I0056509

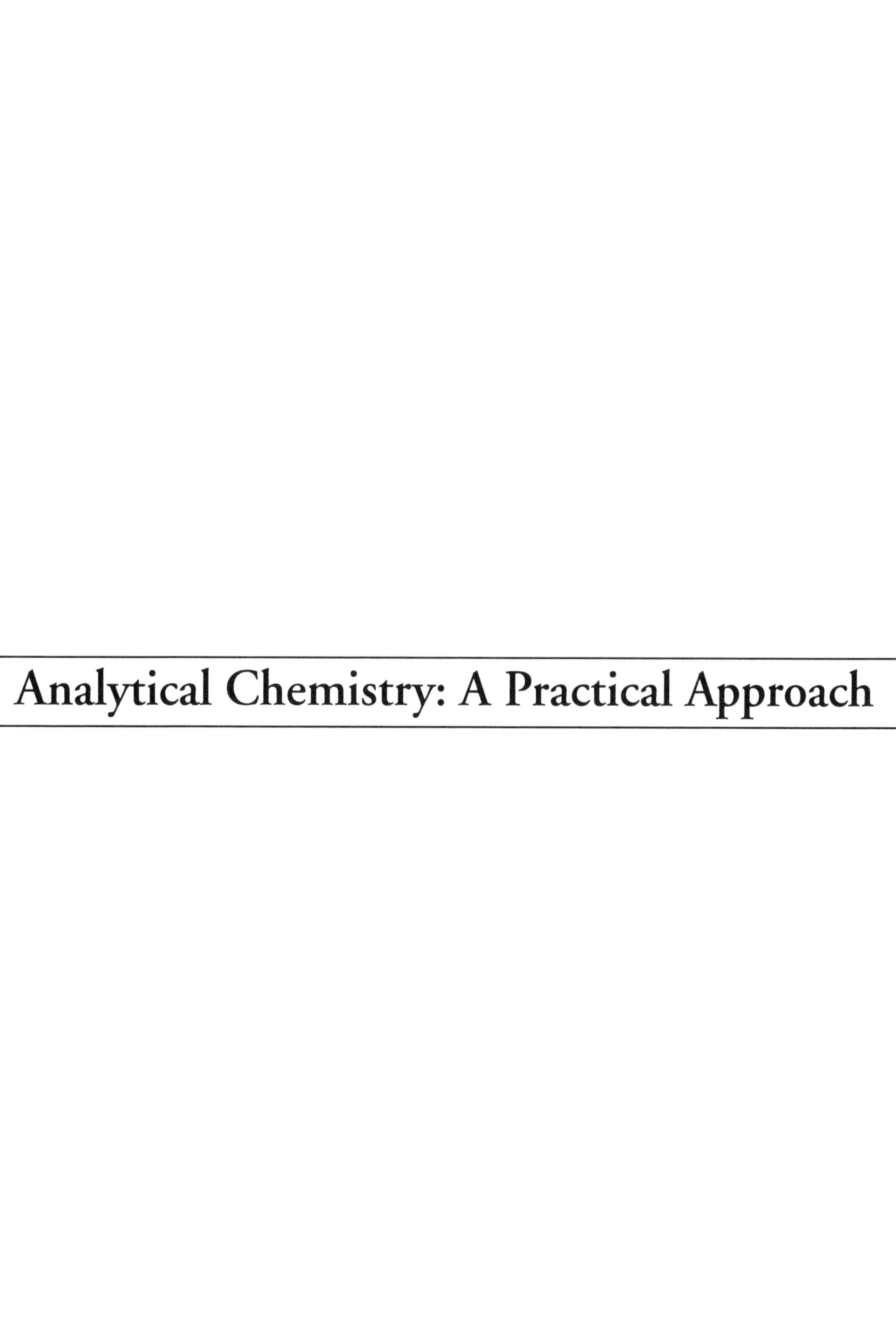

Analytical Chemistry: A Practical Approach

Analytical Chemistry: A Practical Approach

Editor: Lucio Price

New York

Published by NY Research Press
118-35 Queens Blvd., Suite 400,
Forest Hills, NY 11375, USA
www.nyresearchpress.com

Analytical Chemistry: A Practical Approach
Edited by Lucio Price

International Standard Book Number: 978-1-63238-834-6 (Hardback)

Cataloging-in-Publication Data

Analytical chemistry : a practical approach / edited by Lucio Price.
p. cm.
Includes bibliographical references and index.
ISBN 978-1-63238-834-6
1. Chemistry, Analytic. 2. Chemistry. I. Price, Lucio.
QD75.22 .A53 2022
543--dc23

Contents

Preface

This book has been a concerted effort by a group of academicians, researchers and scientists, who have contributed their research works for the realization of the book. This book has materialized in the wake of emerging advancements and innovations in this field. Therefore, the need of the hour was to compile all the required researches and disseminate the knowledge to a broad spectrum of people comprising of students, researchers and specialists of the field.

The field of analytical chemistry is concerned with the separation, identification and quantification of matter. It consists of a diverse range of methods, such as classical wet chemical methods and instrumental methods. Separation methods of extraction, precipitation and distillation are qualitative classical methods. Identification is generally achieved on the basis of differences in melting point, color, odor, boiling point, reactivity or radioactivity. Instrumental methods separate samples using electrophoresis, chromatography or field flow fractionation. Analytical chemistry also emphasizes on chemometrics, improvements in experimental design and creation of new measurement tools. Analytical chemistry has applications in medicine, forensics, science and engineering. This field also plays an important role in the development of nanotechnology. This book provides significant information of this discipline to help develop a good understanding of analytical chemistry and related fields. It is a valuable compilation of topics, ranging from the basic to the most complex advancements in this field. This book is a complete source of knowledge on the present status of this important field.

At the end of the preface, I would like to thank the authors for their brilliant chapters and the publisher for guiding us all-through the making of the book till its final stage. Also, I would like to thank my family for providing the support and encouragement throughout my academic career and research projects.

Editor

1

LC-MS/MS Quantification of Tramadol and Gabapentin Utilizing Solid Phase Extraction

Pappula Nagaraju[1] **Balaji Kodali,**[2] **Peda Varma Datla,**[3] **and Surya Prakasarao Kovvasu**[4]

[1] *Department of Pharmaceutical Analysis, Hindu College of Pharmacy, Guntur 522002, Andhra Pradesh, India*
[2] *College of Pharmaceutical Sciences, Acharya Nagarjuna University, Nagarjuna Nagar 522510, Guntur, Andhra Pradesh, India*
[3] *Clinical Pharmacology and Bio Sciences Division, RA Chem Pharma, Hyderabad, India*
[4] *College of Pharmacy, Western University of Health Sciences, Pomona, CA 91766, USA*

Correspondence should be addressed to Pappula Nagaraju; pappulanagaraju@gmail.com

Academic Editor: Barbara Bojko

An accurate, highly sensitive, and precise method for quantitative analysis of tramadol (TMD) and gabapentin (GBP) by high performance liquid chromatography and tandem mass spectrometry in human plasma was proposed and validated successfully using venlafaxine and pregabalin as internal standards (ISTDs), respectively. An aliquot of 200 μL of plasma was mixed with internal standard dilution and extraction was performed by using solid phase extraction (SPE) technique. Peak resolution was achieved on Phenomenex PFP column (50×4.6 mm, 2.6 μm). The total analytical run time was 3.8 min. Both analytes were monitored using multiple reaction monitoring (MRM) scan and the mass spectrometer was operated in positive polarity mode. The method was validated for specificity, sensitivity, precision, accuracy, and other analytical parameters. The results found were satisfactory over the linear calibration range of 1-500 ng/mL and 10-6000 ng/mL for TMD and GBP, respectively. The developed method can be ready to use by scientific community for quantification of analytes in plasma samples from various clinical studies of different dose strengths.

1. Introduction

Tramadol hydrochloride (TMD), chemically (+)-trans-2-[(dimethyl-amino) methyl]-1-(3-methoxyphenyl) cyclohexanol, is a central analgesic agent for the treatment of severe moderate to chronic pain. Tramadol is also considered as an alternate to opiates for neuropathic pains. Tramadol also proves to produce antitussive, antidepressant, anti-inflammatory, and immune stimulatory effects [1, 2]. In humans, TMD metabolized by cytochrome P4502D6 to its phase 1 metabolites, namely, O-desmethyltramadol and N-desmethyltramadol. These are again metabolized to N,N-didesmethyltramadol, N,N,O-tridesmethyltramadol, and N, O-desmethyltramadol then further produce sulfate and glucuronic acid conjugates before excretion via kidneys in urine [3–5]. TMD is selective opiate agonist at μ-opioid receptors and inhibits reuptake of norepinephrine and serotonin [6]. TMD has plasma protein binding of about 20% and is rapidly absorbed with bioavailability of 65-70% after oral administration [7, 8]. As per the literature search, the analytical methods available for estimation of TMD with its desmethylates in plasma including liquid chromatography coupled to ultraviolet (UV) detector [9–11], fluorescence detector [12, 13], and tandem mass spectrometry (MS/MS) [14–22] are well reported.

Gabapentin (GBP), 1-(aminomethyl-1-cyclohexyl) acetic acid, is a structural analog of the inhibitory neurotransmitter amino butyric acid (GABA) which is a new generation effective antiepileptic drug for partial epileptic seizures with or without secondary generalization [23–25]. The GBP mechanism of action was not clearly defined, but described cellular actions are likely to be related to multiple concentration-dependent actions resulting in supremacy over seizure control [26]. It has been observed that GBP bioavailability varies greatly (inter- and intrasubjects) due to its active absorption by gut and renal excretion of unchanged drug. The bioavailability of a 600 mg oral dose was 49%; individual subjects may vary greatly from 5% to 74% [27, 28]. Gabapentin was proved to be beneficial in the treatment of neuropathic pain as well as postoperative pain following spinal surgery and hysterectomy [29]. Gabapentin in neuropathic pain models prevents mechanical and thermal allodynia and mechanical hyperalgesia. Though the mechanism of action of gabapentin in the treatment of neuropathic pain is not clear, it does not influence the same pathways as opioids or tricyclic depressants. Current evidence indicates that gabapentin affects voltage-gated calcium channels in the CNS [30, 31]. It was also reported that GBP was also effective in pain management because of neuralgia, diabetic neuropathy, multiple sclerosis, and neuropathic cancer pain in miscellaneous reports [32]. Several analytical methods were reported for the determination of GBP that includes high performance liquid chromatography (HPLC) coupled ultraviolet (UV) [33, 34], fluorescence detection [35], and mass spectrometry (MS) [36–42].

Fixed dose combination (FDC) of TMD with paracetamol for pain management in patients was available in market. The TMD combination with GBP is the present choice for doctors to treat pain carried by healthy nerves because of damaged tissues and damaged nerves (neuropathic). In present days, different combinations of TMD and GBP along with other analgesics like ibuprofen are under investigation [43]. The individual dosage forms for TMD and GBP were available around the globe but the FDC (TMD+ GBP) is commonly available in Latin America. The phase-IV clinical or bioequivalence studies are necessary for FDC approvals. Since no method was reported so far for simultaneous determination of TMD and GBP in human plasma, hence we aimed to develop specific and selective achiral assay for quantification of TMD and GBP in human plasma as per USFDA [44] and EMEA [45] bioanalytical method validation guidelines. The biological TMD metabolites (desmethlyates) measurement was not required to prove bioequivalence as per major health regulatory bodies; hence only parent drugs (TMD, GBP) are considered for method development. Finally, highly sensitive and repeatable method was developed for quantification of analytes in human plasma, useful to assess either efficacy or toxicity of both TMD and GBP (particularly) in various clinical situations. The present method is able to quantify the TMD and GBP at very low level (i.e., LLOQ 1 ng/mL and 10 ng/mL), which means that the established linear range is suitable to monitor TMD and GBP circulating levels across the relevant clinical range up to five terminal half-lives (t1/2), right from administration to approximate elimination (trough and subclinical concentrations) from the body [21, 22, 38].

2. Experimentation

2.1. Reference Standards and Reagents. The high purity reference standards of TMD, GBP, venlafaxine (VFX), and pregabalin (PGB) were procured from Clearsynth Labs Pvt. Ltd. (Mumbai, India). The HPLC grade methanol and acetonitrile are purchased from Thermo Fisher Scientific India Pvt. Ltd. (Mumbai, India). GR grade ammonium formate and ammonium acetate reagents were procured from Merck Specialities Pvt. Ltd. (Mumbai, India). Milli-Q water was collected from Milli-Q A10 gradient water purification system (Millipore, Bedford, MA, USA). Strata-X polymeric extraction cartridges (30 mg, 1cc) for solid phase extraction (SPE) are purchased from Phenomenex India Pvt. Ltd.

2.2. Analytical Instrumentation. An ultra flow prominence high performance liquid chromatography (UF-HPLC) coupled with tandem mass spectrometer (MS/MS-3200 model, Sciex, Canada) was used for analysis. The mass spectrometer was assembled with electro spray ionization (ESI) interface. The HPLC was supplied with LC-20AD binary pumps, 20A3 solvent degasser, column oven, and high-throughput SIL HTC auto sampler. After chromatographic separation, the positive polarity MS detection was performed in multiple reaction monitoring (MRM) mode. Analyst software 1.5.1 platform was used for data collection and hardware controlling.

2.3. Chromatographic Conditions. Analytical peak resolution was achieved on a Phenomenex, Kinetex PFP column (C18, 50 × 4.6 mm, 5 μm) pumped with isocratic mobile phase consisting a mixture of 5 mM ammonium formate buffer (pH 3.0 ± 0.3), acetonitrile, and methanol in the ratio of 25: 50: 25 v/v. The flow rate was 0.8 mL/min. The auto sampler and column oven were programmed to maintain the set temperatures at 5°C and 35°C, respectively. Sample volume of 10 μL was injected into the LC-MS/MS system. The total analytical run was 3.8 min.

2.4. MS/MS Compound and Source Dependent Conditions. The mass spectrometer was operated in positive mode to monitor parent⟶product ion (m/z) transitions of analytes (TMD, GBP) and their internal standards (ISTDs) (VFX, PGB). The specific details of MRM transitions and their respective mass spectrometer voltage values like declustering potential (DP), entrance potential (EP), collision energy (CE), and collision exit potential (CXP) used for quantification of respective analytes and ISTDs are summarized in Table 1. Manual tuning was performed to optimize the source dependent and compound dependent parameters to get highest credible intensities. The source dependent parameters like drying gas (GSI) and nebulizer gas (GS2) were set at 35 psi, 45 Psi duly. The turbo ion spray (TIS) temperature and ion spray voltage were set at 500°C and 4,500 V, respectively. The curtain gas (CUR) and collision associated dissociation gas (CAD) pressure were maintained at 30 psi and 8 psi. The unit resolution mode was employed in Q1 and Q3 (quadrupoles) with a dwell time of 300 milliseconds.

2.5. Standard Curve and Control Samples. Stock solutions of TMD and GBP were prepared in methanol and respective working (spiking) dilutions were made using diluent solution of methanol: water mixture (50:50,v/v). Separate stock weighing was done for preparation of calibration curve and quality control stock solutions. Calibration curves in range

Table 1: MRM and mass spectrometer voltage details (TMD, GBP) and IS (VFX, PGB).

Name of the molecule	MRM Transition (Q1/Q3)	DP	EP	CE	CXP
TMD	264.2/58.1	50	10	22	15
GBP	172.2/154.2	38	10	27	10
VFX	278.3/121.1	70	10	27	8
PGB	160.2/97.1	90	10	23	11

of 1-500 ng/mL and 10-6000 ng/mL were prepared for TMD and GBP, respectively. Quality control samples were made at concentration of 1 ng/mL lower limit of quality control (LLOQQC), 3 ng/mL lower quality control (LQC), 212 ng/mL middle quality control (MQC), 380 ng/mL high quality control (HQC), 1000 ng/mL diluted quality control (DQC) for TMD and 10 ng/mL (LLOQQC), 30 ng/mL (LQC), 2500 ng/mL (MQC), 4500 ng/mL (HQC), and 12000 ng/mL (DQC) for GBP. The 1% of respective working dilution was spiked into the total volume of plasma (for example, 10μL of working solution was added to 990μL of plasma, which is 1% to the total volume) to get above-mentioned concentrations for both the analytes. The long-term plasma stability samples at LQC and HQC level were prepared and stored at -70°C in ploy propylene tubes. The spiked samples were prepared freshly based on the validation experimentation plan. All the stock solutions and working dilutions were stored in refrigerator maintained at 2-8°C.

2.6. Bio Analytical Extraction Procedure. 200 μL of plasma sample was aliquoted using micropipette in to a 6mL polypropylene tube containing 100 μL of ISTD solution (containing each 500 ng/mL of VFX and PGB) and then 0.2 mL of 100 mM ammonium acetate buffer as pretreatment solution was added. The resultant sample was briefly mixed and subjected to positive pressure solid phase extraction procedure using strata-X cartridges (30 mg/1 cc). The samples were loaded on cartridges which were already preconditioned with 1mL methanol and 1 mL Milli-Q water. Followed by loading, cartridge was washed with 1 mL 0.1% formic acid, 1 mL n-Hexane, and 1 mL methanol: water (5:95 v/v) solution step by step. Allow the cartridges to dry for about 3 min and then elute with 1 mL of 2% ammoniated methanol solution. The eluent solution was evaporated to dryness under gentle stream of nitrogen at a pressure of 20psi and at temperature of 50°C. The residue was reconstituted with 400 μL of mobile phase and 10 μL was injected into chromatographic system for analysis.

3. Method Validation

The developed method was validated to ensure method performance. The method was validated as per USFDA and EMEA guidelines. Method sensitivity, selectivity, linearity, precision, accuracy, recovery, matrix effect, dilution integrity, and analyte stability in biological matrix were evaluated. Each analytical run in validation begins with calibration curve and evenly distributed quality control samples at different levels based on standard experimental requirements.

3.1. System Suitability. Two injections of low standard solution and six injections of high standard solution containing both analytes (TMD, GBP) were injected to ensure system conditions. The low standard solution was injected to check the peak shape. The % CV for area ratio (analyte/ ISTD for both TMD, GBP) of high standard solution should be less than 4.

3.2. Biological Matrix Screening and Selectivity. The percentage of interference due to exogenous and endogenous components at retention times of analytes and ISTD was evaluated by processing eight different lots of blank plasma along with each two lots of hemolytic and lipemic plasma. The interference due to concomitant medication at retention time was also investigated by spiking paracetamol, ibuprofen, ranitidine, and ondansetron into drug free plasma at concentration equal to their available literature Cmax values. The interference observed at the retention times of analytes and ISTDs in blank plasma lots was compared against mean response of extracted LLOQ (n=6) samples. The observed interference should be less than 20% and 5% at analyte and ISTD retention times, respectively, when compared to mean response of extracted LLOQ samples.

3.3. Reproducibility (Precision) and Accuracy. At four different quality control levels (LLOQQC, LQC, MQC, and HQC, n=12) within day (intrabatch) and between day (interbatch, n=24) precision and accuracy of TMD, GBP was evaluated by calculating the %CV and %accuracy. In together six reproducibility batches were performed on two different days by two different analysts.

3.4. Effect of Matrix. The signal suppression or enhancement via ionization should be studied in mass spectrometric detection methods. To prove that, the method is free from matrix effect, postextraction response from 10 different lots (including each two lots of hemolytic and lipemic plasma) were compared with response of aqueous samples. The matrix effect was evaluated at LQC, HQC levels by calculating matrix factor of analyte and ISTD. Later ISTD normalized matrix factor was calculated by using matrix factor of analyte and ISTD. If ISTD normalized matrix factor value is 1, that indicates there is no suppression or enhancement due to the presence of matrix. If the value is less than 1, that indicates ion suppression or more than 1, that indicates ion enhancement. The acceptable limits for ISTD normalized matrix factor are 0.85-1.15.

3.5. Linearity of Analytes. The method linearity was assessed by constructing three eight-point calibration curves. A linear

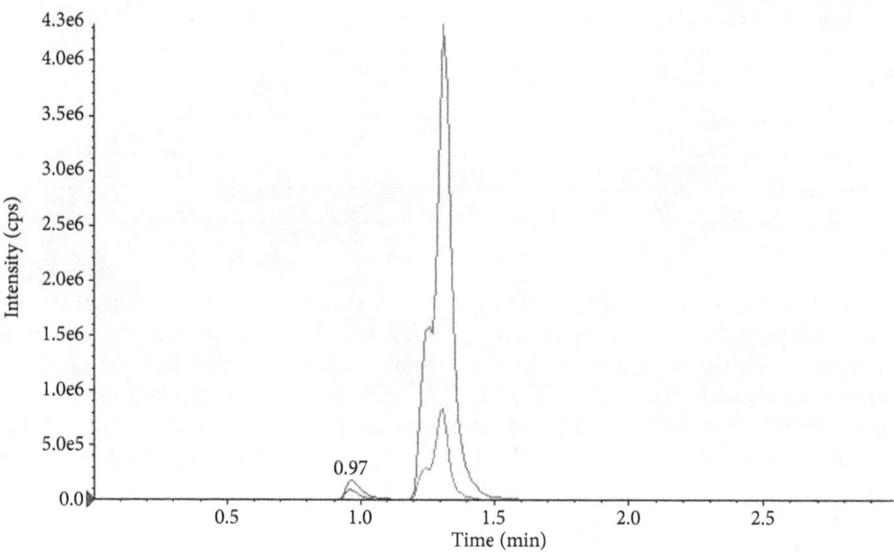

FIGURE 1: Chromatogram of analytes/ISTDs (LLOQ level) injected normal gemini C_{18} column (50 × 4.6 mm, 5 μm).

least-square regression analysis was applied for back calculated concentrations using weighing factors, none, 1/x, 1/x2. The weighing factor with least regression value is 1/x2; therefore 1/x2 was further used as weighing factor for constructing the calibration curves throughout the validation.

3.6. Extraction Recovery/Efficiency. Good extraction recovery was needed for accurate and reproducible results. Stable and consistent recovery was the basic requirement to achieve method sensitivity at limit of quantification (LOQ) level. The analyte recovery might be low or medium or 100% but it should be steady at all levels (LQC, MQC, HQC). Care should be taken while optimizing the procedure to achieve good extraction recovery. Relative recovery (RR) was evaluated at three different levels LQC, MQC, HQC (n=6) by comparing response in postspiked samples versus extracted samples. To evaluate true effect of matrix on recovery of analyte and ISTD (absolute recovery-AR), the response of extracted samples was also compared with aqueous samples. The recovery of analyte should not be more than 115%.

3.7. Stability of Analytes/ISTD. Stability of analytes (TMD, GBP) was evaluated in different experimental conditions based on the requirement of real time unknown sample analysis conditions like freeze and thaw stability (at −70°C), dry extract stability, spiked sample room temperature stability, auto sampler stability, long-term stability (at −70°C) and stability in whole human blood. For all the stability experiments six replicates of LQC, HQC samples were processed and analyzed against fresh calibration curve. The back calculated concentrations are compared to nominal concentration. Stability of aqueous samples was assessed by comparing the responses from high standard solutions prepared from stored aqueous stock solutions/working dilutions (at 2-8°C) with freshly prepared stock solutions/working dilutions.

4. Results and Discussion

4.1. Method Development. For efficient quantification and reliable results, it is prerequisite to give equal importance to optimize the chromatographic conditions, extraction procedure and mass spectrometric conditions. All analytes dissolved in methanol, individually infused into MS (mass spectrometer) source for tuning and then selected positive mode because of better intensity. The Q1 scan was performed to select the parent ion. The declustering potential (DP), entrance potential (EP) voltage values were further optimized to get highest intensity for parent ion. After that, collision energy (CE), collision cell exit potential (CXP) values were optimized in MSMS scan to select product ion for TMD, GBP, PGB and VFX. The observed $[M+H]^+$ peaks (parent ion) and respective consistent product ions were selected for mass spectrometric transitions (Q1/Q3) in MRM (multiple reaction monitoring) mode for quantification. The selected transitions and optimized voltage values were shown in Table 1. The unit resolution mode with a dwell time of 300 milliseconds was used for each MRM transition channel.

Several analytical bonded stationary phases of C_{18} and C_8 were checked and retention times of analytes are overlapped. Initially, aqueous solution of LLOQ level was injected into normal gemini C_{18} (50 × 4.6 mm, 5 μm) column, but the observed peak resolution was not good and peak intensity is very low, the identical chromatogram of LLOQ solution in gemini column was shown in Figure 1. Then sample solution was injected into thermo high purity C_{18} (100 × 4.6 mm, 3.5 μm), column to improve the peak shape. The observed peak resolution was comparatively good with low intensity. The representative chromatogram was shown in Figure 2. The better peak shape and resolution with required sensitivity was achieved on Phenomenex, PFP (50 × 4.6 mm, 2.6 μm) column may be because of its combining C_{18} retention properties and unique aromatic PFP selectivity. A medium

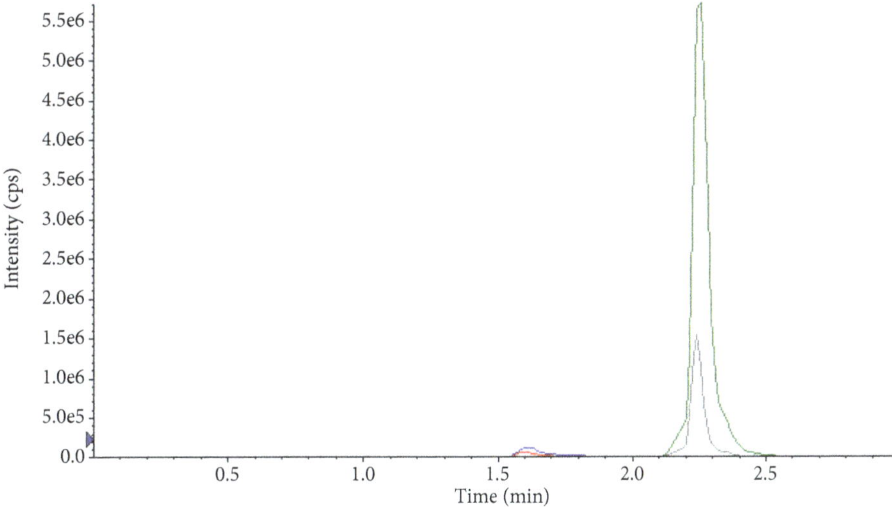

FIGURE 2: Chromatogram of analytes/ISTDs injected on high purity C_{18} column (100 × 4.6 mm, 3.5 μm).

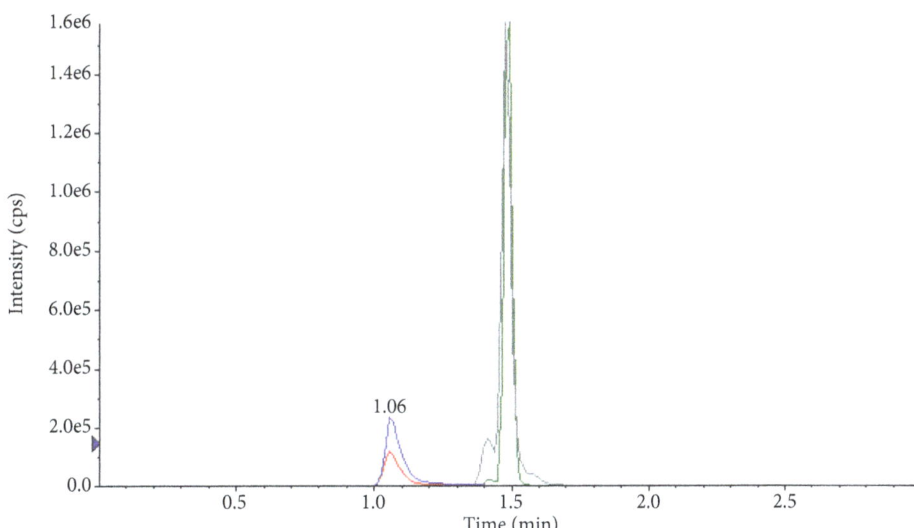

FIGURE 3: Chromatogram of analytes/ISTDs with high base line noise (liquid-liquid extraction).

strength buffer 5 mM ammonium formate gives high signal to noise ratio with negligible baseline noise at LLOQ level.

In sample extraction, liquid-liquid and solid phase extraction techniques were investigated. In liquid-liquid extraction high base line was observed because of possible matrix contaminants. The representative chromatogram is shown in Figure 3. Finally, solid phase extraction was selected due to its high consistent extraction recoveries with no matrix effect and cleaner extracts. Phenomenex Strata-X cartridges with 5% methanol wash produced side peaks in chromatography and low recovery was observed with GBP. An acidic wash with 0.1 % formic acid increases the GBP recovery, followed by n-hexane wash to eliminate nonpolar interferences prior to the elution resolved the side peaks issue. Method was strictly optimized to get similar recoveries for analytes and ISTDs. The nearly same % recovery results for analytes and ISTDs with acceptable ISTD normalized factor values of the method assure reproducible quantification.

4.2. Selectivity. Eight plasma lots along with each two different lots of hemolytic and lipemic plasma were processed and injected for LC-MS/MS analysis. Similar chromatography was observed with no significant interference at the retention times of analytes and ISTDs in all analyzed blank lots, which indicates that the developed method was highly selective.

4.3. Linearity. Three calibration curves were generated by plotting the area ratios (analyte response/ ISTD response) on y-axis and concentration on x-axis. The plot was linear throughout the established calibration ranges, 1-500 ng/mL for TMD and 10-6000 ng/mL for GBP. The slope values are consistent and regression values were

TABLE 2: Precision and accuracy results of TMD.

QC name/nominal concentration	TMD			
	Intra batch (n=12)		Inter batch (n=24)	
	% Accuracy	% CV	% Accuracy	% CV
LLOQQC/ 1 ng/mL	92.3	10.9	91.8	9.8
LQC/ 3 ng/mL	97.3	8.6	95.2	6.1
MQC/ 212 ng/mL	94.9	3.8	96.7	5.5
HQC/ 380 ng/mL	99.2	5.8	97.6	4.2

TABLE 3: Precision and accuracy results of GBP.

QC name/nominal concentration	GBP			
	Intra batch (n=12)		Inter batch (n=24)	
	% Accuracy	% CV	% Accuracy	% CV
LLOQQC/ 10 ng/mL	96.7	6.7	98.4	3.8
LQC/ 30 ng/ mL	93.5	5.8	97.9	6.6
MQC/ 2500 ng/mL	99.1	3.2	94.1	2.9
HQC/ 4500 ng/mL	95.6	2.9	100.9	5.3

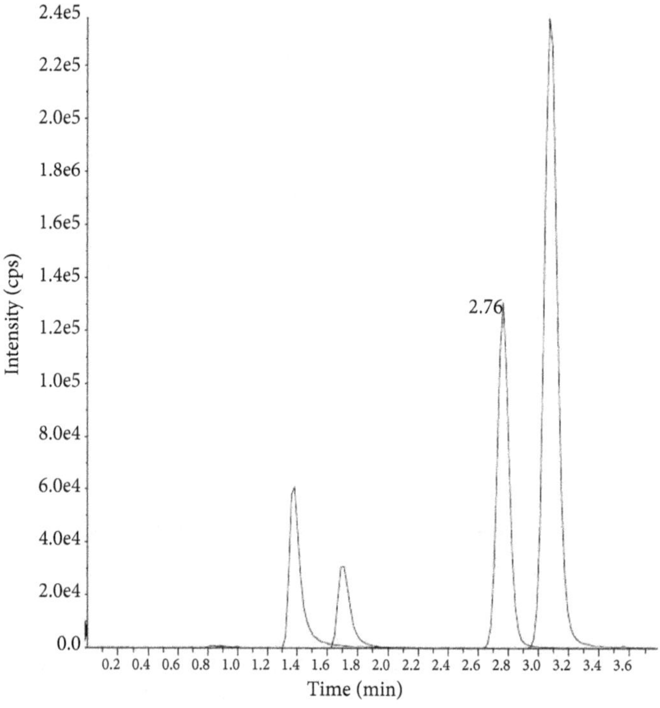

FIGURE 4: Chromatogram of analytes/ISTDs at LLOQQC level.

found to be more than 0.99. The back calculated concentrations for individual calibration standards are meeting acceptance criteria for accuracy (±15 %) and precision (≤ 15 %).

4.4. Sensitivity. Six replicates of LLOQ samples were processed and analyzed against calibration curve. The accuracy and precision values were 91.3 % and 3.8 % for TMD and 98.6 % and 2.9 % for GBP. The observed signal to noise ratio is more than 5:1 for both the analytes.

4.5. Precision and Accuracy. Accuracy and reproducibility results of intra- and interbatches of TMD and GBP were reported in Tables 2 and 3, respectively. The intra- and interbatch accuracy values were in the range of 92.3%-99.2% and intra and interbatch precision were found to be less than 10.9 % and 6.7 % for TMD and GBP. The chromatogram at LLOQ level was shown in Figure 4.

4.6. Effect of Matrix. In general consideration, effect of matrix does not influence peak resolution due to MS selectivity.

TABLE 4: Matrix effect results of TMD and GBP.

Blank plasma lots	TMD (ISNMF)		GBP (ISNMF)	
	LQC	HQC	LQC	HQC
LOT-1	0.99	0.96	0.93	1.02
LOT-2	0.98	0.97	0.94	1.05
LOT-3	0.98	0.96	0.92	0.94
LOT-4	0.97	0.95	0.96	0.98
LOT-5	1.02	0.99	0.99	0.96
LOT-6	1.01	1.06	0.97	0.97
LOT-7 hemolytic	0.99	1.01	0.98	0.99
LOT-8 hemolytic	1.03	0.99	0.96	1.04
LOT-9 lipemic	0.98	0.94	0.96	0.94
LOT-10 lipemic	0.99	0.97	0.94	0.96
Mean	0.994	0.980	0.955	0.985
SD	0.0196	0.0350	0.0222	0.0395
%CV	2.0	3.6	2.3	4.0

ISTDNMF: internal standard normalised matrix factor.

TABLE 5: Recovery results of TMD and GBP.

Sample name	% Mean recovery absolute		% Mean recovery relative	
	TMD	GBP	TMD	GBP
LQC	81.5	83.5	83.8	86.4
MQC	78.4	80.5	81.2	82.1
HQC	84.7	79.9	78.9	82.2
VFX at MQC level	85.4		79.1	
PGB at MQC level	81.9		82.4	

However, in this method sufficient resolution between the analytes (TMD, GBP) was established chromatographically. Matrix effect was evaluated in 10 different lots. The obtained ISTD normalized matrix factor values for both the analytes were in the range of 0.93 to 1.06. The precision values for ISTD normalized factor at LQC and HQC level were 2.0% and 3.6% for TMD and 2.3% and 4.0 % for GBP. The results are presented in Table 4.

4.7. Recovery. Absolute and relative recovery of analytes and ISTDs was evaluated. The mean recovery results of TMD, GBP, VFX, and PGB are represented in Table 5.

4.8. Dilution Integrity. Precision and accuracy of diluted plasma samples were assessed at 1:4 dilution. The DQC (dilution quality control) was prepared by spiking at a concentration equal to two times of high level calibration standard of proposed range for TMD and GBP, respectively. Then $1/4^{th}$ volume of plasma aliquot was diluted with drug free plasma and analyzed against calibration curve. The accuracy values were in the range of 92.4%-106.5 % and %CV was less than 3.2% for TMD and GBP.

4.9. Stability. All the stock solutions and stock dilutions were stable for 21 days at refrigerated storage maintained at 2-8°C. The processed stability samples in plasma at LQC and HQC levels were analyzed against freshly prepared calibration curve. The stability data results are given in Table 6. TMD and GBP were stable in plasma at room temperature for about 16 h and for 5 freeze and thaw cycles. The established stability time for TMD and GBP was 41 h and 52 h for auto sampler and dry extract stabilities. The analytes were found to be stable for 2.5 h in blood. The long-term stability was evaluated and analytes were stable for 32 days at −70°C.

5. Conclusion

Full method validation was carried out using screened and pooled human plasma to ensure that developed procedure is accurate and precise for estimation of TMD and GBP simultaneously. The high-throughput LC-ESI-MS/MS method is sensitive and specific. The recovery, precision, and accuracy results were reproducible over the proposed calibration ranges for TMD and GBP. The shorter runtime allows the analysis of more samples (~300) per day. The method can be readily used by scientific community for the application of sample analysis for therapeutic monitoring/pharmacokinetic or bioequivalence studies.

Table 6: Stability results of TMD and GBP.

Stability experiment	Stability condition	%Mean stability			
		TDL		FNS	
		LQC	HQC	LQC	HQC
Auto sampler stability	41 h at 5°C	98.1	96.4	96.1	106.7
Free and thaw stability	5 cycles at -70°C ± 15°C	97.1	100.1	99.9	96.1
Dry extract stability	52 h at 2-8°C	94.8	96.8	101.2	96.6
Room temperature stability	16 h at room temperature at 25°C ± 5°C	94.1	102.8	91.7	97.2
Long term stability	32 days at -70°C ± 15°C	99.6	95.7	98.8	95.5
Stability in blood	2.5 h room temperature at 25°C ± 5°C	96.4	99.1	103.2	97.6

Acknowledgments

The authors are thankful to Praveen Datla, Clinical Research and Bio Sciences, Hyderabad, India, for their continuous support, motivation, and assistance during the course of this project.

References

[1] B. Kukanich and M. G. Papich, "Pharmacokinetics of tramadol and the metabolite O-desmethyltramadol in dogs," *Journal of Veterinary Pharmacology and Therapeutics*, vol. 27, pp. 239–246, 2004.

[2] I. Yalein, F. Aksu, and C. Belzung, "Effects of desipramine and tramadol in a chronic mild stress model in mice are altered by yohimbine but not by pindolol," *European Journal of Pharmacology*, vol. 514, pp. 165–174, 2005.

[3] K. Miotto, A. K. Cho, M. A. Khalil, K. Blanco, J. D. Sasaki, and R. Rawson, "Trends in tramadol: pharmacology, metabolism, and misuse," *Anesthesia & Analgesia*, vol. 124, no. 1, pp. 44–51, 2017.

[4] W. S. James and D. N. Young, "Tramadol," *Cmaj*, vol. 185, no. 8, p. E352, 2013.

[5] W. Leppert, "CYP2D6 in the metabolism of opioids for mild to moderate pain," *Pharmacology*, vol. 87, no. 5-6, pp. 274–285, 2011.

[6] L. Poulsen, L. Arendt-Nielsen, K. Brosen, and S. H. Sindrup, "The hypoalgesic effect of tramadol in relation to CYP2D6," *Clinical Pharmacology & Therapeutics*, vol. 60, no. 6, pp. 636–644, 1996.

[7] S. Groud and A. Sablotzki, "Clinical Pharmacology of Tramadol," *Clinical Pharmacokinetics*, vol. 43, no. 13, pp. 879–923, 2004.

[8] W. Lintz, H. Barth, R. Becker, E. Frankus, and E. Schmidt-Bothelt, "Pharmacokinetics of tramadol and bioavailability of enteral tramadol formulations-2nd communication: Drops with ethanol," *Arzneimittel-Forschung*, vol. 48, no. 5, pp. 436–445, 1998.

[9] L. Qu, S. Feng, Y. Wu, and S. Da, "HPLC method for determination of tramadol hydrochloride in human plasma," *Journal of Sichuan University. Medical Science Edition = Sichuan Daxue Xuebao (Yixue Ban)*, vol. 34, no. 3, pp. 574-575, 2003.

[10] G. C. Yeh, M. T. Sheu, C. L. Yen, Y. W. Wang, C. H. Liu, and H. O. Ho, "High-performance liquid chromatographic method for determination of tramadol in human plasma," *Journal of Chromatography B: Biomedical Sciences and Applications*, vol. 723, pp. 247–253, 1999.

[11] S. H. Gan, R. Ismail, W. A. W. Adnan, and Z. Wan, "Method development and validation of a high-performance liquid chromatographic method for tramadol in human plasma using liquid–liquid extraction," *Journal of Chromatography B*, vol. 772, pp. 123–129, 2002.

[12] H. Ebrahimzadeh, Y. Yamini, A. Sedighi, and M. R. Rouini, "Determination of tramadol in human plasma and urine samples using liquid phase microextraction with back extraction combined with high performance liquid chromatography," *Journal of Chromatography B*, vol. 863, no. 2, pp. 229–234, 2008.

[13] A. Curticapean, D. Muntean, M. Curticapean, M. Dogaru, and C. Vari, "Optimized HPLC method for tramadol and O-desmethyl tramadol determination in human plasma," *Journal of Biochemical and Biophysical Methods*, vol. 70, no. 6, pp. 1304–1312, 2008.

[14] M. J. Bogusz, R.-D. Maier, K.-D. Krüger, and U. Kohls, "Determination of common drugs of abuse in body fluids using one isolation procedure and liquid chromatography-atmospheric-pressure chemical-ionization mass spectrometry," *Journal of Analytical Toxicology*, vol. 22, no. 7, pp. 549–558, 1998.

[15] A. Ceccato, F. Vanderbist, and B. Streel, "Enantiomeric determination of tramadol and its main metabolite O-desmethyltramadol in human plasma by liquid chromatography–tandem mass spectrometry," *Journal of Chromatography B: Biomedical Sciences and Applications*, vol. 748, no. 1, pp. 65–76, 2000.

[16] B. N. Patel, N. Sharma, M. Sanyal, and P. S. Shrivastav, "An accurate, rapid and sensitive determination of tramadol and its active metabolite O-desmethyltramadol in human plasma by LC–MS/MS," *Journal of Pharmaceutical and Biomedical Analysis*, vol. 49, no. 2, pp. 354–366, 2009.

[17] M. Gergov, P. Nokua, E. Vuori, and I. Ojanperä, "Simultaneous screening and quantification of 25 opioid drugs in post-mortem blood and urine by liquid chromatography-tandem mass spectrometry," *Forensic Science International*, vol. 186, pp. 36–43, 2009.

[18] M. De Leo, M. Giorgi, G. Saccomanni, C. Manera, and A. Braca, "Evaluation of tramadol and its main metabolites in horse plasma by high-performance liquid chromatography/fluorescence and liquid chromatography/electrospray ionization tandem mass spectrometry techniques," *Rapid Communications in Mass Spectrometry*, vol. 23, no. 2, pp. 228–236, 2009.

[19] N. V. de Moraes, G. R. Lauretti, M. N. Napolitano et al., "Enantioselective analysis of unbound tramadol, O-desmethyltramadol and N-desmethyltramadol in plasma by ultrafiltration and LC–MS/MS: Application to clinical pharmacokinetics," *Journal of Chromatography B*, vol. 880, pp. 140–147, 2012.

[20] G. Saccomanni, S. Del Carlo, M. Giorgi, C. Manera, A. Saba, and M. Macchia, "Determination of tramadol and metabolites by HPLC-FL and HPLC-MS/MS in urine of dogs," *Journal of Pharmaceutical and Biomedical Analysis*, vol. 53, pp. 194–199, 2010.

[21] L. Vlase, S. E. Leucuta, and S. Imre, "Determination of tramadol and O-desmethyltramadol in human plasma by high-performance liquid chromatography with mass spectrometry detection," *Talanta*, vol. 75, no. 4, pp. 1104–1109, 2008.

[22] T. Hironari, N. Takafumi, M. Yasuaki, and J. Kawakami, "Validated determination method of tramadol and its desmethylates in human plasma using an isocratic LC-MS/MS and its clinical application to patients with cancer pain or non-cancer pain," *Journal of Pharmaceutical Health Care and Sciences*, vol. 2, no. 25, pp. 1–9, 2016.

[23] M. C. Walker and P. N. Patsalos, "Clinical pharmacokinetics of new antiepileptic drugs," *Pharmacology & Therapeutics*, vol. 67, no. 3, pp. 351–384, 1995.

[24] P. Gareri, T. Gravina, G. Ferreri, and G. De Sarro, "Treatment of epilepsy in the elderly Prog Neurobiol," *Journal of the Korean Medical Association*, vol. 58, no. 5, pp. 389–407, 1999.

[25] R. K. Berlin, P. M. Butler, and M. D. Perloff, "Gabapentin therapy in psychiatric disorders: A systematic review," *Primary Care Companion to the Journal of Clinical Psychiatry*, vol. 17, no. 5, 2015.

[26] M. J. McLean, "Gabapentin in the management of convulsive disorders," *Epilepsia*, vol. 40, no. 6, pp. S39–S50, 1999.

[27] B. E. Gidal, L. L. Radulovic, S. Kruger, P. Rutecki, M. Pitterle, and H. N. Bockbrader, "Inter- and intra-subject variability in gabapentin absorption and absolute bioavailability," *Epilepsy Research*, vol. 40, no. 2-3, pp. 123–127, 2000.

[28] D. Ouellet, H. N. Bockbrader, D. L. Wesche, D. Y. Shapiro, and E. Garofalo, "Population pharmacokinetics of gabapentin in infants and children," *Epilepsy Research*, vol. 47, no. 3, pp. 229–241, 2001.

[29] C. W. Goodman and A. S. Brett, "Gabapentin and Pregabalin for Pain—Is Increased Prescribing a Cause for Concern?" *The New England Journal of Medicine*, vol. 377, pp. 411–414, 2017.

[30] M. A. Rose and P. C. Kam, "Gabapentin: pharmacology and its use in pain management," *Anaesthesia*, vol. 57, no. 5, Article ID 451462, pp. 451–462, 2002.

[31] C. Y. Chang, C. K. Challa, J. Shah, and J. D. Eloy, "Gabapentin in Acute Postoperative Pain Management," *BioMed Research International*, vol. 2014, Article ID 631756, 7 pages, 2014.

[32] J. Mao and L. L. Chen, "Gabapentin in Pain Management," *Anesthesia & Analgesia*, vol. 91, pp. 680–687, 2000.

[33] G. L. Lensmeyer, T. Kempf, B. Gidal, and D. Weibe, "Optimized method for determination of gabapentin in serum by HPLC," *Ther Drug Monit*, vol. 17, pp. 251–258, 1995.

[34] Z. Zhu and L. Neirinck, "High-performance liquid chromatographic method for the determination of gabapentin in human plasma," *Journal of Chromatography B*, vol. 779, pp. 307–312, 2002.

[35] Q. Jiang and S. Li, "Rapid high-performance liquid chromatographic determination of serum gabapentin," *Journal of Chromatography B: Biomedical Sciences and Applications*, vol. 727, no. 1-2, pp. 119–123, 1999.

[36] D. C. Borrey, K. O. Godderis, D. R. Engelrelst Bernard, and M. R. Langlois, "Quantitative determination of vigabatrin and gabapentin in human serum by gas chromatography-mass spectrometry," *Clinica Chimica Acta*, vol. 354, no. 1-2, pp. 147–151, 2005.

[37] T. Wattananat and W. Akarawut, "Validated LC-MS-MS method for the determination of gabapentin in human plasma: Application to a bioequivalence study," *Journal of Chromatographic Science (JCS)*, vol. 47, no. 10, pp. 868–871, 2009.

[38] J.-H. Park, O.-H. Jhee, S.-H. Park et al., "Validated LC-MS/MS method for quantification of gabapentin in human plasma: application to pharmacokinetic and bioequivalence studies in Korean volunteers," *Biomedical Chromatography*, vol. 21, no. 8, pp. 829–835, 2007.

[39] N. V. S. Ramakrishna, K. N. Vishwottam, M. Koteshwara, S. Manoj, M. Santosh, and J. Chidambara, "Rapid quantification of gabapentin in human plasma by liquid chromatography/tandem mass spectrometry," *Journal of Pharmaceutical and Biomedical Analysis*, vol. 40, no. 2, pp. 360–368, 2006.

[40] H. Y. Ji, D. W. Jeong, Y. H. Kim, H. H. Kim, Y. S. Yoon, and K. C. Lee, "Determination of gabapentin in human plasma using hydrophilic interaction liquid chromatography with tandem mass spectrometry," *Rapid Communications in Mass Spectrometry*, vol. 20, pp. 2127–2132, 2006.

[41] Z. L. Xiong, J. Yu, J. F. He, F. Qin, and F. M. Li, "LC-MS/MS method for quantification and pharmacokinetic study of gabapentin in human plasma," *Yao Xue Xue Bao*, vol. 46, no. 10, pp. 1246–1250, 2011.

[42] S. Muratović, K. Durić, E. Veljović et al., "Synthesis of biscoumarin derivatives as antimicrobial agents," *Asian Journal of Pharmaceutical and Clinical Research*, vol. 6, no. 3, pp. 213–216, 2013.

[43] https://www.ncbi.nlm.nih.gov/pubmedhealth/PMH0014677/.

[44] Guidance for Industry: Bioanalytical method validation, U.S. Department of Health and Human services, Food and Drug administration, Rockville, MD, USA, 2001.

[45] Guidance for Industry: Bioanalytical method validation, European Medicines Agency, European Union, 2012.

2

Determination of Chlorpromazine, Haloperidol, Levomepromazine, Olanzapine, Risperidone, and Sulpiride in Human Plasma by Liquid Chromatography/ Tandem Mass Spectrometry (LC-MS/MS)

Abderrezak Khelfi,[1,2] Mohammed Azzouz,[3] Rania Abtroun,[1] Mohammed Reggabi,[3] and Berkahoum Alamir[1,2]

[1]*Department of Toxicology, Bab-El-Oued hospital, Avenue Mohamed Lamine Debaghine, 16009, Algiers, Algeria*
[2]*National Center of Toxicology, Avenue Petit Staouali Delly Brahim, 16062, Algiers, Algeria*
[3]*Department of Biology and Toxicology, Ait-Idir hospital, Avenue Abderrezak Hahad Casbah, 16017, Algiers, Algeria*

Correspondence should be addressed to Abderrezak Khelfi; khelfi_ar@hotmail.com

Academic Editor: Mohamed Abdel-Rehim

Background and Objective. In this study, turbo-ion spray as an interface of tandem mass spectrometry (MS/MS) was performed for sensitive and accurate quantification of chlorpromazine, haloperidol, levomepromazine, olanzapine, risperidone, and sulpiride in plasma samples. *Methods.* Separation was performed by gradient reversed phase high-performance liquid chromatography using a mobile phase containing ammonium formiate 2 mM, pH 2.7, and acetonitrile flowing through a Restek PFP Propyl C18 analytical column (50 mm×2.1 mm i.d.) with particle size of 5 μm, at a flow rate of 800 μL/min. Positive ion fragments were detected in multiple reaction monitoring (MRM) mode. Sample preparation was achieved by solid phase extraction (SPE) (Oasis HLB). *Results.* Mean extraction recoveries ranged from 82.75% to 100.96%. The standard calibration curves showed an excellent linearity, covering subtherapeutic, therapeutic, and toxic ranges. Intraday and interday validation using quality control (QC) samples were performed. The inaccuracy and imprecision were below 12% at all concentration levels. The limits of detection (LOD) and quantification (LOQ) for all analytes were under therapeutic ranges for all tested analytes. Thus, the proposed method was sensitive enough for the detection and determination of subtherapeutic levels of these antipsychotics in plasma samples. No interference of endogenous or exogenous molecules was observed and no carryover effects were recorded. *Conclusion.* According to the results, the proposed method is simple, specific, linear, accurate, and precise and can be applied for antipsychotic analysis in clinical routine. This method was applied for the determination of the tested antipsychotics in plasma samples taken from 71 individuals.

1. Introduction

Antipsychotics, also named neuroleptics, are widely used to treat several mental disorders. These drugs are most often prescribed for schizophrenia, hallucinations, mania, sleeping disorders, dementia, and bipolar disorders [1]. Finding the right therapy in such pathologies is difficult and complex therapeutic schemes are common. Therefore, therapeutic drug monitoring (TDM) of antipsychotics can aid in optimizing therapy, nonresponse, pharmacokinetic interactions, or noncompliance [1, 2]. Many antipsychotic drugs have high pharmacokinetic variability and small therapeutic range, so the antipsychotics are administered at relatively low daily dosages. As defined by the "The AGNP-TDM Expert Group Consensus Guidelines: Therapeutic Drug Monitoring in Psychiatry", therapeutic ranges are narrow and low plasma concentrations are often seen with antipsychotics. Sensitive and specific analytical methods are required for their reliable, accurate, and precise quantification [3].

All antipsychotics may cause unpleasant side effects and severe poisoning after overdose. Suicide and suicide attempts are also frequent in populations using antipsychotics [4] and several intoxications have been published [5, 6]. In order to contribute to optimal drug therapy, a well-organised TDM service with fast turn-around times is very important.

Apart from the target drugs, plasma samples contain numerous endogenous compounds (proteins, acids, bases, and salts). Therefore, preparation of plasma samples prior to analysis is essential to concentrate the drugs (at trace levels) and to remove the proteins and other macromolecules from the matrix. Simple, fast, and universal sample preparation procedure is advantageous, particularly if suitable for different analysis methods. The endogenous compounds could impair the performance of the analytical column, increase the column backpressure, and suppress or intensify the signals during electrospray ionization (ESI) LC-MS/MS analysis. Sample preparation for antipsychotic analyses was mostly performed by liquid-liquid extraction (LLE) [7–13] and SPE [14–17].

At present, determination of some of these drugs is established by high-performance liquid chromatography with UV detection [8, 17–19], coulometric detection [20, 21], and fluorescence detection [7]. Also there are some reports on gas chromatography-mass spectrometry (GC-MS) methods [17] for the determination of antipsychotic drugs, which, however, require derivatization steps. Capillary electrophoresis methods were reported to detect the antipsychotic drugs, but they are not sensitive and robust enough for biological samples [22].

The usefulness of LC-ESI-MS/MS has been demonstrated for a wide range of applications in the bioanalytical filed. Several LC-MS/MS methods have been reported for the quantification of antipsychotic drugs in biological fluids [10, 12, 23–32]. Different sources of ionization have been used (ESI and APCI) although the ESI seems most interesting for the antipsychotic determination in biological samples. All of the analytes were detected in positive ion mode using MRM. The main analyzer used in these works was triple quadrupole.

Considering limited sample volumes, multianalyte procedures for screening and quantification of analytes using mass spectrometry in different biological matrices have become more and more popular in the field of TDM as well as in clinical and forensic toxicology. Multianalyte procedures are also preferable because they make the analytical process much simpler, faster, and cheaper. For unambiguous identification, it is recommended to monitor two or more ion transitions per compound in combination with acceptable tolerance ratios for these transitions.

One of the most important problems when using ESI is the possible reduction or increase of analyte ionization by coeluting compounds. Ionization influence results from the presence of compounds that can change the efficiency of droplet formation or droplet evaporation, which in turn affects the amount of charged ions in the gas phase that ultimately reach the detector. Such effects (suppression or enhancement of ionization) possibly influence the sensitivity, linearity, accuracy, and precision of the assay in quantitative LC-ESI-MS. Sample preparation could reduce (clean-up) or enhance (preconcentrate) matrix effects. Bioanalytical procedures using LC-ESI-MS should only be used routinely and only be accepted if ion suppression studies by sample preparation and/or chromatographic condition optimization have been performed.

The aim of this study is to develop a high throughput LC-MS/MS method for simultaneous identification and quantification of the most commonly used antipsychotics in human plasma. The focus is also on drugs that often occur in poisonings cases. The quantification procedure was fully validated and proved to be suitable for TDM and clinical toxicology. During method development and validation, recovery, matrix effect, linearity, accuracy, precision, LOD, LOQ, selectivity, specificity, carryover, and stability were tested.

2. Materials and Methods

2.1. Apparatus. An API BioSystem 3200 tandem mass spectrometer, equipped with turbo-ion spray interface was used for measurements. The HPLC system consisted of Perkin-Elmer 200 series autosampler and binary pump, Restek PFP Propyl precolumn (10 mm×2.1 mm) and C18 analytical column (50 mm × 2.1 mm i.d.) with particle size of 5 μm. The mobile phase was degassed using vacuum degasser (Perkin-Elmer 200 series). Data acquisition and processing were achieved using Analyst 1.6 software (Applied Bio-systems).

2.2. Reagents. Chlorpromazine, haloperidol, levomepromazine, olanzapine, risperidone, sulpiride, and repaglinide were obtained from the National Laboratory of Pharmaceutical Products of Algiers. HPLC grade acetonitrile and methanol were obtained from Panreac and Sigma-Aldrich, respectively. Ammonium formate and formic acid used as buffer system were obtained from Analar Normapur and Panreca, respectively.

High-purity water for preparative purpose was produced by double distillation for ultrapure deionised water (18.2 MΩ-cm, type I) by bidistillation apparatus (PureLab Option-Q).

2.3. Sample and Calibration Standard Preparation. The stock solution was prepared by dissolving 10 mg of chlorpromazine, haloperidol, levomepromazine, olanzapine, risperidone, and sulpiride in methanol to obtain a final concentration of 100 mg/L, which was kept at 4°C until analysis. Work solutions were freshly prepared by an adequate dilution of the stock solution with methanol. The calibration standards were obtained by diluting (1: 20) the corresponding work solution with free blank plasma.

2.4. Sample Preparation. In this study, the sample preparation was successfully used to measure antipsychotic concentrations in human plasma samples. Blood samples were collected from patients. After centrifugation at 5,000 rpm for 5 minutes, the obtained plasma samples were transferred to cleaned tubes and kept at 4°C until analysis. 20 μL of the internal standard (repaglinide 1,000 ng/ml) were added to 500 μl of calibration standards, plasma samples, and QC.

Sample preparation consisted of SPE with Oasis HLB cartridges. This extraction includes the following steps:

(1) Conditioning of the cartridge with 1 ml of methanol

(2) Equilibration with 1ml of distilled water

TABLE 1: Optimal instrumental settings.

Injection	
Injection volume	20 μl
Injection temperature	40°C
Flush volume	250 μl
Pre-inject Flush	2
Post-inject Flush	2
Flush speed	Medium
Scan	
Type of scan	MRM: Multiple reaction monitoring
Mode of scan	Positive
Source/Gas parameter	
TEM (temperature)	500 K
GS1 (gas source 1)	40 psi
GS2 (gas source 2)	50 psi
CUR (curtain gas)	10 psi
CAD (collision gas)	Medium
IS (ion spray voltage)	5,000 v
IHE (interface heater)	On

(3) Plasma loading

(4) Rinsing with 1ml distilled water solution containing 5% methanol

(5) Elution with 1 ml of methanol.

2.5. Chromatographic Conditions. An aliquot of 20 μl of each sample and calibration standard was loaded on the column. Gradient reversed phase high-performance liquid chromatography was performed by mobile phase consisting of 90% solvent A (water ammonium formiate 2 mM; pH 2.7) + 10% solvent B (acetonitrile) for 3 minutes and subsequently decreased linearly to 10% solvent A over 4 minutes. The system was reequilibrated to the initial condition over 2 minutes. The reequilibrated condition remained for 1 minute. For all separation process, the mobile phase was set at a flow rate of 800 μL/min.

2.6. MS-MS Conditions. The spectrometric measurements were made in positive mode and operated using MRM mode. The optimal instrumental settings are given in Table 1. All given values (compounds and source/gas parameters) are the averages of three measurements (Tables 1 and 2).

2.7. Validation Experiments. The proposed method was validated for recovery, matrix effect, LOD, LOQ, selectivity, specificity, carryover, linearity, and stability. In addition, an intra- and interday validation were performed to evaluate the accuracy and precision of the measurements. All these validation experiments were carried out to allow a bioanalytical application of the present method.

3. Results and Discussion

3.1. Extraction Experiment. In this study, the extraction efficiency (recovery) was evaluated by comparing detector signals (peak areas) obtained from extracts of QC samples at low, medium, and high levels of all tested antipsychotics with those obtained with the corresponding standard solutions added to extracted matrices (Table 3). For all tested antipsychotics, the mean recoveries were more than 80% showing the higher efficiency of the proposed SPE procedure for some analytes compared to those obtained with other methods in previous studies (Table 6). To our knowledge, this is the first work that uses SPE-HLB cartridges for antipsychotic extraction. Although this work uses a higher amount of plasma compared to other methods, the obtained extract was clear and prevented rapid clogging of injection and ESI needles often encountered with simple LLE.

3.2. Chromatogram. Chromatograms with MRM profiles obtained from human plasma containing the tested antipsychotics are shown in Figure 1. Distinct peaks appeared for all compounds with different retention times (Table 4). The total chromatographic run time for analyte separation was 10 minutes, which is suitable for routine analysis as reported in previous works [23, 27, 31]. Representative chromatogram of drug free blank plasma is shown in Figure 2(a).

3.3. Linearity. The calibration curve was established with six points of standard solutions with all tested antipsychotics. Each point was determined by five calibration runs. The first-order regression equations with correlation coefficients are shown in Table 4. The linearity of the calibration curves was evaluated by the lack of fit test at 5% level of significance. According to F_{reg} values, the regression explains the observed variations. The F_{nl} values indicate nonsignificant lack of fit of the calibration curve. Therefore, the regression equations established a linear relationship between antipsychotic concentrations and detector signals in the tested ranges. Thus, subtherapeutic, therapeutic, and toxic levels of all

Table 2: Optimal compound related settings.

Antipsychotics	Q1 (m/z)	Q3 (m/z)	DP (v)	EP (v)	CE (v)	CXP (v)
Chlorpromazine	319.1	86.1	65	10	32	4
	319.1	214.1	60	15	30	3
	319.1	246.1	57	10	25	4
Haloperidol	376.1	123.0	55	15	55	4
	376.1	165.2	60	10	35	3
	376.1	358.2	47	10	35	2
Levomepromazine	329.2	58.0	50	10	45	2
	329.2	100.1	70	15	38	3
	329.2	210.2	50	15	30	4
Olanzapine	313.2	198.2	45	20	30	3
	313.2	256.0	62	15	35	2
	313.2	169.1	53	20	40	2
Risperidone	411.2	191.2	80	20	40	3
	411.2	110.1	55	20	30	2
	411.2	82.2	54	10	38	3
Sulpiride	342.2	112.1	72	15	35	5
	342.2	214.1	45	15	45	3
	342.2	98.1	43	10	45	2
Repaglinide	453.3	230.2	32	10	43	3
	453.3	174.2	35	15	45	3

Table 3: Recovery values at low, medium, and high levels.

Antipsychotics	Concentration levels (ng/ml)	Recovery (%)	Mean recovery (%)
Chlorpromazine	15	83.75	82.75
	150	82.69	
	450	81.82	
Haloperidol	2	96.22	99.69
	20	100.73	
	60	102.12	
Levomepromazine	5	95.24	93.66
	50	94.53	
	150	91.20	
Olanzapine	5	77.77	85.23
	50	84.90	
	150	93.01	
Risperidone	5	98.75	95.96
	50	94.75	
	150	94.39	
Sulpiride	40	100.28	100.96
	400	98.37	
	1,200	104.22	

tested psychotics can easily be determined without constantly needing of dilution procedures (Table 6).

3.4. Accuracy and Precision. In this study, precision and accuracy were determined by intraday and interday validation using QC at four levels (LOQ, 3 LOQ, 50% and 75% of the calibration curve). The intraday validation was performed by five replicate analysis of QC on the same day. For interday validation, five replicate measurements on three different days were performed. Analyte concentrations in QC were calculated using the regression equation of the calibration curve. Accuracy and precision of the analytical method were calculated and the expressed values are summarized in Table 5.

All the coefficients of variation (CV%) of intraday and interday measurements were not greater than 9% and 11%,

TABLE 4: Retention times, calibration curve linearity, LOD, and LOQ.

Antipsychotics	Rt(min)	Therapeutic range (ng/ml)	Concentration range (ng/ml)	Regression equations	R^2	LOD (ng/ml)	LOQ (ng/ml)
Chlorpromazine	2.45±0.04	30-300	15-450 (15, 30, 75, 150, 300 and 450)	y = 0.06762x + 0.17059	0.99965	3.95 (0.53)	13.17 (1.70∗)
Haloperidol	5.38±0.05	5-17	2-60 (2, 4, 10, 20, 40 and 60)	y = 0.18126x + 0.06470	0.99902	0.36 (0.11)	1.19 (0.49∗)
Levomepromazine	5.62±0.08	15-60	5-150 (5, 10, 25, 50, 100 and 150)	y = 0.03452x + 0.01757	0.99927	1.50 (0.80)	4.99 (2.26∗)
Olanzapine	2.34±0.05	20-80	5-150 (5, 10, 25, 50, 100 and 150)	y = 0.03241x - 0.01622	0.99915	0.87 (0.33)	2.89 (1.01∗)
Risperidone	3.88±0.05	20-60	5-150 (5, 10, 25, 50, 100 and 150)	y = 0.08304x + 0.01389	0.99975	1.378 (0.15)	4.59 (0.52∗)
Sulpiride	5.07±0.05	200-1,000	40-1,200 (40, 80, 200, 400, 800 and 1,200)	y = 0.04711x + 0.05818	0.99984	2.11 (0.41)	7.04 (1.82∗)

(∗): LOD and LOQ using S/N approach.

Table 5: Precision and accuracy values.

Antipsychotics	Concentration level (ng/ml)	Precision (CV%)		Accuracy (relative bias%)	
		Intraday assay (n=5)	Interday assay (n = 5 within 3 days)	Intraday assay (n=5)	Interday assay (n = 5 within 3 days)
Chlorpromazine	15	7.15	1.09	5.97	-4.87
	30	7.75	9.16	9.60	-2.40
	75	4.23	1.65	-1.12	-1.21
	150	4.01	4.69	-3.93	0.48
	300	4.58	1.50	0.53	1.67
	450	4.09	0.63	0.18	-0.74
	CQ1 (13)	4.01	8.50	8.62	6.50
	CQ2 (39)	4.36	1.20	8.67	6.12
	CQ3 (160)	5.15	7.71	0.14	4.80
	CQ4 (250)	3.18	1.91	1.06	2.42
Haloperidol	2	8.58	1.10	-9.02	-5.15
	4	2.20	1.25	-3.00	-0.99
	10	3.98	1.35	-4.36	-3.53
	20	1.98	0.61	2.39	1.59
	40	3.63	0.22	2.93	1.65
	60	4.37	1.26	-1.42	-0.80
	CQ1 (1)	3.66	6.54	4.52	9.65
	CQ2 (3)	3.52	8.56	-4.69	-0.13
	CQ3 (30)	1.13	0.62	7.16	6.00
	CQ4 (45)	1.78	0.87	0.35	-0.66
Levomepromazine	5	3.35	6.10	0.78	11.92
	10	3.12	6.31	8.77	10.55
	25	2.69	2.85	-1.05	-2.18
	50	6.08	1.95	-5.12	-6.56
	100	6.88	2.86	1.85	2.50
	150	4.82	0.96	-0.27	-0.40
	CQ1 (2)	2.16	7.17	-3.57	11.01
	CQ2 (6)	3.54	9.77	9.73	11.26
	CQ3 (80)	2.41	2.14	7.82	3.01
	CQ4 (120)	3.21	1.43	0.67	-1.90
Olanzapine	5	1.56	2.22	-5.66	5.13
	10	5.79	3.47	3.19	10.40
	25	5.51	3.59	-6.41	-0.18
	50	2.80	1.12	-7.91	-5.97
	100	2.72	2.33	0.64	1.82
	150	3.15	0.79	-0.74	-0.22
	CQ1 (3)	1.79	7.25	-5.05	9.78
	CQ2 (9)	3.87	3.31	-8.60	1.54
	CQ3 (80)	5.10	1.68	-0.09	3.14
	CQ4 (120)	4.72	2.52	3.37	-1.13

Table 5: Continued.

Antipsychotics	Concentration level (ng/ml)	Precision (CV%)		Accuracy (relative bias%)	
		Intraday assay (n=5)	Interday assay (n = 5 within 3 days)	Intraday assay (n=5)	Interday assay (n = 5 within 3 days)
Risperidone	5	7.58	2.78	6.48	2.00
	10	4.28	1.77	-0.88	-6.12
	25	5.01	4.69	-5.73	-4.09
	50	4.70	1.04	2.49	2.52
	100	2.24	0.62	0.42	1.48
	150	2.63	0.56	-0.30	-0.79
	CQ1 (5)	3.58	2.81	4.36	-1.17
	CQ2 (15)	4.36	1.17	4.54	5.04
	CQ3 (80)	1.72	0.23	-1.04	0.41
	CQ4 (120)	1.75	0.57	-2.01	0.05
Sulpiride	40	3.99	5.91	9.60	9.10
	80	1.82	1.30	7.34	7.01
	200	5.08	1.37	-4.52	-4.55
	400	3.73	2.01	-1.12	-0.34
	800	2.63	1.22	0.09	-0.10
	1,200	1.17	1.79	0.18	0.17
	CQ1 (7)	4.62	5.15	-7.46	-4.13
	CQ2 (21)	2.67	10.28	-8.20	-1.40
	CQ3 (600)	2.42	2.16	-0.54	0.13
	CQ4 (800)	2.23	0.53	0.15	-1.09

Table 6: Recovery, concentration range, LOD, and LOQ from different mass spectrometry methods.

Antipsychotics	Extraction recovery (%)	Concentration range (ng/ml)	LOD (ng/ml)	LOQ (ng/ml)
Chlorpromazine	67 [10]	30- 300 [10]	11.3 [24]	15 [24]
	84.5 [28]	15-600 [28]	7.5 [28]	15 [28]
	94.13 [32]	10-1,000 [30]	0.3 [32]	1 [29]
		1-50 [32]		1 [32]
Haloperidol	65 [10]	5-17 [10]	3.8 [24]	5 [24]
	93.97 [23]	1-60 [26]	<0.5 [27]	1 [26]
	86.95 [12]	2.5-30 [28]	1 [28]	1 [27]
	88.5 [33]	1-20 [30]	0.3 [32]	2.5 [28]
	83 [28]	1-50 [32]		0.5 [29]
				0.23 [30]
				1 [32]
Levomepromazine	70 [10]	15-60 [10]	5 [28]	7.5 [24]
	81 [28]	10-300 [28]		10 [28]
		10-1,000 [30]		0.47 [30]
Olanzapine	92.02 [23]	2-200 [25]	1 [27]	2 [25]
	96 [25]	10-160 [28]	2 [28]	5 [27]
	93.95 [12]	10-1,000 [30]		10 [28]
	102 [33]			0.5 [29]
	77 [28]			1.83 [30]
	86 [31]			
Risperidone	69 [10]	5- 60 [10]	1.9 [24]	2.5 [24]
	89.6 [23]	1.5-60 [28]	<1 [27]	5 [27]
	94 [25]	1-50 [30]	0.8 [28]	1.5 [28]
	88.85 [12]			0.67 [30]
	84 [33]			
	80.5 [28]			
	88.5 [31]			
Sulpiride	12 [10]	200-1,000 [10]	75 [24]	100 [24]
	105 [28]	100-1,500 [28]	2 [27]	20 [27]
		100-10,000 [30]	80 [28]	100 [28]
				8.3 [30]

Sample volume (μl): 50 [10], 200 [12], 250 [23], 500 [24], 200 [25], 100 [26], 500 [27], 500 [28], 200 [29], 100 [30], 250 [31], 200 [32], and 500 [33].
Sample preparation technique: LLE [10, 12, 23–26, 28, 30, 31, 33]; SPE [27, 29, 32].

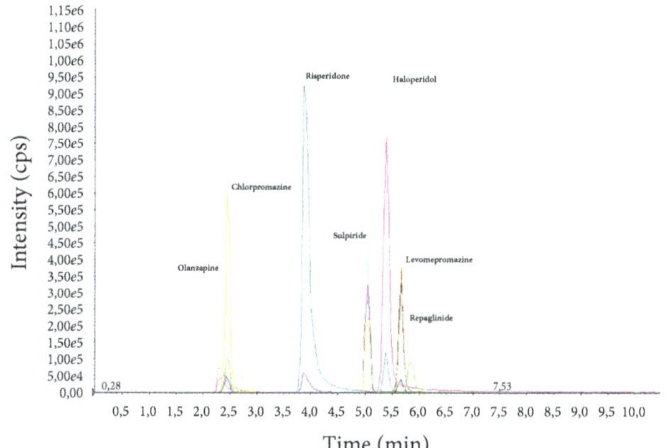

Figure 1: MRM chromatogram of blank plasma sample spiked with antipsychotics (chlorpromazine, haloperidol, levomepromazine, olanzapine, risperidone, and sulpiride) at concentration of 75, 40, 100, 100, 100, and 80 ng/ml, respectively.

respectively. These results indicate a multilevel high precision of the present method regardless of time factor. All accuracy measures of intraday and interday validation were less than 10% and 12% (absolute values), respectively, which indicate a multilevel high accuracy of the present method regardless of time factor. Moreover, the intraday precision in all calibration range (repeatability) was acceptable at 5% level of significance according to Cochran test (C (5%)=0.727). The interday precision (reproducibility) in all calibration range was also acceptable at 5% level of significance according to Grubbs test (G (5%)=1.115).

3.5. Limits of Detection and Quantification. The LOD and LOQ were calculated using S/N (S/N ≥ 3 for LOD and S/N ≥ 10 for LOQ (Figure 2(b))) as well as the slope and the standard deviation of calibration curve intercepts (3 s(b0)/b and 10 s(b0)/b values, respectively) (Table 4). The highest values between the two approaches were considered as LOD

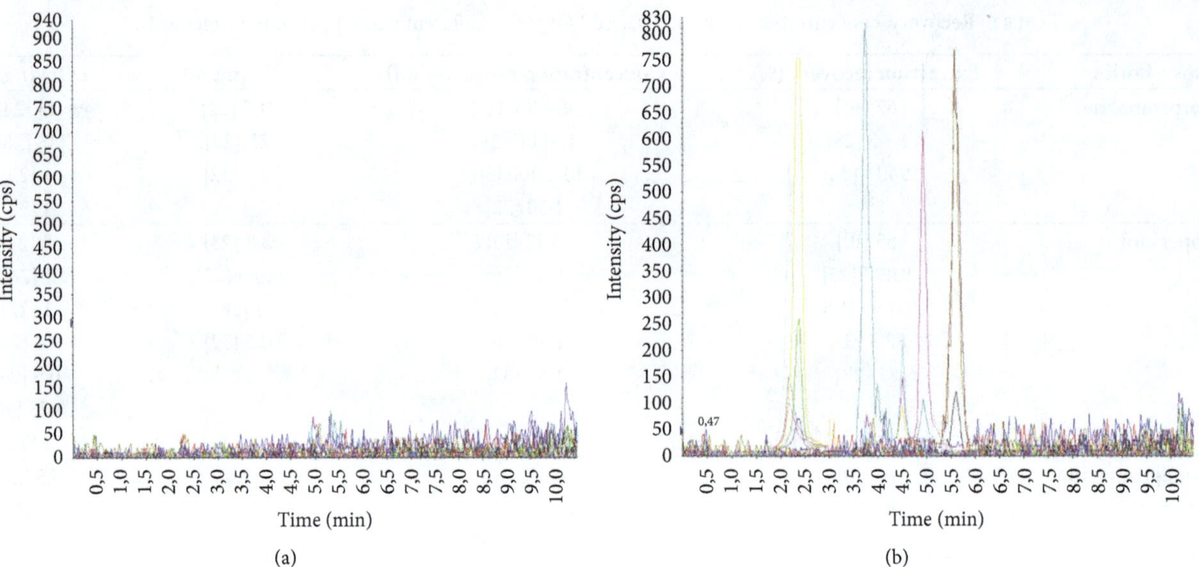

FIGURE 2: MRM chromatogram of free blank plasma (a) and blank plasma sample spiked with antipsychotics at LOQ (b) (chlorpromazine (1.70 ng/ml), haloperidol (0.49 ng/ml), levomepromazine (2.26 ng/ml), olanzapine (1.01 ng/ml), risperidone (0.52 ng/ml), and sulpiride (1.82 ng/ml)).

and LOQ for routine analyses. In general, the values of LOD and LOQ using S/N approach were below of those obtained in previous studies and demonstrate that concentrations below therapeutic ranges can be reached and determined (Tables 4 and 6).

3.6. Matrix Effect, Selectivity, Specificity, and Carryover. With respect to matrix effect and selectivity, no major interferences (0.2 times the response of the LOQ) were detected at the retention times of all tested antipsychotics or the internal standard in 10 batches of free blank plasma and no suppression effect was found to all tested antipsychotics. In addition, common drugs, when injected into the mass spectrometer, do not generate any interfering ions with those selected for antipsychotic quantification.

Due to the large range of the calibration curve in this method, the carryover was assessed by measuring detector signals (peak areas) of blank samples after the higher calibration point. The accepted limit for carry over was that the detector signals of blank samples must be less than 20% of the LOQ signal. The obtained carryover in this method complied with the acceptable limits.

3.7. Stability. The stability of the tested antipsychotics in plasma matrix was estimated using QC stored in different conditions (Table 7). The stability was expressed by the relative bias of the found concentrations to the nominal concentrations. The tested antipsychotics were considered stable when less than 15% of the nominal concentration is obtained using relative bias.

The results indicate that the signs of deterioration of all tested antipsychotics were within the acceptance limits in different conditions (Table 7).

3.8. Application. The present method was successfully used in our laboratory for routine analysis of plasma samples taken from patients receiving these antipsychotics in TDM context. The method was also used for clinical analyses of plasma samples obtained from intoxicated individuals. Overall, plasma samples from 71 patients (48 males and 23 females) were analyzed and some of the patients were examined repeatedly. A summary of the obtained results is given in Table 8. Three samples which gave results above the calibration range were diluted with calf serum and reanalyzed.

The average age of the patients was 29.3 ± 7.4 years (14-69) for males and 37.1 ± 6.9 years (16-72) for females. Examples of representative MRM chromatograms of plasma samples from six different patients taking one of the studied drugs are shown in Figures 3(a)–3(f).

According to our experience, TDM could help to enhance the therapeutic response, design optimal dosing regimens, and avoid the build-up of excessively high and potentially toxic drug concentrations, as well as monitor patient's adherence to treatment.

4. Conclusion

A simple method was developed for antipsychotic determination in plasma samples by tandem mass spectrometry detection using turbo-ion spray as an interface. SPE procedure with good recovery was performed in order to render this analytical method relevant in routine clinical diagnosis. For separation, a gradient reversed phase high-performance liquid chromatography with low time consumption was performed.

Validation experiments showed good results in terms of accuracy, precision, and sensitivity in all concentration levels.

TABLE 7: Evaluation of sample storage procedure.

Antipsychotics	Nominal concentrations of QC (ng/ml)	Stability: Storage condition of samples
Chlorpromazine	13	Four freeze/thaw cycles
	39	Bench top (6 h)
	160	Autosampler stability (24 h)
	250	Preserved during 1 month at -20°C
Haloperidol	1	Three freeze/thaw cycles
	3	Bench top (6 h)
	30	Autosampler stability (12 h)
	45	Preserved during 1 month at -20°C
Levomepromazine	2	Four freeze/thaw cycles
	6	Bench top (6 h)
	80	Autosampler stability (24 h)
	120	Preserved during 1 month at -20°C
Olanzapine	3	Four freeze/thaw cycles
	9	Bench top (6 h)
	80	Autosampler stability (24 h)
	120	Preserved during 1 month at -20°C
Risperidone	5	Three freeze/thaw cycles
	15	Bench top (6 h)
	80	Autosampler stability (24 h)
	120	Preserved during 1 month at -20°C
Sulpiride	7	Four freeze/thaw cycles
	21	Bench top (6 h)
	600	Autosampler stability (24 h)
	800	Preserved during 1 month at -20°C

TABLE 8: Overview of antipsychotic determination in plasma samples taken from 71 patients.

Antipsychotics	Number of positive cases	Mean (ng/ml)	Minimum (ng/ml)	Maximum (ng/ml)	< therapeutic range	> therapeutic range
Chlorpromazine	13	270.76	21.44	440.77	3	4
Haloperidol	30	27.42	1.29	131.46	4	21
Levomepromazine	8	79.10	17.08	375.27	0	4
Olanzapine	9	141.33	18.31	168.46	1	3
Risperidone	6	116.70	12.34	280.93	1	1
Sulpiride	13	800.61	10.37	1431.72	2	7

From the viewpoints of calibration range, this analytical method seems recommendable for TDM and toxicological diagnosis without constantly needing of dilution procedures. The LOD and LOQ appear to be sufficiently low to evaluate subtherapeutic concentrations, which is very useful in therapeutic monitoring.

As an application, this method was applied for antipsychotic determination in 71 plasma samples collected from different cases.

Additional Points

Highlights. This method includes the following highlights: (i) enhanced extraction recoveries; (ii) excellent linearity; (iii)

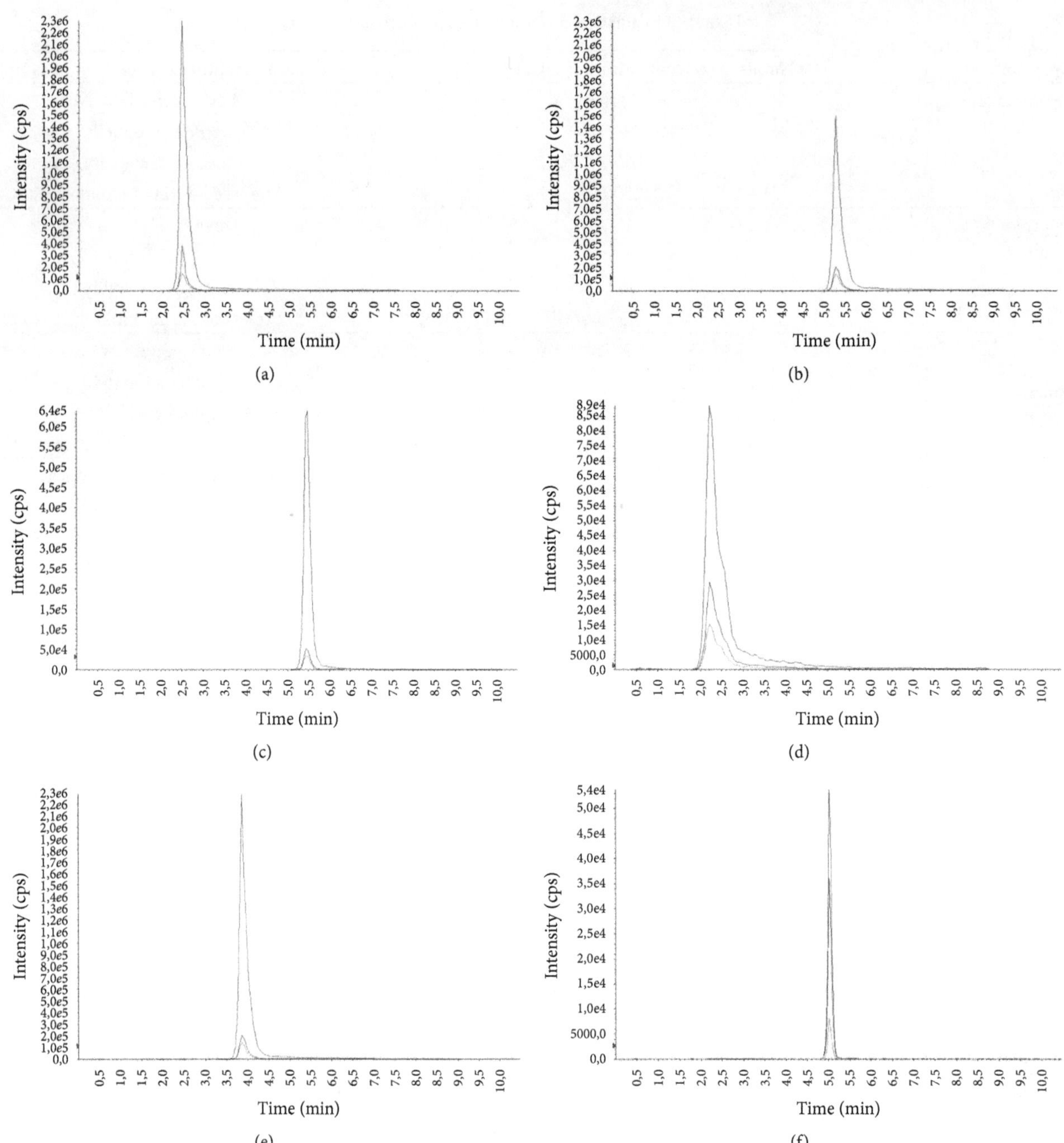

FIGURE 3: Representative chromatograms of patients' plasma samples containing antipsychotics. (a) Chlorpromazine (322.52 ng/ml), (b) haloperidol (7.31 ng/ml), (c) levomepromazine (161. 83 ng/ml), (d) olanzapine (28.58 ng/ml), (e) risperidone (280.93 ng/ml), and (f) sulpiride (10.37 ng/ml).

high precision and accuracy; (iv) low limit of detection and quantification for all tested antipsychotics; (v) no matrix effect detected; (vi) application in routine analysis for 71 individuals.

Acknowledgments

The authors wish to express their deep gratitude to other members of the National Center of Toxicology for their skillful cooperation.

References

[1] L. Patteet, M. Morrens, K. E. Maudens, P. Niemegeers, B. Sabbe, and H. Neels, "Therapeutic drug monitoring of common antipsychotics," *Therapeutic Drug Monitoring*, vol. 34, no. 6, pp. 629–651, 2012.

[2] P. Baumann, C. Hiemke, S. Ulrich et al., "The AGNP-TDM expert group consensus guidelines: Therapeutic Drug Monitoring in psychiatry," *Pharmacopsychiatry*, vol. 37, no. 6, pp. 243–265, 2004.

[3] C. Hiemke, P. Baumann, N. Bergemann et al., "AGNP consensus guidelines for therapeutic drug monitoring in psychiatry:

update 2011," *Pharmacopsychiatry*, vol. 44, no. 6, pp. 195–235, 2011.

[4] E. C. Harris and B. Barraclough, "Suicide as an outcome for mental disorders. A meta-analysis," *The British Journal of Psychiatry*, vol. 170, pp. 205–228, 1997.

[5] D. S. Ravin and J. W. Levenson, "Fatal cardiac event following initiation of risperidone therapy," *Annals of Pharmacotherapy*, vol. 31, no. 7-8, pp. 867–870, 1997.

[6] A. C. Springfield and E. Bodiford, "An overdose of risperidone," *Journal of Analytical Toxicology*, vol. 20, no. 3, pp. 202-203, 1996.

[7] N. Kishikawa, C. Hamachi, Y. Imamura et al., "Determination of haloperidol and reduced haloperidol in human serum by liquid chromatography after fluorescence labeling based on the Suzuki coupling reaction," *Analytical and Bioanalytical Chemistry*, vol. 386, no. 3, pp. 719–724, 2006.

[8] L. Garay Garcia, I. Forfar-Bares, F. Pehourcq, and C. Jarry, "Simultaneous determination of four antipsychotic drugs in plasma by high-performance liquid chromatography: Application to management of acute intoxications," *Journal of Chromatography B*, vol. 795, no. 2, pp. 257–264, 2003.

[9] E. Tanaka, T. Nakamura, M. Terada et al., "Simple and simultaneous determination for 12 phenothiazines in human serum by reversed-phase high-performance liquid chromatography," *Journal of Chromatography B*, vol. 854, no. 1-2, pp. 116–120, 2007.

[10] D. Remane, M. R. Meyer, F. T. Peters, D. K. Wissenbach, and H. H. Maurer, "Fast and simple procedure for liquid-liquid extraction of 136 analytes from different drug classes for development of a liquid chromatographic-tandem mass spectrometric quantification method in human blood plasma," *Analytical and Bioanalytical Chemistry*, vol. 397, no. 6, pp. 2303–2314, 2010.

[11] J. Gradinaru, A. Vullioud, C. B. Eap, and N. Ansermot, "Quantification of typical antipsychotics in human plasma by ultra-high performance liquid chromatography tandem mass spectrometry for therapeutic drug monitoring," *Journal of Pharmaceutical and Biomedical Analysis*, vol. 88, pp. 36–44, 2014.

[12] L. Patteet, K. E. Maudens, B. Sabbe, M. Morrens, M. De Doncker, and H. Neels, "High throughput identification and quantification of 16 antipsychotics and 8 major metabolites in serum using ultra-high performance liquid chromatography-tandem mass spectrometry," *Clinica Chimica Acta*, vol. 429, pp. 51–58, 2014.

[13] R. Uřinovská, H. Brozmanová, P. Šištík et al., "Liquid chromatography-tandem mass spectrometry method for determination of five antidepressants and four atypical antipsychotics and their main metabolites in human serum," *Journal of Chromatography B*, vol. 907, pp. 101–107, 2012.

[14] M. C. Sampedro, N. Unceta, A. Gómez-Caballero et al., "Screening and quantification of antipsychotic drugs in human brain tissue by liquid chromatography-tandem mass spectrometry: Application to postmortem diagnostics of forensic interest," *Forensic Science International*, vol. 219, no. 1-3, pp. 172–178, 2012.

[15] J. Wang, H. Huang, Q. Yao et al., "Simple and accurate quantitative analysis of 16 antipsychotics and antidepressants in human plasma by ultrafast high-performance liquid chromatography/tandem mass spectrometry," *Therapeutic Drug Monitoring*, vol. 37, no. 5, pp. 649–660, 2015.

[16] T. Kumazawa, C. Hasegawa, S. Uchigasaki, X.-P. Lee, O. Suzuki, and K. Sato, "Quantitative determination of phenothiazine derivatives in human plasma using monolithic silica solid-phase extraction tips and gas chromatography-mass spectrometry," *Journal of Chromatography A*, vol. 1218, no. 18, pp. 2521–2527, 2011.

[17] B. M. da Fonseca, I. E. D. Moreno, M. Barroso, S. Costa, J. A. Queiroz, and E. Gallardo, "Determination of seven selected antipsychotic drugs in human plasma using microextraction in packed sorbent and gas chromatography-tandem mass spectrometry," *Analytical and Bioanalytical Chemistry*, vol. 405, no. 12, pp. 3953–3963, 2013.

[18] L. J. Dusci, L. Peter Hackett, L. M. Fellows, and K. F. Ilett, "Determination of olanzapine in plasma by high-performance liquid chromatography using ultraviolet absorbance detection," *Journal of Chromatography B*, vol. 773, no. 2, pp. 191–197, 2002.

[19] H. Aboul-Enein, I. Ali, and H. Hoenen, "Rapid determination of haloperidol and its metabolites in human plasma by HPLC using monolithic silica column and solid-phase extraction," *Biomedical Chromatography*, vol. 20, no. 8, pp. 760–764, 2006.

[20] C. Sabbioni, M. A. Saracino, R. Mandrioli, L. Albers, G. Boncompagni, and M. A. Raggi, "Rapid analysis of olanzapine and desmethylolanzapine in human plasma using high-performance liquid chromatography with coulometric detection," *Analytica Chimica Acta*, vol. 516, no. 1-2, pp. 111–117, 2004.

[21] D. S. Schatz and A. Saria, "Simultaneous determination of paroxetine, risperidone and 9-hydroxyrisperidone in human plasma by high-performance liquid chromatography with coulometric detection," *Pharmacology*, vol. 60, no. 1, pp. 51–56, 2000.

[22] V. Pucci, M. Raggi, and E. Kenndler, "Separation of eleven central nervous system drugs by capillary zone electrophoresis," *Journal of Chromatography B: Biomedical Sciences and Applications*, vol. 728, no. 2, pp. 263–271, 1999.

[23] G. Zhang, A. V. Terry Jr., and M. G. Bartlett, "Liquid chromatography/tandem mass spectrometry method for the simultaneous determination of olanzapine, risperidone, 9-hydroxyrisperidone, clozapine, haloperidol and ziprasidone in rat plasma," *Rapid Communications in Mass Spectrometry*, vol. 21, no. 6, pp. 920–928, 2007.

[24] D. Remane, M. R. Meyer, D. K. Wissenbach, and H. H. Maurer, "Ultra high performance liquid chromatographic-tandem mass spectrometric multi-analyte procedure for target screening and quantification in human blood plasma: Validation and application for 31 neuroleptics, 28 benzodiazepines, and Z-drugs," *Analytical and Bioanalytical Chemistry*, vol. 401, no. 4, pp. 1341–1352, 2011.

[25] D. S. Fisher, S. J. Partridge, S. A. Handley, L. Couchman, P. E. Morgan, and R. J. Flanagan, "LC-MS/MS of some atypical antipsychotics in human plasma, serum, oral fluid and haemolysed whole blood," *Forensic Science International*, vol. 229, no. 1-3, pp. 145–150, 2013.

[26] J. M. Juenke, P. I. Brown, F. M. Urry, K. L. Johnson-Davis, and G. A. McMillin, "Simultaneous UPLC-MS/MS assay for the detection of the traditional antipsychotics haloperidol, fluphenazine, perphenazine, and thiothixene in serum and plasma," *Clinica Chimica Acta*, vol. 423, pp. 32–34, 2013.

[27] C. Kratzsch, F. T. Peters, T. Kraemer, A. A. Weber, and H. H. Maurer, "Screening, library-assisted identification and validated quantification of fifteen neuroleptics and three of their metabolites in plasma by liquid chromatography/mass spectrometry with atmospheric pressure chemical ionization," *Journal of Mass Spectrometry*, vol. 38, no. 3, pp. 283–295, 2003.

[28] D. Montenarh, M. Hopf, H. H. Maurer, P. Schmidt, and A. H. Ewald, "Development and validation of a multi-analyte LC-MS/MS approach for quantification of neuroleptics in whole blood, plasma, and serum," *Drug Testing and Analysis*, vol. 8,

no. 10, pp. 1080–1089, 2016.

[29] M. A. L. Pinto, I. D. de Souza, and M. E. C. Queiroz, "Determination of drugs in plasma samples by disposable pipette extraction with C18-BSA phase and liquid chromatography–tandem mass spectrometry," *Journal of Pharmaceutical and Biomedical Analysis*, vol. 139, pp. 116–124, 2017.

[30] H. Kirchherr and W. N. Kühn-Velten, "Quantitative determination of forty-eight antidepressants and antipsychotics in human serum by HPLC tandem mass spectrometry: a multi-level, single-sample approach," *Journal of Chromatography B*, vol. 843, no. 1, pp. 100–113, 2006.

[31] N. Ansermot, M. Brawand-Amey, A. Kottelat, and C. B. Eap, "Fast quantification of ten psychotropic drugs and metabolites in human plasma by ultra-high performance liquid chromatography tandem mass spectrometry for therapeutic drug monitoring," *Journal of Chromatography A*, vol. 1292, pp. 160–172, 2013.

[32] L. Zhang, P. Wu, Q. Jin, Z. Hu, and J. Wang, "Multi-residue analysis of sedative drugs in human plasma by ultra-high performance liquid chromatography tandem mass spectrometry," *Journal of Chromatography B*, vol. 1072, pp. 305–314, 2018.

[33] D. Montenarh, M. Hopf, S. Warth, H. H. Maurer, P. Schmidt, and A. H. Ewald, "A simple extraction and LC-MS/MS approach for the screening and identification of over 100 analytes in eight different matrices," *Drug Testing and Analysis*, vol. 7, no. 3, pp. 214–240, 2015.

3

The Use of Dried Blood Spots for the Quantification of Antihypertensive Drugs

Alexander Chernonosov

Institute of Chemical Biology and Fundamental Medicine, Siberian Branch of Russian Academy of Sciences, Academician Lavrentiev Avenue 8, Novosibirsk 630090, Russia

Correspondence should be addressed to Alexander Chernonosov; alexander.chernonosov@niboch.nsc.ru

Academic Editor: Jan Åke Jönsson

Hypertension or high blood pressure is a harbinger of cardiovascular diseases. There are several classes of drugs used to treat hypertension. This review discusses the use of dried blood spots (DBSs) for the quantification by mass spectrometry (MS), tandem mass spectrometry (MS/MS), or, in some cases, by fluorescence detection methods the following antihypertensive medications: angiotensin-converting enzyme inhibitors (ramipril, ramiprilat, captopril, and lisinopril); angiotensin II receptor antagonists (valsartan, irbesartan, losartan, and losartan carboxylic acid); calcium channel blockers (verapamil, amlodipine, nifedipine, pregabalin, and diltiazem); α blockers (guanfacine, doxazosin, and prazosin); β blockers (propranolol, bisoprolol, atenolol, and metoprolol); endothelin receptor antagonists (bosentan and ambrisentan); and statins (simvastatin, atorvastatin, and rosuvastatin).

1. Introduction

Cardiovascular disease is the number one cause of death in the world [1]. Hypertension is the main attributable risk factor of death and is responsible for approximately half of the cases of cerebrovascular and ischemic heart diseases [2, 3]. Moreover, blood pressure over 140/90 mmHg is a cause of heart failure and chronic kidney disease in 20% and 23% of cases, respectively [3–5]. As a rule, high blood pressure is widespread among the elderly [3]. Blood pressure above 140/90 mmHg is observed in more than 50% of people aged 60 years and 75% over the age of 70 years [6, 7], and the treatment of hypertension is becoming an increasingly urgent problem with the ageing of society [3]. On the other hand, blood pressure >140/90, despite treatment with a diuretic and two other antihypertensive drugs of various classes, is defined as drug-resistant hypertension [8, 9]. In the USA, patients who need to take more than four antihypertensive drugs to achieve a normal blood pressure level are also considered resistant hypertensives [10].

Nevertheless, most people with hypertension require more than one drug for effective control of the disease. Because hypertension can develop in different biochemical ways, various classes of antihypertensive drugs have been developed. Diuretics, calcium channel blockers (CCBs), α blockers, β blockers (BBs), and inhibitors of the renin-angiotensin system are used for initial antihypertension therapy [11]. In addition, angiotensin-converting enzyme (ACE) inhibitors, angiotensin II receptor antagonists (ARAs), and endothelin receptor antagonists (ERAs) have been used for antihypertensive therapy over the past 20 years [12, 13]. CCBs, ARAs, and BBs have been the most prescribed antihypertensive medications [14–16]. Treatment of most patients (67.92 %) includes more than one drug. The most commonly used combination of medications has been CCB + BB + α blocker (7.55 %) [13]. For effective treatment, patients should follow the indications for taking a given drug and the prescribed dose [17]. The big problem is that ~50% of all patients with heart disease do not adhere to their prescribed regimen [18–20]. Often, such patients are referred to as patients with apparently resistant hypertension [9]. Patients, who ignore or do not adhere to their prescribed medication [21], more often as a result show considerable morbidity, higher costs of care, and mortality [22–26]. Many clinicians fail to assess blood pressure regularly or effectively titrate and regulate the dose of drugs; these shortcomings result in ineffective treatment [27].

The use of dried blood spots (DBSs) can simplify the methods for determining the concentrations of drugs. The DBS sampling technique is minimally invasive, and capillary blood can be obtained from a finger prick with a lancet by the patients themselves or guardians with minimal training. Such a sampling technique is also suitable even for small children [28] and is ideal for routine clinical testing [29] or helps with recruitment of subjects for preclinical or clinical studies [30]. Besides, DBSs reduce to a minimum the risk of infection with HIV and other infectious pathogens [31]. Moreover, DBSs offer a simpler storage and easier transfer by mail to the assigned laboratory, preventing unnecessary costs [32–34]. They should be well desiccated after sampling (2–3 hours minimum). The combined advantage of the above benefits coupled with improved analytical instrumental capability [35] has been recognized for the use of this methodology for various applications including therapeutic drug monitoring [36–38], toxicokinetic studies [39], and preclinical or clinical pharmacokinetic studies [30, 39–43].

In the current paper, quantification of major classes of antihypertensive drugs and statins in DBS by liquid chromatography with mass spectrometry (LC-MS), liquid chromatography with tandem mass spectrometry (LC-MS/MS), and fluorescence detection methods in recent years are reviewed. Parameters of analysis that can influence sensitivity were analyzed: types of used DBS cards, mass-spectrometers and detection modes, and the diameter of a disk punched for analysis, linear concentration range, elution solvents, extraction procedures, recovery, and stability of DBS samples.

2. DBS Analysis

The technique of DBS was first suggested by Dr. Robert Guthrie in the 1960s for diagnosis of phenylketonuria by neonatal screening [44]. In this sampling technique, a small volume of blood (30–50 μL) is collected on filter paper or DBS cards based on a cellulose matrix with specific pore size and thickness [45, 46].

Several types of DBS cards are usually used for blood sampling in quantitative drug analysis. The FTA DMPK-A/ FTA card and FTA DMPK-B/ FTA elute card are coated and contain additives for analysis of nucleic acids [47]. Four types of DBS cards are uncoated: FTA DMPK-C/ 31ETF, Whatman 903, and two cards manufactured by Ahlstrom (226 and 237) [45]. After drying, a DBS sample is stored under conditions suitable for the analyte, usually at room temperature (RT) or 4°C. The sample could be sent by mail to a remote analytical laboratory if required.

A disk of a certain diameter is then punched and subjected to an extraction procedure. This procedure is simple: a suitable solvent is applied to dissolve the analyte in the punched disk of a DBS. Sometimes a second step, involving solid phase extraction (SPE) or liquid extraction (LLE), is required for further analysis [45]. The analysis involves a small amount of an analyte (as a consequence of the small volume of blood in a DBS): nanogram or even picogram amounts are suitable for fluorescence, MS, or MS/MS detection methods. Thus, the combination of the DBS technique with such methods allows for analyzing low concentrations of drugs in blood samples [30].

MS/MS includes selected-reaction monitoring (SRM) and multiple-reaction monitoring (MRM) modes, when only the transition between a selected precursor and product ions is detected. This approach enables excluding other possible ions and thus significantly improves sensitivity and selectivity of the assay [74].

At present, triple quadrupole (QQQ) mass-spectrometers remain the gold standard for quantification [74, 75], but quadrupole time-of-flight (QTOF) mass-spectrometers are also employed in many DBS studies.

Some analytical parameters are important for comparing assays of the same medication. First of all, there is the type of DBS card, size of the disk punched for analysis, and recovery of the analyte from a DBS; these directly affect the lower limit of quantitation (LLOQ). Assay sensitivity also depends on equipment type and detection mode. The extraction procedure with a suitable elution solvent is often characterized by simplicity. The linear concentration range and stability of DBS samples describe the assay in general. Table 1 summarizes the methods of quantification of antihypertensive drugs in DBS, published from 2009 to 2017. The search for the articles was performed in the PubMed database by means of the following keywords: DBS or "dried blood spots" with the name of medications.

DBS sampling was reported for several classes of antihypertensive drugs, such as ACE inhibitors, ARAs, CCBs, alpha blockers, BBs, and ERAs as well as for statins. Although statins are not antihypertensive drugs themselves, they are often used as part of a combination therapy and enhance the antihypertensive effect of ACE inhibitors or CCBs, but not BBs or diuretics [76]. In most of the reviewed studies, statins have been simultaneously analyzed along with antihypertensive drugs.

3. ACE Inhibitors

These drugs block the peripheral conversion of angiotensin I to angiotensin II [77]. The DBS technique has predominantly been used by Tanna and coauthors for quantification of ramipril [48, 50, 51], captopril [53], and lisinopril [50]. Ramipril is the most studied medication among ACE inhibitors and is stable up to 84 days at RT with good recovery (~90%) from Whatman 903 and Ahlstrom 226 cards. Ramipril has been detected by QTOF mass-spectrometer with 417.2384 m/z. The only difference in the elution procedure was the extraction solvent: 300 μL of methanol (MeOH) [50] or 150 μL of MeOH/H_2O (70:30, v/v) [48, 49, 51], and the latter showed better recovery. Lisinopril was analyzed in the same assay [50] in TOF mode at 406.2336 m/z. Captopril has been quantitated with similar results in SRM and TOF modes [53]. For this analysis, the Whatman 903 card was pretreated by dithiothreitol (DDT) and extraction solvent containing 10% v/v 200 mM DTT. It was necessary to reconstitute captopril from its disulfide dimer in blood

Table 1: Published DBS sampling techniques for analysis of antihypertensive medications and statins.

Drug analyzed	DBS card/paper used	Diameter of disk punched for analysis, mm	Calibration range, ng/mL	Stability of DBS samples	Elution procedure (min)	V, Extraction solvent	Recovery (%)	Mass-spectrometer	Detection mode	Mass or transmission, m/z	Internal standard	Ref.
Angiotensin converting enzyme inhibitors												
Ramipril	Whatman 903	8	0.5 - 100	84 days, RT	mixed (1 min) + sonicated (30 min)	150 μL, MeOH/H_2O (70:30, v/v)	≥92	Agilent G6530A QTOF	TOF	417.2384	Atenolol	[48]
	Whatman 903	8	-	84 days, RT	mixed (1 min) + sonicated (30 min)	150 μL, MeOH/H_2O (70:30, v/v)	-	Agilent 6530 QTOF	TOF	417.2384	Atenolol	[49]
	Whatman 903	8	0.1 - 100	70 days, RT	mixed (1 min) + sonicated (30 min)	300 μL, MeOH	≥87	Agilent G6530A QTOF	TOF	417.2384	Atenolol-d7	[50]
	Ahlstrom 226	8	See [48]	-	mixed (1 min) + sonicated (30 min)	150 μL, MeOH/H_2O (70:30, v/v)	See [48]	Agilent G6530A QTOF	TOF	417.2384	Atenolol	[51]
Ramiprilat	Whatman 903	Whole DBS	5 - 250	21 days, 25°C 21 days, 37°C 21 days,-20°C	mixed (1 min) + sonicated (10 min) + mixed (45 min)	300 μL, MeOH/0.1% FA (90:10, v/v)	65-81	Agilent 6460 QQQ	MRM	389.2 ⟶ 206.1	Ramiprilat-d5	[52]
	Ahlstrom 226	8	See [48]	-	mixed (1 min) + sonicated (30 min)	150 μL, MeOH/H_2O (70:30, v/v)	See [48]	Agilent G6530A QTOF	TOF	389.2071	Atenolol	[51]
Captopril	Whatman 903	8	10 - 400	84 days, RT	mixed (1 min) + sonicated (30 min)	200 μL, MeOH/H_2O (60:40, v/v) with 10% v/v 200 mM DTT	90 ± 10	Agilent 1100 Ion Trap Agilent 6530 TOF	SRM TOF	218.0 218.0845	Enalapril maleate	[53]
Lisinopril	Whatman 903	8	0.1 - 100	70 days, RT	mixed (1 min) + sonicated (30 min)	300 μL, MeOH	≥87	Agilent G6530A QTOF	TOF	406.2336	Atenolol-d7	[50]
Angiotensin II receptor antagonists												
Valsartan	Whatman 903	8	50 - 4000	70 days, RT	mixed (1 min) + sonicated (30 min)	300 μL MeOH	≥87	Agilent G6530A QTOF	TOF	436.2343	Atenolol-d7	[50]
	Whatman 903	8	-	84 days, RT	mixed (1 min) + sonicated (30 min)	150 μL, MeOH/H_2O (70:30, v/v)	-	Agilent 6530 QTOF	TOF	436.2343	Atenolol	[49]
	Whatman DMPK-A	10	6 - 2000	30 days, RT	mixed (15 min)	1000 μL, MeOH	98	-	fluorescence	-	-	[54]
Irbesartan	Whatman 903	3	20	30 days, RT	mixed (5 min) + sonicated (10 min)	500 μL, MeOH/H_2O (50:50, v/v)		Agilent 1100 Ion Trap	MRM	427.3 ⟶ 193.0	Irbesartan	[55]
	Whatman DMPK-A	10	6 - 2000	30 days, RT	mixed (15 min)	1000 μL, MeOH	99	-	fluorescence	-	-	[54]
Losartan	Whatman 903	8	5 - 1000	70 days, RT	mixed (1 min) + sonicated (30 min)	300 μL, MeOH	≥87	Agilent G6530A QTOF	TOF	423.1695	Atenolol-d7	[50]
	Whatman 903	3	1 - 200	30 days, RT	mixed (5 min) + sonicated (10 min)	500 μL, MeOH/H_2O (50:50, v/v)	89-94	Agilent 1100 Ion Trap	MRM	421.2 ⟶ 179.0	Irbesartan	[55]
	Whatman DMPK-A	10	12 - 4000	30 days, RT	mixed (15 min)	1000 μL, MeOH	98	-	fluorescence	-	-	[54]
Losartan carboxylic acid	Whatman 903	3	5 - 1000	30 days, RT	mixed (5 min) + sonicated (10 min)	500 μL, MeOH/H_2O (50:50, v/v)	93-98	Agilent 1100 Ion Trap	MRM	421.2 ⟶ 179.0	Irbesartan	[55]

Table 1: Continued.

Drug analyzed	DBS card/paper used	Diameter of disk punched for analysis, mm	Calibration range, ng/mL	Stability of DBS samples	Elution procedure (min)	V, Extraction solvent	Recovery (%)	Mass-spectrometer	Detection mode	Mass or transmission, m/z	Internal standard	Ref.
Calcium channel blockers												
Verapamil	Whatman 903 Whatman DMPK-C Ahlstrom 237 Whatman DMPK-A Whatman DMPK-B	3	500		mixed (60 min)	200 μL, ACN/H_2O (75:25, v/v)		AB Sciex API 4000	SRM	455 ⟶ 165	Verapamil	[56]
	Whatman DMPK-C	2 *	1 - 1000		Used SPE cartridge	1000 μL, 0.2% FA in H_2O	81	AB Sciex API 4000	MRM	455.3⟶ 165.2	Imipramine	[57]
	Whatman filter paper, Grade CF 12	Whole DBS	50 - 500	-	On-line desorption	100% ACN	-	Agilent 1100 MSD single Q VL version	SRM	-	Trimipramine-d3	[58]
	Whatman grade SG81 ion exchange paper	Whole DBS	0.01 - 10000	-	Paper spray	$MeCl_2$/isopropanol (90:10, v/v)	-	Thermo TSQ Quantum Access Max QQQ	SRM	455 ⟶ 303	-	[59]
Amlodipine	Whatman 903	8	-	84 days, RT	mixed (1 min) + sonicated (30 min)	150 μL, MeOH/H_2O (70:30, v/v)	-	Agilent 6530 QTOF	TOF	431.1344	Atenolol	[49]
Nifedipine	Ahlstrom 226	4	10 - 10000	1 day, RT	-	200 μL, ACN/H_2O (75:25, v/v)	-	AB Sciex API 4000	MRM	347⟶ 315	[^{13}C,$^{15}N_2$]-nifedipine nitrophenyl pyridine	[60]
Pregabalin	Whatman 903	4	200 - 20000	120 days, 4°C	mixed (10 min) + Derivatization	100 μL, 0.1 mol/L HCl	98-102	Thermo TSQ Quantum Access MAX QQQ	SRM	288.0 ⟶ 142.1 288.0 ⟶ 228.0	4-aminocyclohexanecarboxylic acid	[61]
Diltiazem	Whatman 903	8	0.5 - 600	70 days, RT	mixed (1 min) + sonicated (30 min)	300 μL, MeOH	≥87	Agilent G6530A QTOF	TOF	415.1686	Atenolol-d7	[50]
α blockers												
Guanfacine	Whatman DMPK-C	5	0.05 - 25	76 days, RT	mixed (10 min)	600 μL, TBME	~80	AB SCIEX 5500 QTRAP	SRM	246.1⟶ 60.1	[^{13}C,$^{15}N_3$]-guanfacine hydrochloride	[62]
	Whatman DMPK-C	4 *	0.01 - 25	-	On-line DBS-SPE system	1000 μL, ACN/0.1% FA (80:20, v/v) + 1000 μL, 0.1% FA	-	AB SCIEX 5500 QTRAP	SRM	246.1 ⟶ 60.1	[^{13}C,$^{15}N_3$]-guanfacine hydrochloride	[63]
	two-layered polymeric membrane	2 *	0.25 - 250	-	On-line DBS-SPE system	1000 μL, ACN/0.1% FA (80:20, v/v) + 1000 μL, 0.1% FA	-	AB SCIEX 5500 QTRAP	SRM	246.1 ⟶ 60.1	[^{13}C,$^{15}N_3$]-guanfacine hydrochloride	[64]
Doxazosin	Whatman 903	8	-	84 days, RT	mixed (1 min) + sonicated (30 min)	150 μL, MeOH/H_2O (70:30, v/v)	-	Agilent 6530 QTOF	TOF	452.1928	Atenolol	[49]
	Whatman 903	8	0.1 - 100	70 days, RT	mixed (1 min) + sonicated (30 min)	300 μL, MeOH	≥87	Agilent G6530A QTOF	TOF	452.1928	Atenolol-d7	[50]
Prazosin	Whatman 903 Whatman DMPK-C Ahlstrom 237 Whatman DMPK-A Whatman DMPK-B	3	100 - 10000	-	mixed (60 min) + sonicated (30 min)	200 μL, ACN/H_2O (75:25, v/v) or MeOH for DMPK-B	-	AB Sciex API 4000	SRM	384 ⟶ 247	Verapamil	[56]

Table 1: Continued.

Drug analyzed	DBS card/paper used	Diameter of disk punched for analysis, mm	Calibration range, ng/mL	Stability of DBS samples	Elution procedure (min)	V, Extraction solvent	Recovery (%)	Mass-spectrometer	Detection mode	Mass or transmission, m/z	Internal standard	Ref.
β blockers												
Propranolol	Whatman 903	Whole DBS	5 - 250	21 days, 25°C 21 days, 37°C 21 days,-20°C	mixed (1 min) + sonicated (10 min)+ mixed (45 min)	300 μL, MeOH/0.1% FA (90:10, v/v)	91-99	Agilent 6460 QQQ	MRM	260.2 ⟶ 116.1	Propranolol-d7	[52]
	Whatman 903	3.2	2.5 - 200	30 days, RT 30 days, 4°C 30 days,-20°C	mixed (25 min)	200 μL, MeOH/0.1% FA (95:5, v/v)	94-100	AB SCIEX 5500 QTRAP	MRM	260.1 ⟶ 116.1	Propranolol-d7	[65]
	Whatman DMPK-C	2 *	1 - 1000	-	Used SPE cartridge	1000 μL, 0.2% FA in H_2O	83	AB Sciex API 4000	MRM	260.1⟶ 116.1	Imipramine	[57]
	Sartorius TFN, Munktell TFN, Whatman DMPK-C	Whole DBS	0 - 20	7 days, 4°C	sonicated (45 + 30 min)	100 μL, MeOH + 400 μL, TBME + 300 μL, acetone	71	Thermo Q Exactive	scan-to-scan	260.1645	Cocaine-d3	[66]
	Watman DMPK-B	3	1 - 500	-	mixed (10 min)	300 μL, MeOH	-	AB Sciex API 4000	MRM	-	Propranolol-d3	[67]
	Whatman filter paper	4 *	1 - 500	-	Heating	25 μL, MeOH with 0.1% FA	70–82	AB SCIEX 3200 QTRAP	MRM	260 ⟶ 183	Propranolol-d7	[68]
Bisoprolol	Whatman 903	8	0.1 - 100	70 days, RT	mixed (1 min) + sonicated (30 min)	300 μL, MeOH	≥87	Agilent G6530A QTOF	TOF	326.2326	Atenolol-d7	[50]
	Whatman 903	8	0.1 - 100	84 days, RT	mixed (1 min) + sonicated (30 min)	150 μL, MeOH/H_2O (70:30, v/v)	≥92	Agilent G6530A QTOF	TOF	326.2326	Atenolol	[48]
	Whatman 903	8	-	84 days, RT	mixed (1 min) + sonicated (30 min)	150 μL, MeOH/H_2O (70:30, v/v)	~98	Agilent 6530 QTOF	TOF	326.2326	Atenolol	[49]
	Whatman 903	Whole DBS	5 - 250	21 days, 25°C 21 days, 37°C 21 days,-20°C	mixed (1 min) + sonicated (10 min)+ mixed (45 min)	300 μL, MeOH/0.1% FA (90:10, v/v)	74-83	Agilent 6460 QQQ	MRM	326.2 ⟶ 116.1	Bisoprolol-d5	[52]
	Sartorius TFN, Munktell TFN, Whatman DMPK-C	Whole DBS	0 - 20	7 days, 4°C	sonicated (45 + 30 min)	100 μL, MeOH + 400 μL, TBME + 300 μL, acetone	60	Thermo Q Exactive	scan-to-scan	326.2326	Cocaine-d3	[66]
	Ahlstrom 226	8	See [48]	See [48]	mixed (1 min) + sonicated (30 min)	150 μL, MeOH/H_2O (70:30, v/v)	See [48]	Agilent G6530A QTOF	TOF	326.2326	Atenolol	[51]
Atenolol	Whatman 903	8	-	84 days, RT	mixed (1 min) + sonicated (30 min)	150 μL, MeOH/H_2O (70:30, v/v)	-	Agilent 6530 QTOF	TOF	267.1703	Atenolol	[49]
	Whatman 903	8	10 - 1500	70 days, RT	mixed (1 min) + sonicated (30 min)	300 μL, MeOH	≥87	Agilent G6530A QTOF	TOF	267.1703	Atenolol-d7	[50]
	Whatman 903	5	25 - 1500	70 days, 25°C	mixed (1 min) + sonicated (30 min)	200 μL, MeOH/H_2O (60:40, v/v)	96-100	Agilent 6530 QTOF	TOF	267.1703	Atenolol-d7	[29]
Metoprolol	Whatman DMPK-A	5	2.5 - 2500	14 days, RT	mixed (3 min)	200 μL, 2% ammonium hydroxide + 700 μL, ethyl acetate	77-86	AB Sciex API 5000	MRM	268.2 ⟶ 116.2	Metoprolol-d7	[69]
	Sartorius TFN, Munktell TFN, Whatman DMPK-C	Whole DBS	0 - 20	7 days, 4°C	sonicated (45 + 30 min)	100 μL, MeOH + 400 μL, TBME + 300 μL, acetone	68	Thermo Q Exactive	scan-to-scan	268.1907	Cocaine-d3	[66]

TABLE 1: Continued.

Drug analyzed	DBS card/paper used	Diameter of disk punched for analysis, mm	Calibration range, ng/mL	Stability of DBS samples	Elution procedure (min)	V, Extraction solvent	Recovery (%)	Mass-spectrometer	Detection mode	Mass or transmission, m/z	Internal standard	Ref.
Endothelin receptor antagonists												
Bosentan	Whatman DMPK-A	-	2 - 1500	105 days, RT	Used Sample Card And Prep DBS System	$MeOH/H_2O$ (50:50, v/v)	83-92	AB Sciex API 4000	MRM	552.2 ⟶ 202.1	Bosentan-d4	[70]
	Whatman DMPK-A	-	2 - 3000	105 days, RT	Used Sample Card And Prep DBS System	$MeOH/H_2O$ (50:50, v/v)	83-92	AB Sciex API 4000	MRM	552.2 ⟶ 202.1	Bosentan-d4	[71]
	Whatman DMPK-C	6.2	2.5 - 4000	147 days, RT 14 days, 37°C 14 days,-20°C	sonicated (20 min) + mixed (20 min)	400 μL, H_2O + 4000 μL, TBME	69-83	Thermo TSQ7000 QQQ	MRM	552.4 ⟶ 202.1	Bosentan-d4	[72]
Ambrisentan	Whatman DMPK-C	6.2	2.5 - 1000	147 days, RT 14 days, 37°C 14 days,-20°C	sonicated (20 min) + mixed (20 min)	400 μL, H_2O + 4000 μL, TBME	88-94	Thermo TSQ7000 QQQ	MRM	347.3 ⟶ 125	Ambrisentan-d10	[72]
Statins												
Simvastatin	Whatman 903	8	1 - 100	84 days, RT	mixed (1 min) + sonicated (30 min)	150 μL, $MeOH/H_2O$ (70:30, v/v)	~43	Agilent G6530A QTOF	TOF	441.261 Na adduct	Atenolol	[48]
	Whatman 903	8	0.1 - 100	70 days, RT	mixed (1 min) + sonicated (30 min)	300 μL, MeOH	68	Agilent G6530A QTOF	TOF	441.2611	Atenolol-d7	[50]
	Whatman 903	8	-	84 days, RT	mixed (1 min) + sonicated (30 min)	150 μL, $MeOH/H_2O$ (70:30, v/v)	~40	Agilent 6530 QTOF	TOF	441.2611	Atenolol	[49]
	Ahlstrom 226	8	See [70]	See [70]	mixed (1 min) + sonicated (30 min)	150 μL, $MeOH/H_2O$ (70:30, v/v)	See [70]	Agilent G6530A QTOF	TOF	418.2719 / 441.2611 NA adduct	Atenolol	[51]
Atorvastatin	Whatman 903	8	0.5 - 100	70 days, RT	mixed (1 min) + sonicated (30 min)	300 μL, MeOH	≥87	Agilent G6530A QTOF	TOF	559.2610	Atenolol-d7	[50]
Rosuvastatin	Whatman DMPK-B	3	0.5 - 80		mixed (60 min)	100 μL, MeOH		Waters Xevo TQ-S QQQ	MRM	482 ⟶ 258	Rosuvastatin-d6	[73]

ACN: acetonitrile, FA: formic acid, DTT: dithiothreitol, $MeCl_2$: dichloromethane, MeOH: methanol, MSX: multiplex targeted SRM mode, and TBME: *tert*-butyl methyl ether.
∗: elution area

samples. Ramiprilat is an active metabolite of ramipril and can be analyzed alone [52] or together with ramipril [51]. These studies cannot be completely compared because the assay was not described in detail in [51] and simply cited a paper where method of ramipril analysis was reported [48].

4. ARAs

ARAs specifically and selectively block type I angiotensin, a receptor of the renin-angiotensin system, by displacing angiotensin II from it [78, 79]. The presence of fluorescent functional groups in the molecular structure of ARA has allowed researchers to develop fluorescent assays for simultaneous quantification of valsartan, irbesartan, and losartan in DBS [54]. In this case, 1000 μL of MeOH served as an extraction solvent with 98–99 % recovery. The assay sensitivity for valsartan was better than that reported in studies on MS detection in TOF mode [49, 50]. Irbesartan has been detected in MRM mode only as an internal standard (IS), but losartan detection in the MRM mode enables investigators to achieve LOQ of 1 ng/mL [55]. Losartan carboxylic acid is a more potent active metabolite of losartan [80], which has also been detected in MRM mode [55] with excellent recovery (93–98%). DBS samples of valsartan, irbesartan, losartan, and losartan carboxylic acid are stable up to 84, 30, 70, and 30 days at RT, respectively.

5. CCBs

Antihypertensive properties of CCBs are mediated by their ability to disrupt the movement of calcium ions through membrane channels. Several classes of CCBs can be distinguished based on their chemical structure: 1.4-dihydropyridines (amlodipine, pregabalin, and nifedipine), phenylalkylamine (verapamil), benzothiazepines (diltiazem), and others.

Verapamil has been used as a model analyte to develop and improve methods of analysis in DBS. For example, the following are direct assays via on-line desorption of DBS [58]: flow-through desorption of DBS without an LC column [57], the paper spray ambient ionization method for direct analysis of biological samples [59], and desorption electrospray ionization (DESI) that operates at atmospheric pressure [56].

In these works, verapamil has been detected by MS/MS with transitions 455 $\longrightarrow$ 303 or 455 $\longrightarrow$ 165 m/z. The paper spray method involving Whatman grade SG81 ion exchange paper shows the best sensitivity: down to 0.01 ng/mL. In this case, in contrast to other studies, the spot was cut out as a triangle (10 × 5 mm) [59]. Different solvents have been applied for elution: acetonitrile (ACN)/H_2O, 100% ACN, and dichloromethane/isopropanol (90:10, v/v), but only for the flow-through desorption method was the recovery (81%) reported [57]. The stability of DBS samples in these studies has not been evaluated.

Amlodipine [49] and diltiazem [50] have been simultaneously analyzed along with ramipril and other medications under the conditions described above for ramipril.

To improve LOQ, the derivatization protocol with n-propyl chloroformate [81] has been developed for pregabalin quantitation in DBS [61]. The simple elution procedure with 0.1 M HCl yielded approximately 100% recovery. Nevertheless, LOQ was still high: 200 ng/mL.

The DBS technique has been applied to a study on the ways to increase stability of photosensitive compounds [60]. Nifedipine was used as a model analyte for this purpose. Extraction of nifedipine was performed on the Ahlstrom 226 card using micronic 96-well plates and silicone mats with a polytetrafluoroethylene film [60] by means of 200 μL of ACN/H_2O (70:30, v/v). It was shown that photosensitive compounds are more stable in DBSs than in blood or plasma under exposure to light.

6. Alpha Blockers

These are a class of antihypertensive drugs that has a vasodilating effect as a result of predominantly blocking α-adrenergic receptors. Guanfacine, like verapamil, has been chosen as a model analyte in several studies on improvement of DBS assays [62–64]. The analyte has been detected on the same equipment in SRM mode with transition 246.1 $\longrightarrow$ 60.1 m/z. At first, the standard DBS technique was developed with good stability of DBS samples (76 days at RT), a simple extraction procedure (10 min mixing) involving 600 μL of *tert*-butyl methyl ether (TBME), with ~80% recovery from a Whatman DMPK-C card and the concentration range from 0.05 to 25 ng/mL [62]. Then, methods of on-line desorption from a Whatman DMPK-C card [63] and from a two-layered polymeric membrane [64] were developed. The best sensitivity was achieved with the on-line DBS-SPE system based on Whatman DMPK-C cards. Doxazosin was simultaneously analyzed along with ramipril and other medications under the conditions described above for ramipril [49, 50]. Prazosin was analyzed by the DESI technique [56]. Several types of filter card were tested in that work: Whatman 903, FTA DMPK-A, FTA DMPK-B, and FTA DMPK-C. Nonetheless, the LOQ shown by the DESI technique was 100 ng/mL: not as low as in a comparable electrospray ionization assay [56].

7. BBs

This is a large group of drugs that block the binding sites on β-adrenergic receptors. Antihypertensive medications of this class are most often subjected to the DBS technique, for example, propranolol and bisoprolol. These two have been tested on different types of filter paper, including Sartorius TFN and Munktell TFN [66]. The DBS samples were stable on a Whatman 903 card for up to 30 and 84 days for propranolol [65] and bisoprolol [48, 49], respectively. The extraction procedure for bisoprolol was a standard one—mixing and sonication—whereas propranolol was extracted via on-line desorption [57] and paper heating [68]. The best recovery rates, up to 98–99%, for propranolol and bisoprolol were achieved with an extraction solvent consisting of MeOH or MeOH/H_2O mixtures in different combination with or without 0.1% formic acid (FA). In contrast, extraction with 100 μL of MeOH, 400 μL of TBME, and 300 μL of acetone resulted in 71% and 60% recovery for propranolol and bisoprolol, respectively [66]. The difference lies in the main detection mode: MRM for propranolol and TOF for bisoprolol. In two studies, atenolol was simultaneously assayed along with ramipril and other medications under conditions described above for ramipril [49, 50]. In the third paper, the DBS technique was generally the same and yielded similar results [29]. Metoprolol was extracted in two ways: by means of 100 μL of MeOH, 400 μL of TBME, and 300 μL of acetone or with 200 μL of 2% ammonium hydroxide and 700 μL of ethyl acetate. The last solvent with simple 3 min mixing showed better recovery, up to 86%. In this case, the concentration range was 2.5–2500 ng/mL, with 14-day stability of DBS samples at RT.

8. ERAs

These drugs block endothelin receptors; ambrisentan acts on endothelin A receptors, whereas bosentan affects both endothelin A and B receptors. Only these two medications have been analyzed by the DBS sampling technique, and detection has been carried out in MRM mode for both. Elution of bosentan was performed using a Sample Card and Prep DBS System by means of MeOH/H_2O (50:50, v/v) [70, 71] or simultaneously with ambrisentan in a standard assay with 400 μL H_2O and 4 mL of TBME [72]. The recovery in these two approaches was approximately the same, but bosentan recovery in the case of MeOH/H_2O (50:50, v/v) was a little better. DBSs of both analytes were found to be stable for up to 147 days at RT.

9. Statins

These agents inhibit 3-hydroxy-3-methyl-glutaryl-coenzyme A reductase; as a result, the blood levels of total cholesterol, low-density lipoproteins, and triglycerides decrease, while the concentration of antiatherogenic high-density lipoproteins increases. The combination of statins with antihypertensive drugs increases the activity of the latter. Therefore, statins are often quantified simultaneously with antihypertensive drugs. Just as the analytes above, simvastatin [48–51] and atorvastatin [50] have been analyzed simultaneously with other medications under conditions described in a previous section for ramipril. The only problem with simvastatin is recovery, which did not exceed 68%. Rosuvastatin was assayed by 2-dimensional analytical-scale chromatography with at-column dilution enabling the injection of large sample volumes of organic extracts of DBS [73]. Analysis of rosuvastatin was performed in MRM mode with simple mixing extraction by means of 100 μL of MeOH in the concentration range 5–80 ng/mL.

10. Effects of Blood Properties

Aside from the parameters of analysis described above, a researcher should take into account the effects of blood properties, such as hematocrit, blood volume, and blood distribution in a DBS as well as sample quality, which may have a direct impact on quantitative analysis. Hematocrit is the most important parameter influencing the accuracy and precision of DBS analysis [82, 83]. Hematocrit varies with gender, health status, and age and slightly with ethnicity [74]. Moreover, capillary blood tends to have higher hematocrit (e.g., 61%) than venous blood does [30]. Reference ranges slightly vary among sources but are typically 40–50% for adult men and 35–45% for adult women [30, 84, 85]. In a hospital population, 95% of routine hematocrit measurements yield values between 23% and 48% [74]. Hematocrit affects the spot size and formation, homogeneity of the spot, drying time, recovery of the analyte, and reproducibility and robustness of the assay [46, 74]. Therefore, it is important to evaluate the hematocrit effect during validation of an assay [46]. In several research papers, investigators have considered the influence of hematocrit on quantitative determination of analyte concentrations. No significant impact of hematocrit on the quantification was found in the hematocrit range 35–65% for bosentan [70], 38–45% for bosentan and ambrisentan [72], 23–43% for propranolol [65], 41–48% for guanfacine [62], and 20–50% for losartan and losartan carboxylic acid [55] and the hematocrit range of 35–55% for ramipril, lisinopril, valsartan, losartan, diltiazem, doxazosin, bisoprolol, atenolol, simvastatin, and atorvastatin [50]. The impact of hematocrit was evaluated in a follow-up to [64], where unacceptable bias was detected for guanfacine at hematocrit values of 30% and 60%, but acceptable accuracy and precision results were obtained at hematocrit 45% [86]. Besides, alternative approaches have been used for correction of hematocrit values: a mathematical equation for pregabalin [61], a number of consecutive analyses of the same blood spot for verapamil and propranolol [57], and analysis of the entire blood spot sample for ramiprilat, propranolol, and bisoprolol [52]. In some cases, the hematocrit effect was not studied because the authors assumed that hematocrit variation is likely to be within the "normal" hematocrit range in adults [29, 48]. In other studies, the hematocrit effect has not been considered.

Because it has been observed that the DBS area decreases nearly linearly when hematocrit increases, the spot size and homogeneity of spots also influence quantification results [74, 84]. Usually, the spotted blood volume is between 15 and 40 μL. Accordingly, the impact of the blood spotting volume on DBS analysis has been investigated in several works: blood volume of 20, 25, and 30 μL [70]; 20, 30, and 40 μL [29, 48, 50, 53]; 10 to 40 μL [69]; 15, 20, and 25 μL [62]; and 10, 15, and 20 μL [55]. In all cases, accuracy and precision were less than 15% within the tested range, and therefore accurate pipetting during preparation of DBS may not be necessary. To minimize the influence of spot homogeneity on measurement, the same location of a spot (center) [29, 48–51, 53, 54, 61–63, 69, 73] or the whole spot [52, 66] has been punched. No difference between punching the central area and punching an offcenter site of DBS has been observed for pregabalin [61] and guanfacine [62].

In many cases, analyte concentration can differ between capillary and venous blood [87] as well as between plasma and DBS. Thus, a comparison is recommended as part of the assay validation [30]. Such a study has been conducted for guanfacine, showing on average 20% higher concentrations in DBS than in plasma, but an 11.6% to 23.2% decrease in concentration was observed when whole blood rather than plasma was spiked with guanfacine [62]. The analyte concentrations in plasma show a significant correlation with DBS concentrations with a conversion factor (slope) of 1.73 for propranolol [65], 1.58 for ambrisentan, and 1.52 for bosentan [72]. Bosentan DBS and plasma concentrations have also been compared in a Bland–Altman plot, and a limited systematic difference between the measurements was found [71]. For rosuvastatin, quantification methods for either plasma or DBS have been developed, but comparison experiments for plasma and DBS samples have not been carried out [73]. For nifedipine, only photostability has been compared between DBS and liquid aqueous and biological matrices [60].

Samples during validation assays are collected in the laboratory. If samples are collected in the laboratory from volunteers [29, 48–51] or rats [52, 54], they are usually spotted correctly. For home sample collection, the proportion of unsatisfactory samples is 19% [49]. This figure can reach more than 30% [74, 88] and may pose a problem for clinical application because the assays are validated by a correct technique for spotting of samples.

11. Conclusions

Interest in the DBS technique as an easy blood sampling method for monitoring of antihypertensive drug concentrations is increasing continuously. The main limitation of the DBS technique is sensitivity, which is expected to improve with the growing availability of MS and MS/MS equipment in clinical and scientific laboratories for analysis of antihypertensive drugs. Among the reviewed works, medications have in general been analyzed in MRM mode on QQQ mass-spectrometers or in TOF mode on QTOF mass-spectrometers. The elution procedure in many cases consists of mixing and sonication in an extraction solvent with suitable recovery. Most popular types of filter paper are Whatman 903 or Whatman DMPK-C/-A cards. Almost all antihypertensive drugs show great stability in DBS samples for weeks or even months. An automated DBS technique and on-line desorption of DBS have been devised in some studies to enable analysis of a large number of samples. The major variables of the DBS technique are hematocrit and differences in drug concentration measured in blood, blood plasma, and DBS. Validation experiments are necessary for new DBS assays, but not all the published drug assays involving DBS include full validation; this situation can be considered the main obstacle for their broad application.

Acknowledgments

This work was supported by a Program of RAS, "Basic Research for Biomedical Technology" 2018–2020.

References

[1] World Health Organization, "Cardiovascular Diseases (CVDs)," 2017, http://www.who.int/mediacentre/factsheets/fs317/en/.

[2] C. M. Lawes, S. V. Hoorn, and A. Rodgers, "Global burden of blood-pressure-related disease, 2001," *The Lancet*, vol. 371, no. 9623, pp. 1513–1518, 2008.

[3] D. J. Campbell, M. Mcgrady, D. L. Prior et al., "Most individuals with treated blood pressures above target receive only one or two antihypertensive drug classes," *Internal Medicine Journal*, vol. 43, no. 2, pp. 137–143, 2013.

[4] S. M. Dunlay, S. A. Weston, S. J. Jacobsen, and V. L. Roger, "Risk factors for heart failure: a population-based case-control study," *American Journal of Medicine*, vol. 122, no. 11, pp. 1023–1028, 2009.

[5] M. K. Haroun, B. G. Jaar, S. C. Hoffman, G. W. Comstock, M. J. Klag, and J. Coresh, "Risk factors for chronic kidney disease: a prospective study of 23,534 men and women in Washington County, Maryland," *Journal of the American Society of Nephrology*, vol. 14, no. 11, pp. 2934–2941, 2003.

[6] R. S. Vasan, A. Beiser, S. Seshadri et al., "Residual lifetime risk for developing hypertension in middle-aged women and men," *Journal of the American Medical Association*, vol. 287, no. 8, pp. 1003–1010, 2002.

[7] N. D. Wong, V. A. Lopez, G. L'Italien, R. Chen, S. E. J. Kline, and S. S. Franklin, "Inadequate control of hypertension in US adults with cardiovascular disease comorbidities in 2003-2004," *JAMA Internal Medicine*, vol. 167, no. 22, pp. 2431–2436, 2007.

[8] G. Mancia, R. Fagard, K. Narkiewicz et al., "2013 ESH/ESC Guidelines for the management of arterial hypertension: The Task Force for the management of arterial hypertension of the European Society of Hypertension (ESH) and of the European Society of Cardiology (ESC)," *Journal of hypertension*, vol. 31, no. 7, pp. 1281–1357, 2013.

[9] E. Berra, M. Azizi, A. Capron et al., "Evaluation of adherence should become an integral part of assessment of patients with apparently treatment-resistant hypertension," *Hypertension*, vol. 68, no. 2, pp. 297–306, 2016.

[10] A. V. Chobanian, G. L. Bakris, and H. R. Black, "Seventh report of the Joint National Committee on prevention, detection, evaluation, and treatment of high blood pressure," *Hypertension*, vol. 42, no. 6, pp. 1206–1252, 2003.

[11] G. Rimoy, M. Justin-Temu, and C. Nilay, "Prescribing Patterns and cost of antihypertensive drugs in private hospitals in Dar es Salaam, Tanzania," *East and Central African Journal of Pharmaceutical Sciences*, vol. 11, no. 3, 2009.

[12] M. C. Cáceres, P. Moyano, H. Fariñas et al., "Trends in antihypertensive drug use in spanish primary health care (1990-2012)," *Advances in Pharmacoepidemiology & Drug Safety*, vol. 04, no. 01, 2015.

[13] N. Jarari, N. Rao, J. R. Peela et al., "A review on prescribing patterns of antihypertensive drugs," *Clinical Hypertension*, vol. 22, article 7, 2015.

[14] H. Xu, Y. He, L. Xu, X. Yan, and H. Dai, "Trends and patterns of five antihypertensive drug classes between 2007 and 2012 in China using hospital prescription data," *International Journal of*

Clinical Pharmacology and Therapeutics, vol. 53, no. 6, pp. 430-437, 2015.

[15] P.-H. Liu and J.-D. Wang, "Antihypertensive medication prescription patterns and time trends for newly-diagnosed uncomplicated hypertension patients in Taiwan," *BMC Health Services Research*, vol. 8, article no. 133, 2008.

[16] M. Beg, S. Dutta, A. Varma et al., "Study on drug prescribing pattern in hypertensive patients in a tertiary care teaching hospital at Dehradun, Uttarakhand," *International Journal of Medical Science and Public Health*, vol. 3, no. 8, pp. 922-926, 2014.

[17] S. Tanna and G. Lawson, *Analytical Chemistry for Assessing Medication Adherence*, Elsevier, Amsterdam, Netherlands, 2016.

[18] A. Shroufi and J. W. Powles, "Adherence and chemoprevention in major cardiovascular disease: A simulation study of the benefits of additional use of statins," *Journal of Epidemiology and Community Health*, vol. 64, no. 2, pp. 109-113, 2010.

[19] S. Baroletti and H. Dell'Orfano, "Medication adherence in cardiovascular disease," *Circulation*, vol. 121, no. 12, pp. 1455-1458, 2010.

[20] I. M. Kronish and S. Ye, "Adherence to cardiovascular medications: Lessons learned and future directions," *Progress in Cardiovascular Diseases*, vol. 55, no. 6, pp. 590-600, 2013.

[21] P. M. Ho, J. A. Spertus, F. A. Masoudi et al., "Impact of medication therapy discontinuation on mortality after myocardial infarction," *JAMA Internal Medicine*, vol. 166, no. 17, pp. 1842-1847, 2006.

[22] P. M. Ho, C. L. Bryson, and J. S. Rumsfeld, "Medication adherence: its importance in cardiovascular outcomes," *Circulation*, vol. 119, no. 23, pp. 3028-3035, 2009.

[23] J. B. Garner, "Problems of nonadherence in cardiology and proposals to improve outcomes," *American Journal of Cardiology*, vol. 105, no. 10, pp. 1495-1501, 2010.

[24] H. B. Bosworth, B. B. Granger, P. Mendys et al., "Medication adherence: A call for action," *American Heart Journal*, vol. 162, no. 3, pp. 412-424, 2011.

[25] A. La Caze, G. Gujral, and W. N. Cottrell, "How do we better translate adherence research into improvements in patient care?" *International Journal of Clinical Pharmacy*, vol. 36, no. 1, pp. 10-14, 2014.

[26] I. Barat, F. Andreasen, and E. M. S. Damsgaard, "Drug therapy in the elderly: What doctors believe and patients actually do," *British Journal of Clinical Pharmacology*, vol. 51, no. 6, pp. 615-622, 2001.

[27] T. A. Kotchen, "The search for strategies to control hypertension," *Circulation*, vol. 122, no. 12, pp. 1141-1143, 2010.

[28] H. C. Pandya, N. Spooner, and H. Mulla, "Dried blood spots, pharmacokinetic studies and better medicines for children," *Bioanalysis*, vol. 3, no. 7, pp. 779-786, 2011.

[29] G. Lawson, E. Cocks, and S. Tanna, "Quantitative determination of atenolol in dried blood spot samples by LC-HRMS: A potential method for assessing medication adherence," *Journal of Chromatography B*, vol. 897, pp. 72-79, 2012.

[30] W. Li and F. L. S. Tse, "Dried blood spot sampling in combination with LC-MS/MS for quantitative analysis of small molecules," *Biomedical Chromatography*, vol. 24, no. 1, pp. 49-65, 2010.

[31] S. P. Parker and W. D. Cubitt, "The use of the dried blood spot sample in epidemiological studies," *Journal of Clinical Pathology*, vol. 52, no. 9, pp. 633-639, 1999.

[32] R. G. Boy, J. Henseler, R. Mattern, and G. Skopp, "Determination of morphine and 6-acetylmorphine in blood with use of dried blood spots," *Therapeutic Drug Monitoring*, vol. 30, no. 6, pp. 733-739, 2008.

[33] G. la Marca, S. Malvagia, L. Filippi, F. Luceri, G. Moneti, and R. Guerrini, "A new rapid micromethod for the assay of phenobarbital from dried blood spots by LC-tandem mass spectrometry.," *Epilepsia*, vol. 50, no. 12, pp. 2658-2662, 2009.

[34] A. Chernonosov, "Quantification of warfarin in dried rat plasma spots by high-performance liquid chromatography with tandem mass spectrometry," *Journal of Pharmaceutics*, vol. 2016, Article ID 6053295, 6 pages, 2016.

[35] S. Tanna and G. Lawson, "Analytical methods used in conjunction with dried blood spots," *Analytical Methods*, vol. 3, no. 8, pp. 1709-1718, 2011.

[36] P. M. Edelbroek, J. V. D. Heijden, and L. M. L. Stolk, "Dried blood spot methods in therapeutic drug monitoring: Methods, assays, and pitfalls," *Therapeutic Drug Monitoring*, vol. 31, no. 3, pp. 327-336, 2009.

[37] S. AbuRuz, J. Millership, and J. McElnay, "Dried blood spot liquid chromatography assay for therapeutic drug monitoring of metformin," *Journal of Chromatography B*, vol. 832, no. 2, pp. 202-207, 2006.

[38] J. van der Heijden, Y. de Beer, K. Hoogtanders et al., "Therapeutic drug monitoring of everolimus using the dried blood spot method in combination with liquid chromatography-mass spectrometry," *Journal of Pharmaceutical and Biomedical Analysis*, vol. 50, no. 4, pp. 664-670, 2009.

[39] M. Barfield, N. Spooner, R. Lad, S. Parry, and S. Fowles, "Application of dried blood spots combined with HPLC-MS/MS for the quantification of acetaminophen in toxicokinetic studies," *Journal of Chromatography B*, vol. 870, no. 1, pp. 32-37, 2008.

[40] P. Wong, R. Pham, B. A. Bruenner, and C. A. James, "Increasing efficiency for dried blood spot analysis: Prospects for automation and simplified sample analysis," *Bioanalysis*, vol. 2, no. 11, pp. 1787-1789, 2010.

[41] N. Spooner, R. Lad, and M. Barfield, "Dried blood spots as a sample collection technique for the determination of pharmacokinetics in clinical studies: considerations for the validation of a quantitative bioanalytical method," *Analytical Chemistry*, vol. 81, no. 4, pp. 1557-1563, 2009.

[42] P. Patel, S. Tanna, H. Mulla, V. Kairamkonda, H. Pandya, and G. Lawson, "Dexamethasone quantification in dried blood spot samples using LC-MS: The potential for application to neonatal pharmacokinetic studies," *Journal of Chromatography B*, vol. 878, no. 31, pp. 3277-3282, 2010.

[43] M. F. Suyagh, G. Iheagwaram, P. L. Kole et al., "Development and validation of a dried blood spot-HPLC assay for the determination of metronidazole in neonatal whole blood samples," *Analytical and Bioanalytical Chemistry*, vol. 397, no. 2, pp. 687-693, 2010.

[44] R. Guthrie and A. Susi, "A simple phenylalanine method for detecting phenylketonuria in large populations of newborn infants," *Pediatrics*, vol. 32, pp. 338-343, 1963.

[45] I. Taneja, M. Erukala, K. S. R. Raju, S. P. Singh, and Wahajuddin, "Dried blood spots in bioanalysis of antimalarials: Relevance and challenges in quantitative assessment of antimalarial drugs," *Bioanalysis*, vol. 5, no. 17, pp. 2171-2186, 2013.

[46] P. Timmerman, S. White, S. Globig, S. Lüdtke, L. Brunet, and J. Smeraglia, "EBF recommendation on the validation of bioanalytical methods for dried blood spots," *Bioanalysis*, vol. 3, no. 14, pp. 1567-1575, 2011.

[47] T. K. Majumdar and D. R. Howard, "The use of dried blood spots for concentration assessment in pharmacokinetic evaluations," *Pharmacokinetics in Drug Development*, vol. 3, pp. 91–114, 2011.

[48] G. Lawson, E. Cocks, and S. Tanna, "Bisoprolol, ramipril and simvastatin determination in dried blood spot samples using LC-HRMS for assessing medication adherence," *Journal of Pharmaceutical and Biomedical Analysis*, vol. 81-82, pp. 99–107, 2013.

[49] S. Tanna, D. Bernieh, and G. Lawson, "LC-HRMS analysis of dried blood spot samples for assessing adherence to cardiovascular medications," *Journal of Bioanalysis & Biomedicine*, vol. 07, no. 01, 2015.

[50] D. Bernieh, G. Lawson, and S. Tanna, "Quantitative LC–HRMS determination of selected cardiovascular drugs, in dried blood spots, as an indicator of adherence to medication," *Journal of Pharmaceutical and Biomedical Analysis*, vol. 142, pp. 232–243, 2017.

[51] S. Tanna and G. Lawson, "Cardiovascular drug medication adherence assessed by dried blood spot analysis," *Journal of Analytical & Bioanalytical Techniques*, vol. S12, article 006, 2014.

[52] K. Cvan Trobec, J. Trontelj, J. Springer, M. Lainscak, and M. Kerec Kos, "Liquid chromatography-tandem mass spectrometry method for simultaneous quantification of bisoprolol, ramiprilat, propranolol and midazolam in rat dried blood spots," *Journal of Chromatography B*, vol. 958, pp. 29–35, 2014.

[53] G. Lawson, H. Mulla, and S. Tanna, "Captopril determination in dried blood spot samples with LC-MS and LC-HRMS: A potential method for neonate pharmacokinetic studies," *Journal of Bioanalysis & Biomedicine*, vol. 4, no. 2, pp. 16–25, 2012.

[54] R. N. Rao, S. Bompelli, and P. K. Maurya, "High-performance liquid chromatographic determination of anti- hypertensive drugs on dried blood spots using a fluorescence detector - method development and validation," *Biomedical Chromatography*, vol. 25, no. 11, pp. 1252–1259, 2011.

[55] R. N. Rao, S. S. Raju, R. M. Vali, and G. G. Sankar, "Liquid chromatography-mass spectrometric determination of losartan and its active metabolite on dried blood spots," *Journal of Chromatography B*, vol. 902, pp. 47–54, 2012.

[56] J. M. Wiseman, C. A. Evans, C. L. Bowen, and J. H. Kennedy, "Direct analysis of dried blood spots utilizing desorption electrospray ionization (DESI) mass spectrometry," *Analyst*, vol. 135, no. 4, pp. 720–725, 2010.

[57] J. A. Ooms, L. Knegt, and E. H. M. Koster, "Exploration of a new concept for automated dried blood spot analysis using flow-through desorption and online SPE-MS/MS," *Bioanalysis*, vol. 3, no. 20, pp. 2311–2320, 2011.

[58] J. Déglon, A. Thomas, A. Cataldo, P. Mangin, and C. Staub, "On-line desorption of dried blood spot: A novel approach for the direct LC/MS analysis of μ-whole blood samples," *Journal of Pharmaceutical and Biomedical Analysis*, vol. 49, no. 4, pp. 1034–1039, 2009.

[59] Z. Zhang, W. Xu, N. E. Manicke, R. G. Cooks, and Z. Ouyang, "Silica coated paper substrate for paper-spray analysis of therapeutic drugs in dried blood spots," *Analytical Chemistry*, vol. 84, no. 2, pp. 931–938, 2012.

[60] C. L. Bowen, M. D. Hemberger, J. R. Kehler, and C. A. Evans, "Utility of dried blood spot sampling and storage for increased stability of photosensitive compounds," *Bioanalysis*, vol. 2, no. 11, pp. 1823–1828, 2010.

[61] N. Kostić, Y. Dotsikas, N. Jović, G. Stevanović, A. Malenović, and M. Medenica, "Quantitation of pregabalin in dried blood spots and dried plasma spots by validated LC-MS/MS methods," *Journal of Pharmaceutical and Biomedical Analysis*, vol. 109, pp. 79–84, 2015.

[62] Y. Li, J. Henion, R. Abbott, and P. Wang, "Dried blood spots as a sampling technique for the quantitative determination of guanfacine in clinical studies," *Bioanalysis*, vol. 3, no. 22, pp. 2501–2514, 2011.

[63] Y. Li, J. Henion, R. Abbott, and P. Wang, "Semi-automated direct elution of dried blood spots for the quantitative determination of guanfacine in human blood," *Bioanalysis*, vol. 4, no. 12, pp. 1445–1456, 2012.

[64] Y. Li, J. Henion, R. Abbott, and P. Wang, "The use of a membrane filtration device to form dried plasma spots for the quantitative determination of guanfacine in whole blood," *Rapid Communications in Mass Spectrometry*, vol. 26, no. 10, pp. 1208–1212, 2012.

[65] M. L. Della Bona, S. Malvagia, F. Villanelli et al., "A rapid liquid chromatography tandem mass spectrometry-based method for measuring propranolol on dried blood spots," *Journal of Pharmaceutical and Biomedical Analysis*, vol. 78-79, pp. 34–38, 2013.

[66] A. Thomas, H. Geyer, W. Schänzer et al., "Sensitive determination of prohibited drugs in dried blood spots (DBS) for doping controls by means of a benchtop quadrupole/Orbitrap mass spectrometer," *Analytical and Bioanalytical Chemistry*, vol. 403, no. 5, pp. 1279–1289, 2012.

[67] G. T. Clark, G. Giddens, L. Burrows, and C. Strand, "Utilization of dried blood spots within drug discovery: Modification of a standard DiLab® AccuSampler® to facilitate automatic dried blood spot sampling," *Laboratory Animals*, vol. 45, no. 2, pp. 124–126, 2011.

[68] H. Huang, Q. Wu, L. Zeng et al., "Heating paper spray mass spectrometry for enhanced detection of propranolol in dried blood samples," *Analytical Methods*, vol. 9, no. 29, pp. 4282–4287, 2017.

[69] X. Liang, H. Jiang, and X. Chen, "Human DBS sampling with LC-MS/MS for enantioselective determination of metoprolol and its metabolite O-desmethyl metoprolol," *Bioanalysis*, vol. 2, no. 8, pp. 1437–1448, 2010.

[70] N. Ganz, M. Singrasa, L. Nicolas et al., "Development and validation of a fully automated online human dried blood spot analysis of bosentan and its metabolites using the Sample Card And Prep DBS System," *Journal of Chromatography B*, vol. 885-886, pp. 50–60, 2012.

[71] M. Géhin, P. N. Sidharta, and J. Dingemanse, "Bosentan Pharmacokinetics in Pediatric Patients with Pulmonary Arterial Hypertension: Comparison of Dried Blood Spot and Plasma Analysis," *Pharmacology*, vol. 98, no. 3-4, pp. 111–114, 2016.

[72] Y. Enderle, A. D. Meid, J. Friedrich et al., "Dried Blood Spot Technique for the Monitoring of Ambrisentan, Bosentan, Sildenafil, and Tadalafil in Patients with Pulmonary Arterial Hypertension," *Analytical Chemistry*, vol. 87, no. 24, pp. 12112–12120, 2015.

[73] P. D. Rainville, J. L. Simeone, D. S. Root, C. R. Mallet, I. D. Wilson, and R. S. Plumb, "A method for the direct injection and analysis of small volume human blood spots and plasma extracts containing high concentrations of organic solvents using revered-phase 2D UPLC/MS," *Analyst*, vol. 140, no. 6, pp. 1921–1931, 2015.

[74] M. V. Antunes, M. F. Charão, and R. Linden, "Dried blood spots analysis with mass spectrometry: Potentials and pitfalls

in therapeutic drug monitoring," *Clinical Biochemistry*, vol. 49, no. 13-14, pp. 1035–1046, 2016.

[75] J. Déglon, L. A. Leuthold, and A. Thomas, "Potential missing steps for a wide use of dried matrix spots in biomedical analysis," *Bioanalysis*, vol. 7, no. 18, pp. 2375–2382, 2015.

[76] C. Borghi, A. Dormi, M. Veronesi, V. Immordino, and E. Ambrosioni, "Use of Lipid-Lowering Drugs and Blood Pressure Control in Patients With Arterial Hypertension," *The Journal of Clinical Hypertension*, vol. 4, no. 4, pp. 277–285, 2002.

[77] J. E. Rodgers and J. H. Patterson, "The role of the renin-angiotensin-aldosterone system in the management of heart failure," *Pharmacotherapy*, vol. 20, no. 11, part 2, pp. 368S–378S, 2000.

[78] J. H. Bauer, "The angiotensin II type 1 receptor antagonists. A new class of antihypertensive drugs," *JAMA Internal Medicine*, vol. 155, no. 13, pp. 1361–1368, 1995.

[79] J. Nie, M. Zhang, Y. Fan, Y. Wen, B. Xiang, and Y.-Q. Feng, "Biocompatible in-tube solid-phase microextraction coupled to HPLC for the determination of angiotensin II receptor antagonists in human plasma and urine," *Journal of Chromatography B*, vol. 828, no. 1-2, pp. 62–69, 2005.

[80] M. Polinko, K. Riffel, H. Song, and M.-W. Lo, "Simultaneous determination of losartan and EXP3174 in human plasma and urine utilizing liquid chromatography/tandem mass spectrometry," *Journal of Pharmaceutical and Biomedical Analysis*, vol. 33, no. 1, pp. 73–84, 2003.

[81] N. D. S. Kostić, Y. Dotsikas, A. D. S. Malenović, and M. Medenica, "Effects of derivatization reagents consisting of n-alkyl chloroformate/n-alcohol combinations in LC-ESI-MS/MS analysis of zwitterionic antiepileptic drugs," *Talanta*, vol. 116, pp. 91–99, 2013.

[82] R. J. W. Meesters and G. P. Hooff, "State-of-the-art dried blood spot analysis: An overview of recent advances and future trends," *Bioanalysis*, vol. 5, no. 17, pp. 2187–2208, 2013.

[83] Y. Enderle, K. Foerster, and J. Burhenne, "Clinical feasibility of dried blood spots: Analytics, validation, and applications," *Journal of Pharmaceutical and Biomedical Analysis*, vol. 130, pp. 231–243, 2016.

[84] M. Wagner, D. Tonoli, E. Varesio, and G. Hopfgartner, "The use of mass spectrometry to analyze dried blood spots," *Mass Spectrometry Reviews*, vol. 35, no. 3, pp. 361–368, 2016.

[85] P. Denniff and N. Spooner, "The effect of hematocrit on assay bias when using DBS samples for the quantitative bioanalysis of drugs," *Bioanalysis*, vol. 2, no. 8, pp. 1385–1395, 2010.

[86] R. Sturm, J. Henion, R. Abbott, and P. Wang, "Novel membrane devices and their potential utility in blood sample collection prior to analysis of dried plasma spots," *Bioanalysis*, vol. 7, no. 16, pp. 1987–2002, 2015.

[87] B. G. Keevil, "The analysis of dried blood spot samples using liquid chromatography tandem mass spectrometry," *Clinical Biochemistry*, vol. 44, no. 1, pp. 110–118, 2011.

[88] T. Panchal, N. Spooner, and M. Barfield, "Ensuring the collection of high-quality dried blood spot samples across multisite clinical studies," *Bioanalysis*, vol. 9, no. 2, pp. 209–213, 2017.

4

Determination of Aflatoxin B_1 in Feedstuffs without Clean-Up Step by High-Performance Liquid Chromatography

Sasiprapa Choochuay,[1,2] Jutamas Phakam,[1,2] Prakorn Jala,[3] Thanapoom Maneeboon,[4] and Natthasit Tansakul[1,2]

[1] *Department of Veterinary Pharmacology and Toxicology, Faculty of Veterinary Medicine, Kasetsart University, 10900, Bangkok, Thailand*
[2] *Center for Advanced Studies for Agriculture and Food, KU Institute for Advanced Studies, Kasetsart University (CASAF, NRU-KU), 10900, Bangkok, Thailand*
[3] *Faculty of Veterinary Medicine, Kamphaengsaen Campus, 73140, Nakhon Pathom, Thailand*
[4] *Kasetsart University Research and Development Institute, 10900, Bangkok, Thailand*

Correspondence should be addressed to Natthasit Tansakul; natthasitt@yahoo.com

Academic Editor: Mohamed Abdel-Rehim

A reliable and rapid method has been developed for the determination of aflatoxin B_1 (AFB_1) in four kinds of feedstuffs comprising broken rice, peanuts, corn, and fishmeal. A sample preparation was carried out based on the QuEChERS method with the exclusion of the clean-up step. In this study, AFB_1 was extracted using acetonitrile/methanol (40/60 v/v), followed by partitioning with sodium chloride and magnesium sulfate. High-performance liquid chromatography with precolumn derivatization and fluorescence detection was performed. The coefficients of determination were greater than 0.9800. Throughout the developed method, the recovery of all feedstuffs achieved a range of 82.50-109.85% with relative standard deviation lower than 11% for all analytes at a concentration of 20-100 ng/g. The limit of detection (LOD) ranged from 0.2 to 1.2 ng/g and limit of quantitation (LOQ) ranged from 0.3 to 1.5 ng/g. The validated method was successfully applied to a total of 120 samples. The occurrence of AFB_1 contamination was found at the following concentrations: in broken rice (0.44-2.33ng/g), peanut (3.97-106.26ng/g), corn (0.88-50.29 ng/g), and fishmeal (1.06-10.35 ng/g). These results indicate that the proposed method may be useful for regularly monitoring AFB_1 contamination in feedstuffs.

1. Introduction

Aflatoxins (AFs) are secondary metabolites of fungi (e.g., *Aspergillus flavus* and *A. parasiticus*). AFs occur naturally and can be found in common food and feedstuffs such as rice, peanuts, corn, fishmeal, and soybean meal [1–3]. The four major analogues—aflatoxins B_1 (AFB_1), B_2 (AFB_2), G_1 (AFG_1), and G_2 (AFG_2)—are the most important members because they all pose a potential risk to human and animal health if food and feedstuffs have been contaminated. In particular, the toxicity of AFB_1 can range from levels that may cause immune system suppression to the induction of teratogenic, mutagenic, and carcinogenic activities [4, 5], which is collectively classified as a carcinogen (Group 1) by the International Agency for Research on Cancer (IARC) [6]. In addition, feedstuffs comprise the first link of the food chain; therefore, there is a risk of AFB_1 carryover from feedstuffs into animal tissues and/or biological fluids such as meat, milk, and eggs, which may eventually be hazardous for human consumption [7–9]. As a result of such adverse health effects of the toxin, it is necessary to have a sensitive, reliable, and accurate method for monitoring AFB_1 level in feedstuffs.

High-performance liquid chromatography with fluorescence detection (HPLC-FLD) is considered as the most reliable instrument for the quantification of AFs due to its accuracy and high sensitivity. However, it requires clean-up steps involving immune affinity columns (IAC) to remove interferences and preconcentration of AFB_1. IAC is also quite expensive, difficult to use, and time-consuming [10]. In this regard, a QuEChERS (Quick, Easy, Cheap, Effective, Rugged, and Safe) method, which was originally developed for pesticides and veterinary drug residue analysis, has been applied for the measurement of mycotoxins in a variety of matrices [10–12]. Commonly, it involves two steps: the first step is the extraction step based on a partitioning phase, and the second step is the dispersive-solid phase extraction step (d-SPE) using a combination of magnesium sulfate with different sorbents. The advantages of this method are that it is simple and reduces time consumption [13, 14].

In order to reduce sample handling and increase sample output efficiency, the aim of this study is focused on the optimization of sample preparation using QuEChERS-based extraction and optimizing the chromatographic conditions of an HPLC-FLD method with pre-column derivatization for quantification of AFB_1 in the feedstuffs as well as a cross-sectional investigation of the contamination of AFB_1 content in the feedstuffs using a QuEChERS-based method.

2. Materials and Methods

2.1. Chemicals and Reagents. AFB_1 standard was purchased from Sigma-Aldrich (Madrid, Spain). Stock solution of AFB_1 was prepared in methanol at a concentration of 10 mg/mL. From the stock standard solution, working solution was prepared by diluting AFB_1 in methanol at a concentration of 1 mg/mL and stored at 4°C. Before use, the solution was brought to laboratory room temperature (25°C) and thoroughly shaken. Acetonitrile and methanol were of HPLC grade and supplied by ACILabscan (Thailand). Sodium chloride was obtained from Ajax Finechem Pty. Ltd. (New Zealand) and anhydrous magnesium sulfate was from Applicati Chem Panreac ITW Companies (Germany). Trifluoroacetic acid was purchased from ACROS Organics (Belgium) and water was purified (18 MΩ) on a Milli-Q Plus apparatus.

2.2. Optimization of the HPLC-FLD Analytical Method. To fulfill the purposes, several conditions, including a mobile phase, extraction solvents combined with the QuEChERS salt, volume of extraction, and precolumn derivatization solvents, were optimized to ensure an efficient determination of AFB_1. Different combinations of the mobile phase solution, including methanol/water (50/50 v/v), acetonitrile/methanol/water (20/20/60 v/v/v), acetonitrile/methanol/water (16/22/62 v/v/v), and acetonitrile/methanol/water (15/15/70 v/v/v), were evaluated in order to optimize the retention time of the AFB_1 and to enhance the sensitivity. In addition, the different mixtures of extract solvents that consisted of acetonitrile (100%), acetonitrile/methanol (40/60 v/v), and acetonitrile/water (50/50 v/v) were investigated. Rather than optimizing the ratio of extraction solvents alone, this study simultaneously evaluated the addition of different drying agents or salt; these included sodium chloride/magnesium sulfate (1/4 w/w) or sodium acetate/magnesium sulfate (1/4 w/w). The effect of extraction volume of the satisfied extraction solvent was tested by two different volumes of 10 and 20 mL. Moreover, the critical points of reconstitution of the precolumn derivatization with suitable solvent between 10% acetonitrile and the selected mobile phase were compared.

2.3. Optimization of Sample Preparation. The optimization of sample preparation was carried out using broken rice as a representative matrix. Briefly, the effects of different extraction solvents (acetonitrile 100%, acetonitrile/methanol 40/60 v/v, and acetonitrile/water 50/50 v/v) and salts (sodium chloride and sodium acetate) on the signal response of AFB_1 were compared. After milling and homogenization, a 10 g sample was weighed into the extraction tube. Then, the sample was spiked by adding AFB_1 standard solution and left at laboratory room temperature (25°C) for approximately 30 min to allow the solvent to evaporate. After leaving the samples to undergo a process of equilibration, the extraction solvents were added and shaken at 3000 g for 3 min to ensure that the solvents had mixed thoroughly with the entire sample. Thereafter, 1 g of the selected salt (sodium chloride) with 4 g of anhydrous magnesium sulfate was added to the mixture and the shaking procedure was repeated in order to induce phase separation and AFB_1 partitioning. Subsequently, an aliquot of the supernatant layer (1 mL) was evaporated until dry under nitrogen gas. Following that, the precolumn derivatization of AFB_1 was done in terms of how to enhance its sensitivity. The residue was reconstituted in 900 μl of 10% acetonitrile or in the mobile phase that had been optimized in this study. After that, 100 μl of trifluoroacetic acid was added, followed by an incubation period at 50°C for 15 min. The derivatized solution was then centrifuged at 1000 g for 5 min before HPLC-FLD analysis took place.

The influence of the volume of extraction solvent on the signal response of AFB_1 was optimized (in this case, a mixture of acetonitrile/methanol 40/60 v/v was used as the extraction solvent). The volumes of extraction solvent compared were 10 mL and 20 mL. Subsequent extraction procedures were done as described above.

2.4. Validation of Methods

2.4.1. Final HPLC Analysis Used in Validation Experiments. The apparatus used for the analysis was a reverse phase HPLC (Shimadzu, Kyoto, Japan) equipped with a fluorescence detector. The analytical column used was a Symmetry® C-18 3.9x150 mm with 5 μm particle size from Waters (Massachusetts) and the guard column was Silfilter STD C-18 3.0x10 mm. The column was maintained at 40°C. Analysis was run at a flow rate of 1 ml/min by an isocratic mobile phase using a mixture of acetonitrile/methanol/water (15/15/70 v/v/v). The total run time was 20 min. An aliquot of a 10 μl sample extract was injected into the chromatographic system and detection was carried out by a fluorescence detector (excitation and emission wavelengths were 360 and 440 nm, respectively). Chromatograms were displayed with Class VP LC software.

2.4.2. Final Sample Preparation Method Used in Validation Experiments. 10 g of the sample and 20 mL of acetonitrile/methanol (40/60 v/v) were added to an extraction tube. Extraction was achieved through shaking the sample at

3000 g for 3 min. Thereafter, phase partition was induced using 1 g of sodium chloride and 4 g of anhydrous magnesium sulfate with shaking at 3000 g for 3 min. Subsequently, 1 ml of the organic layer was evaporated until dry under nitrogen gas. After evaporation, a precolumn derivatization of AFB_1 was prepared by reconstituting the residue in 900 μl of 10% acetonitrile solution and adding 100 μl of trifluoroacetic acid followed by incubation at 50°C for 15 min. Finally, the derivatized solution was centrifuged at 1000 g for 5 min and an aliquot was transferred to the vials for HPLC-FLD analysis which was described previously.

The method was validated according to SANCO/12571/2013 which demonstrates the conformity of the analytical performances with criteria established in regulation (EC) no. 178/2010 [15]. The guidelines recommend the validation procedure for linearity, specificity, limit of detection (LOD), limit of quantitation (LOQ), accuracy, and precision. This study used the maximum permitted levels (MPLs) for the legislated mycotoxin in animal feed under the Animal Feed Quality Control Act B.E. 2525 of Thailand at 100 ng/g to consider as a maximum fortified concentration for the samples.

Linearity was tested by external standardization using matrix calibration curves constructed from AFB_1 standard solutions at 6 different concentrations within the range of 5-100 ng/g (5, 10, 20, 40, 60, and 100 ng/g). Analytical curves were established by plotting the peak areas which were used as the analytical signal response (y) versus the concentration of AFB_1 (x).

The specificity of the method was evaluated by comparing the retention times in the blank sample matrices and the samples were spiked with 100 ng/g of AFB_1 to ensure there was no interference in the retention time of the target analyte.

The sensitivity of the method was considered according to the LOD and LOQ. The LOD was calculated as the lowest concentration of the AFB_1 giving a signal response 3 times greater than the average of the baseline noise obtained from 10 independent blank samples of each matrix (S/N 3:1) and the LOQ was defined as an AFB_1 signal response 10 times greater than the average of the baseline noise obtained from 10 independent blank samples of each matrix (S/N 10:1).

The accuracy was tested through recovery studies by spiked AFB_1 standard solution at 3 different concentration levels (equivalent to 20, 40, and 100 ng/g) into blank sample matrices. Six replicates of each concentration were prepared for each matrix. The level of precision, which is expressed as relative standard deviations (RSDs), was estimated by performing intraday repeatability, expressed as RSD_r, and interday reproducibility, expressed as RSD_R, by spiked blank sample with AFB_1 standard solution at 3 concentration levels (20, 40, and 100 ng/g). Again, 6 replicates of each concentration were prepared for each matrix and determined within 1 day and in 3 consecutive days.

2.5. Limited Survey of AFB_1 Contamination in Selected Feedstuffs. A total of 120 samples, including 30 samples of each kind of feedstuffs (broken rice, peanuts, corn, and fishmeal), were randomly collected from swine farms in the western region of Thailand. All samples were sealed in plastic bags and stored at -20°C until analysis.

3. Results and Discussion

3.1. Optimization of HPLC-FLD Conditions. To optimize the chromatographic separation, the study was carried out using four different solutions of mobile phase solvent as described above. The results showed that using methanol/water (50/50 v/v), acetonitrile/methanol/water (20/20/60 v/v/v), and acetonitrile/methanol/water (16/22/62 v/v/v) obtained poor signal responses and the AFB_1 was not separated from the unretained compound. The best mobile phase which gave the highest signal response was obtained and that offered adequate chromatographic separation with acetonitrile/methanol/water (15/15/70 v/v/v). This might be due to the fact that a high polarity of the mobile phase combination was able to exclude nonpolar substances from the C-18 analytical column [16]. However, broadened peak may occur when the mobile phase is composed of water over 70% [10].

HPLC-FLD has been the most popular analytical instrument for AFB_1 analysis. Nevertheless, fluorescent detection sensitivity of AFB_1 requires a conversion to more highly fluorescent derivatives (AFB_{2a}). Several derivatization methods are available, including precolumn treatment with trifluoroacetic acid or postcolumn derivatization with pyridinium hydrobromide [2, 17]. With an existence instrument, the method used the precolumn derivatization throughout the quantitative determinations. Generally, the residues were reconstituted using a mobile phase to reduce the solvent effects. However, in this study, the chromatographic response signal of the AFB_1 derivative was unsatisfied when dissolved with mobile phase (acetonitrile/methanol/water 15/15/70 v/v/v). This might be due to the fact that methanol affects the trifluoroacetic acid derivatization and AFB_{2a} could potentially be degraded by methanol, which would then result in a poor signal response [18, 19].

3.2. Optimization of Sample Preparation. In quantifying the substance, the sample preparation has a crucial impact on the accuracy and precision of the results, especially when the sample was contaminated at very low level. Usually, the sample was extracted and partitioned using acidified acetonitrile and salt as a drying agent followed by the purification of extractant with d-SPE [11, 12]. To reduce sample handling, in this study, a simple extraction QuEChERS-based procedure without d-SPE clean-up step was applied.

Following the current protocol, the optimization method was carried out using 10 g of broken rice as a representative. The conventional extraction procedures of AFB_1 from different matrices used a mixture of water and organic solvents such as acetonitrile, methanol, or acetone [20]. Therefore, different mixtures of extract solvents, which were acetonitrile (100%), acetonitrile/methanol (40/60 v/v), and methanol/water (50/50 v/v), were evaluated. An important characteristic of the QuEChERS procedure is the addition of salts, such as sodium chloride, sodium acetate, or sodium sulfate, which the typical QuEChERS method uses to separate interfering compounds from matrix and increase the solubility of the analyte into the organic phase [10, 14, 21]. Therefore, all of the extraction solvents were tested by adding either sodium chloride/magnesium sulfate (1/4 w/w)

TABLE 1: The results of validation method with mean value∗(accuracy and precision), n=6.

Matrices	Spiked level (ng/g)	%Recovery∗			Intra-day∗ (%RSD_r)			Inter-day∗(%RSD_R)	LOD (ng/g)	LOQ (ng/g)
		Day 1	Day 2	Day 3	Day 1	Day 2	Day 3			
Broken rice	20	86.83	91.83	96.83	3.52	5.48	3.16	5.44	0.2	0.3
	40	82.50	89.83	90.83	3.73	4.30	4.25	4.12		
	100	84.00	93.66	92.50	3.61	3.35	2.34	5.77		
Peanut	20	106.57	109.37	109.37	3.34	5.89	1.98	1.42	0.6	0.9
	40	106.65	104.13	109.41	1.02	2.45	5.42	2.47		
	100	101.64	93.89	107.10	4.06	1.83	3.97	6.58		
Corn	20	90.10	90.98	91.72	6.77	5.63	4.89	0.89	0.6	0.8
	40	91.84	95.06	84.95	5.05	5.27	1.69	5.70		
	100	86.36	85.64	85.43	3.28	4.20	6.33	0.57		
Fish meal	20	109.85	108.88	102.31	5.02	2.96	6.18	6.33	1.2	1.5
	40	100.35	89.13	97.92	4.26	2.99	11.0	6.16		
	100	86.63	86.70	99.49	2.76	2.34	2.38	8.29		

or sodium acetate/magnesium sulfate (1/4 w/w). The results clearly showed that the highest efficiency to extract AFB_1 was achieved when using acetonitrile/methanol (40/60 v/v) combined with sodium chloride/magnesium sulfate (1/4 w/w). Generally, the initial QuEChERS method was developed for the determination of pesticide residue in fruits and vegetables which contain a high concentration of water (>80%), while feedstuffs have lower percentages of water composition than in those matrices. An earlier report suggested that the sample should be soaked in water prior to extraction [22]. However, the extraction procedure in this study was done without this soaking step.

Moreover, the present results showed that using acetonitrile (100%) and a mixture of acetonitrile/water (50/50 v/v) as an extraction solvent resulted in a low recovery of determination which is in accordance with Capriotti [23]. This might be due to the fact that AFB_1 is normally soluble in moderately polar organic solvents, e.g., methanol or chloroform, and scarcely soluble in water. Consequently, AFB_1 could not be completely extracted and remained in matrix [24]. Although acetonitrile/water (50/50 v/v) offered high signal response, its broadened tailing chromatogram overlapped with interference peaks. Moreover, the presence of a large amount of water in the solvent was adsorbed by matrices and tended to aggregate in the tube. Generally, the drying agents are anhydrous inorganic salts that acquire water for hydration. Therefore, the crystal form with large clumps of drying agent commonly occurred when exposed to water. Then, the results created a high viscosity of the extractant, reduced volume of supernatant layer, and extended duration for the evaporation step and a low signal response as well as percentage of recovery [10, 20].

The volume of extraction solvent (10 mL and 20 mL) was optimized. The results clearly showed that using 20 mL of extraction volume gave more satisfying data than using 10 mL. This is in agreement with an earlier study that used an adequate volume for extraction that increased the extraction efficiency by decreasing the matrix effect and enhanced the release of the AFB_1 from the matrix resulting in high recoveries [10]. Therefore, 20 mL of acetonitrile/methanol (40/60 v/v) combined with sodium chloride (1g) and magnesium sulfate (4g) without any clean-up step was chosen as an extraction solution for AFB_1 analysis.

3.3. Method Performance. Performance characteristics of the optimized method were established by a validation procedure with spiked feeding stuff samples to test linearity, specificity, accuracy, precision, LOD, and LOQ. As a result, the analytical curves showed good linearity within the working range (5-100 ng/g), with coefficients (R^2) of determination higher than 0.9800. The discrimination between AFB_1 and interference compounds was defined as specificity. The absence of any signal response close to the retention time of AFB_1 in all matrices indicated that there were no matrix interferences in spite of the high complexity of the matrices. The obtained results have no coeluting peaks close to the retention time of the AFB_1 (Figure 1), demonstrating that it was a specification method.

The sensitivity of the method was considered according to the LOD and LOQ (Table 1). As presented, the calculated values in all matrices were found to have the LOD between 0.2 and 1.2 ng/g. The LOQ was in the range of 0.3-1.5 ng/g. Both the LOD and LOQ values were satisfactory for the AFB_1 detection in the feedstuffs, being far below the MPLs legislated values.

In order to check the accuracy of the method, recovery experiments were tested with a spiked AFB_1 standard solution at 3 different concentration levels (20, 40, and 100 ng/g). Average recoveries for all feedstuffs were in the range of 82.50-109.85% (Table 1) which demonstrated the conformity of the guidelines for the recovery of analysis range of 70–110%, according to SANCO/12571/2013. RSDs were calculated under RSD_r and RSD_R conditions to evaluate the precision of the method. In the present study, the RSD_r and RSD_R of most samples were 1.02-6.77% and 0.57-6.58%, respectively, except for the fishmeal matrix which obtained

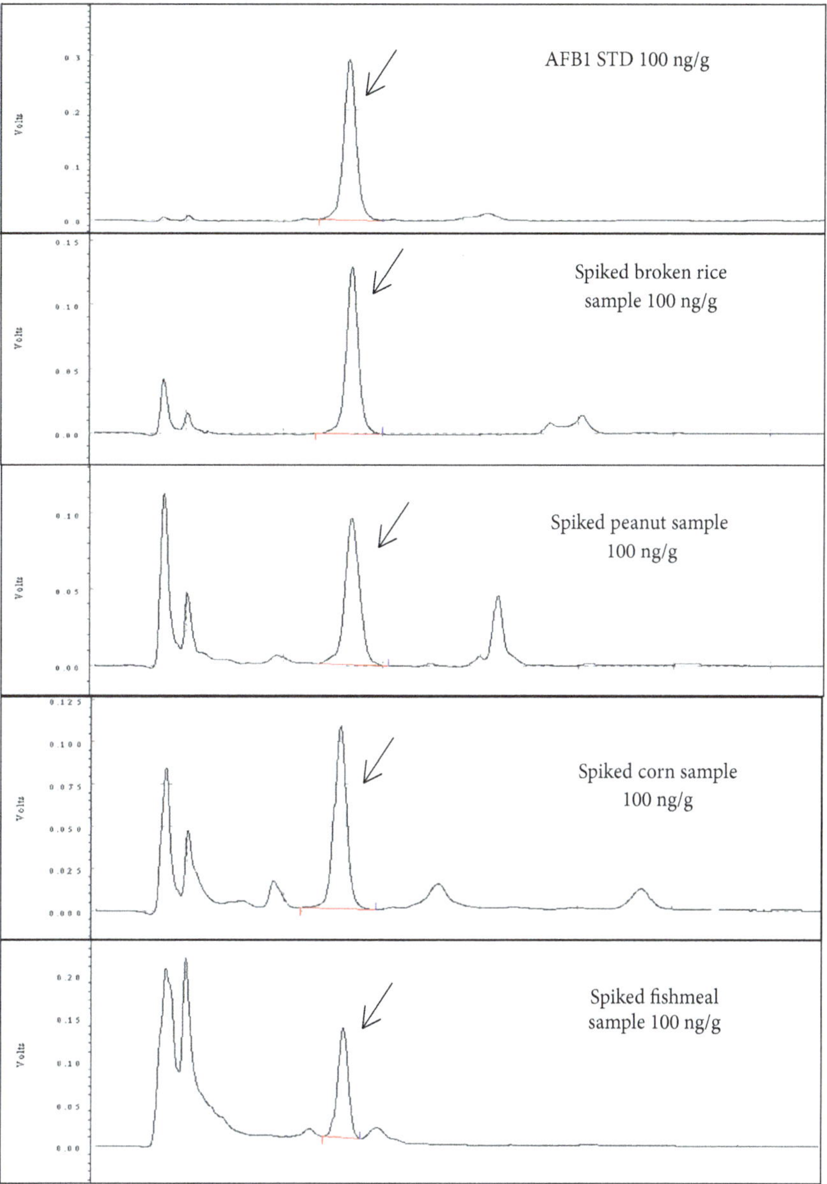

FIGURE 1: Chromatograms of AFB_1 standard solution containing 100 ng/g and sample matrices (broken rice, peanut, corn, and fishmeal) were spiked with AFB_1 standard solution at 100 ng/g.

RSD_r at 11% on the third day and RSD_R was found to be between 6.16 and 8.29% (Table 1). However, the performed method was in the acceptable range of RSDs of less than 20% with the performance criteria requirement of the EC [15]. Additionally, the obtained results comply well with the Thai regulations for the detection of AFB_1 in feedstuff.

3.4. Occurrence of AFB1 from Collected Samples. The developed method was applied for the analysis of a total of 120 feedstuff samples including 30 samples of each type of feedstuffs of broken rice, peanuts, corn, and fishmeal which were randomly collected from swine farms in the western region of Thailand where there is a high density of livestock available. Several types of AFs analogues are naturally produced in food and feed. Obviously, AFB_1 contamination in animal feed is the most predominant compared to other metabolites. Currently, of four major AFs analogues, AFB_1 is the only metabolite with maximum permitted levels (MPLs) for animal feed set under Directive 2002/32. A guidance value of the AFs contamination by the European Food Safety Authority (EFSA) in animal feed is limited between 5 ng/g for compound feed of dairy and young animals and 20 ng/g for all other feed materials. In Thailand, the safe limits of AFB_1 at 30-100 ng/g in animal feed (depending on the type of feed and animal species) have been regulated under the Animal Feed Quality Control Act B.E. 2525 of Thailand.

With regard to the data (Table 2), a low incidence (23.33%) with AFB_1 trace level contamination in broken rice samples was determined by the validated method. It is possible that AFB_1 is present mainly at the fungal invading

Table 2: Occurrence and levels of AFB_1 contamination in feedstuffs.

Sample matrices (n=120)	n.d.	<4	4-20	21-50	51-100	>100	Concentration range (ng/g) (Mean±SD)
Broken rice (30)	23	7	-	-	-	-	0.44-2.33 (0.27±0.61)
Peanut (30)	18	1	-	4	6	1	3.97-106.26 (23.29±33.76)
Corn (30)	7	10	7	5	1	-	0.88-50.29 (10.67±11.72)
Fishmeal (30)	24	4	2	-	-	-	1.06-10.35 (0.82±2.20)

area and remains on the outer layer of the grain which was removed by the degree of milling. As a result, AFB_1 in broken rice is usually found at trace level [25–27].

However, postharvest factors, for instance, storage under inappropriate conditions and being exposed to rain, may lead to increased AFB_1 contamination levels [28]. Likewise, Anjum [1] reported that AFB_1 contamination in broken rice is affected by rain. In addition, Thieu [25] has reported that broken rice could be contaminated with AFB_1 during harvesting and via a dehumidification process through sun-drying technique.

A high incidence of AFB_1 contamination in peanuts (40%) with a wide range from 3.97 to 106.27 ng/g was found in this study. Among those, 11 samples were contaminated with an AFB_1 level that exceeded the maximum levels set in Directive 2003/100/EC, amended Directive 2002/32 (20 ng/g), and 7 samples were contaminated with AFB_1 level higher than the permissible limits as established by the Animal Feed Quality Control Act B.E. 2525 of Thailand (50 ng/g). In contrast, our current study showed that the contamination level was higher than that found in peanuts from Thailand in a study conducted by Thanida [29]. Moreover, 93-99% of the 200 peanut samples collected from 2012 to 2015 were found to have lower than 4 ng/g (unpublished in-house data). This may be due to underestimation detected by the ELISA method used in previous survey.

Corn (maize) is one of the major ingredients used in animal feed which susceptible to infection by mycotoxin-producing fungi and, consequently, has the potential to become contaminated with mycotoxins [30]. The current results showed that the incidence of AFB_1 remained in 23 out of 30 corn samples (76.67%) with a wide range of concentrations (0.88-50.29 ng/g). Similarly, previous reports have shown that corn can become contaminated with AFB_1 in various ranges. Recently, AFs in corn ingredient were found with a high prevalence [31]. Thanida [29] reported AFB_1 contamination in corn samples from Thailand found ranging from 0.275 to 40 ng/g. As compared to the in-house data, 15% of the total 400 corn feed samples collected during 2012 to 2015 were contaminated with AFB_1 ranging from 51 to 100 ng/g while 7% of the samples contained AFB_1 over 100 ng/g (unpublished data). Our current study presented the notion that the AFB_1 contents were within the safe limits under the Animal Feed Quality Control Act B.E. 2525 of Thailand (for feedstuff set as 100 ng/g) and the FDA guidance (100-300 ng/g).

Typically, AFB_1 is not found in fishmeal because its matrix is unsuitable for the production of mycotoxin. It has been presented that AFB_1 was not detected in any fishmeal samples [32]. However, fishmeal contaminated with AFB_1 has been reported. Anjum [1] found AFB_1 24 ng/g from Pakistan during rainy season which was in accordance with the results of Mayahi [33] who found AFB_1 contamination at an average of 15 ng/g. As detected in the present study, 20% of the sample was found to be contaminated with AFB_1 in the range of 1.06-10.35 ng/g. This coincides with earlier results reported by Thanida [29] that AFB_1 contamination in fishmeal was found in the range of 1.67-11.90 ng/g. This indicates that fishmeal as a feedstuff presents no risk of AFB_1 contamination to animal health as the detected AFB_1 was not above the MPLs.

4. Conclusion

A modified method based on the QuEChERS procedure by a single extraction step without employing a clean-up step was developed for AFB_1 determination in feedstuffs. Good analytical results were obtained, including good linearity, specificity, accuracy, precision (repeatability and reproducibility), and analytical limits (LOD and LOQ). Therefore, it can be suggested as an alternative to expensive and time-consuming methods by using immune affinity columns or two steps of liquid/solid extraction procedure. Finally, an efficient and sensitive method was developed here that was applied to four different kinds of matrices which were detectable at low level. Considering that, the usefulness of the method could be applied as a regular monitoring method of AFB_1 in broken rice, peanuts, corn, fishmeal, and their related matrices.

Acknowledgments

This work was partially supported by the Kasetsart University Research and Development Institute (KURDI) and Center for Advanced Studies for Agriculture and Food, Institute for Advanced Studies (CASAF), Kasetsart University, under the Higher Education Research Promotion and National Research University Project of Thailand, Office of the Higher Education Commission, Ministry of Education, Thailand. The authors gratefully acknowledge the staffs of the diagnostic and laboratory service units (Mycotoxin Unit) in Kamphaengsaen Campus for their technical and equipment support.

References

[1] M. A. Anjum, S. H. Khan, A. W. Sahota, and R. Sardar, "Assessment of aflatoxin B1 in commercial poultry feed and feed ingredients," *Journal of Animal and Plant Sciences*, vol. 22, no. 2, pp. 268–272, 2012.

[2] N. Tansakul, S. Limsuwan, J. Böhm, M. Hollmann, and E. Razzazi-Fazeli, "Aflatoxins in selected Thai commodities," *Food Additives & Contaminants: Part B. Surveillance*, vol. 6, no. 4, pp. 254–259, 2013.

[3] M. Oplatowska-Stachowiak, N. Sajic, Y. Xu et al., "Fast and sensitive aflatoxin B1 and total aflatoxins ELISAs for analysis of peanuts, maize and feed ingredients," *Food Control*, vol. 63, pp. 239–245, 2016.

[4] N. Arroyo-Manzanares, L. Gámiz-Gracia, A. M. Gámiz-Gracia, J. J. Soto-Chinchilla, A. M. García-Campaña, and L. E. García-Ayuso, "On-line preconcentration for the determination of aflatoxins in rice samples by micellar electrokinetic capillary chromatography with laser-induced fluorescence detection," *Electrophoresis*, vol. 31, no. 13, pp. 2180–2185, 2010.

[5] W. S. Khayoon, B. Saad, T. P. Lee, and B. Salleh, "High performance liquid chromatographic determination of aflatoxins in chilli, peanut and rice using silica based monolithic column," *Food Chemistry*, vol. 133, no. 2, pp. 489–496, 2012.

[6] International Agency for Research on Cancer, "Some traditional herbal medicines, some mycotoxins, naphthalene and styrene," *IARC Monograph on the evaluation of carcinogenic risk to human*, vol. 82, no. 2-3, pp. 171–300, 2002.

[7] S. A. Aly and W. Anwer, "Effect of naturally contaminated feed with aflatoxins on performance of laying hens and the carryover of aflatoxin B1 residues in table eggs," *Pakistan Journal of Nutrition*, vol. 8, no. 2, pp. 181–186, 2009.

[8] S. M. Herzallah, "Determination of aflatoxins in eggs, milk, meat and meat products using HPLC fluorescent and UV detectors," *Food Chemistry*, vol. 114, no. 3, pp. 1141–1146, 2009.

[9] R. W. Han, N. Zheng, J. Q. Wang, Y. P. Zhen, X. M. Xu, and S. L. Li, "Survey of aflatoxin in dairy cow feed and raw milk in China," *Food Control*, vol. 34, no. 1, pp. 35–39, 2013.

[10] A. Y. Sirhan, G. H. Tan, A. Al-Shunnaq, L. Abdulra'uf, and R. C. S. Wong, "QuEChERS-HPLC method for aflatoxin detection of domestic and imported food in Jordan," *Journal of Liquid Chromatography & Related Technologies*, vol. 37, no. 3, pp. 321–342, 2014.

[11] N. Arroyo-Manzanares, A. M. García-Campaña, and L. Gámiz-Gracia, "Multiclass mycotoxin analysis in *Silybum marianum* by ultra high performance liquid chromatography–tandem mass spectrometry using a procedure based on QuEChERS and dispersive liquid–liquid microextraction," *Journal of Chromatography A*, vol. 1282, pp. 11–19, 2013.

[12] U. Koesukwiwat, K. Sanguankaew, and N. Leepipatpiboon, "Evaluation of a modified QuEChERS method for analysis of mycotoxins in rice," *Food Chemistry*, vol. 153, pp. 44–51, 2014.

[13] N. Arroyo-Manzanares, J. F. Huertas-Pérez, A. M. García-Campaña, and L. Gámiz-Gracia, "Simple methodology for the determination of mycotoxins in pseudocereals, spelt and rice," *Food Control*, vol. 36, no. 1, pp. 94–101, 2014.

[14] M. Á. González-Curbelo, B. Socas-Rodríguez, A. V. Herrera-Herrera, J. González-Sálamo, J. Hernández-Borges, and M. Á. Rodríguez-Delgado, "Evolution and applications of the QuEChERS method," *TrAC - Trends in Analytical Chemistry*, vol. 71, pp. 169–185, 2015.

[15] EC, "Commission regulation (EU) No 178/2010 of 2 March 2010 amending regulation (EC) No 401/2006 of 23 February 2006 laying down the methods of sampling and analysis for the official control of the levels of mycotoxins in foodstuffs," *Official Journal of the European Union*, 2010.

[16] A. Cismileanu, G. Voicu, V. Ciuca, and M. Ionescu, "Determination of aflatoxin B1 in cereal-based feed by a high-performance chromatographic method," *LucrariStiintificMedicinaverinara*, vol. 41, pp. 565–569, 2008.

[17] M. Muscarella, M. Iammarino, D. Nardiello et al., "Validation of a confirmatory analytical method for the determination of aflatoxins B1, B2, G1 and G2 in foods and feed materials by HPLC with on-line photochemical derivatization and fluorescence detection," *Food Additives and Contaminants - Part A Chemistry, Analysis, Control, Exposure and Risk Assessment*, vol. 26, no. 10, pp. 1402–1410, 2009.

[18] W. T. Kok, "Derivatization reactions for the determination of aflatoxins by liquid chromatography with fluorescence detection," *Journal of Chromatography B: Biomedical Sciences and Applications*, vol. 659, no. 1-2, pp. 127–137, 1994.

[19] S. M. Pearson, A. A. G. Candlish, K. E. Aidoo, and J. E. Smith, "Determination of aflatoxin levels in pistachio and cashew nuts using immunoaffinity column clean-up with HPLC and fluorescence detection," *Biotechnology Techniques*, vol. 13, no. 2, pp. 97–99, 1999.

[20] E. Reiter, J. Zentek, and E. Razzazi, "Review on sample preparation strategies and methods used for the analysis of aflatoxins in food and feed," *Molecular Nutrition & Food Research*, vol. 53, no. 4, pp. 508–524, 2009.

[21] E. A. Orlando and A. V. C. Simionato, "Extraction of tetracyclinic antibiotic residues from fish filet: Comparison and optimization of different procedures using liquid chromatography with fluorescence detection," *Journal of Chromatography A*, vol. 1307, pp. 111–118, 2013.

[22] S. C. Cunha and J. O. Fernandes, "Development and validation of a method based on a QuEChERS procedure and heart-cutting GC-MS for determination of five mycotoxins in cereal products," *Journal of Separation Science*, vol. 33, no. 4-5, pp. 600–609, 2010.

[23] A. L. Capriotti, C. Cavaliere, S. Piovesana, R. Samperi, and A. Laganà, "Multiclass screening method based on solvent extraction and liquid chromatography-tandem mass spectrometry for the determination of antimicrobials and mycotoxins in egg," *Journal of Chromatography A*, vol. 1268, pp. 84–90, 2012.

[24] W. S. Khayoon, B. Saad, C. B. Yan et al., "Determination of aflatoxins in animal feeds by HPLC with multifunctional column clean-up," *Food Chemistry*, vol. 118, no. 3, pp. 882–886, 2010.

[25] N. Q. Thieu, B. Ogle, and H. Pettersson, "Screening of Aflatoxins and Zearalenone in feedstuffs and complete feeds for pigs in Southern Vietnam," *Tropical Animal Health and Production*, vol. 40, no. 1, pp. 77–83, 2008.

[26] S. Z. Iqbal, M. R. Asi, A. Arino, N. Akram, and M. Zuber, "Aflatoxin contamination in different fractions of rice from Pakistan and estimation of dietary intakes," *Mycotoxin Research*, vol. 28, no. 3, pp. 175–180, 2012.

[27] S. Choi, H. Jun, J. Bang et al., "Behaviour of Aspergillus flavus and Fusarium graminearum on rice as affected by degree of milling, temperature, and relative humidity during storage," *Food Microbiology*, vol. 46, pp. 307–313, 2015.

[28] K. R. N. Reddy, C. S. Reddy, and K. Muralidharan, "Detection of Aspergillus spp. and aflatoxin B1 in rice in India," *Food Microbiology*, vol. 26, no. 1, pp. 27–31, 2009.

[29] W. Thanita, C. Surapol, K. Saowanit, and W. Sutha, "Aflatoxin contamination in feedstuffs and complete feed for layer, broiler and swine in Songkhla Province," *Thaksin*, vol. 11, no. 1, pp. 8–18, 2008.

[30] F. Wu and G. P. Munkvold, "Mycotoxins in ethanol co-products: Modeling economic impacts on the livestock industry and management strategies," *Journal of Agricultural and Food Chemistry*, vol. 56, no. 11, pp. 3900–3911, 2008.

[31] F. Granados-Chinchilla, A. Molina, G. Chavarría, M. Alfaro-Cascante, D. Bogantes-Ledezma, and A. Murillo-Williams, "Aflatoxins occurrence through the food chain in Costa Rica: Applying the One Health approach to mycotoxin surveillance," *Food Control*, vol. 82, pp. 217–226, 2017.

[32] B. M. Kokic, I. S. Cabarkapa, I. D. Levicet, and etal., "Screening of mycotoxins in animal feed from the region of Vojodina," *Proc Nat SciMaticaSrpska Novi Sad*, vol. 117, pp. 87–96, 2009.

[33] M. Mayahi, R. J. Mohammad, and S. Negin, "Isolation of Aspergillus spp. and determination of aflatoxin level in fish meal, maize and soya meal," *ShahidChamran University Journal of Science*, vol. 17, pp. 95–105, 2007.

5

A Data Mining Approach to Improve Inorganic Characterization of *Amanita ponderosa* Mushrooms

Cátia Salvador,[1] M. Rosário Martins,[1,2] Henrique Vicente,[2,3] and A. Teresa Caldeira[1,2]

[1]*HERCULES Laboratory, Évora University, Largo Marquês de Marialva 8, 7000-809 Évora, Portugal*
[2]*Departamento de Quimica, School of Sciences and Technology, Évora University, Rua Romão Ramalho 59, 7000-671 Évora, Portugal*
[3]*Centro ALGORITMI, Universidade do Minho, Braga, Portugal*

Correspondence should be addressed to A. Teresa Caldeira; atc@uevora.pt

Academic Editor: Günther K. Bonn

Amanita ponderosa are wild edible mushrooms that grow in some microclimates of Iberian Peninsula. Gastronomically this species is very relevant, due to not only the traditional consumption by the rural populations but also its commercial value in gourmet markets. Mineral characterisation of edible mushrooms is extremely important for certification and commercialization processes. In this study, we evaluate the inorganic composition of *Amanita ponderosa* fruiting bodies (Ca, K, Mg, Na, P, Ag, Al, Ba, Cd, Cr, Cu, Fe, Mn, Pb, and Zn) and their respective soil substrates from 24 different sampling sites of the southwest Iberian Peninsula (e.g., Alentejo, Andalusia, and Extremadura). Mineral composition revealed high content in macroelements, namely, potassium, phosphorus, and magnesium. Mushrooms showed presence of important trace elements and low contents of heavy metals within the limits of RDI. Bioconcentration was observed for some macro- and microelements, such as K, Cu, Zn, Mg, P, Ag, and Cd. *A. ponderosa* fruiting bodies showed different inorganic profiles according to their location and results pointed out that it is possible to generate an explanatory model of segmentation, performed with data based on the inorganic composition of mushrooms and soil mineral content, showing the possibility of relating these two types of data.

1. Introduction

Mushrooms are known from ancient times for their medicinal properties and gastronomic properties. Therefore the consumption of edible wild-growing mushrooms has been very popular. The demand for the commercialization of edible wild mushrooms has proved to be a widely expanding business with increasing economic importance in many rural areas of some countries. In recent years, the consumption of edible mushrooms has been increasing and gaining prominence due to their gastronomic potential, also for their both organoleptic properties (texture and pleasant aroma) [1–4], their chemical composition, mineral content, and nutraceutical value [4–8]. Mushrooms are an important source of proteins, dietary fibres, and vitamins (B, C, D, E) containing low levels of sugar and fats. They can assimilate large amounts of water and minerals such as phosphorus, iron, potassium, cadmium, magnesium, copper, and zinc, due to the large area of mycelium overgrowing the surface layer of soil [9]. This mycelia network is ideally suited to penetrate and access soil pore spaces and an extensive surface area of fungal hyphae and physiology enable for many species on effective absorption and bioconcentration of various metallic elements, metalloids, and nonmetals [10]. Bioconcentration factor (BCF), the ratio of the element content in fruiting body to the content in underlying substrate, can express the ability of fungi to accumulate elements from substrate, and this capacity of the mushroom is affected by fungal lifestyle, age of fruiting body, specific species and element, and environment such as pH, organic matter, and pollution [9]. Moreover, the symbiotic relationships that some mushrooms species, namely, *A. ponderosa*, can establish with some plants of their habitats allowing the accumulation of high concentrations of some metals. Therefore, mineral content and organic composition of edible mushrooms are dependent on the species and the characteristics of the ecosystems in

Figure 1: Geographic representation of sampling location areas of the *A. ponderosa* fruiting bodies of forest area of the Alentejo, Andalusia, and Extremadura regions. In detail, we observe the sampling site of Évora region (Alentejo, Portugal), showing the surrounding vegetation of *Quercus suber*, *Cistus ladanifer* and a sample of the fruiting body collected at this local.

which they are inserted [11, 12]. Some minerals are essential elements for the correct human body function, although others may present toxicity [5, 6, 9, 11, 13, 14]. There are many species of edible mushrooms growing wild, and some of them are slightly characterized about minerals content and the potential of heavy metals bioaccumulation such as lead, mercury, cadmium, and silver [5, 13, 15–17]. On the other hand, due to the great diversity of wild mushrooms, the similarity between some species and the lack of knowledge of others may lead to intoxication when harvesting wild species, leading in some cases to death [18].

The genus *Amanita* is one of the best known from Agaricales order and comprises edible and poisonous mushrooms distributed worldwide, occupying mainly a mycorrhizal habitat and playing a significant role in forest ecosystems [19, 20]. This genus includes important species of edible mushrooms, as is the case of *Amanita ponderosa* [18, 20–24]. The south of Portugal, due to its Mediterranean characteristics and diversity of flora, is one of the regions of Europe with a greater predominance of *A. ponderosa* wild edible mushrooms [23]. These robust basidiomata grow during spring in mounted areas with acid soils, in forests of holm oaks and cork trees like *Quercus ilex* and *Quercus suber*, and with shrubs like *Cistus ladanifer*, *Cistus laurifolius*, and *Lavandula stoechas* establishing a symbiotic relationship with them (Figure 1). Gastronomically, this species is very relevant, not only due to the traditional consumption in the rural populations, but also due to its commercial value in the gourmet markets having high exportation potential in Portugal, thus, the chemical and mineral characterization of numerous species of edible mushrooms for certification purposes and further commercialization process becoming of extreme importance.

In recent years, some artificial intelligence based tools, namely, Data Mining, Artificial Neural Networks (ANNs), and Decision Trees (DTs) have been applied for fungal environment systems [18, 25, 26]. Data mining tools were used in latest studies about *A. ponderosa* mushrooms in order to establish a bridge of inorganic contents, molecular fingerprints, and geographical sites. In one study a segmentation model based on the molecular analysis was developed, which allowed relating the clusters obtained to the geographical site of sampling. There were also developed explanatory models of the segmentation, using Decision Trees, to relate the molecular and inorganic data, following two different strategies: one based on DNA profile and another based on the mineral composition [22]. Another study based on ANNs exposed the selected model and can predict molecular profile based on inorganic composition with a good match between the observed and predicted values [18].

The aim of this study was to determine the inorganic composition of fruiting bodies of *A. ponderosa* mushrooms from different sampling sites in the southwest of the Iberian Peninsula, analysing the variations in mineral content. Additionally, the mineral composition of the corresponding soil substrates was analysed in order to correlate with the mineral composition of the fruiting bodies. In the present work, the k-means clustering method was used to study the mineral composition of *A. ponderosa* fruiting bodies. To explain the segmentation model, that is, in order to obtain rules to assign a case to a cluster DTs were used.

2. Materials and Methods

2.1. Sampling Collection. Fruiting bodies of the *Amanita ponderosa* mushrooms were collected in spring, between

TABLE 1: Sampling sites description of *A. ponderosa* mushrooms and soil substrates.

Sampling sites	GPS coordinates	Region/country
(1) Almendres	38°33′57″N 8°02′40″O	
(2) Azaruja	38°42′10″N 7°45′58″O	
(3) Baleizão	38°01′34″N 7°42′38″O	Alentejo, Portugal
(4) Beja	38°02′44″N 7°50′56″O	
(5) Cabeça Gorda	37°55′19″N 7°48′47″O	
(6) Cabezas Rubias	37°43′50″N 7°05′11″O	Andaluzia, Spain
(7) Évora	38°35′01″N 7°51′46″O	
(8) Evoramonte	38°46′22″N 7°42′45″O	
(9) Her^{de} da Mitra	38°31′35″N 8°00′51″O	
(10) Mértola	37°37′44″N 7°39′30″O	
(11) Mina S. Domingos	37°41′02″N 7°28′49″O	Alentejo, Portugal
(12) M^{te} da Borralha	37°58′31″N 7°37′22″O	
(13) M^{te} Novo	38°30′39″N 7°43′06″O	
(14) Montejuntos	38°32′24″N 7°19′49″O	
(15) N. S^{ra} Guadalupe	38°33′47″N 8°01′23″O	
(16) N. S^{ra} Machede	38°35′21″N 7°48′19″O	
(17) Rosal de la Frontera	37°57′21″N 7°13′12″O	Andaluzia, Spain
(18) S^{to} Aleixo da Restauração	38°04′01″N 7°09′46″O	
(19) S. Miguel de Machede	38°37′34″N 7°42′33″O	
(20) Serpa	37°55′59″N 7°35′05″O	Alentejo, Portugal
(21) Ve Rocins	37°52′15″N 7°44′41″O	
(22) Valverde	38°32′24″N 8°01′18″O	
(23) V. N. S. Bento	37°56′27″N 7°23′51″O	
(24) Villanueva del Fresno	38°22′49″N 7°11′35″O	Extremadura, Spain

February and April, from 24 different sampling sites, in the southwest of the Iberian Peninsula, namely, 11 samples collected from Alto Alentejo region and 10 from Baixo Alentejo, Portugal, 2 samples collected from Andalusia and 1 from Extremadura, Spain (Table 1). These mushrooms were collected in acid soils, in forests of *Quercus suber*, *Q. ilex* ssp. *ballota*, *Cistus ladanifer*, and *Cistus laurifolius* (Figure 1).

Three individuals in the same growth stage were selected per each sampling site to avoid the effect of size. Fruiting bodies were identified by a specialist, based on morphological features according to taxonomic description of *A. ponderosa* [27]. During the collection process fruiting bodies were placed in wicker baskets and samples were transported in refrigerated containers and deposited in the Biotechnology Laboratory of Chemistry Department of University of Évora (Portugal). In parallel, soil samples were randomly collected from the surrounding soil of fruiting bodies of each site.

Fruiting bodies samples were weighed and carefully washed with double distilled water and representative samples of each sampling site were catalogued, stored in sterile bags, and preserved at −20°C for its inorganic study. The sheath was eliminated during the pretreatment of fruiting bodies samples, since it is not usually consumed.

Soil substrates were sampled (0–15 cm), after removing some visible organisms, small stones, sticks, and leaves. Samples were air-dried in room temperature for 1-2 weeks and, subsequently, sieved through a pore size of 2 mm and stored at −4°C.

2.2. Inorganic Characterization. Mineral composition was determined in *A. ponderosa* fruiting bodies and respective soil collected in 24 different sampling sites described, using different analytical techniques: Flame Atomic Absorption Spectrometry (FAAS), Flame Emission Photometry (FEP), and UV-Vis molecular absorption spectrometry.

2.2.1. Treatment of Samples. Mineral composition analysis of fruiting bodies was performed by the dry mineralization method [28]. Three samples of mushrooms from each sampling site (25 g, fresh weight) were dried in a furnace at 100°C to constant weight (48 h) from which the moisture content was calculated. Samples were homogenized, transferred to porcelain crucibles, and incinerated in a muffle furnace (Termolab) at 460°C for 14 h. In order to determine the organic matter and mineral content in mushrooms samples, ashes were weighted at constant weight. Afterwards, ashes were bleached after cooling by adding 2 M nitric acid, drying them on thermostatic hotplates, and maintaining them at 460°C for 1 h. Ashes recovery was performed with 15 mL of 0.1 M nitric acid and stored at 4°C until analyses.

Three soil samples (5 g dw) collected from each sampling site were cold treated with an extracting solution of 25%

nitric acid (HNO_3) (40 ml) and shaking with orbital agitation for 24 h at room temperature. The extracts obtained after filtering through *Whatman* No. 42 filter paper were stored into polyethylene bottles and stored at 4°C until analyses [11].

2.2.2. Analytical Determinations. Aluminium (Al), barium (Ba), calcium (Ca), cadmium (Cd), lead (Pb), copper (Cu), chromium (Cr), iron (Fe), magnesium (Mg), manganese (Mn), silver (Ag), and zinc (Zn) contents were quantified by flame atomic absorption spectrometry (Perkin-Elmer 3100 model) with atomization in an air-acetylene flame and single element hollow cathode lamps and background correction with deuterium lamp for manganese. For the determinations of calcium and magnesium, strontium chloride ($SrCl_2$) was added to make up a final concentration of 0.1125% of the sample, in order to prevent anionic interferences which might modify the result [29].

Sodium (Na) and potassium (K) were quantified by flame emission photometry with a Jenway model PFP7 flame emission photometer [30, 31]. Phosphorus (P) was quantified by vanadomolybdophosphoric acid colorimetric method, using a spectrophotometer (Hitachi, U-3010 model) [32].

All sample determinations were performed in triplicate for each one of the three independent ash extracts of mushroom samples from 24 sampling sites ($n = 9$) and for three extracts of each soil substrate ($n = 9$). Concentration values were calculated through the standard curves for each element expressed in mg/kg dw. The bioconcentration factor (BCF) value, which is the quotient of the concentration in fruiting bodies divided by the concentration in the soil substrates, was determined for all minerals in each sampling site.

2.3. Data Mining

2.3.1. Cluster Analysis. In the present study a cluster analysis was carried out. The technique applied in order to build up clusters was the k-means clustering method [33] and the software used was WEKA [34]. In WEKA Simple k-means algorithm the normalization of the numerical attributes is carried out automatically when the Euclidean distance is computed. A more detailed description of the mentioned algorithm can be found in Witten [35].

2.3.2. Decision Trees. In the k-means clustering method clusters were formed without information about the groups and their characteristics. Therefore, it is necessary to understand how the clusters were formed. To attain such a purpose Decision Trees (DTs) were used. In this study the algorithm used to generate DTs was the WEKA J48 [34], corresponding to the 8th revision of the C4.5 algorithm. The detailed description of the J48 algorithm can be found in Witten et al. [35].

2.4. Statistical Analysis. Data were evaluated statistically using the SPSS® *21.0* software *Windows Copyright©, (Microsoft Corporation)*, by descriptive parameters and by *one-way* ANOVA in order to determine statistically significant differences at the 95% confidence level ($p < 0.05$). The homogeneity of the population variances was confirmed by the *Levene* test and the multiple comparisons of media were evaluated by *Tukey's* test.

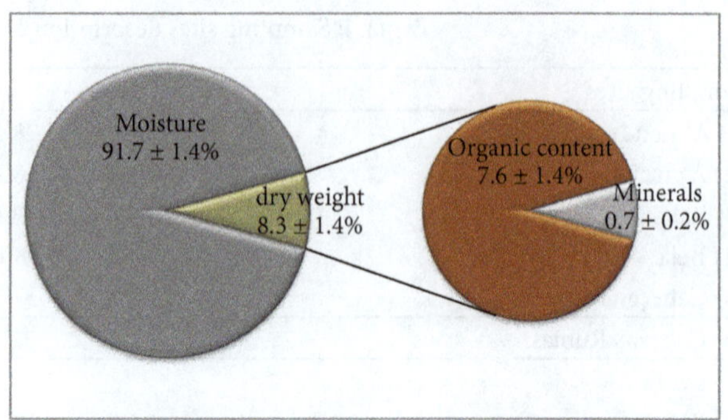

Figure 2: Moisture, dry weight, and organic and minerals content of *A. ponderosa* fruiting bodies from 24 different sampling sites.

3. Results

3.1. Fruiting Bodies and Soil Substrates' Mineral Composition. Mineral analysis of *A. ponderosa* mushrooms samples collected from 24 sampling sites of Alentejo (Portugal), Andalusia, and Extremadura (Spain) regions was evaluated.

Table 2 shows the contents in moisture, organic content, and minerals present in the samples of *A. ponderosa* collected in the different places. Moisture content ranged from 89.5 ± 0.0 to 93.8 ± 0.5% (Table 2). The analyses of variance (one-way ANOVA) showed that *A. ponderosa* fruiting bodies samples collected in Almendres (1) and Serpa (20) presented significantly higher values compared to the other samples, while the sample from Mina S. Domingos (11) showed moisture content significantly lower than the remaining samples. Dry weight values, for several samples, ranged between 6.2 ± 0.5 and 10.5 ± 0.1%, consisting in organic content values between 5.8±0.3 and 9.8±0.1% and minerals between 0.5±0.0 and 1.4± 0.1%. Fruiting bodies samples collected in Almendres (1) and Serpa (20) presented values of organic content significantly lower and the sample collected in the Mina S. Domingos (11) significantly higher than all analysed samples ($p < 0.05$). The mineral content of the fruiting bodies collected in Cabezas Rubias (6) was 1.4 ± 0.1%, a value significantly higher than the remaining samples. However, the samples from Serpa (20) and Villanueva del Fresno (24) showed the lowest values, significantly different from the another samples ($p < 0.05$).

Figure 2 shows the average contents of water, dry weight, organic content, and minerals present in the 24 samples of *A. ponderosa* analysed. The fruiting bodies presented water content, corresponding to 90.3 to 93.1% of their fresh weight. The organic matter content ranged between 6.2 and 9% and mineral content between 0.5 and 0.9%. Thus, 100 g of edible *A. ponderosa* mushrooms corresponds to a maximum of 9 g of macronutrients, such as lipids, carbohydrates, and proteins, and less than 1 g of mineral content. These values were similar to those observed in a study with mushrooms harvested in some regions of Andalusia (water content 87.8%, organic content 11.8%, carbohydrates 6.6%, proteins 3.2%, lipids 0.5%, fibre 1.5%, and mineral content 0.4%) [24] and these fruiting

TABLE 2: Moisture, organic content, and minerals contents of *A. ponderosa* fruiting bodies from different sampling sites.

Fruiting bodies	Composition in fresh weight		
Sampling sites	Moisture (%)	Organic content (%)	Minerals (%)
(1) Almendres	93.6 ± 0.3^{a}	5.8 ± 0.3^{a}	$0.6 \pm 0.1^{a,b,c}$
(2) Azaruja	$92.4 \pm 0.5^{a,b,c,d}$	$6.7 \pm 0.5^{a,b,c}$	$0.9 \pm 0.0^{b,c}$
(3) Baleizão	$90.7 \pm 0.3^{b,c,d,e}$	$8.6 \pm 0.4^{b,c,d}$	$0.7 \pm 0.1^{a,b,c}$
(4) Beja	$91.0 \pm 0.0^{a,b,c,d,e}$	$8.4 \pm 0.1^{a,b,c,d}$	$0.6 \pm 0.0^{a,b,c}$
(5) Cabeça Gorda	$91.6 \pm 0.7^{a,b,c,d,e}$	$7{,}8 \pm 0.7^{a,b,c,d}$	$0.5 \pm 0.0^{a,b}$
(6) Cabezas Rubias	$91.8 \pm 0.4^{a,b,c,d,e}$	$6.8 \pm 0.1^{a,b,c}$	1.4 ± 0.1^{d}
(7) Évora	$90.2 \pm 1.6^{c,d,e}$	$9.2 \pm 1.4^{c,d}$	$0.7 \pm 0.2^{a,b,c}$
(8) Evoramonte	$90.4 \pm 0.4^{b,c,d,e}$	$8.8 \pm 0.7^{b,c,d}$	$0.8 \pm 0.3^{a,b,c}$
(9) Her[de] da Mitra	$90.5 \pm 0.3^{b,c,d,e}$	$8.9 \pm 0.3^{c,d}$	$0.6 \pm 0.1^{a,b,c}$
(10) Mértola	$93.2 \pm 0.3^{a,b}$	$6.0 \pm 0.3^{a,b}$	$0.8 \pm 0{,}1^{a,b,c}$
(11) Mina S. Domingos	89.5 ± 0.0^{e}	9.8 ± 0.1^{d}	$0.8 \pm 0.1^{a,b,c}$
(12) M[te] da Borralha	$91.5 \pm 1.5^{a,b,c,d,e}$	$7.8 \pm 1.5^{a,b,c,d}$	$0.7 \pm 0.1^{a,b,c}$
(13) M[te] Novo	$91.7 \pm 1.2^{a,b,c,d,e}$	$7.6 \pm 1.1^{a,b,c,d}$	$0.7 \pm 0.1^{a,b,c}$
(14) Montejuntos	$92.7 \pm 0.4^{a,b,c,d}$	$6.7 \pm 0.4^{a,b,c}$	$0.5 \pm 0.1^{a,b}$
(15) N. S[ra] Guadalupe	$91.4 \pm 1.1^{a,b,c,d,e}$	$7.9 \pm 1.3^{a,b,c,d}$	$0.7 \pm 0.2^{a,b,c}$
(16) N. S[ra] Machede	$91.4 \pm 1.5^{a,b,c,d,e}$	$8.0 \pm 1.4^{a,b,c,d}$	$0.7 \pm 0.2^{a,b,c}$
(17) Rosal de la Frontera	$90.4 \pm 1.7^{b,c,d,e}$	$8.8 \pm 1.6^{b,c,d}$	$0.8 \pm 0.1^{a,b,c}$
(18) S[to] Aleixo da Restauração	$91.1 \pm 1.1^{a,b,c,d,e}$	$8.1 \pm 1.1^{a,b,c,d}$	$0.8 \pm 0.1^{a,b,c}$
(19) S. Miguel de Machede	$92.7 \pm 0.4^{a,b,c,d}$	$6.5 \pm 0.3^{a,b,c}$	$0.7 \pm 0.1^{a,b,c}$
(20) Serpa	93.8 ± 0.5^{a}	5.8 ± 0.5^{a}	0.5 ± 0.0^{a}
(21) Ve Rocins	$93.0 \pm 0.7^{a,b,c}$	$6.4 \pm 0.7^{a,b,c}$	$0.6 \pm 0.1^{a,b,c}$
(22) Valverde	$92.0 \pm 1.6^{a,b,c,d,e}$	$7.1 \pm 1.6^{a,b,c,d}$	0.9 ± 0.0^{c}
(23) V. N. S. Bento	$90.0 \pm 0.2^{d,e}$	$9.1 \pm 0.2^{c,d}$	$0.8 \pm 0.0^{a,b,c}$
(24) Villanueva del Fresno	$93.1 \pm 0.6^{a,b}$	$6.4 \pm 0.5^{a,b,c}$	0.5 ± 0.1^{a}

Values of each determination represents mean ± SD ($n = 9$). Different letters following the values indicate significant differences ($p < 0.05$).

bodies have a caloric content of 42 kcal/100 g of mushroom, similar to other edible mushroom species, characterized as low calorie foods [8, 28, 36].

Tables 3–6 show the mineral content of 24 sampling sites of *A. ponderosa* fruiting bodies as well as their corresponding soil samples. The studied *A. ponderosa* fruiting bodies showed higher mineral content in macroelements calcium (Ca), magnesium (Mg), sodium (Na), potassium (K), phosphorus (P), and microelements aluminium (Al), copper (Cu), and iron (Fe) (Tables 3 and 4). Potassium was present in higher concentrations and was higher in Cabezas Rubias (6) fruiting bodies, 69565 ± 362 mg/kg dw. The samples showed lower values of the trace elements silver (Ag), barium (Ba), cadmium (Cd), chromium (Cr), manganese (Mn), lead (Pb), and zinc (Zn) (Tables 5 and 6).

Minerals contents of *A. ponderosa* fruiting bodies and their corresponding soil samples presented significant differences for all the studied elements ($p < 0.05$). Cabezas Rubias (6) presented significantly higher aluminium and calcium contents ($p < 0.05$). The cadmium content was significantly higher ($p < 0.05$) in the fruiting bodies collected in N. S[ra] Machede (16) and Almendres (1). Samples from Évora (7), Cabezas Rubias (6), and Villanueva del Fresno (24) showed similar chromium contents ($p > 0.05$), which were significantly higher than the others *A. ponderosa* fruiting bodies ($p < 0.05$). Results obtained for soil mineral content showed some significant differences between sampling sites ($p < 0.05$). The sample collected in N. S[ra] Machede (16) presented significant different contents of silver, aluminium, and magnesium ($p < 0.05$). The Montejuntos sample (14) presented significantly differences of aluminium, iron, phosphorus, and magnesium contents ($p < 0.05$) and Villanueva del Fresno (24) presented significant differences for barium and manganese ($p < 0.05$). Significant differences in copper content were observed for Her[de] da Mitra sample (9). Mértola sample (10) presented significant differences in zinc and Baleizão sample (3) in potassium contents ($p < 0.05$). Serpa samples (20), presented significant differences in calcium, chromium, iron, magnesium, sodium, and zinc ($p < 0.05$). Almendres (1), Évora (7), M[te] Novo (13) and Valverde (22) samples did not present significant differences for the iron content ($p > 0.05$). Samples of Cabezas Rubias (6), Serpa (20), and Rosal de la Frontera (17) did not present significant differences in lead content ($p > 0.05$). Concerning phosphorus content, two groups with no significant differences ($p > 0.05$) were observed, one including Baleizão (3) and Rosal de la Frontera (17) and the other group including S[to] Aleixo Restauração (18), S. Miguel de Machede (19), and

TABLE 3: Ca, Mg, Na, and K mineral content of *A. ponderosa* fruiting bodies and corresponding soil substrates from different sampling sites.

Sampling site	Sample	Minerals (mg/Kg dry weight)			
		Ca	K	Mg	Na
(1) Almendres	FB	757 ± 156^{a}	$26677 \pm 915^{a,b,c,d,e}$	$774 \pm 130^{a,b,c}$	1408 ± 146^{a}
	SS	$106 \pm 11^{a,b}$	$363 \pm 9^{a,b}$	$286 \pm 14^{a,b,c,d}$	$70 \pm 1^{a,b,c,d}$
(2) Azaruja	FB	139 ± 13^{b}	4589 ± 103^{a}	194 ± 10^{a}	216 ± 135^{f}
	SS	$102 \pm 12^{a,b}$	981 ± 37^{g}	$324 \pm 29^{a,b}$	$75 \pm 4^{a,b,h,i}$
(3) Baleizão	FB	606 ± 61^{a}	$25758 \pm 4728^{a,b,c,d,e}$	$856 \pm 107^{a,b,c,d}$	$2249 \pm 231^{g,h}$
	SS	$162 \pm 11^{c,d}$	1099 ± 24^{h}	$312 \pm 38^{a,b}$	109 ± 4^{k}
(4) Beja	FB	615 ± 1^{a}	$6090 \pm 34^{a,b}$	$734 \pm 1^{a,b,c}$	$1540 \pm 17^{a,g}$
	SS	$189 \pm 14^{d,e}$	$619 \pm 19^{d,e}$	$277 \pm 33^{a,b,c,d,e}$	$69 \pm 2^{a,b,c,d,e}$
(5) Cabeça Gorda	FB	554 ± 21^{a}	$38652 \pm 1607^{c,d,e}$	$848 \pm 159^{a,b,c}$	$889 \pm 38^{a,b,c,d,e,f}$
	SS	$108 \pm 6^{a,b}$	$384 \pm 16^{a,b,c}$	$186 \pm 8^{e,f,g}$	$69 \pm 5^{a,b,c,d,e}$
(6) Cabezas Rubias	FB	352 ± 2^{c}	69565 ± 362^{f}	$1265 \pm 1^{c,d}$	2645 ± 18^{h}
	SS	$118 \pm 16^{a,b,c}$	443 ± 9^{c}	$277 \pm 19^{a,b,c,d,e}$	$62 \pm 2^{c,d,e,f}$
(7) Évora	FB	605 ± 1^{a}	$34630 \pm 186^{c,d,e}$	$713 \pm 1^{a,b,c}$	$1241 \pm 16^{a,b}$
	SS	91 ± 23^{a}	651 ± 9^{e}	$171 \pm 25^{f,g}$	$52 \pm 3^{f,g}$
(8) Évoramonte	FB	733 ± 15^{a}	$25148 \pm 4556^{a,b,c,d,e}$	$760 \pm 101^{a,b,c}$	$614 \pm 131^{b,c,d,e,f}$
	SS	$102 \pm 8^{a,b}$	576 ± 16^{d}	156 ± 11^{g}	51 ± 2^{g}
(9) Her^{de} da Mitra	FB	579 ± 111^{a}	$31852 \pm 5364^{a,b,c,d,e}$	$685 \pm 89^{a,b,c}$	$930 \pm 58^{a,b,c,d,e,f}$
	SS	$159 \pm 8^{c,d}$	$229 \pm 9^{j,k}$	$298 \pm 29^{a,b,c}$	51 ± 3^{g}
(10) Mértola	FB	$370 \pm 57^{a,b}$	$18021 \pm 1806^{a,b,c}$	$658 \pm 33^{a,b,c}$	$484 \pm 171^{c,d,e,f}$
	SS	$112 \pm 8^{a,b,c}$	731 ± 19^{f}	$193 \pm 18^{d,e,f,g}$	$82 \pm 2^{h,i,j}$
(11) Mina S. Domingos	FB	582 ± 1^{a}	$33607 \pm 173^{b,c,d,e}$	$737 \pm 1^{a,b,c}$	$1097 \pm 14^{a,b,c,d}$
	SS	$143 \pm 25^{b,c,d}$	$256 \pm 16^{i,j}$	365 ± 43^{a}	$59 \pm 5^{e,f,g}$
(12) M^{te} da Borralha	FB	$448 \pm 33^{a,b}$	$26099 \pm 1618^{a,b,c,d,e}$	$696 \pm 29^{a,b,c}$	$368 \pm 70^{e,f}$
	SS	70 ± 14^{a}	155 ± 9^{l}	$172 \pm 23^{f,g}$	$67 \pm 5^{b,c,d,e}$
(13) M^{te} Novo	FB	$436 \pm 72^{a,b}$	$25763 \pm 1608^{a,b,c,d,e}$	$562 \pm 114^{a,b}$	$379 \pm 25^{e,f}$
	SS	$148 \pm 19^{b,c,d}$	$603 \pm 56^{d,e}$	$217 \pm 34^{c,d,e,f,g}$	$53 \pm 4^{f,g}$
(14) Montejuntos	FB	$412 \pm 50^{a,b}$	$41285 \pm 982^{c,d,e}$	$742 \pm 53^{a,b,c}$	$1146 \pm 304^{a,b,c}$
	SS	$224 \pm 16^{e,f}$	$256 \pm 16^{i,j}$	910 ± 25^{j}	$85 \pm 2^{i,j}$
(15) N. S^{ra} Guadalupe	FB	$408 \pm 11^{a,b}$	$34814 \pm 1763^{c,d,e}$	$930 \pm 158^{b,c,d}$	$551 \pm 52^{b,c,d,e,f}$
	SS	$148 \pm 8^{b,c,d}$	656 ± 16^{e}	$255 \pm 14^{b,c,d,e,f}$	$73 \pm 1^{a,b,h}$
(16) N. S^{ra} de Machede	FB	632 ± 111^{a}	$24717 \pm 1881^{a,b,c,d,e}$	$843 \pm 120^{a,b,c}$	$2185 \pm 438^{g,h}$
	SS	$143 \pm 3^{b,c,d}$	912 ± 48^{g}	804 ± 48^{i}	$80 \pm 2^{a,h,i,j}$
(17) Rosal de la Frontera	FB	584 ± 25^{a}	$51996 \pm 2660^{e,f}$	1540 ± 77^{d}	$964 \pm 453^{a,b,c,d,e}$
	SS	83 ± 9^{a}	$160 \pm 16^{k,l}$	365 ± 11^{a}	$80 \pm 1^{a,h,i,j}$
(18) S^{to} Aleixo da Restauração	FB	366 ± 1^{a}	$18586 \pm 106^{d,e,f}$	$481 \pm 1^{a,b,c}$	$1098 \pm 103^{a,b,c,d}$
	SS	79 ± 8^{a}	773 ± 24^{f}	477 ± 14^{h}	$70 \pm 2^{a,b,c}$
(19) S. Miguel de Machede	FB	649 ± 74^{a}	$45011 \pm 1152^{c,d,e,f}$	$549 \pm 140^{a,b}$	$897 \pm 11^{a,b,c,d,e,f}$
	SS	$338 \pm 35^{a,b,c}$	$315 \pm 9^{k,l}$	$1466 \pm 68^{a,b,c,d,e}$	$158 \pm 5^{c,d,e,f,g}$
(20) Serpa	FB	$698 \pm 148^{a,b}$	$47715 \pm 5220^{a,b,c}$	$736 \pm 86^{a,b}$	$1078 \pm 174^{a,b,c,d,e}$
	SS	119 ± 16^{g}	$171 \pm 9^{a,i}$	272 ± 15^{k}	60 ± 5^{l}
(21) Ve Rocins	FB	$448 \pm 269^{a,b}$	$21788 \pm 1237^{a,b,c,d}$	$418 \pm 203^{a,b}$	$419 \pm 212^{d,e,f}$
	SS	$113 \pm 8^{a,b,c}$	$320 \pm 16^{a,i}$	$284 \pm 14^{a,b,c,d}$	86 ± 2^{j}
(22) Valverde	FB	$521 \pm 146^{a,b}$	$27256 \pm 2180^{a,b,c,d,e}$	$586 \pm 35^{a,b,c}$	$895 \pm 65^{a,b,c,d,e,f}$
	SS	93 ± 23^{a}	$405 \pm 24^{b,c}$	$265 \pm 41^{b,c,d,e,f}$	111 ± 3^{k}
(23) V. N. S. Bento	FB	599 ± 1^{a}	$6228 \pm 30^{a,b}$	$657 \pm 1^{a,b,c}$	1467 ± 15^{a}
	SS	$233 \pm 11^{e,f}$	$192 \pm 16^{j,k,l}$	536 ± 37^{h}	$60 \pm 2^{d,e,f,g}$
(24) Villanueva del Fresno	FB	$459 \pm 13^{a,b}$	$25713 \pm 5847^{a,b,c,d,e}$	$738 \pm 49^{a,b,c}$	1451 ± 147^{a}
	SS	240 ± 26^{f}	$368 \pm 16^{a,b}$	$267 \pm 27^{b,c,d,e}$	$78 \pm 4^{a,h,i,j}$
Total	FB	523 ± 143	29648 ± 14908	738 ± 261	1092 ± 620
	SS	143 ± 63	484 ± 273	381 ± 296	75 ± 24

FB, fruiting bodies; SS, soil substrate. Mean values ($n = 9$) ± SD. Different letters for each element indicate significant differences with the confidence level of $p < 0.05$ (ANOVA, Tukey's test).

TABLE 4: Al, Cu, Fe, and P mineral content of *A. ponderosa* fruiting bodies and corresponding soil substrates from different sampling sites.

Sampling site	Sample	Minerals (mg/Kg dry weight)			
		Al	Cu	Fe	P
(1) Almendres	FB	349 ± 124^{b,c,d,e,f}	80 ± 1^{a,b}	66 ± 14^{a,b,c}	319 ± 37^{a,b,c,d,e}
	SS	277 ± 8^{a,b}	1 ± 0^{a}	1547 ± 92^{a}	72 ± 2^{a,b}
(2) Azaruja	FB	58 ± 8^{a}	30 ± 2^{a}	18 ± 4^{a}	44 ± 18^{h,i}
	SS	243 ± 30^{a}	2 ± 0^{c}	2240 ± 160^{b}	74 ± 4^{a,b}
(3) Baleizão	FB	357 ± 29^{b,c,d,e,f}	356 ± 37^{a,b,c}	27 ± 8^{a}	525 ± 62^{j,k}
	SS	397 ± 4^{c,d,e}	7 ± 0^{k}	3093 ± 92^{d,e,f}	156 ± 4^{g}
(4) Beja	FB	449 ± 13^{d,e,f,g}	373 ± 1^{b,c}	227 ± 1^{f,g}	212 ± 1^{d,e,f,g}
	SS	254 ± 15^{a}	9 ± 1^{l}	4107 ± 92^{i,j}	133 ± 8^{f}
(5) Cabeça Gorda	FB	151 ± 25^{a,b}	161 ± 10^{a,b}	51 ± 12^{a,b,c}	447 ± 2^{a,j}
	SS	196 ± 4^{k}	1 ± 0^{a}	5067 ± 244^{k}	130 ± 3^{e,f}
(6) Cabezas Rubias	FB	932 ± 77^{h}	140 ± 1^{a,b}	193 ± 1^{e,f}	73 ± 1^{g,h,i}
	SS	375 ± 8^{c,d}	5 ± 0^{g}	3520 ± 160^{e,f,g,h}	93 ± 4^{c}
(7) Évora	FB	208 ± 11^{a,b,c,d}	198 ± 1^{a,b}	292 ± 1^{g,h}	201 ± 1^{d,e,f,g,h}
	SS	368 ± 18^{c}	2 ± 0^{b,c}	1653 ± 92^{a}	82 ± 5^{b,c}
(8) Évoramonte	FB	423 ± 49^{c,d,e,f,g}	124 ± 20^{a,b}	62 ± 7^{a,b,c}	469 ± 83^{a,j}
	SS	311 ± 6^{b}	2 ± 0^{c,d}	2400 ± 160^{b,c}	94 ± 1^{c}
(9) Her^{de} da Mitra	FB	373 ± 26^{b,c,d,e,f}	193 ± 19^{a,b}	35 ± 10^{a,b}	277 ± 90^{b,c,d,e,f}
	SS	157 ± 6^{l}	10 ± 0^{m}	4747 ± 92^{k}	115 ± 3^{d,e}
(10) Mértola	FB	333 ± 36^{b,c,d,e}	43 ± 15^{a,b}	22 ± 3^{a}	61 ± 21^{g,h,i}
	SS	160 ± 8^{l}	5 ± 0^{f,g}	4853 ± 244^{k}	119 ± 2^{d,e,f}
(11) Mina S. Domingos	FB	299 ± 14^{a,b,c,d,e}	334 ± 1^{a,b,c}	62 ± 1^{a,b,c}	149 ± 1^{f,g,h,i}
	SS	417 ± 6^{e,f}	3 ± 0^{d,e}	3573 ± 92^{f,g,h}	74 ± 2^{a,b}
(12) M^{te} da Borralha	FB	209 ± 42^{a,b,c,d}	232 ± 63^{a,b}	41 ± 38^{a,b}	417 ± 65^{a,b,c,j}
	SS	159 ± 6^{l}	2 ± 0^{c}	3840 ± 160^{h,i}	75 ± 5^{a,b}
(13) M^{te} Novo	FB	423 ± 59^{c,d,e,f,g}	121 ± 16^{a,b}	29.3 ± 1^{a}	316 ± 76^{a,b,c,d,e,f}
	SS	300 ± 15^{b}	3 ± 0^{c,d}	1653 ± 244^{a}	84 ± 4^{b,c}
(14) Montejuntos	FB	545 ± 99^{e,f,g}	170 ± 21^{a,b}	62 ± 10^{a,b,c}	442 ± 2^{a,b,j}
	SS	813 ± 11^{j}	7 ± 0^{j,k}	8099 ± 251^{l}	294 ± 11^{i}
(15) N. S^{ra} Guadalupe	FB	209 ± 73^{a,b,c,d}	264 ± 23^{a,b,c}	110 ± 27^{b,c,d}	344 ± 74^{a,b,c,d}
	SS	200 ± 6^{k}	8 ± 0^{l}	4960 ± 160^{k}	131 ± 5^{e,f}
(16) N. S^{ra} de Machede	FB	313 ± 47^{a,b,c,d,e}	148 ± 26^{a,b}	48 ± 13^{a,b,c}	542 ± 48^{j,k}
	SS	632 ± 19^{i}	4 ± 0^{e,f}	3738 ± 258^{g,h,i}	78 ± 1^{a,b,c}
(17) Rosal de la Frontera	FB	659 ± 36^{g}	584 ± 51^{c}	119 ± 97^{c,d,e}	641 ± 98^{k}
	SS	381 ± 4^{c,d}	9 ± 0^{l}	3360 ± 160^{e,f,g,h}	154 ± 4^{g}
(18) S^{to} Aleixo da Restauração	FB	65 ± 12^{d,e,f,g}	129 ± 1^{a,b}	51 ± 1^{a,b,c}	34 ± 1^{a,j}
	SS	449 ± 16^{f,g}	6 ± 0^{g,h,i}	2827 ± 92^{c,d}	185 ± 1^{h}
(19) S. Miguel de Machede	FB	610 ± 183^{f,g}	207 ± 39^{a,b}	311 ± 10^{h}	270 ± 60^{c,d,e,f}
	SS	497 ± 6^{c,d}	5 ± 0^{g,h}	9040 ± 139^{d,e,f,g}	111 ± 2^{h}
(20) Serpa	FB	453 ± 66^{a}	143 ± 14^{a,b}	43 ± 18^{a,b,c}	460 ± 16^{i}
	SS	378 ± 4^{h}	5 ± 0^{f,g}	3253 ± 92^{m}	179 ± 7^{d}
(21) Ve Rocins	FB	181 ± 24^{a,b,c}	84 ± 7^{a,b}	36 ± 24^{a,b}	165 ± 85^{e,f,g,h,i}
	SS	391 ± 9^{c,d,e}	9 ± 0^{l}	3040 ± 160^{d,e}	75 ± 5^{a,b}
(22) Valverde	FB	183 ± 54^{a,b,c}	62 ± 3^{a,b}	57 ± 21^{a,b,c}	152 ± 34^{f,g,h,i}
	SS	453 ± 6^{g}	1 ± 0^{a,b}	1440 ± 160^{a}	65 ± 2^{a}
(23) V. N. S. Bento	FB	303 ± 7^{a,b,c,d,e}	172 ± 1^{a,b}	182 ± 1^{d,e,f}	197 ± 1^{d,e,f,g,h,i}
	SS	409 ± 6^{d,e}	6 ± 0^{h,i,j}	4587 ± 244^{j,k}	116 ± 5^{d,e}
(24) Villanueva del Fresno	FB	602 ± 66^{f,g}	100 ± 9^{a,b}	53 ± 17^{a,b,c}	310 ± 83^{a,b,c,d,e,f}
	SS	494 ± 4^{h}	6 ± 0^{i,j}	2347 ± 92^{b,c}	179 ± 12^{h}
Total	FB	362 ± 204	185 ± 125	92 ± 85	294 ± 171
	SS	363 ± 155	5 ± 3	3708 ± 1870	120 ± 53

FB, fruiting bodies; SS, soil substrate. Mean values (n = 9) ± SD. Different letters for each element indicate significant differences with the confidence level of $p < 0.05$ (ANOVA, Tukey's test).

TABLE 5: Ag, Ba, Cd, and Cr mineral content of *A. ponderosa* fruiting bodies and corresponding soil substrates from different sampling sites.

Sampling site	Sample	Minerals (mg/kg dry weight)			
		Ag	Ba	Cd	Cr
(1) Almendres	FB	$2.6 \pm 0.5^{a,b,c,d,e}$	2.9 ± 1.1^{a}	2.2 ± 0.8^{a}	$1.3 \pm 0.6^{a,b,c}$
	SS	0.1 ± 0.0^{a}	$2.0 \pm 0.3^{a,b}$	$0.3 \pm 0.0^{a,b}$	1.0 ± 0.2^{a}
(2) Azaruja	FB	6.6 ± 2.2^{f}	$0.9 \pm 0.2^{d,e,f,g}$	0.3 ± 0.0^{b}	$0.6 \pm 0.1^{c,d,e}$
	SS	$0.1 \pm 0.0^{a,b,c,d}$	1.6 ± 0.2^{a}	0.3 ± 0.1^{a}	0.9 ± 0.1^{a}
(3) Baleizão	FB	$5.1 \pm 1.8^{a,b,c,d}$	$0.6 \pm 0.3^{d,e,f,g}$	$1.1 \pm 0.3^{b,c,d,e}$	$1.0 \pm 0.2^{d,e}$
	SS	$0.1 \pm 0.0^{a,b,c,d}$	5.3 ± 0.9^{i}	0.2 ± 0.0^{e}	$2.4 \pm 0.1^{b,c,d}$
(4) Beja	FB	$1.1 \pm 0.2^{a,b}$	$0.8 \pm 0.0^{b,c,d,e,f,g}$	$0.4 \pm 0.0^{b,c,d,e}$	0.8 ± 0.0^{a}
	SS	$0.1 \pm 0.0^{a,b,c,d,e,f}$	$2.6 \pm 0.7^{a,b,c,d}$	0.1 ± 0.0^{f}	$4.4 \pm 0.2^{i,j}$
(5) Cabeça Gorda	FB	$2.1 \pm 0.7^{a,b,c}$	$0.3 \pm 0.0^{d,e,f,g}$	$0.7 \pm 0.1^{c,d,e}$	$0.8 \pm 0.0^{b,c,d,e}$
	SS	$0.2 \pm 0.0^{b,c,d,e,f}$	$2.3 \pm 0.1^{a,b,c,d}$	$0.2 \pm 0.0^{c,d,e}$	$4.2 \pm 0.0^{g,h,i,j}$
(6) Cabezas Rubias	FB	$1.5 \pm 0.2^{a,b,c}$	$2.0 \pm 0.3^{d,e,f,g}$	$0.8 \pm 0.1^{c,d,e}$	0.9 ± 0.1^{f}
	SS	$0.2 \pm 0.0^{c,d,e,f,g}$	$2.6 \pm 0.4^{a,b,c,d}$	$0.1 \pm 0.0^{e,f}$	$3.1 \pm 0.5^{c,d,e,f,g,h}$
(7) Évora	FB	$0.5 \pm 0.3^{a,b,c,d}$	$0.8 \pm 0.4^{f,g}$	$0.4 \pm 0.3^{b,c,d,e}$	0.9 ± 0.5^{f}
	SS	$0.2 \pm 0.0^{g,h,i}$	$4.3 \pm 1.1^{f,g,h,i}$	$0.1 \pm 0.0^{e,f}$	$2.8 \pm 0.5^{c,d,e,f}$
(8) Évoramonte	FB	$1.0 \pm 0.1^{d,e,f}$	$0.9 \pm 0.1^{b,c,d,e}$	$1.2 \pm 0.2^{b,c,d,e}$	$2.9 \pm 0.1^{b,c,d,e}$
	SS	$0.1 \pm 0.0^{a,b,c,d,e,f}$	$2.8 \pm 0.1^{a,b,c,d,e}$	$0.1 \pm 0.0^{e,f}$	$2.6 \pm 0.0^{c,d,e,f}$
(9) Her^{de} da Mitra	FB	$0.7 \pm 0.1^{a,b,c,d}$	1.2 ± 0.1^{g}	$1.0 \pm 0.0^{b,c,d,e}$	$1.8 \pm 0.1^{b,c,d,e}$
	SS	$0.2 \pm 0.0^{e,f,g,h}$	$2.3 \pm 0.1^{a,b,c}$	$0.1 \pm 0.0^{e,f}$	$3.8 \pm 0.1^{f,g,h,i,j}$
(10) Mértola	FB	$4.3 \pm 0.8^{a,b}$	$1.5 \pm 0.2^{b,c,d,e,f}$	$1.0 \pm 0.3^{b,c,d}$	$0.9 \pm 0.3^{d,e}$
	SS	$0.1 \pm 0.0^{a,b}$	$3.0 \pm 0.4^{b,c,d,e,f}$	$0.1 \pm 0.0^{e,f}$	$3.0 \pm 0.2^{c,d,e,f,g}$
(11) Mina S. Domingos	FB	$0.6 \pm 0.1^{a,b,c}$	$0.6 \pm 0.0^{c,d,e,f,g}$	$1.4 \pm 0.2^{d,e}$	$0.7 \pm 0.1^{a,b,c,d,e}$
	SS	0.3 ± 0.0^{i}	$1.9 \pm 0.0^{a,b}$	$0.1 \pm 0.0^{e,f}$	$5.1 \pm 0.7^{j,k}$
(12) M^{te} da Borralha	FB	$1.5 \pm 0.1^{a,b,c}$	$0.9 \pm 0.1^{f,g}$	$1.1 \pm 0.2^{b,c}$	$0.9 \pm 0.1^{c,d,e}$
	SS	$0.2 \pm 0.0^{d,e,f,g}$	$2.3 \pm 0.0^{a,b,c,d}$	$0.1 \pm 0.0^{e,f}$	$3.7 \pm 0.0^{d,e,f,g,h,i}$
(13) M^{te} Novo	FB	0.6 ± 0.1^{a}	$1.3 \pm 0.2^{e,f,g}$	$0.6 \pm 0.1^{b,c,d,e}$	0.5 ± 0.1^{e}
	SS	$0.1 \pm 0.0^{a,b,c}$	$3.4 \pm 0.2^{c,d,e,f,g}$	$0.2 \pm 0.0^{b,c,d,e}$	$2.6 \pm 0.2^{c,d,e,f}$
(14) Montejuntos	FB	$1.5 \pm 0.2^{a,b,c}$	$1.1 \pm 0.1^{b,c,d}$	$1.3 \pm 0.1^{b,c,d,e}$	$1.0 \pm 0.1^{a,b,c,d}$
	SS	$0.2 \pm 0.1^{h,i}$	$3.6 \pm 0.3^{d,e,f,g,h}$	$0.1 \pm 0.0^{e,f}$	5.7 ± 0.2^{k}
(15) N. S^{ra} Guadalupe	FB	$2.5 \pm 0.3^{e,f}$	$0.7 \pm 0.2^{d,e,f,g}$	$0.9 \pm 0.2^{b,c,d,e}$	$0.5 \pm 0.1^{a,b,c,d,e}$
	SS	$0.2 \pm 0.0^{b,c,d,e,f,g}$	$2.4 \pm 0.1^{a,b,c,d}$	$0.2 \pm 0.0^{c,d,e}$	$4.3 \pm 0.1^{h,i,j}$
(16) N. S^{ra} Machede	FB	$0.4 \pm 0.2^{a,b,c,d,e}$	$0.6 \pm 0.1^{b,c,d,e,f,g}$	0.7 ± 0.1^{f}	$0.4 \pm 0.1^{b,c,d,e}$
	SS	0.4 ± 0.0^{j}	$4.0 \pm 0.2^{e,f,g,h,i}$	$0.1 \pm 0.0^{e,f}$	$3.8 \pm 0.3^{e,f,g,h,i,j}$
(17) Rosal de la Frontera	FB	$1.6 \pm 0.3^{c,d,e}$	$1.5 \pm 0.1^{b,c,d,e,f}$	$1.0 \pm 0.1^{b,c}$	1.2 ± 0.0^{a}
	SS	$0.2 \pm 0.0^{c,d,e,f,g}$	$2.3 \pm 0.1^{a,b,c,d}$	$0.1 \pm 0.0^{e,f}$	$2.9 \pm 0.2^{c,d,e,f}$
(18) S^{to} Aleixo da Restauração	FB	$1.3 \pm 0.3^{a,b,c}$	$1.1 \pm 0.1^{d,e,f,g}$	$0.7 \pm 0.1^{b,c}$	$0.9 \pm 0.0^{b,c,d,e}$
	SS	$0.2 \pm 0.0^{e,f,g,h}$	$2.8 \pm 0.2^{a,b,c,d,e}$	$0.3 \pm 0.0^{a,b,c,d}$	$2.8 \pm 0.9^{c,d,e,f}$
(19) S. Miguel de Machede	FB	$2.6 \pm 0.3^{b,c,d,e}$	$1.3 \pm 0.3^{b,c,d,e,f,g}$	$3.9 \pm 0.5^{b,c,d,e}$	$0.9 \pm 0.1^{a,b}$
	SS	$0.2 \pm 0.0^{e,f,g}$	$3.4 \pm 0.1^{c,d,e,f,g}$	$0.1 \pm 0.0^{e,f}$	$2.3 \pm 0.2^{b,c}$
(20) Serpa	FB	$1.7 \pm 0.3^{a,b,c}$	$0.4 \pm 0.1^{a,b,c}$	$0.4 \pm 0.1^{b,c,d,e}$	$0.6 \pm 0.1^{b,c,d,e}$
	SS	$0.2 \pm 0.0^{f,g,h}$	$2.4 \pm 0.3^{a,b,c,d}$	0.0 ± 0.0^{f}	7.9 ± 0.4^{l}
(21) Ve Rocins	FB	3.3 ± 2.2^{a}	$1.3 \pm 0.5^{d,e,f,g}$	$0.5 \pm 0.2^{b,c}$	$1.8 \pm 0.6^{b,c,d,e}$
	SS	$0.1 \pm 0.0^{a,b,c,d,e}$	$4.8 \pm 0.2^{h,i}$	$0.2 \pm 0.0^{d,e}$	$2.5 \pm 0.1^{b,c,d,e}$
(22) Valverde	FB	$1.8 \pm 0.6^{a,b}$	$0.5 \pm 0.0^{d,e,f,g}$	1.0 ± 0.1^{e}	$2.9 \pm 0.1^{b,c,d,e}$
	SS	$0.2 \pm 0.0^{c,d,e,f,g}$	$1.7 \pm 0.3^{a,b}$	$0.2 \pm 0.0^{b,c,d,e}$	$1.2 \pm 0.1^{a,b}$
(23) V. N. S. Bento	FB	$0.4 \pm 0.3^{a,b,c}$	$2.1 \pm 0.2^{c,d,e,f,g}$	$0.7 \pm 0.1^{b,c,d,e}$	$2.8 \pm 0.4^{b,c,d,e}$
	SS	$0.1 \pm 0.0^{a,b,c,d,e,f}$	$4.5 \pm 0.4^{g,h,i}$	$0.3 \pm 0.1^{a,b,c}$	$1.8 \pm 1.3^{a,b,c}$
(24) Villanueva del Fresno	FB	3.0 ± 0.3^{a}	$1.1 \pm 0.2^{a,b}$	$0.8 \pm 0.3^{b,c,d,e}$	1.5 ± 0.4^{f}
	SS	$0.1 \pm 0.0^{a,b,c,d,e}$	11.3 ± 1.0^{j}	$0.3 \pm 0.0^{a,b,c}$	$4.9 \pm 0.6^{i,j,k}$
Total	FB	2.0 ± 1.6	1.1 ± 0.6	1.0 ± 0.7	1.2 ± 0.7
	SS	0.2 ± 0.1	3.3 ± 2.0	0.2 ± 0.1	3.3 ± 1.6

FB, fruiting bodies; SS, soil substrate. Mean values ($n = 9$) $\pm$ SD. Different letters for each element indicate significant differences with the confidence level of $p < 0.05$ (ANOVA, Tukey's test).

TABLE 6: Mn, Pb, and Zn mineral content of *A. ponderosa* fruiting bodies and corresponding soil substrates from different sampling sites.

Sampling site	Sample	Minerals (mg/kg dry weight)		
		Mn	Pb	Zn
(1) Almendres	FB	$28 \pm 6^{a,b,c,d}$	$0.7 \pm 0.5^{a,b}$	$59 \pm 1^{a,b,c,d}$
	SS	22 ± 1^{a}	4.0 ± 0.4^{a}	$4 \pm 0^{a,b}$
(2) Azaruja	FB	6 ± 1^{a}	0.4 ± 0.1^{a}	16 ± 2^{a}
	SS	$131 \pm 9^{a,b}$	$7.0 \pm 1.0^{a,b,c,d,e}$	$5 \pm 1^{a,b,c}$
(3) Baleizão	FB	$91 \pm 14^{g,h}$	$1.7 \pm 0.5^{c,d,e,f}$	$97 \pm 2^{c,d,e}$
	SS	$893 \pm 17^{f,g}$	$9.2 \pm 1.1^{d,e,f}$	9 ± 0^{d}
(4) Beja	FB	$84 \pm 1^{f,g,h}$	$2.8 \pm 0.1^{a,b,c,d}$	$104 \pm 1^{d,e}$
	SS	507 ± 45^{c}	$6.9 \pm 0.6^{a,b,c,d,e}$	13 ± 0^{g}
(5) Cabeça Gorda	FB	$66 \pm 6^{c,d,e,f,g,h}$	$1.9 \pm 0.2^{c,d,e,f}$	$81 \pm 1^{b,c,d,e}$
	SS	$695 \pm 62^{d,e}$	$4.5 \pm 0.2^{a,b}$	10 ± 0^{d}
(6) Cabezas Rubias	FB	$81 \pm 1^{e,f,g,h}$	$5.1 \pm 0.8^{h,i}$	132 ± 1^{e}
	SS	$735 \pm 30^{d,e,f}$	16.5 ± 0.6^{g}	$11 \pm 1^{d,e}$
(7) Évora	FB	$59 \pm 1^{b,c,d,e,f,g,h}$	1.9 ± 0.9^{i}	$75 \pm 1^{b,c,d,e}$
	SS	$175 \pm 3^{a,b}$	10.9 ± 0.2^{f}	$5 \pm 1^{b,c}$
(8) Évoramonte	FB	$79 \pm 11^{e,f,g,h}$	$4.2 \pm 0.2^{g,h,i}$	$59 \pm 7^{a,b,c,d}$
	SS	212 ± 12^{b}	$7.4 \pm 0.3^{a,b,c,d,e,f}$	6 ± 0^{c}
(9) Her^{de} da Mitra	FB	$47 \pm 9^{a,b,c,d,e,f,g}$	$1.4 \pm 0.2^{b,c,d,e,f}$	$61 \pm 2^{a,b,c,d}$
	SS	$725 \pm 17^{d,e}$	$5.4 \pm 0.3^{a,b,c}$	$11 \pm 0^{e,f}$
(10) Mértola	FB	$29 \pm 6^{a,b,c,d}$	$3.7 \pm 0.9^{a,b,c,d,e}$	$52 \pm 8^{a,b,c,d}$
	SS	$725 \pm 17^{d,e}$	$8.0 \pm 0.7^{b,c,d,e,f}$	15 ± 1^{h}
(11) Mina S. Domingos	FB	104 ± 1^{h}	$3.1 \pm 0.7^{e,f,g,h}$	$91 \pm 1^{c,d,e}$
	SS	290 ± 6^{b}	$10.3 \pm 0.4^{e,f}$	4 ± 1^{a}
(12) M^{te} da Borralha	FB	$39 \pm 16^{a,b,c,d,e,f}$	$2.2 \pm 0.3^{a,b,c}$	$62 \pm 4^{a,b,c,d}$
	SS	199 ± 9^{b}	$5.4 \pm 0.3^{a,b,c}$	$10 \pm 1^{d,e}$
(13) M^{te} Novo	FB	$34 \pm 4^{a,b,c,d}$	1.6 ± 0.3^{a}	$52 \pm 4^{a,b,c,d}$
	SS	$139 \pm 3^{a,b}$	$8.2 \pm 0.7^{c,d,e,f}$	$5 \pm 0^{a,b,c}$
(14) Montejuntos	FB	$49 \pm 9^{a,b,c,d,e,f,g}$	$3.0 \pm 0.1^{e,f,g,h}$	$60 \pm 3^{a,b,c,d}$
	SS	$649 \pm 27^{c,d,e}$	$10.2 \pm 1.5^{e,f}$	$10 \pm 0^{d,e}$
(15) N. S^{ra} Guadalupe	FB	$38 \pm 5^{a,b,c,d,e}$	$2.1 \pm 0.3^{a,b,c,d,e,f}$	$63 \pm 10^{a,b,c,d}$
	SS	$794 \pm 232^{e,f}$	$4.5 \pm 0.2^{a,b}$	$12 \pm 0^{f,g}$
(16) N. S^{ra} Machede	FB	$50 \pm 13^{a,b,c,d,e,f,g}$	$0.4 \pm 0.1^{a,b,c}$	$70 \pm 2^{a,b,c,d}$
	SS	$123 \pm 4^{a,b}$	$8.4 \pm 0.1^{c,d,e,f}$	9 ± 0^{d}
(17) Rosal de la Frontera	FB	$90 \pm 56^{g,h}$	$3.0 \pm 0.1^{c,d,e,f,g}$	$95 \pm 6^{c,d,e}$
	SS	$725 \pm 17^{d,e}$	16.5 ± 0.6^{g}	$6 \pm 0^{b,c}$
(18) S^{to} Aleixo da Restauração	FB	$47 \pm 15^{c,d,e,f,g,h}$	3.9 ± 0.3^{i}	$47 \pm 1^{a,b,c,d}$
	SS	193 ± 7^{b}	$6.4 \pm 0.9^{a,b,c,d}$	$11 \pm 0^{d,e}$
(19) S. Miguel de Machede	FB	$91 \pm 1^{g,h}$	$1.4 \pm 0.1^{f,g,h}$	$58 \pm 8^{a,b,c,d}$
	SS	576 ± 17^{g}	$5.8 \pm 0.2^{a,b,c,d}$	9 ± 0^{d}
(20) Serpa	FB	$69 \pm 19^{a,b,c,d,e,f,g}$	$1.0 \pm 0.2^{d,e,f,g,h}$	$61 \pm 9^{a,b,c,d}$
	SS	$1002 \pm 30^{c,d}$	19.5 ± 3.7^{g}	30 ± 1^{j}
(21) Ve Rocins	FB	$15 \pm 10^{a,b}$	$2.3 \pm 0.8^{b,c,d,e,f}$	$31 \pm 2^{a,b}$
	SS	$893 \pm 17^{f,g}$	$9.2 \pm 1.1^{d,e,f}$	$5 \pm 1^{a,b,c}$
(22) Valverde	FB	$26 \pm 7^{a,b,c}$	$5.1 \pm 0.2^{f,g,h}$	$42 \pm 2^{a,b,c}$
	SS	21 ± 1^{a}	$4.4 \pm 0.4^{a,b}$	$5 \pm 0^{a,b,c}$
(23) V. N. S. Bento	FB	$72 \pm 1^{d,e,f,g,h}$	$1.6 \pm 0.1^{h,i}$	$86 \pm 1^{b,c,d,e}$
	SS	$745 \pm 45^{e,f}$	$8.4 \pm 3.0^{c,d,e,f}$	27 ± 1^{i}
(24) Villanueva del Fresno	FB	$39 \pm 6^{a,b,c,d,e,f}$	$3.1 \pm 0.6^{a,b,c,d,e}$	$68 \pm 4^{a,b,c,d}$
	SS	1518 ± 45^{h}	$7.4 \pm 0.5^{a,b,c,d,e,f}$	$5 \pm 1^{a,b,c}$
Total	FB	56 ± 27	2.4 ± 1.3	68 ± 25
	SS	529 ± 378	8.5 ± 4.0	10 ± 6

FB, fruiting bodies; SS, soil substrate. Mean values ($n = 9$) ± SD. Different letters for each element indicate significant differences with the confidence level of $p < 0.05$ (ANOVA, Tukey's test).

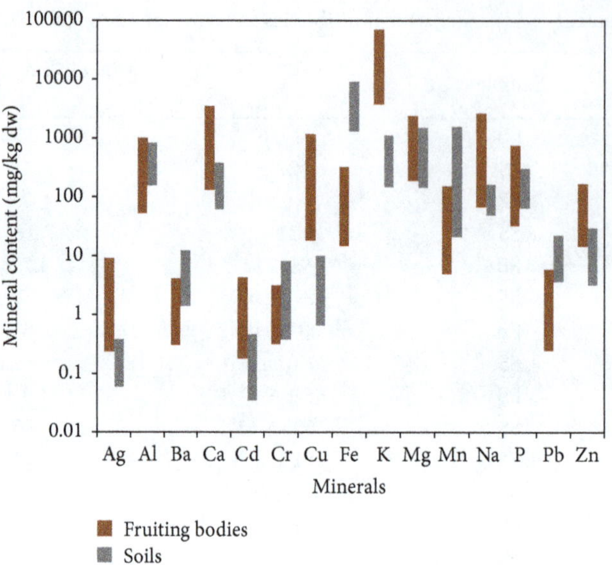

FIGURE 3: Concentration of mineral content of *A. ponderosa* fruiting bodies and soil substrates. The values are represented in logarithmic scale.

Vila Nueva del Fresno (24) samples. Finally, samples from Baleizão (3) and Valverde (22) did not present significant differences in sodium and potassium content ($p > 0.05$).

In fact, mineral content of *A. ponderosa* fruiting bodies and their soil samples (Figure 3) varied very similarly for almost all the analysed elements, indicating the influence of the substrate characteristics on the composition of mushrooms samples collected at the different sites.

Mushrooms have a specialized mechanism to accumulate nutrients and minerals in their fruiting bodies. The age of the fruiting body or its size may contribute to mineral composition. Thus, *A. ponderosa* mushrooms analysed in this study were obtained at the same development stage, in order to eliminate possible size interferences in the comparison of their mineral content. Macroelements content, such as Ca, Mg, Na, K, and P, in fruiting bodies were similar to those reported in literature for *A. ponderosa* [28] and for other edible mushrooms species [5, 36–38]. The calcium and phosphorus contents for the different samples studied were 523 ± 143 and 294 ± 171 mg/kg dw. Potassium and sodium, minerals responsible for the hydroelectrolytic balance maintenance and important enzyme cofactors, have high RDIs (Recommended Daily Intakes), which are 4700 mg/day and 2000 mg/day, respectively, for an adult [39]. Potassium and sodium levels of *A. ponderosa* mushrooms studied were 29648 ± 14908 and 1092 ± 620 mg/kg dw, respectively. Potassium content ranged from 18021 ± 1806 to 69565 ± 362 mg/Kg dw, presenting values similar to those described by Moreno-Rojas et al. [28], that report potassium levels ranging between 22410 ± 211 and 60890 ± 23950 mg/Kg dw. However, the low K levels observed for fruiting bodies from three sampling sites can be correlated with the large differences observed for potassium content in the surrounding substrates. Magnesium also plays an important role in large number of biological functions, particularly linked to energy metabolism, it is required for the proper function of certain enzymes as cofactor, and structural functions, the recommended magnesium intakes for adult (19–65 years), are 220–260 mg/day [40]. The mean magnesium content in the fruiting bodies was 738 ± 261 mg/kg dw, similar to that described in the literature [28].

A. ponderosa fruiting bodies presented values of trace minerals smaller than those found in other species of edible mushrooms [6, 11, 14, 38, 41]. Concentrations of trace elements in fruiting bodies are generally species-dependent [36]; however only one study of *A. ponderosa* mushrooms is reported [28]. The existence of higher metal concentrations in younger fruiting bodies can be explained by the transport of metals from the mycelium to the fruiting body during the beginning of fructification [36]. Trace elements like Fe, Cu, Zn, Cr, and Mn are essential metals since they play an important role in biological systems; however, these trace metals can also produce toxic effects when intake is excessively amount [38, 42, 43]. Copper is the third most abundant trace element found in the human body, and being an important element of several enzymes, it is also one of the agents involved in iron metabolism [42]. Iron is a component of haemoglobin and myoglobin, proteins responsible for transporting oxygen to tissues. It also participates in protein metabolism, energy production in cells and in various enzymatic reactions [44]. The copper content was 185 ± 125 mg/kg dw, and the highest value was found in *A. ponderosa* samples collected from Rosal de la Frontera (17) (584 ± 51 mg/kg dw). The iron content in *A. ponderosa* mushrooms was 92±85 mg/kg dw. Copper levels were similar and iron levels are slightly lower than those described for *Amanita* spp. [5, 28]. The adult RDI is 2 mg/day for Cu and 18 mg/day for Fe [12, 39], so the concentrations of copper and iron present in these edible mushrooms are not considered to be a health risk. Manganese and zinc are important trace elements for the human organism, participating in macronutrients and nucleic acids metabolism and promoting

Table 7: Bioconcentration factor (BCF) values for *A. ponderosa* fruiting bodies from different sampling sites.

Sampling site	Al	Ca	Cu	Fe	K	Mg	Na	P
(1) Almendres	1.3 ± 0.4	7.1 ± 0.7	123 ± 2	0.04 ± 0.01	74 ± 1	2.7 ± 0.3	20 ± 2	4.4 ± 0.4
(2) Azaruja	0.2 ± 0.0	1.4 ± 0.0	16 ± 1	0.01 ± 0.00	5 ± 1	0.6 ± 0.0	3 ± 2	0.6 ± 0.2
(3) Baleizão	0.9 ± 0.1	3.7 ± 0.1	48 ± 3	0.01 ± 0.00	23 ± 4	2.7 ± 0.0	21 ± 1	3.4 ± 0.3
(4) Beja	1.8 ± 0.1	3.3 ± 0.2	43 ± 2	0.06 ± 0.00	10 ± 0	2.7 ± 0.3	22 ± 0	1.6 ± 0.1
(5) Cabeça Gorda	0.8 ± 0.1	5.1 ± 0.1	248 ± 16	0.01 ± 0.00	101 ± 0	4.5 ± 0.7	13 ± 0	3.4 ± 0.1
(6) Cabezas Rubias	2.5 ± 0.2	3.0 ± 0.4	28 ± 1	0.05 ± 0.00	157 ± 2	4.6 ± 0.3	43 ± 1	0.8 ± 0.0
(7) Évora	0.6 ± 0.0	7.0 ± 1.8	122 ± 19	0.18 ± 0.01	53 ± 0	4.2 ± 0.6	24 ± 1	2.5 ± 0.1
(8) Évoramonte	1.4 ± 0.1	7.2 ± 0.4	52 ± 1	0.03 ± 0.00	44 ± 7	4.9 ± 0.3	12 ± 2	5.0 ± 0.8
(9) Her[de] da Mitra	2.4 ± 0.1	3.6 ± 0.5	20 ± 1	0.01 ± 0.00	139 ± 18	2.3 ± 0.1	18 ± 0	2.4 ± 0.7
(10) Mértola	2.1 ± 0.1	3.3 ± 0.3	9 ± 3	0.00 ± 0.00	25 ± 2	3.4 ± 0.2	6 ± 2	0.5 ± 0.2
(11) Mina S. Domingos	0.7 ± 0.0	4.2 ± 0.7	114 ± 10	0.02 ± 0.00	132 ± 8	2.0 ± 0.2	19 ± 1	2.0 ± 0.0
(12) M[te] da Borralha	1.3 ± 0.2	6.5 ± 0.9	119 ± 32	0.01 ± 0.01	168 ± 1	4.1 ± 0.4	5 ± 1	5.5 ± 0.5
(13) M[te] Novo	1.4 ± 0.1	2.9 ± 0.1	48 ± 2	0.02 ± 0.00	43 ± 1	2.6 ± 0.1	7 ± 0	3.7 ± 0.7
(14) Montejuntos	0.7 ± 0.1	1.8 ± 0.1	25 ± 1	0.01 ± 0.00	162 ± 6	0.8 ± 0.0	13 ± 3	1.5 ± 0.1
(15) N. S[ra] Guadalupe	1.0 ± 0.3	2.8 ± 0.1	31 ± 1	0.02 ± 0.00	53 ± 10	3.6 ± 0.4	8 ± 1	2.6 ± 0.5
(16) N. S[ra] Machede	0.5 ± 0.1	4.4 ± 0.7	39 ± 7	0.01 ± 0.00	27 ± 1	1.1 ± 0.1	27 ± 5	6.9 ± 0.5
(17) Rosal de la Frontera	1.7 ± 0.1	7.1 ± 0.5	68 ± 4	0.03 ± 0.03	326 ± 16	4.2 ± 0.1	12 ± 6	4.2 ± 0.5
(18) S[to] Aleixo da Restauração	0.1 ± 0.0	4.7 ± 0.5	23 ± 1	0.02 ± 0.00	24 ± 1	1.0 ± 0.0	16 ± 1	0.2 ± 0.0
(19) S. Miguel de Machede	1.2 ± 0.4	1.9 ± 0.0	38 ± 4	0.03 ± 0.00	142 ± 33	0.4 ± 0.1	6 ± 0	2.4 ± 0.5
(20) Serpa	1.2 ± 0.2	5.8 ± 0.5	31 ± 1	0.01 ± 0.01	278 ± 16	2.7 ± 0.2	18 ± 1	2.6 ± 0.0
(21) Ve Rocins	0.5 ± 0.1	3.9 ± 2.1	10 ± 1	0.01 ± 0.01	67 ± 34	1.5 ± 0.6	5 ± 2	2.2 ± 1.0
(22) Valverde	0.4 ± 0.1	5.6 ± 0.2	83 ± 24	0.04 ± 0.01	65 ± 50	2.2 ± 0.2	8 ± 0	2.3 ± 0.5
(23) V. N. S. Bento	0.7 ± 0.0	2.6 ± 0.1	28 ± 1	0.04 ± 0.00	33 ± 3	1.2 ± 0.1	24 ± 1	1.7 ± 0.1
(24) Villanueva del Fresno	1.2 ± 0.1	1.9 ± 0.2	16 ± 1	0.02 ± 0.01	70 ± 13	2.8 ± 0.1	19 ± 1	1.7 ± 0.4

Values of each determination represents mean ± SD ($n = 3$).

several enzyme activity processes [14, 45, 46]. These elements can be accumulated by mushrooms and the recommended daily intakes were 2 mg/day and 15 mg/day, for manganese and zinc, respectively [12, 39]. In this study, mushrooms presented a mean value of 56 ± 27 mg/kg dw for manganese and 68 ± 25 mg/kg dw for zinc content, similar to values described in literature [5, 6, 11, 36, 47]. Chromium biological functions are not known precisely; it seems to participate in the metabolism of lipids and carbohydrates, as well as in the insulin action. The RDI for this metal is 120 μg/day. The mean chromium content obtained was 1,19 ± 0,74 mg/kg dw, lower than those reported in the literature for other species of *Amanita* [5] and similar to some species of edible mushrooms described [11, 36]. Aluminium is one of the few abundant elements in nature but no significant biological function is known, although there are some evidences of toxicity when ingested in large quantities. Most acidic soils are saturated in aluminium instead of hydrogen ions, and this acidity is the result of aluminium compounds hydrolysis [48]. *A. ponderosa* fruiting bodies from the different sampling sites showed a high range of aluminium content with a medium value of 362±204 mg/Kg dw. High aluminium levels were reported for some *Amanita* species, for example, *A. rubescens* that showed values around 262 mg/kg dw [15] and *A. strobiliformis* and *A. verna* presented aluminium levels of 72 and 343 mg/kg dw, respectively [5]. Other studies report different levels of aluminium in *Amanita rubescens*: 293, 75, and 512 mg/kg dw for the whole fruiting body, cap, and stipe, respectively [37]. The large range in aluminium content was also reported in a study of *A. fulva* that showed aluminium levels ranging from 40 to 500 mg/Kg dw in the stipe and 40 to 200 mg/Kg dw in cap [10]. Regarding cadmium and lead, these elements have the highest toxicological significance. Cadmium has probably been the most damaging metal found in mushrooms; some studies point out the existence of accumulating species, which, in polluted areas, accumulate this metal. The mean values of cadmium and lead were 1.00 ± 0.73 mg/kg dw and 2.41 ± 1.34 mg/kg dw, respectively, showing that the *A. ponderosa* had no toxicity due to the presence of these two elements [5, 11]. The contents of heavy metals barium and silver in the studied mushrooms were 1.10 ± 0.59 mg/kg dw and 2.01 ± 1.56 mg/kg dw, respectively. These values are similar to those described in the literature for other species of edible mushrooms [11] and lower than those described for species of the genus *Amanita* [5, 13].

Bioconcentration factor (BCF) allows estimating the mushroom potential for the bioextraction of elements from the substratum (soil). Values of BCF of *A. ponderosa* fruiting bodies are summarized in Tables 7 and 8. The macroelements, K and Na, exhibited the highest values of BCF but Ca, Mg, and P also presented BCF > 1. Trace elements, Ag, Cd, Cu, and Zn, presented BCF > 1 with higher values for Cu. The remaining elements (Fe, Mn, Ba, Cr, and Pb) are bioexcluded showing lower values (BCF < 1). For copper, BCF values ranged from

Table 8: Bioconcentration factor (BCF) values for *A. ponderosa* fruiting bodies from different sampling sites.

Sampling site	Ag	Ba	Cd	Cr	Mn	Pb	Zn
(1) Almendres	38 ± 4	1.41 ± 0.33	7 ± 2	1.30 ± 0.39	1.27 ± 0.22	0.15 ± 0.10	13 ± 0
(2) Azaruja	72 ± 8	0.57 ± 0.04	1 ± 0	0.68 ± 0.03	0.05 ± 0.00	0.06 ± 0.00	3 ± 0
(3) Baleizão	55 ± 8	0.12 ± 0.04	7 ± 0	0.42 ± 0.07	0.10 ± 0.01	0.19 ± 0.03	10 ± 0
(4) Beja	8 ± 1	0.33 ± 0.08	15 ± 1	0.19 ± 0.00	0.17 ± 0.01	0.41 ± 0.02	8 ± 0
(5) Cabeça Gorda	14 ± 4	0.14 ± 0.01	4 ± 0	0.18 ± 0.01	0.09 ± 0.00	0.41 ± 0.03	8 ± 0
(6) Cabezas Rubias	10 ± 1	0.77 ± 0.01	8 ± 1	0.29 ± 0.00	0.11 ± 0.00	0.31 ± 0.04	13 ± 1
(7) Évora	2 ± 1	0.19 ± 0.05	4 ± 2	0.30 ± 0.13	0.34 ± 0.00	0.17 ± 0.08	15 ± 2
(8) Évoramonte	7 ± 0	0.32 ± 0.02	9 ± 1	1.14 ± 0.02	0.37 ± 0.03	0.56 ± 0.00	10 ± 1
(9) Her^{de} da Mitra	4 ± 0	0.53 ± 0.02	11 ± 2	0.47 ± 0.00	0.06 ± 0.01	0.26 ± 0.03	5 ± 0
(10) Mértola	61 ± 2	0.50 ± 0.01	8 ± 1	0.32 ± 0.07	0.04 ± 0.01	0.46 ± 0.07	3 ± 0
(11) Mina S. Domingos	2 ± 0	0.30 ± 0.01	13 ± 1	0.13 ± 0.00	0.36 ± 0.00	0.30 ± 0.06	25 ± 3
(12) M^{te} da Borralha	9 ± 0	0.40 ± 0.04	12 ± 1	0.24 ± 0.02	0.19 ± 0.07	0.41 ± 0.03	6 ± 0
(13) M^{te} Novo	6 ± 0	0.39 ± 0.05	3 ± 0	0.20 ± 0.02	0.24 ± 0.02	0.20 ± 0.02	10 ± 0
(14) Montejuntos	6 ± 1	0.29 ± 0.00	10 ± 2	0.18 ± 0.01	0.08 ± 0.01	0.30 ± 0.04	6 ± 0
(15) N. S^{ra} Guadalupe	16 ± 1	0.31 ± 0.08	5 ± 1	0.12 ± 0.01	0.05 ± 0.01	0.48 ± 0.06	5 ± 1
(16) N. S^{ra} Machede	1 ± 1	0.14 ± 0.02	8 ± 1	0.10 ± 0.00	0.40 ± 0.09	0.05 ± 0.02	7 ± 0
(17) Rosal de la Frontera	11 ± 2	0.65 ± 0.01	9 ± 1	0.42 ± 0.02	0.12 ± 0.07	0.18 ± 0.00	16 ± 1
(18) S^{to} Aleixo da Restauração	7 ± 0	0.39 ± 0.00	3 ± 0	0.36 ± 0.10	0.24 ± 0.07	0.62 ± 0.04	4 ± 0
(19) S. Miguel de Machede	16 ± 1	0.38 ± 0.08	38 ± 1	0.41 ± 0.03	0.16 ± 0.00	0.23 ± 0.01	6 ± 1
(20) Serpa	9 ± 0	0.18 ± 0.02	14 ± 3	0.07 ± 0.01	0.07 ± 0.02	0.05 ± 0.00	2 ± 0
(21) Ve Rocins	35 ± 21	0.27 ± 0.10	3 ± 1	0.71 ± 0.24	0.02 ± 0.01	0.25 ± 0.06	6 ± 0
(22) Valverde	12 ± 2	0.31 ± 0.04	5 ± 0	2.38 ± 0.03	1.23 ± 0.28	1.16 ± 0.05	9 ± 0
(23) V. N. S. Bento	4 ± 2	0.46 ± 0.01	2 ± 0	2.46 ± 2.01	0.10 ± 0.00	0.21 ± 0.07	3 ± 0
(24) Villanueva del Fresno	32 ± 6	0.10 ± 0.01	3 ± 1	0.30 ± 0.04	0.03 ± 0.00	0.42 ± 0.06	14 ± 1

Values of each determination represents mean ± SD (n = 3).

9 to 248, showing bioaccumulation of this metal. The same heterogeneous behaviour is observed for Ag, Cd, and Zn with values ranging from 1–72, 1–38, to 2–25, respectively. Ag bioaccumulation was found in other *Amanita* species with much higher values of BCF [13]. The high levels of aluminium observed for some *A. ponderosa* fruiting bodies can be related to soil content and to their different ability to accumulate this mineral, with BCF > 1 for some stands. Some works with different species such as *Leccinum scabrum*, *Amanita rubescens*, and *Xerocomus chrysenteron* reported different aluminium accumulation [15]. Some species of Basidiomycetes can be useful for assessing the environmental pollution levels [38].

Metal concentrations were usually assumed to be species-dependent, but soil composition is also an important factor in mineral content [12, 36, 38, 49]. In order to clarify this association between inorganic composition of *A. ponderosa* mushrooms and soil, a data mining approach was developed.

3.2. Segmentation Models Based on Mushrooms and Soil Mineral Content: Interpretation and Assessment. The k-Means Clustering Method is a segmentation algorithm that uses unsupervised learning. The input variables used in the segmentation approach are macroelements content (Na, K, Ca, P, and Mg), trace elements content (Al, Fe, Cu, Zn, Cr, and Mn), and heavy metals (Ag, Ba, Cd, and Pb). The algorithm input parameter is the number of clusters, 3 in this study. Table 9 shows the clusters centre of gravity, in order to characterize the clusters formed.

The analysis of Table 9 shows that cluster 1 is characterized by high values of Ag, Ba, Cd, K, Mg, and P. Cluster 2 is characterized by high content of Cr, Fe, Mn, Na, Pb, and Zn and Cluster 3 showed lower mineral content. In order to evaluate the relationships between clusters and fruiting bodies sampling sites the graph presented Figure 4 was conceived.

The analysis of Figure 4 shows that cluster 1 is formed by the samples collected at Almendres (1), Baleizão (3), Cabeça Gorda (5), Évoramonte (8), Mina S. Domingos (11), Montejuntos (14), N. S^{ra} Guadalupe (15), N. S^{ra} Machede (16), Rosal de la Frontera (17), Serpa (20), and Villanueva del Fresno (24). Cluster 2 includes the samples collected at Beja (2), Évora (7), S. Miguel de Machede (19), and V. N. S. Bento (23). Finally, cluster 3 is composed by the samples collected at Azaruja (2), Her^{de} da Mitra (9), Mértola (10), M^{te} da Borralha (12), M^{te} Novo (13), S^{to} Aleixo da Restauração (18), Ve de Rocins (21), and Valverde (22).

In order to generate an explanatory model of segmentation (i.e., seek to establish rules for assigning a new case to a cluster), Decisions Trees (DT) were used. Two different strategies were followed: one of them based on the mineral

TABLE 9: Clusters center of gravity.

Variable	Cluster 1	Cluster 2	Cluster 3
Ag	2.433 ± 1.593	1.690 ± 1.248	1.690 ± 2.309
Al	396.503 ± 210.361	392.625 ± 226.842	227.812 ± 187.093
Ba	1.442 ± 0.962	0.996 ± 0.616	0.719 ± 0.616
Ca	583.148 ± 246.929	617.110 ± 151.838	413.341 ± 260.514
Cd	1.316 ± 1.120	0.854 ± 0.322	0.630 ± 0.526
Cr	1.213 ± 0.933	1.753 ± 1.122	0.656 ± 0.486
Cu	224.094 ± 168.207	238.065 ± 108.827	112.320 ± 90.198
Fe	63.880 ± 41.865	252.718 ± 74.423	36.283 ± 27.911
K	34192.173 ± 13478.898	23008.328 ± 23670.094	21750.972 ± 12089.189
Mg	864.049 ± 419.246	664.112 ± 268.732	533.816 ± 354.098
Mn	64.046 ± 29.957	76.347 ± 18.841	30.254 ± 19.582
Na	1240.238 ± 624.901	1286.973 ± 354.879	598.937 ± 391.228
P	422.333 ± 167.110	219.741 ± 66.174	182.971 ± 173.598
Pb	2.440 ± 1.468	3.389 ± 1.805	1.640 ± 1.256
Zn	72.936 ± 29.664	80.972 ± 33.836	45.118 ± 29.973

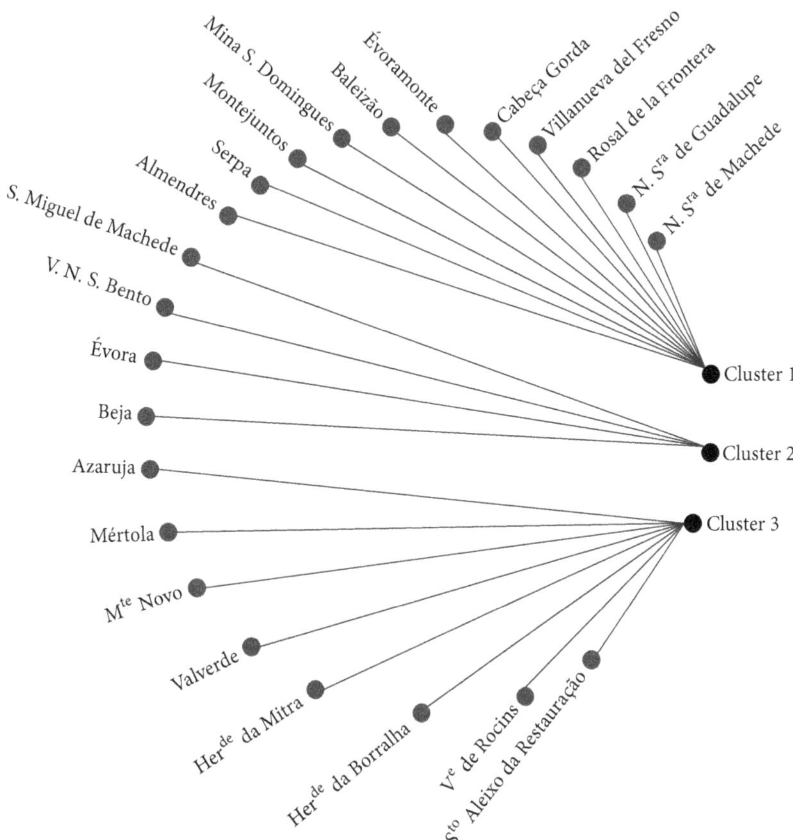

FIGURE 4: Relationships between clusters of mushrooms and different sampling sites.

mushrooms content (strategy 1) and the other one based on the soil mineral composition (strategy 2).

To ensure statistical significance of the attained results, 25 (twenty-five) runs were applied in all tests, the accuracy estimates being achieved using the Holdout method [50]. In each simulation, the available data are randomly divided into two mutually exclusive partitions: the training set, with about 2/3 of the available data and used during the modelling phase, and the test set, with the remaining examples, being used after training, in order to compute the accuracy values.

The DT obtained using the strategy 1 is shown in Figure 5. The minerals that contribute to this explanatory model are Fe,

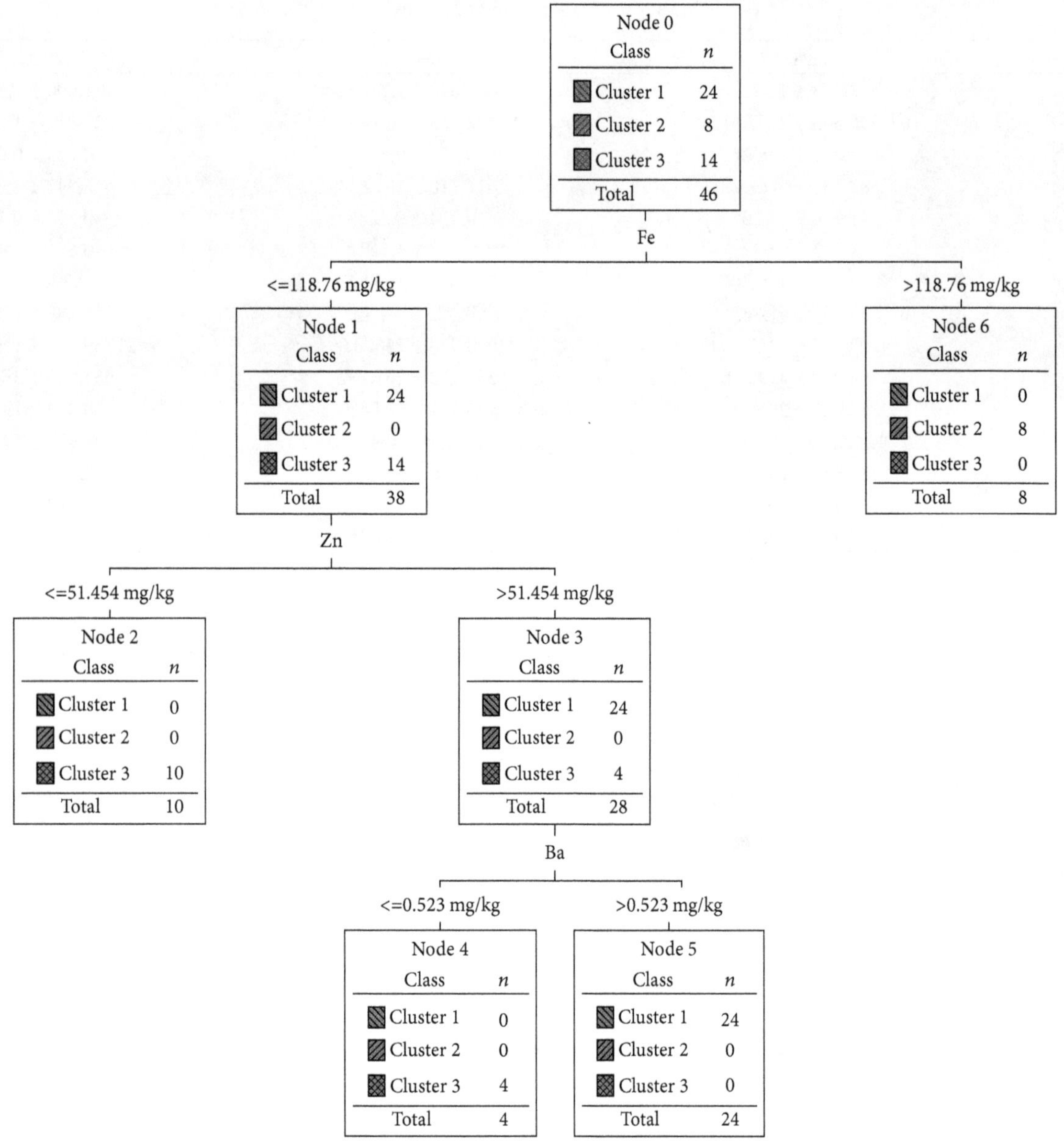

FIGURE 5: The Decision Tree explanatory of segmentation model based in mineral content of mushrooms samples (strategy 1).

Zn, and Ba and the rules to assign a case to a cluster are as follows:

(i) **If** Fe ≤ 118.76 mg/Kg
and Zn > 51.454 mg/Kg
and Ba > 0.523 mg/Kg
Then → Cluster 1

(ii) **If** Fe > 118.76 mg/Kg
Then → Cluster 2

(iii) **If** Fe ≤ 118.76 mg/Kg
and Zn ≤ 51.454 mg/Kg
Then → Cluster 3

(iv) **If** Fe ≤ 118.76 mg/Kg
and Zn > 51.454 mg/Kg
and Ba ≤ 0.523 mg/Kg
Then → Cluster 3

A common tool for classification analysis is the coincidence matrix [51], a matrix of size $L \times L$, where L denotes the number of possible classes. This matrix is created by matching the predicted and actual values. L was set to 3 (three) in the present case. Table 10 presents the coincidence matrix. The results reveal that the model accuracy is 100% both in the training set and in test set.

In order to relate the mineral mushrooms content to the soil mineral composition an explanatory model of the clusters formed was made, using the soil content of the same minerals (i.e., Fe, Zn, and Ba). Since the model accuracy is only 74% a new explanation model was built up. In this attempt all

TABLE 10: The coincidence matrix for Decision Tree model presented in Figure 5.

Variables	Training set			Test set		
	Cluster 1	Cluster 2	Cluster 3	Cluster 1	Cluster 2	Cluster 3
Cluster 1	24	0	0	9	0	0
Cluster 2	0	8	0	0	4	0
Cluster 3	0	0	14	0	0	10

TABLE 11: The coincidence matrix for Decision Tree model presented in Figure 6.

Variables	Training Set			Test Set		
	Cluster 1	Cluster 2	Cluster 3	Cluster 1	Cluster 2	Cluster 3
Cluster 1	22	2	0	8	1	0
Cluster 2	1	7	0	1	3	0
Cluster 3	2	0	12	2	0	8

inorganic soil mineral content was available. The referred model is shown in Figure 6 and the respective coincidence matrix is presented in Table 11.

The model accuracy was 89.1% and 82.6%, respectively, for training and test sets. The soil minerals that contribute to this explanatory model are Cr, Ba, and Zn. The rules to assign the cases to each cluster are as follows:

(i) **If** Cr (soil) > 3.765 mg/Kg
 Then → Cluster 1

(ii) **If** Cr (soil) ≤ 3.765 mg/Kg
 and Ba (soil) ≤ 3.012 mg/Kg
 and Zn (soil) ≤ 5.944 mg/Kg
 Then → Cluster 1

(iii) **If** Cr (soil) ≤ 3.765 mg/Kg
 and Ba (soil) > 3.012 mg/Kg
 and Zn (soil) > 5.033 mg/Kg
 Then → Cluster 2

(iv) **If** Cr (soil) ≤ 3.765 mg/Kg
 and Ba (soil) ≤ 3.012 mg/Kg
 and Zn (soil) > 5.944 mg/Kg
 Then → Cluster 3

(v) **If** Cr (soil) ≤ 3.765 mg/Kg
 and Ba (soil) > 3.012 mg/Kg
 and Zn (soil) ≤ 5.033 mg/Kg
 Then → Cluster 3

This study observed that the mineral content was influenced by the location area in which the mushroom samples were collected, possibly due to the soil composition and by environmental factors, such as medium temperature, vegetation, and rainfall. Indeed the inorganic composition of *A. ponderosa* allows group mushrooms according to the location area, based mainly in their Fe, Zn, and Ba content. On the other hand, it is possible to predict the same mushroom clustering taking into account the mineral soil content, based in Cr, Ba, and Zn soil composition although with a lower model accuracy.

Results of mineral composition do not reveal a direct correlation between inorganic composition of *A. ponderosa* mushrooms and their corresponding soil substrate; nevertheless, mushrooms are agents that play an important role in the continuous changes that occur in their habitats, and indeed they present a very effective mechanism for accumulating metals from the environment.

Moreno-Rojas et al. (2004), in the study of mineral content evaluation of *A. ponderosa* samples from Andalusia, also verified that variations occurred in the mineral composition according to the sample collection site, particularly in relation to Fe, K, and Na contents. Other authors also report that the main cause of variation of mineral composition between samples of different species of *Amanita* is the character and composition of the substrates (e.g., sand and wood) and may be influenced by the presence or absence of ability of the different species for a specific accumulation of some metals, namely, copper and zinc [5]. In a study carried out with a species of Boletaceae *(Suillus grevillei)*, it is also mentioned that the variations in mineral composition of the different samples are related to the composition of the substrates and the geochemistry of the soils of each site [11].

4. Conclusions

A. ponderosa mushrooms collected from different sites showed fruiting bodies with water content of 90–93%, dry mass ranging from 6.9 to 9.7%, contents of organic matter between 6.2 and 9.0%, and minerals between 0.5 and 0.9%. Mineral composition revealed high content in macroelements, such as potassium, phosphorus, and magnesium. Copper, chromium, iron, manganese, and zinc, essential microelements in biological systems, can also be found in fruiting bodies of *A. ponderosa*, within the limits of RDI. Bioconcentration was observed for some macro- and microelements, such as K, Na, Cu, Zn, Mg, P, Ag, Ca, and Cd. The presence of heavy metals, such as barium, cadmium,

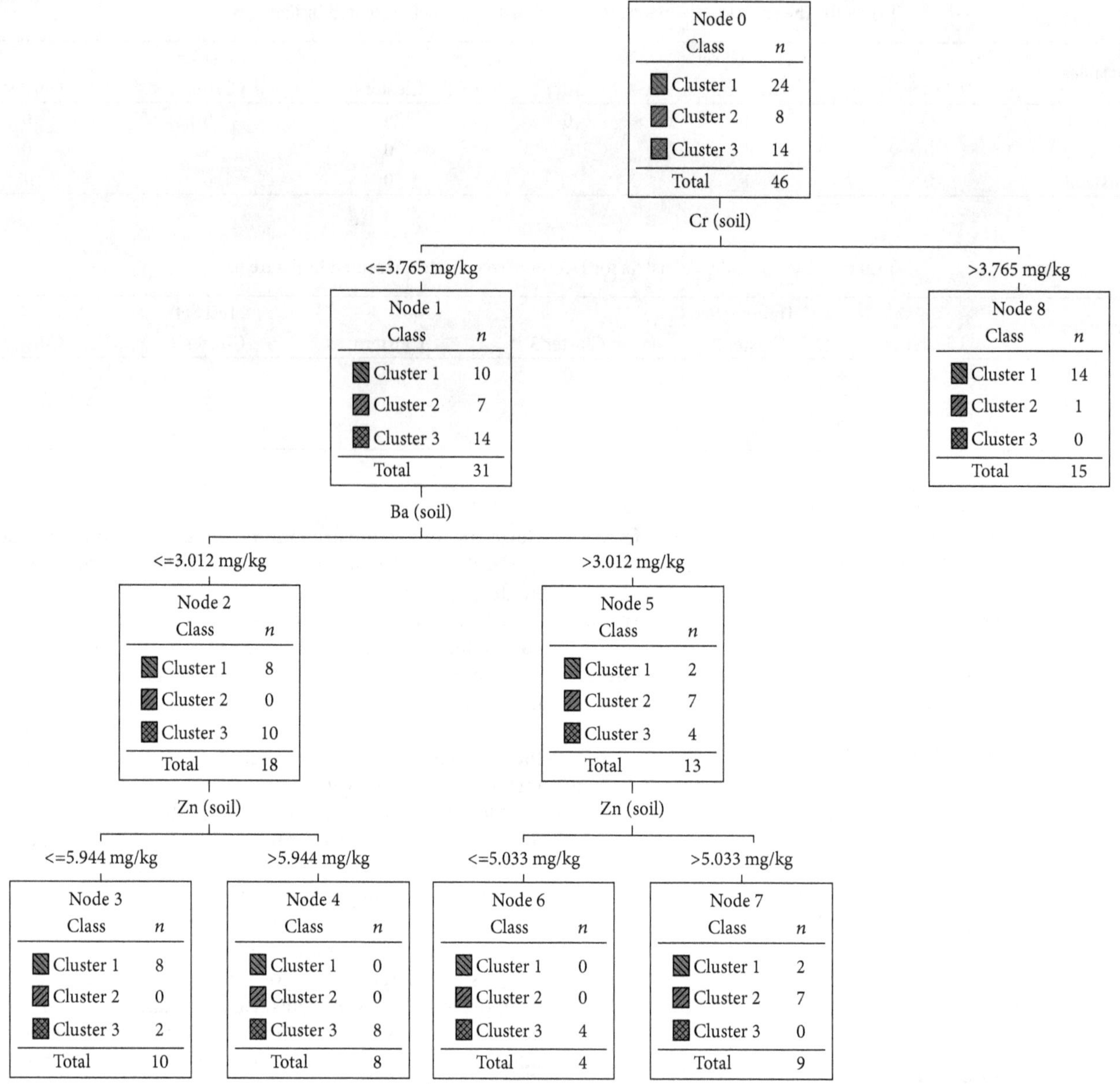

FIGURE 6: The Decision Tree explanatory of segmentation model based in mineral content of substrates.

lead and silver, was quite low, within the limits of RDI, and did not constitute a risk to human health.

Our results pointed out that it is possible to generate an explanatory model of segmentation performed with data based on the inorganic composition of mushrooms and soil mineral content, showing that it may be possible to relate these two types of data. The inorganic analysis provides evidence that mushrooms mineral composition variation is according to the collecting location, indicating the influence of the substrate characteristics in the fruiting bodies. The relationship between mineral elements in mushrooms and soils from the different sampling sites can be an important contribution to the certification process and seem to be related to the substrate effects from interindividual or interstrain differences.

Acknowledgments

The authors would like to acknowledge the HIT3CH Project-HERCULES Interface for Technology Transfer and Teaming in Cultural Heritage (ALT20-03-0246-FEDER-000004), MEDUSA Project-Microorganisms Monitoring and Mitigation-Developing and Unlocking Novel Sustainable Approaches (ALT20-03-0145-FEDER-000015), cofinanced by the European Regional Development Fund (ERDF), and ALENTEJO 2020 (Alentejo Regional Operational Programme).

References

[1] D. Agrahar-Murugkar and G. Subbulakshmi, "Nutritional value of edible wild mushrooms collected from the Khasi hills of Meghalaya," *Food Chemistry*, vol. 89, no. 4, pp. 599–603, 2005.

[2] P. G. De Pinho, B. Ribeiro, R. F. Gonçalves et al., "Correlation between the pattern volatiles and the overall aroma of wild edible mushrooms," *Journal of Agricultural and Food Chemistry*, vol. 56, no. 5, pp. 1704–1712, 2008.

[3] R. Zawirska-Wojtasiak, M. Siwulski, S. Mildner-Szkudlarz, and E. Wąsowicz, "Studies on the aroma of different species and strains of Pleurotus measured by GC/MS, sensory analysis and electronic nose," *ACTA Scientiarum Polonorum Technologia Alimentaria*, vol. 8, no. 1, pp. 47–61, 2009.

[4] E. Guillamón, A. García-Lafuente, M. Lozano et al., "Edible mushrooms: role in the prevention of cardiovascular diseases," *Fitoterapia*, vol. 81, no. 7, pp. 715–723, 2010.

[5] J. Vetter, "Mineral composition of basidiomes of Amanita species," *Mycological Research*, vol. 109, no. 6, pp. 746–750, 2005.

[6] P. K. Ouzouni, P. G. Veltsistas, E. K. Paleologos, and K. A. Riganakos, "Determination of metal content in wild edible mushroom species from regions of Greece," *Journal of Food Composition and Analysis*, vol. 20, no. 6, pp. 480–486, 2007.

[7] S. Beluhan and A. Ranogajec, "Chemical composition and non-volatile components of Croatian wild edible mushrooms," *Food Chemistry*, vol. 124, no. 3, pp. 1076–1082, 2011.

[8] F. S. Reis, L. Barros, A. Martins, and I. C. F. R. Ferreira, "Chemical composition and nutritional value of the most widely appreciated cultivated mushrooms: an inter-species comparative study," *Food and Chemical Toxicology*, vol. 50, no. 2, pp. 191–197, 2012.

[9] X.-M. Wang, J. Zhang, T. Li, Y.-Z. Wang, and H.-G. Liu, "Content and bioaccumulation of nine mineral elements in ten mushroom species of the genus boletus," *Journal of Analytical Methods in Chemistry*, vol. 2015, Article ID 165412, 2015.

[10] J. Falandysz, M. Drewnowska, M. Chudzińska, and D. Barałkiewicz, "Accumulation and distribution of metallic elements and metalloids in edible Amanita fulva mushrooms," *Ecotoxicology and Environmental Safety*, vol. 137, pp. 265–271, 2017.

[11] K. Chudzyński and J. Falandysz, "Multivariate analysis of elements content of Larch Bolete (*Suillus grevillei*) mushroom," *Chemosphere*, vol. 73, no. 8, pp. 1230–1239, 2008.

[12] H. Gençcelep, Y. Uzun, Y. Tunçtürk, and K. Demirel, "Determination of mineral contents of wild-grown edible mushrooms," *Food Chemistry*, vol. 113, no. 4, pp. 1033–1036, 2009.

[13] J. Borovička, Z. Řanda, E. Jelínek, P. Kotrba, and C. E. Dunn, "Hyperaccumulation of silver by Amanita strobiliformis and related species of the section Lepidella," *Mycological Research*, vol. 111, no. 11, pp. 1339–1344, 2007.

[14] J. Borovička and Z. Řanda, "Distribution of iron, cobalt, zinc and selenium in macrofungi," *Mycological Progress*, vol. 6, no. 4, pp. 249–259, 2007.

[15] P. Kalač, "Trace element contents in European species of wild growing edible mushrooms: a review for the period 2000–2009," *Food Chemistry*, vol. 122, no. 1, pp. 2–15, 2010.

[16] J. Falandysz and M. Drewnowska, "Distribution of mercury in Amanita fulva (Schaeff.) Secr. mushrooms: Accumulation, loss in cooking and dietary intake," *Ecotoxicology and Environmental Safety*, vol. 115, pp. 49–54, 2015.

[17] J. Falandysz, "Mercury accumulation of three Lactarius mushroom species," *Food Chemistry*, vol. 214, pp. 96–101, 2017.

[18] C. Salvador, M. R. Martins, H. Vicente, J. Neves, J. M. Arteiro, and A. T. Caldeira, "Modelling molecular and inorganic data of Amanita ponderosa mushrooms using artificial neural networks," *Agroforestry Systems*, vol. 87, no. 2, pp. 295–302, 2013.

[19] V. González, F. Arenal, G. Platas, F. Esteve-Raventós, and F. Peláez, "Molecular typing of Spanish species of Amanita by restriction analysis of the ITS region of the DNA," *Mycological Research*, vol. 106, no. 8, pp. 903–910, 2002.

[20] G. Moreno, G. Platas, F. Peláez et al., "Molecular phylogenetic analysis shows that Amanita ponderosa and A. curtipes are distinct species," *Mycological Progress*, vol. 7, no. 1, pp. 41–47, 2008.

[21] A. T. Caldeira, C. Salvador, F. Pinto, J. M. Arteiro, and M. R. Martins, "MSP-PCR and RAPD molecular biomarkers to characterize Amanita ponderosa mushrooms," *Annals of Microbiology*, vol. 59, no. 3, pp. 629–634, 2009.

[22] M. R. Martins, C. Salvador, H. Vicente, J. Neves, J. M. Arteiro, and A. T. Caldeira, A Data Mining approach to characterize *Amanita ponderosa* mushrooms using inorganic profile and M13-PCR molecular Data. In Vasco Cadavez & Daniel Thiel Eds, Eurosis - ETI Publication, Ghent, Belgium, FOOD-SIM'2010, 5–12, 2010.

[23] C. Salvador, M. R. Martins, J. M. Arteiro, and A. T. Caldeira, "Molecular evaluation of some Amanita ponderosa and the fungal strains living in association with these mushrooms in the southwestern Iberian Peninsula," *Annals of Microbiology*, vol. 64, no. 3, pp. 1179–1187, 2014.

[24] A. T. Caldeira, C. Salvador, F. Pinto, J. M. Arteiro, and M. R. Martins, "Molecular biomarkers to characterize Amanita ponderosa mushrooms," in *Proceedings of the 10th International Chemical and Biological Engineering Conference*, pp. 2094–2099, Braga, Portugal, 2008.

[25] A. Teresa Caldeira, J. M. Arteiro, J. C. Roseiro, J. Neves, and H. Vicente, "An artificial intelligence approach to Bacillus amyloliquefaciens CCMI 1051 cultures: Application to the production of anti-fungal compounds," *Bioresource Technology*, vol. 102, no. 2, pp. 1496–1502, 2011.

[26] A. T. Caldeira, J. C. Roseiro, J. M. Arteiro, J. Neves, and H. Vicente, "Production of bioactive compounds against wood contaminant fungi: an artificial intelligence approach," in *Minimizing the Environmental Impact of the Forest Products Industries*, 137, p. 131, University Fernando Pessoa Edition, Portugal, 2011.

[27] G. Malençon, and R. Heim, "Notes critiques sur quelques hymΦnomycetes dEurope et dAfrique du Nord I. Les amanites blanches meridionales," *Bulletin trimestriel de la Société mycologique de France*, vol. 58, pp. 14–34, 1942.

[28] R. Moreno-Rojas, M. A. Díaz-Valverde, B. M. Arroyo, T. J. González, and C. J. B. Capote, "Mineral content of gurumelo (Amanita ponderosa)," *Food Chemistry*, vol. 85, no. 3, pp. 325–330, 2004.

[29] A. E. Greenberg and A. D. Eaton, *Standard Methods for the Examination of Water and Wastewater*, American Public Health Association, Washington, Wash, USA, 1992.

[30] S. M. V. Fernandes, A. O. S. S. Rangel, and J. L. F. C. Lima, "Flow Injection Determination of Sodium, Potassium, Calcium, and Magnesium in Beer by Flame Emission and Atomic Absorption Spectrometry," *Journal of Agricultural and Food Chemistry*, vol. 45, no. 4, pp. 1269–1272, 1997.

[31] F. Okumura, É. T. G. Cavalheiro, and J. A. Nóbrega, "Simple flame photometric experiments to teach principles of atomic

spectrometry in undergraduate analytical chemistry courses," *Química Nova*, vol. 27, no. 5, pp. 832–836, 2004.

[32] AOAC, (Association of Official Analytical Chemist) Official methods of analysis, 15th ed, 2nd (suppl 991.25) 101–102, 1991.

[33] J. MacQueen, "Some methods for classification and analysis of multivariate observations," in *Proceedings of the 5th Berkeley Symposium on Mathematical Statistics and Probability*, p. 14, University of California Press, Berkeley, Calif, USA, 1967.

[34] M. Hall, E. Frank, G. Holmes, B. Pfahringer, P. Reutemann, and I. H. Witten, "The WEKA data mining software: an update," *ACM SIGKDD Explorations Newsletter*, vol. 11, no. 1, pp. 10–18, 2009.

[35] I. H. Witten, E. Frank, and M. A. Hall, *Data Mining - Practical Machine Learning Tools and Techniques*, Elsevier, Burlington, USA, 3rd edition, 2011.

[36] P. Kalač and L. Svoboda, "A review of trace element concentrations in edible mushrooms," *Food Chemistry*, vol. 69, no. 3, pp. 273–281, 2000.

[37] M. Rudawska and T. Leski, "Macro- and microelement contents in fruiting bodies of wild mushrooms from the Notecka forest in west-central Poland," *Food Chemistry*, vol. 92, no. 3, pp. 499–506, 2005.

[38] E. Sesli, M. Tuzen, and M. Soylak, "Evaluation of trace metal contents of some wild edible mushrooms from Black sea region, Turkey," *Journal of Hazardous Materials*, vol. 160, no. 2-3, pp. 462–467, 2008.

[39] European Commission, European Commission Opinion of the Scientific Committee on Food on the Tolerable Upper Intake Level of Copper. Health and Consumer Protection Directorate-General, Brussels, Belgium, 2003.

[40] FAO/WHO, Human vitamin and mineral requirements. World Health Organization, Food and Agriculture Organization of United Nations, Rome, Italy, 2002.

[41] E. Sesli and M. Tüzen, "Levels of trace elements in the fruiting bodies of macrofungi growing in the East Black Sea region of Turkey," *Food Chemistry*, vol. 65, no. 4, pp. 453–460, 1999.

[42] D. Afzali, A. Mostafavi, M. A. Taher, and A. Moradian, "Flame atomic absorption spectrometry determination of trace amounts of copper after separation and preconcentration onto TDMBAC-treated analcime pyrocatechol-immobilized," *Talanta*, vol. 71, no. 2, pp. 971–975, 2007.

[43] S. Soriano, A. D. P. Netto, and R. J. Cassella, "Determination of Cu, Fe, Mn and Zn by flame atomic absorption spectrometry in multivitamin/multimineral dosage forms or tablets after an acidic extraction," *Journal of Pharmaceutical and Biomedical Analysis*, vol. 43, no. 1, pp. 304–310, 2007.

[44] N. Abbaspour, R. Hurrell, and R. Kelishadi, "Review on iron and its importance for human health," *Journal of Research in Medical Sciences*, vol. 19, no. 2, pp. 164–174, 2014.

[45] D. Mendil, Ö. D. Uluözlü, E. Hasdemir, and A. Çağlar, "Determination of trace elements on some wild edible mushroom samples from Kastamonu, Turkey," *Food Chemistry*, vol. 88, no. 2, pp. 281–285, 2004.

[46] G. B. Gerber, A. Léonard, and P. Hantson, "Carcinogenicity, mutagenicity and teratogenicity of manganese compounds," *Critical Review in Oncology/Hematology*, vol. 42, no. 1, pp. 25–34, 2002.

[47] M. Tuzen, E. Sesli, and M. Soylak, "Trace element levels of mushroom species from East Black Sea region of Turkey," *Food Control*, vol. 18, no. 7, pp. 806–810, 2007.

[48] J. F. Ma, P. R. Ryan, and E. Delhaize, "Aluminium tolerance in plants and the complexing role of organic acids," *Trends in Plant Science*, vol. 6, no. 6, pp. 273–278, 2001.

[49] L. Cocchi, L. Vescovi, L. E. Petrini, and O. Petrini, "Heavy metals in edible mushrooms in Italy," *Food Chemistry*, vol. 98, no. 2, pp. 277–284, 2006.

[50] J. Souza and N. Japkowicz, "Evaluating data mining models: a pattern language," in *Proceedings of the 9th Conference on Pattern Language of Programs*, pp. 11–23, 2002.

[51] F. Provost and R. Kohavi, "Guest editors' introduction: On applied research in machine learning," *Machine Learning*, vol. 30, no. 2-3, pp. 127–132, 1998.

6

Computations of the M-Polynomials and Degree-Based Topological Indices for Dendrimers and Polyomino Chains

Young Chel Kwun[1], **Adeel Farooq**[2], **Waqas Nazeer**[3], **Zohaib Zahid**[4], **Saba Noreen**[5] and **Shin Min Kang**[6,7]

[1]*Department of Mathematics, Dong-A University, Busan 49315, Republic of Korea*
[2]*Department of Mathematics, COMSATS University Islamabad, Lahore Campus, Lahore 54000, Pakistan*
[3]*Division of Science and Technology, University of Education, Lahore 54000, Pakistan*
[4]*Department of Mathematics, University of Management and Technology, Lahore 54000, Pakistan*
[5]*Department of Mathematics and Statistics, The University of Lahore, Lahore, Pakistan*
[6]*Department of Mathematics and RINS, Gyeongsang National University, Jinju 52828, Republic of Korea*
[7]*Center for General Education, China Medical University, Taichung 40402, Taiwan*

Correspondence should be addressed to Waqas Nazeer; nazeer.waqas@ue.edu.pk and Shin Min Kang; smkang@gnu.ac.kr

Academic Editor: Charles L. Wilkins

Topological indices correlate certain physicochemical properties like boiling point, stability, and strain energy of chemical compounds. In this report, we compute M-polynomials for PAMAM dendrimers and polyomino chains. Moreover, by applying calculus, we compute nine important topological indices of under-study dendrimers and chains.

1. Introduction

The **polyomino chains** constitute a finite 2-connected floor plan, where each inner face (or a unit) is surrounded by a square of length one. We can say that it is a union of cells connected by edges in a planar square lattice. For the origin of dominoes, we quote [1]. The **polyomino chains** have a long history dating back to the beginning of the 20th century, but they were originally promoted by Golomb [2, 3]. Dendrimers [4] are repetitively branched molecules. The name comes from the Greek word, which translates to "trees." Synonymous terms for dendrimers include arborols and cascade molecules. The first dendrimer was made by Fritz Vögtle in [5]. For detailed study about dendrimer structures we refer the reader to [6–9].

Many studies have shown that there is a strong intrinsic link between the chemical properties of chemical compounds and drugs (such as melting point and boiling point) and their molecular structure [10, 11]. The topological index defined on the structure of these chemical molecules can help researchers better understand the physical characteristics, chemical reactivity, and biological activity [12]. Therefore, the study of topological indices of chemical substances and chemical structures of drugs can make up for the lack of chemical experiments and provide theoretical basis for the preparation of drugs and chemical substances. In the previous two decades, a number of topological indices have been characterized and utilized for correlation analysis in pharmacology, environmental chemistry, toxicology, and theoretical chemistry [13]. Hosoya polynomial (Wiener polynomial) [14] plays a pivotal role in finding topological indices that depend on distances. From this polynomial, a long list of distance-based topological indices can be easily evaluated. A similar breakthrough was obtained recently by Klavžar et al. [15], in the context of degree-based indices. In the year 2015, authors in [15] introduced the M-polynomial, which plays similar role "to what Hosoya polynomial does" to determine many topological indices depending on the degree of end vertices [16–20].

In the present paper, we compute M-polynomials for different dendrimer structures and polyomino chains. By applying fundamental calculus, we recover nine degree-based topological indices for these dendrimers and chains.

2. Basic Definitions and Literature Review

In this paper, we fixed G as a connected graph, $V(G)$ is the set of vertices, $E(G)$ is the set of edges, and d_v is the degree of any vertex v. Most of the definitions presented in this section can be found in [17].

Definition 1 (see [15]). The M-polynomial of G is defined as

$$M(G; x, y) = \sum_{\delta \le i \le j \le \Delta} m_{ij}(G)\, x^i y^j \tag{1}$$

where $\delta = \min\{d_v \mid v \in V(G)\}$, $\Delta = \max\{d_v \mid v \in V(G)\}$, and $m_{ij}(G)$ is the edge $vu \in E(G)$ that is $i \le j$.

The very first topological index was the Wiener index, defined by Wiener in 1945, when he was studying boiling point of alkane [21]. For comprehensive details about the applications of Wiener index, see [22, 23]. After that, in 1975, Milan Randić [24] introduced the first degree-based topological index, which is now known as Randić index and is defined as

$$R_{-1/2}(G) = \sum_{uv \in E(G)} \frac{1}{\sqrt{d_u d_v}}. \tag{2}$$

The generalized Randić index is defined as

$$R_\alpha(G) = \sum_{uv \in E(G)} \frac{1}{(d_u d_v)^\alpha}; \tag{3}$$

please see [25–29].

The inverse generalized Randić index is defined as

$$RR_\alpha(G) = \sum_{uv \in E(G)} (d_u d_v)^\alpha. \tag{4}$$

It can be seen easily that the Randić index is particular case of the generalized Randić index and the inverse generalized Randić index. Other oldest degree-based topological indices are Zagreb indices. The first Zagreb index is defined as

$$M_1(G) = \sum_{uv \in E(G)} (d_u + d_v) \tag{5}$$

and the second Zagreb index is defined as

$$M_2(G) = \sum_{uv \in E(G)} (d_u \times d_v). \tag{6}$$

The second modified Zagreb index is defined as

$${}^{m}M_2(G) = \sum_{uv \in E(G)} \frac{1}{d(u)\, d(v)}. \tag{7}$$

For detailed study about Zagreb indices, we refer the reader to [30–32]. There are many other degree-based topological indices, for example, symmetric division index:

$$\mathrm{SDD}(G) = \sum_{uv \in E(G)} \left\{ \frac{\min(d_u, d_v)}{\max(d_u, d_v)} + \frac{\max(d_u, d_v)}{\min(d_u, d_v)} \right\} \tag{8}$$

harmonic index:

$$H(G) = \sum_{vu \in E(G)} \frac{2}{d_u + d_v} \tag{9}$$

inverse sum index:

$$I(G) = \sum_{vu \in E(G)} \frac{d_u d_v}{d_u + d_v} \tag{10}$$

augmented Zagreb index:

$$A(G) = \sum_{vu \in E(G)} \left\{ \frac{d_u d_v}{d_u + d_v - 2} \right\}^3. \tag{11}$$

We refer to [33–45] for detailed survey about the above defined indices and applications. Tables exhibited in [15–19] relate some notable degree-based topological indices with M-polynomial with the following notations [17]:

$$\begin{aligned} D_x &= x \frac{\partial (f(x,y))}{\partial x}, \\ D_y &= y \frac{\partial (f(x,y))}{\partial y}, \\ S_x &= \int_0^x \frac{f(t,y)}{t} dt, \\ S_y &= \int_0^y \frac{f(x,t)}{t} dt, \\ J(f(x,y)) &= f(x,x), \\ Q_\alpha(f(x,y)) &= x^\alpha f(x,y). \end{aligned} \tag{12}$$

3. Computational Results

In this section we give our computational results.

3.1. M-Polynomials and Degree-Based Indices for PAMAM Dendrimers. Polyamidoamine (PAMAM) dendrimers are hyperbranched polymers with unparalleled molecular uniformity, narrow molecular weight distribution, defined size and shape characteristics, and a multifunctional terminal surface. These nanoscale polymers consist of an ethylene-diamine core, a repetitive branching amidoamine internal structure, and a primary amine terminal surface. Dendrimers are "grown" off a central core in an iterative manufacturing process, with each subsequent step representing a new "generation" of dendrimer. Increasing generations (molecular weight) produce larger molecular diameters, twice the number of reactive surface sites and approximately double the molecular weight of the preceding generation. PAMAM dendrimers also assume a spheroidal, globular shape at generation 4 and above (see molecular simulation below). Their functionality is readily tailored, and their uniformity, size, and highly reactive "molecular Velcro" surfaces are the functional keys to their use. Here we consider PD_1, which denote PAMAM dendrimers with trifunctional core unit generated by dendrimer generations G_n with n growth stages, and PD_2, the PAMAM dendrimers with different core generated by dendrimer generators G_n with n growth stages. DS_1 is kinds of PAMAM dendrimers with n growth stages

Theorem 2. *For the PAMAM dendrimers PD_1, we have*

$$M(PD_1, x, y) = 3 \cdot 2^n xy^2 + 3(2^{n+1} - 1)xy^3 + 9(2^{n+1} - 1)x^2y^2 + 3(7 \cdot 2^n - 4)x^2y^3. \quad (13)$$

Proof. Let PD_1 denote PAMAM dendrimers with trifunctional core unit generated by dendrimer generations G_n with n growth stages.

The edge set of PD_1 has following four partitions:

$$\begin{aligned} E_{\{1,2\}} &= \{e = uv \in E(PD_1) \mid d_u = 1, d_v = 2\}, \\ E_{\{1,3\}} &= \{e = uv \in E(PD_1) \mid d_u = 1, d_v = 3\}, \\ E_{\{2,2\}} &= \{e = uv \in E(PD_1) \mid d_u = 2, d_v = 2\}, \\ E_{\{2,3\}} &= \{e = uv \in E(PD_1) \mid d_u = 2, d_v = 3\}. \end{aligned} \quad (14)$$

Now

$$\begin{aligned} |E_{\{1,2\}}| &= 3 \cdot 2^n, \\ |E_{\{1,3\}}| &= 6 \cdot 2^n - 3, \\ |E_{\{2,2\}}| &= 18 \cdot 2^n - 9, \end{aligned} \quad (15)$$

and

$$|E_{\{2,3\}}| = 21 \cdot 2^n - 12.$$

$$\begin{aligned} M(PD_1; x, y) &= \sum_{i \le j} m_{ij}(PD_1) x^i y^j \\ &= \sum_{1 \le 2} m_{12}(PD_1) xy^2 + \sum_{1 \le 3} m_{13}(PD_1) x^1 y^3 + \sum_{2 \le 2} m_{22}(PD_1) x^2 y^2 + \sum_{2 \le 3} m_{23}(PD_1) x^2 y^3 \\ &= \sum_{uv \in E_{\{1,2\}}} m_{12}(PD_1) xy^2 + \sum_{uv \in E_{\{1,3\}}} m_{13}(PD_1) xy^3 + \sum_{uv \in E_{\{2,2\}}} m_{22}(PD_1) x^2 y^2 + \sum_{uv \in E_{\{2,3\}}} m_{23}(PD_1) x^2 y^3 \\ &= |E_{\{1,2\}}| xy^2 + |E_{\{1,3\}}| x^1 y^3 + |E_{\{2,2\}}| x^2 y^2 + |E_{\{2,3\}}| x^2 y^3 \\ &= 3 \times 2^n xy^2 + (6 \times 2^n - 3) xy^3 + (18 \times 2^n - 9) x^2 y^2 + (21 \times 2^n - 12) x^2 y^3 \\ &= 3 \times 2^n xy^2 + 3(2^{n+1} - 1) xy^3 + 9(2^{n+1} - 1) x^2 y^2 + 3(7 \times 2^n - 4) x^2 y^3. \end{aligned} \quad (16)$$

□

Theorem 3. *For the PAMAM dendrimers PD_1, we have*

1. $M_1(G) = 105 \times 2^{n+1} - 108$.
2. $M_2(G) = 111 \times 2^{n+1} - 117$.
3. ${}^m M_2(G) = 2^{n+1} + 19 \times 2^{n-1} - 21/4$.
4. $R_\alpha(G) = 3 \times 2^{n+\alpha} + (3^{\alpha+1} + 2^{2\alpha} \times 9)(2^{n+1} - 1) + 2^\alpha \times 3^{\alpha+1}(7 \cdot 2^n - 4)$.
5. $R_\alpha(G) = 3 \times 2^{n-\alpha} + (1/3^{\alpha-1} + 9/2^{2\alpha})(2^{n+1} - 1) + (1/(3^{\alpha-1} \times 2^\alpha))(7 \cdot 2^n - 4)$.
6. $SSD(G) = 7 \times 2^{n+3} + 3 \times 2^{n+1} + 47 \times 2^n - 54$.
7. $H(G) = (7/5) \times 2^{n+4} - 54/5$
8. $I(G) = (497/20) \times 2^{n+1} - 513/20$.
9. $A(G) = 3 \times 2^{n+3} + 21 \times 2^{n+2} + 369 \times 2^{n-2} - 753/8$.

Proof. Let PD_1 denote PAMAM dendrimers with trifunctional core unit generated by dendrimer generations G_n with n growth stages. Let

$$\begin{aligned} M(G; x, y) &= f(x, y) \\ &= 3 \times 2^n xy^2 + 3(2^{n+1} - 1) xy^3 + 9(2^{n+1} - 1) x^2 y^2 + 3(7 \times 2^n - 4) x^2 y^3. \end{aligned} \quad (17)$$

Then

$$D_x f(x, y) = 3 \times 2^n xy^2 + 3\left(2^{n+1} - 1\right) xy^3 + 18\left(2^{n+1} - 1\right) x^2 y^2 + 6\left(7 \times 2^n - 4\right) x^2 y^3.$$

$$D_y f(x, y) = 3 \times 2^{n+1} xy^2 + 9\left(2^{n+1} - 1\right) xy^3 + 18\left(2^{n+1} - 1\right) x^2 y^2 + 9\left(7 \times 2^n - 4\right) x^2 y^3.,$$

$$D_y D_x f(x, y) = 3 \times 2^{n+1} xy^2 + 9\left(2^{n+1} - 1\right) xy^3 + 36\left(2^{n+1} - 1\right) x^2 y^2 + 18\left(7 \times 2^n - 4\right) x^2 y^3,$$

$$S_x S_y (f(x, y)) = 3 \times 2^{n-1} xy^2 + \left(2^{n+1} - 1\right) xy^3 + \frac{9}{4}\left(2^{n+1} - 1\right) x^2 y^2 + \frac{1}{2}\left(7 \times 2^n - 4\right) x^2 y^3,$$

$$D_x^{\alpha} D_y^{\alpha} (f(x, y)) = 3 \times 2^{n+\alpha} xy^2 + 3^{\alpha+1}\left(2^{n+1} - 1\right) xy^3 + 2^{2\alpha} \times 9\left(2^{n+1} - 1\right) x^2 y^2 + 2^{\alpha} \times 3^{\alpha}\left(7 \times 2^n - 4\right) x^2 y^3,$$

$$S_x^{\alpha} S_y^{\alpha} (f(x, y) = 3 \times 2^{n-\alpha} xy^2 + \frac{1}{3^{\alpha-1}}\left(2^{n+1} - 1\right) xy^3 + \frac{9}{2^{2\alpha}}\left(2^{n+1} - 1\right) x^2 y^2 + \frac{1}{3^{\alpha-1} \times 2^{\alpha}}\left(7 \times 2^n - 4\right) x^2 y^3,$$

$$S_y D_x (f(x, y)) = 3 \times 2^{n-1} xy^2 + \left(2^{n+1} - 1\right) xy^3 + 9\left(2^{n+1} - 1\right) x^2 y^2 + 2\left(7 \times 2^n - 4\right) x^2 y^3,$$

$$S_x D_y (f(x, y)) = 3 \times 2^{n-1} xy^2 + 9\left(2^{n+1} - 1\right) xy^3 + 9\left(2^{n+1} - 1\right) x^2 y^2 + \frac{9}{2}\left(7 \times 2^n - 4\right) x^2 y^3,$$

$$S_x J f(x, y) = 2^n x^3 + 3\left(2^{n+1} - 1\right) x^4 + \frac{3}{5}\left(7 \times 2^n - 4\right) x^5,$$

$$S_x J D_x D_y f(x, y) = 2^{n+1} x^3 + \frac{45}{4}\left(2^{n+1} - 1\right) x^4 + \frac{18}{5}\left(7 \times 2^n - 4\right) x^5,$$

$$S_x^{\ 3} Q_{-2} J D_x^{\ 3} D_y^{\ 3} f(x, y) = 3 \times 2^{n+3} x + \frac{369}{8}\left(2^{n+1} - 1\right) x^2 + 12\left(7 \times 2^n - 4\right) x^3. \quad (18)$$

□

(1) First Zagreb Index

$$M_1(G) = \left(D_x + D_y\right) f(x, y)\big|_{x=y=1} = 105 \times 2^{n+1} - 108. \quad (19)$$

(2) Second Zagreb Index

$$M_2(G) = D_y D_x (f(x, y))\big|_{x=y=1} = 111 \times 2^{n+1} - 117. \quad (20)$$

(3) Modified Second Zagreb Index

$$^{m}M_2(G) = S_x S_y (f(x, y))\big|_{x=y=1} = 2^{n+1} + 19 \times 2^{n-1} - \frac{21}{4}. \quad (21)$$

(4) Generalized Randić Index

$$R_{\alpha}(G) = D_x^{\alpha} D_y^{\alpha} (f(x, y))\big|_{x=y=1} = 3 \times 2^{n+\alpha} + \left(3^{\alpha+1} + 2^{2\alpha} \times 9\right)\left(2^{n+1} - 1\right) + 2^{\alpha} \times 3^{\alpha+1}\left(7 \cdot 2^n - 4\right). \quad (22)$$

(5) Inverse Randić Index

$$RR_{\alpha}(G) = S_x^{\alpha} S_y^{\alpha} (f(x, y))\big|_{x=y=1} = 3 \times 2^{n-\alpha} + \left(\frac{1}{3^{\alpha-1}} + \frac{9}{2^{2\alpha}}\right)\left(2^{n+1} - 1\right) + \frac{1}{3^{\alpha-1} \times 2^{\alpha}}\left(7 \cdot 2^n - 4\right). \quad (23)$$

(6) Symmetric Division Index

$$SSD(G) = \left(S_y D_x + S_x D_y\right)(f(x,y))\big|_{x=y=1} = 7\times 2^{n+3} + 3\times 2^{n+1} + 47\times 2^n - 54. \quad (24)$$

(7) Harmonic Index

$$H(G) = 2S_x J(f(x,y))\big|_{x=1} = H(G) = \frac{7}{5}\times 2^{n+4} - \frac{54}{5}. \quad (25)$$

(8) Inverse Sum Index

$$I(G) = S_x J D_x D_y (f(x,y))_{x=1} = \frac{497}{20}\times 2^{n+1} - \frac{513}{20}. \quad (26)$$

(9) Augmented Zagreb Index

Theorem 4. *For the PAMAM dendrimers PD_2, we have*

$$M(PD_2, x, y) = 2^{n+2}xy^2 + 4\left(2^{n+1}-1\right)xy^3 + \left(24\cdot 2^n - 11\right)x^2y^2 + 14\left(2^{n+1}-1\right)x^2y^3. \quad (27)$$

Proof. Let PD_2 be the PAMAM dendrimers with different core generated by dendrimer generators G_n with n growth stages. Then the edge set of PD_2 has following four partitions:

$$\begin{aligned} E_{\{1,2\}} &= \left\{e = uv \in E(PD_2) \mid d_u = 1, d_v = 2\right\},\\ E_{\{1,3\}} &= \left\{e = uv \in E(PD_2) \mid d_u = 1, d_v = 3\right\},\\ E_{\{2,2\}} &= \left\{e = uv \in E(PD_2) \mid d_u = 2, d_v = 2\right\},\\ E_{\{2,3\}} &= \left\{e = uv \in E(PD_2) \mid d_u = 2, d_v = 3\right\}. \end{aligned} \quad (28)$$

Now

$$\begin{aligned} \left|E_{\{1,2\}}\right| &= 4\cdot 2^n,\\ \left|E_{\{1,3\}}\right| &= 8\cdot 2^n - 4,\\ \left|E_{\{2,2\}}\right| &= 24\cdot 2^n - 11, \end{aligned} \quad (29)$$

and

$$\left|E_{\{2,3\}}\right| = 28\cdot 2^n - 14.$$

$$\begin{aligned} M(PD_2; x, y) &= \sum_{i\le j} m_{ij}(PD_2)\,x^i y^j\\ &= \sum_{1\le 2} m_{12}(PD_2)\,xy^2 + \sum_{1\le 3} m_{13}(PD_2)\,x^1y^3 + \sum_{2\le 2} m_{22}(PD_2)\,x^2y^2 + \sum_{2\le 3} m_{23}(PD_2)\,x^2y^3\\ &= \sum_{uv\in E_{\{1,2\}}} m_{12}(PD_2)\,xy^2 + \sum_{uv\in E_{\{1,3\}}} m_{13}(PD_2)\,xy^3 + \sum_{uv\in E_{\{2,2\}}} m_{22}(PD_2)\,x^2y^2 + \sum_{uv\in E_{\{2,3\}}} m_{23}(PD_2)\,x^2y^3\\ &= \left|E_{\{1,2\}}\right| xy^2 + \left|E_{\{1,3\}}\right| x^1y^3 + \left|E_{\{2,2\}}\right| x^2y^2 + \left|E_{\{2,3\}}\right| x^2y^3\\ &= 4\cdot 2^n xy^2 + (8\cdot 2^n - 4)\,xy^3 + (24\cdot 2^n - 11)\,x^2y^2 + (28\cdot 2^n - 14)\,x^2y^3\\ &= 2^{n+2}xy^2 + 4\left(2^{n+1}-1\right)xy^3 + \left(24\cdot 2^n - 11\right)x^2y^2 + 14\left(2^{n+1}-1\right)x^2y^3. \end{aligned} \quad (30)$$

□

Theorem 5. *For the PAMAM dendrimers PD_2, we have*

1. $M_1(G) = 5(7\times 2^{n+3} - 26)$.
2. $M_2(G) = 37\times 2^{n+3} - 140$.
3. ${}^{m}M_2(G) = (23/3)\times 2^{n+1} - 77/12$.
4. $R_\alpha(G) = 2^{n+\alpha+2} + (4\times 3^\alpha + 2^{\alpha+1}\times 3^\alpha\times 7)(2^{n+1}-1) + 2^{2\alpha}(24\cdot 2^n - 11)$.
5. $R_\alpha(G) = 2^n + (4/3^\alpha + 14/6^\alpha)(2^{n+1}-1) + (1/2^{2\alpha})(24\cdot 2^n - 11)$.
6. $SSD(G) = (109/3)\times 2^{n+2} - 197/3$.
7. $H(G) = (7/5)\times 2^{n+6} - 131/10$.

8. $I(G) = (497/15) \times 2^{n+1} - 154/5$.
9. $A(G) = 475 \times 2^n - 427/2$.

Theorem 6. *For the PAMAM dendrimers DS_1, we have*

$$M(DS_1; x, y) = 4 \cdot 3^n xy^4 + 10(3^n - 1)x^2y^2 + 4(3^n - 1)x^2y^4. \quad (31)$$

Proof. Let DS_1 be kinds of PAMAM dendrimers with n growth stages.

The edge set of DS_1 has the following three partitions:

$$\begin{aligned} E_{\{1,4\}} &= \{e = uv \in E(DS_1) \mid d_u = 1, d_v = 4\}, \\ E_{\{2,2\}} &= \{e = uv \in E(DS_1) \mid d_u = 2, d_v = 2\}, \\ E_{\{2,4\}} &= \{e = uv \in E(DS_1) \mid d_u = 2, d_v = 4\}. \end{aligned} \quad (32)$$

Now

$$\begin{aligned} |E_{\{1,4\}}| &= 4 \cdot 3^n, \\ |E_{\{2,2\}}| &= 10 \cdot 3^n - 10, \end{aligned} \quad (33)$$

and

$$|E_{\{2,4\}}| = 4 \cdot 3^n - 4.$$

$$\begin{aligned} M(DS_1; x, y) &= \sum_{i \le j} m_{ij}(DS_1)x^iy^j \\ &= \sum_{1\le 4} m_{14}(DS_1)xy^4 + \sum_{2\le 2} m_{22}(DS_1)x^2y^2 + \sum_{2\le 4} m_{24}(DS_1)x^2y^4 \\ &= \sum_{uv\in E_{\{1,4\}}} m_{14}(DS_1)xy^4 + \sum_{uv\in E_{\{2,2\}}} m_{22}(DS_1)x^2y^2 + \sum_{uv\in E_{\{2,4\}}} m_{24}(DS_1)x^2y^4 \\ &= |E_{\{1,4\}}|xy^4 + |E_{\{2,2\}}|x^2y^2 + |E_{\{2,4\}}|x^2y^4 \\ &= 4 \cdot 3^n xy^4 + (10 \cdot 3^n - 10)x^2y^2 + (4 \cdot 3^n - 4)x^2y^4 \\ &= 4 \cdot 3^n xy^4 + 10(3^n - 1)x^2y^2 + 4(3^n - 1)x^2y^4. \end{aligned} \quad (34)$$

□

Theorem 7. *For the PAMAM dendrimers DS_1, we have*

1. $M_1(G) = 4(7 \times 3^{n+1} - 16)$.
2. $M_2(G) = 8(11 \times 3^n - 9)$.
3. ${}^mM_2(G) = 4 \times 3^n - 3$.
4. $R_\alpha(G) = 3^n \times 4^{\alpha+1} + (2^{\alpha+1} \times 5 + 2^{3\alpha+2})(3^n - 1)$.
5. $R_\alpha(G) = 3^n/4^{\alpha-1} + (5/2^{2\alpha-1} + 1/2^{3\alpha-2})(3^n - 1)$.
6. $SSD(G) = 47 \times 3^n - 30$
7. $H(G) = (119/15) \times 3^n - 19/5$
8. $I(G) = (278/15) \times 3^n - 46/3$.
9. $A(G) = (3280/27) \times 3^n - 112$.

3.2. M-Polynomials and Degree-Based Indices for Polyomino Chains. From the geometric point of view, a polyomino system is a finite 2-connected plane graph in which each interior cell is encircled by a regular square. In other words, it is an edge-connected union of cells in the planar square lattice. Polyomino chain is a particular polyomino system such that the joining of the centers (set ci as the center of the ith square) of its adjacent regular composes a path $c_1, c_2, c_3, \ldots c_n$.

Let B_n be the set of polyomino chains with n squares. There are *2n+1* edges in every $B_n \in \mathrm{B}_n$, where B_n is named as a linear chain and denoted by L_n if the subgraph of B_n induced by the vertices with *d(v)=3* is a molecular graph with exactly *n-2* squares. Also, B_n can be called a zigzag chain and labelled as Z_n if the subgraph of B_n is induced by the vertices with *d(v)>2* is P_n.

The angularly connected, or branched, squares constitute a link of a polyomino chain. A maximal linear chain (containing the terminal squares and kinks at its end) in the polyomino chains is called a segment of polyomino chain. Let *l(S)* be the length of *S* which is calculated by the number of squares in S. For any segment *S* of a polyomino chain, we get $l(S) \in \{2, 3, 4, \ldots, n\}$. Furthermore, we deduce $l_1 = n$ and *m=1* for a linear chain L_n with n squares and $l_i = 2$ and *m=n-1* for a zigzag chain Z_n with n squares.

In what follows, we always assume that a polyomino chain consists of a sequence of segments $S_1, S_2, S_3, \ldots S_n$ and $L(S_i) = l_i$, where $m \ge 1$ and $i \in \{2, 3, 4, \ldots, m\}$. We derive that $\sum_{i=1}^{m} l_i = n + m - 1$.

Theorem 8. *For a linear polyomino chain L_n, we have $M(L_n; x, y) = 2x^2y^2 + 4x^2y^3 + (3n - 5)x^3y^3$.*

Proof. Let L_n be the polyomino chain with n squares where $l_1 = n$ and *m=1*. L_n is called the linear chain.

The edge set of L_n has the following three partitions:

$$\begin{aligned} E_{\{2,2\}} &= \{e = uv \in E(L_n) \mid d_u = 2, d_v = 2\}, \\ E_{\{2,3\}} &= \{e = uv \in E(L_n) \mid d_u = 2, d_v = 3\}, \\ E_{\{3,3\}} &= \{e = uv \in E(L_n) \mid d_u = 3, d_v = 3\}. \end{aligned} \quad (35)$$

Now

$$|E_{\{2,2\}}| = 2, \quad (36)$$

and

$$\begin{aligned}|E_{\{2,3\}}| &= 4,\\ |E_{\{3,3\}}| &= 3n-5.\end{aligned} \tag{37}$$

$$\begin{aligned}M(L_n;x,y) &= \sum_{i\le j} m_{ij}(L_n)x^iy^j\\ &= \sum_{2\le 2} m_{22}(L_n)x^2y^2 + \sum_{2\le 3} m_{23}(L_n)x^2y^3\\ &\quad + \sum_{3\le 3} m_{33}(L_n)x^3y^3\\ &= \sum_{uv\in E_{\{2,2\}}} m_{22}(L_n)x^2y^2\\ &\quad + \sum_{uv\in E_{\{2,3\}}} m_{23}(L_n)x^2y^3\\ &\quad + \sum_{uv\in E_{\{3,3\}}} m_{33}(L_n)x^3y^3\\ &= |E_{\{2,2\}}|x^2y^2 + |E_{\{2,3\}}|x^2y^3\\ &\quad + |E_{\{3,3\}}|x^3y^3\\ &= 2x^2y^2 + 4x^2y^3 + (3n-5)x^3y^3.\end{aligned} \tag{38}$$

□

Theorem 9. *For a linear polyomino chain L_n, we have the following:*

1. $M_1(G) = 18n - 2$.
2. $M_2(G) = 27n - 13$.
3. ${}^mM_2(G) = (1/3)n - 11/18$.
4. $R_\alpha(G) = 2^{2\alpha+1} + 2^{\alpha+2}\cdot 3^\alpha + 3^{2\alpha}(3n-5)$.
5. $R_\alpha(G) = 1/2^{\alpha-1} + 2^{2-\alpha}/3^\alpha + (1/3^{2\alpha})(3n-5)$.
6. $SSD(G) = 6n + 8/3$.
7. $H(G) = n + 14/15$.
8. $I(G) = (9/2)n - 7/10$.
9. $A(G) = (2187/64)n - 573/64$.

Theorem 10. *Let Z_n be zigzag polyomino chain with n squares such that $l_i = 2$ and $m = n - 1$. Then*

$$\begin{aligned}M(Z_n,x,y) &= 2x^2y^2 + 4x^2y^3 + 2(m-1)x^2y^4\\ &\quad + 2x^3y^4 + (3n-2m-5)x^4y^4.\end{aligned} \tag{39}$$

Proof. Let Z_n be zigzag polyomino chain with n squares such that $l_i = 2$ and $m = n - 1$. Polyomino chain consists of a sequence of segments $S_1, S_2, \ldots S_m$ and $l(S_i) = l_i$ where $m \ge 1$ and $i \in \{1, 2, \ldots, m\}$.

The edge set of Z_n has the following five partitions:

$$\begin{aligned}E_{\{2,2\}} &= \{e = uv \in E(Z_n) \mid d_u = 2, d_v = 2\},\\ E_{\{2,3\}} &= \{e = uv \in E(Z_n) \mid d_u = 2, d_v = 3\},\\ E_{\{2,4\}} &= \{e = uv \in E(Z_n) \mid d_u = 2, d_v = 4\},\\ E_{\{3,4\}} &= \{e = uv \in E(Z_n) \mid d_u = 3, d_v = 4\},\\ E_{\{4,4\}} &= \{e = uv \in E(Z_n) \mid d_u = 4, d_v = 4\}.\end{aligned} \tag{40}$$

Now

$$\begin{aligned}|E_{\{2,2\}}| &= 2,\\ |E_{\{2,3\}}| &= 4,\\ |E_{\{2,4\}}| &= 2(m-1),\\ |E_{\{3,4\}}| &= 2,\end{aligned} \tag{41}$$

and

$$|E_{\{4,4\}}| = 3n - 2m - 5.$$

$$\begin{aligned}M(Z_n;x,y) &= \sum_{i\le j} m_{ij}(Z_n)x^iy^j\\ &= \sum_{2\le 2} m_{22}(Z_n)x^2y^2 + \sum_{2\le 3} m_{23}(Z_n)x^2y^3\\ &\quad + \sum_{2\le 4} m_{24}(Z_n)x^2y^4\\ &\quad + \sum_{3\le 4} m_{34}(Z_n)x^3y^4\\ &\quad + \sum_{4\le 4} m_{44}(Z_n)x^4y^4\\ &= \sum_{uv\in E_{\{2,2\}}} m_{22}(Z_n)x^2y^2\\ &\quad + \sum_{uv\in E_{\{2,3\}}} m_{23}(Z_n)x^2y^3\\ &\quad + \sum_{uv\in E_{\{2,4\}}} m_{24}(Z_n)x^2y^4\\ &\quad + \sum_{uv\in E_{\{3,4\}}} m_{34}(Z_n)x^3y^4\\ &\quad + \sum_{uv\in E_{\{4,4\}}} m_{44}(Z_n)x^4y^4\\ &= |E_{\{2,2\}}|x^2y^2 + |E_{\{2,3\}}|x^2y^3\\ &\quad + |E_{\{2,4\}}|x^2y^4 + |E_{\{3,4\}}|x^3y^4\\ &\quad + |E_{\{4,4\}}|x^4y^4\\ &= 2x^2y^2 + 4x^2y^3 + 2(m-1)x^2y^4\\ &\quad + 2x^3y^4 + (3n-2m-5)x^4y^4.\end{aligned} \tag{42}$$

□

Theorem 11. *For the Zigzag polyomino chain Z_n for $n \geq 2$, we have the following:*

1. $M_1(G) = 24n - 4m - 10$.
2. $M_2(G) = 48n - 16m - 40$.
3. ${}^mM_2(G) = (1/2)n + (1/3)m + 37/18$
4. $R_\alpha(G) = 2^{2\alpha+1} + 2^{\alpha+2} \times 3^\alpha + 2^{3\alpha+1}(m-1) + 3^\alpha \times 2^{2\alpha+1} + 2^{4\alpha}(3n - 2m - 5)$.
5. $R_\alpha(G) = 1/2^{2\alpha-1} + 1/(3^\alpha \cdot 2^{\alpha-2}) + (1/2^{3\alpha-1})(m - 1) + 1/(2^{2\alpha-1} \times 3^\alpha) + (1/2^{4\alpha})(3n - 2m - 5)$.
6. $SSD(G) = 6n + m + 11/6$
7. $H(G) = (3/4)n + (1/6)m + 61/60$.
8. $I(G) = (3/5)n - (4/3)m - 92/35$.
9. $A(G) = (512/9)n - (592/27)m - 118688/3375$.

Theorem 12. *For the polyomino chain with n squares and of m segments S_1 and S_2 satisfying $l_1 = 2$ and $l_2 = n - 1$, B_n^1 $(n \geq 3)$, we have*

$$M\left(B_n^1, x, y\right) = 2x^2y^2 + 5x^2y^3 + x^2y^4 + (3n-10)\,x^3y^3 + 3x^3y^4. \tag{43}$$

Proof. Let B_n^1 $(n \geq 3)$ be the polyomino chain with n squares and of m segments S_1 and S_2 satisfying $l_1 = 2$ and $l_2 = n - 1$. The edge set of B_n^1 $(n \geq 3)$ has the following five partitions:

$$\begin{aligned}
E_{\{2,2\}} &= \left\{e = uv \in E\left(B_n^1\right) \mid d_u = 2, d_v = 2\right\}, \\
E_{\{2,3\}} &= \left\{e = uv \in E\left(B_n^1\right) \mid d_u = 2, d_v = 3\right\}, \\
E_{\{2,4\}} &= \left\{e = uv \in E\left(B_n^1\right) \mid d_u = 2, d_v = 4\right\}, \\
E_{\{3,3\}} &= \left\{e = uv \in E\left(B_n^1\right) \mid d_u = 3, d_v = 3\right\}, \\
E_{\{3,4\}} &= \left\{e = uv \in E\left(B_n^1\right) \mid d_u = 3, d_v = 4\right\}.
\end{aligned} \tag{44}$$

Now

$$\begin{aligned}
\left|E_{\{2,2\}}\right| &= 2, \\
\left|E_{\{2,3\}}\right| &= 5, \\
\left|E_{\{2,4\}}\right| &= 1, \\
\left|E_{\{3,3\}}\right| &= 3n - 10,
\end{aligned} \tag{45}$$

and

$$\left|E_{\{3,4\}}\right| = 3. \tag{46}$$

$$\begin{aligned}
M\left(B_n^1; x, y\right) &= \sum_{i \leq j} m_{ij}\left(B_n^1\right) x^i y^j \\
&= \sum_{2\leq 2} m_{22}\left(B_n^1\right) x^2y^2 + \sum_{2\leq 3} m_{23}\left(B_n^1\right) x^2y^3 \\
&\quad + \sum_{2\leq 4} m_{24}\left(B_n^1\right) x^2y^4 \\
&\quad + \sum_{3\leq 3} m_{33}\left(B_n^1\right) x^3y^3 \\
&\quad + \sum_{3\leq 4} m_{34}\left(B_n^1\right) x^3y^4 \\
&= \sum_{uv \in E_{\{2,2\}}} m_{22}\left(B_n^1\right) x^2y^2 \\
&\quad + \sum_{uv \in E_{\{2,3\}}} m_{23}\left(B_n^1\right) x^2y^3 \\
&\quad + \sum_{uv \in E_{\{2,4\}}} m_{24}\left(B_n^1\right) x^2y^4 \\
&\quad + \sum_{uv \in E_{\{3,3\}}} m_{33}\left(B_n^1\right) x^3y^3 \\
&\quad + \sum_{uv \in E_{\{3,4\}}} m_{34}\left(B_n^1\right) x^3y^4 \\
&= \left|E_{\{2,2\}}\right| x^2y^2 + \left|E_{\{2,3\}}\right| x^2y^3 \\
&\quad + \left|E_{\{2,4\}}\right| x^2y^4 + \left|E_{\{3,3\}}\right| x^3y^3 \\
&\quad + \left|E_{\{3,4\}}\right| x^3y^4 \\
&= 2x^2y^2 + 5x^2y^3 + x^2y^4 \\
&\quad + (3n-10)\,x^3y^3 + 3x^3y^4.
\end{aligned} \tag{47}$$

□

Theorem 13. *For the polyomino chain with n squares and of m segments S_1 and S_2 satisfying $l_1 = 2$ and $l_2 = n - 1$, B_n^1 $(n \geq 3)$, we have the following:*

1. $M_1(G) = 18n$.
2. $M_2(G) = 27n - 8$.
3. ${}^mM_2(G) = n/3 + 43/72$.
4. $R_\alpha(G) = 2^{2\alpha+1} + 6^\alpha \times 5 + 2^{3\alpha} + 3^{2\alpha}(3n-10) + 3^{\alpha+1} \times 4^\alpha$.
5. $R_\alpha(G) = 1/2^{2\alpha-1} + 5/6^\alpha + 1/2^{3\alpha} + (1/3^{2\alpha})(3n - 10) + 1/(2^{2\alpha} \times 3^{\alpha-1})$.
6. $SSD(G) = 6n + 43/12$.
7. $H(G) = n + 6/7$.
8. $I(G) = 27n - 709/10$.
9. $A(G) = (2187/64)n - 33737/4000$.

Theorem 14. *For polyomino chain with n squares and m segments* $S_1, S_2, \ldots, S_m$ $(m \geq 3)$ *satisfying* $l_1 = l_m = 2$ *and* $l_2, \ldots, l_{m-1} \geq 3$, B_n^2 $(n \geq 4)$, *we have*

$$M\left(B_n^2, x, y\right) = 2x^2y^2 + 2mx^2y^3 + 2x^2y^4 + 3(n - 2m + 1)x^3y^3 + 2(2m - 3)x^3y^4. \quad (48)$$

Proof. Let B_n^2 $(n \geq 4)$ be a polyomino chain with n squares and m segments $S_1, S_2, \ldots, S_m$ $(m \geq 3)$ satisfying $l_1 = l_m = 2$ and $l_2, \ldots, l_{m-1} \geq 3$. Then the edge set of B_n^2 $(n \geq 4)$ has the following five partitions:

$$\begin{aligned} E_{\{2,2\}} &= \left\{e = uv \in E\left(B_n^2\right) \mid d_u = 2, d_v = 2\right\}, \\ E_{\{2,3\}} &= n\left\{e = uv \in E\left(B_n^2\right) \mid d_u = 2, d_v = 3\right\}, \\ E_{\{2,4\}} &= \left\{e = uv \in E\left(B_n^2\right) \mid d_u = 2, d_v = 4\right\}, \\ E_{\{3,3\}} &= \left\{e = uv \in E\left(B_n^2\right) \mid d_u = 3, d_v = 3\right\}, \\ E_{\{3,4\}} &= \left\{e = uv \in E\left(B_n^2\right) \mid d_u = 3, d_v = 4\right\}. \end{aligned} \quad (49)$$

Now

$$\begin{aligned} \left|E_{\{2,2\}}\right| &= 2, \\ \left|E_{\{2,3\}}\right| &= 2m, \\ \left|E_{\{2,4\}}\right| &= 2, \\ \left|E_{\{3,3\}}\right| &= 3n - 6m + 3, \end{aligned} \quad (50)$$

and

$$\left|E_{\{4,4\}}\right| = 4m - 6. \quad (51)$$

$$\begin{aligned} M\left(B_n^2; x, y\right) &= \sum_{i \leq j} m_{ij}\left(B_n^2\right)x^iy^j \\ &= \sum_{2 \leq 2} m_{22}\left(B_n^2\right)x^2y^2 + \sum_{2 \leq 3} m_{23}\left(B_n^2\right)x^2y^3 \\ &\quad + \sum_{2 \leq 4} m_{24}\left(B_n^2\right)x^2y^4 + \sum_{3 \leq 3} m_{33}\left(B_n^2\right)x^3y^3 \\ &\quad + \sum_{3 \leq 4} m_{34}\left(B_n^2\right)x^3y^4 \\ &= \sum_{uv \in E_{\{2,2\}}} m_{22}\left(B_n^2\right)x^2y^2 + \sum_{uv \in E_{\{2,3\}}} m_{23}\left(B_n^2\right)x^2y^3 \\ &\quad + \sum_{uv \in E_{\{2,4\}}} m_{24}\left(B_n^2\right)x^2y^4 \sum_{uv \in E_{\{3,3\}}} m_{33}\left(B_n^2\right)x^3y^3 \\ &\quad + \sum_{uv \in E_{\{3,4\}}} m_{34}\left(B_n^2\right)x^3y^4 \\ &= \left|E_{\{2,2\}}\right|x^2y^2 + \left|E_{\{2,3\}}\right|x^2y^3 + \left|E_{\{2,4\}}\right|x^2y^4 \\ &\quad + \left|E_{\{3,3\}}\right|x^3y^3 + \left|E_{\{3,4\}}\right|x^3y^4 \\ &= 2x^2y^2 + 2mx^2y^3 + 2x^2y^4 + (3n - 6m + 3)x^3y^3 \\ &\quad + (4m - 6)x^3y^4 \\ &= 2x^2y^2 + 2mx^2y^3 + 2x^2y^4 + 3(n - 2m + 1)x^3y^3 \\ &\quad + 2(2m - 3)x^3y^4 \end{aligned} \quad (52)$$

□

Theorem 15. *For polyomino chain with n squares and m segments* $S_1, S_2, \ldots, S_m$ $(m \geq 3)$ *satisfying* $l_1 = l_m = 2$ *and* $l_2, \ldots, l_{m-1} \geq 3$, B_n^2 $(n \geq 4)$, *we have the following:*

1. $M_1(G) = 2(9n + m - 2)$.
2. $M_2(G) = 3(9n + 2m - 7)$.
3. ${}^mM_2(G) = (1/3)n + 7/12$
4. $R_\alpha(G) = 3^{2\alpha+1}n + (3^\alpha \times 2^{\alpha+1} - 2 \times 3^{2\alpha+1} + 3^\alpha \times 2^{2\alpha+2})m + (1 + 3^{\alpha+1})2^{2\alpha+1} + 2^{3\alpha+1} + 3^{2\alpha+1}$.
5. $R_\alpha(G) = (1/3^{2\alpha-1})n + (1/(3^\alpha \cdot 2^{\alpha-1}) + 2/3^{2\alpha-1} + 1/(2^{2\alpha-2} \times 3^\alpha))m + (1 - 1/3^{\alpha-1})(1/2^{2\alpha-1}) + 1/2^{3\alpha-1} + 1/3^{2\alpha-1}$.
6. $SSD(G) = 6n + (2/3)m + 5/2$
7. $H(G) = n - (2/35)m + 20/21$.
8. $I(G) = (9/2)n + (9/35)m - 47/42$.
9. $A(G) = (2187/64)n + (11809/4000)m - 134177/8000$.

4. Conclusions

Topological indices calculated in this paper help us to guess biological activities, chemical reactivity, and physical features of under-study dendrimers and polyomino chains. For example, Randić index is useful for determining physiochemical properties of alkanes as noticed by chemist Milan Randić in 1975. He noticed the correlation between the Randić index and several physicochemical properties of alkanes like boiling point, vapor pressure, enthalpies of formation, surface area, and chromatographic retention times. Hence our results are helpful in determination of the significance of PAMAM dendrimers and polyomino chains in pharmacy and industry.

Authors' Contributions

All authors contributed equally to this paper.

Acknowledgments

The first author is funded by Dong-A University, Korea.

References

[1] T. Go, *Handbook of discrete and computational geometry*, J. E. Goodman and J. O'Rourke, Eds., CRC Press Series on Discrete Mathematics and its Applications, CRC Press, Boca Raton, FL, 1997.

[2] S. W. Golomb, *Polyominoes, Puzzles, Patterns, Problems, And Packings*, Princeton University Press, 1996.

[3] L. Alonso and R. Cerf, "The three dimensional polyominoes of minimal area," *Electronic Journal of Combinatorics*, vol. 3, no. 1 R, pp. 1–39, 1996.

[4] D. Astruc, E. Boisselier, and C. Ornelas, "Dendrimers designed for functions: from physical, photophysical, and supramolecular properties to applications in sensing, catalysis, molecular electronics, photonics, and nanomedicine," *Chemical Reviews*, vol. 110, no. 4, pp. 1857–1959, 2010.

[5] F. Vögtle, G. Richardt, and N. Werner, *Dendrimer Chemistry Concepts, Syntheses, Properties, Applications*, 2009.

[6] M. R. Farahani, W. Gao, and M. R. Kanna, "The connective eccentric index for an infinite family of dendrimers," *Journal of Fundamental and Applied Life Sciences*, vol. 5, no. S4, pp. 766–771, 2015.

[7] W. Gao, L. Shi, and M. R. Farahani, "Distance-based indices for some families of dendrimer nanostars," *IAENG International Journal of Applied Mathematics*, vol. 46, no. 2, pp. 168–186, 2016.

[8] W. Gao and W. Wang, "Degree-based indices of polyhex nanotubes and dendrimer nanostar," *Journal of Computational and Theoretical Nanoscience*, vol. 13, no. 3, pp. 1577–1583, 2016.

[9] W. Gao and M. R. Farahani, "The hyper-zagreb index for an infinite family of nanostar dendrimer," *Journal of Discrete Mathematical Sciences & Cryptography*, vol. 20, no. 2, pp. 515–523, 2017.

[10] W. Gao, M. R. Farahani, and L. Shi, "The forgotten topological index of some drug structures," *Acta Medica Mediterranea*, vol. 32, no. 1, pp. 579–585, 2016.

[11] W. Gao, Y. Wang, W. Wang, and L. Shi, "The first multiplication atom-bond connectivity index of molecular structures in drugs," *Saudi Pharmaceutical Journal*, vol. 25, no. 4, pp. 548–555, 2017.

[12] W. Gao, M. Younas, A. Farooq et al., "M-Polynomials and Degree-Based Topological Indices of the Crystallographic Structure of Molecules," *Biomolecules*, vol. 8, no. 4, p. 107, 2018.

[13] H. Zhang and F. Zhang, "The Clar covering polynomial of hexagonal systems," *Discrete Applied Mathematics*, vol. 69, no. 1-2, pp. 147–167, 1996.

[14] H. Wiener, "Structural determination of paraffin boiling points," *Journal of the American Chemical Society*, vol. 69, no. 1, pp. 17–20, 1947.

[15] E. Deutsch and S. Klavžar, "M-Polynomial and degree-based topological indices," *Iranian Journal of Mathematical Chemistry*, vol. 6, pp. 93–102, 2015.

[16] M. Munir, W. Nazeer, S. Rafique, and S. M. Kang, "M-polynomial and related topological indices of nanostar dendrimers," *Symmetry*, vol. 8, no. 9, p. 97, 2016.

[17] M. Munir, W. Nazeer, A. R. Nizami, S. Rafique, and S. M. Kang, "M-polynomials and topological indices of titania nanotubes," *Symmetry*, vol. 8, no. 11, p. 117, 2016.

[18] Y. C. Kwun, M. Munir, W. Nazeer, S. Rafique, and S. Min Kang, "M-Polynomials and topological indices of V-Phenylenic Nanotubes and Nanotori," *Scientific Reports*, vol. 7, no. 1, 2017.

[19] M. Munir, W. Nazeer, S. Rafique, A. R. Nizami, and S. M. Kang, "Some computational aspects of boron triangular nanotubes," *Symmetry*, vol. 9, no. 6, 2017.

[20] M. Munir, W. Nazeer, S. Rafique, and S. M. Kang, "M-polynomial and degree-based topological indices of polyhex nanotubes," *Symmetry*, vol. 8, no. 12, p. 149, 2016.

[21] A. A. Dobrynin, R. Entringer, and I. Gutman, "Wiener index of trees: theory and applications," *Acta Applicandae Mathematicae*, vol. 66, no. 3, pp. 211–249, 2001.

[22] B. Mohar and T. Pisanski, "How to compute the Wiener index of a graph," *Journal of Mathematical Chemistry*, vol. 2, no. 3, pp. 267–277, 1988.

[23] B. Zhou and I. Gutman, "Relations between Wiener, hyper-Wiener and Zagreb indices," *Chemical Physics Letters*, vol. 394, no. 1–3, pp. 93–95, 2004.

[24] M. Randić, "On characterization of molecular branching," *Journal of the American Chemical Society*, vol. 97, no. 23, pp. 6609–6615, 1975.

[25] S. M. Kang, M. A. Zahid, W. Nazeer, and W. Gao, "Calculating the Degree-based Topological Indices of Dendrimers," *Open Chemistry*, vol. 16, no. 1, pp. 681–688, 2018.

[26] B. Bollobás and P. Erdös, "Graphs of extremal weights," *Ars Combinatoria*, vol. 50, pp. 225–233, 1998.

[27] D. Amić, D. Bešlo, B. Lučić, S. Nikolić, and N. Trinajstić, "The Vertex-Connectivity Index Revisited," *Journal of Chemical Information and Computer Sciences*, vol. 38, no. 5, pp. 819–822, 1998.

[28] Y. Hu, X. Li, Y. Shi, T. Xu, and I. Gutman, "On molecular graphs with smallest and greatest zeroth-order general randić index," *MATCH - Communications in Mathematical and in Computer Chemistry*, vol. 54, no. 2, pp. 425–434, 2005.

[29] G. Caporossi, I. Gutman, P. Hansen, and L. Pavlović, "Graphs with maximum connectivity index," *Computational Biology and Chemistry*, vol. 27, no. 1, pp. 85–90, 2003.

[30] I. Gutman and K. C. Das, "The first Zagreb index 30 years after," *MATCH Communications in Mathematical and in Computer Chemistry*, vol. 50, no. 1, pp. 83–92, 2004.

[31] K. C. Das and I. Gutman, "Some properties of the second Zagreb index," *MATCH Communications in Mathematical and in Computer Chemistry*, vol. 52, no. 1, pp. 1–3, 2004.

[32] I. Gutman, "An exceptional property of first Zagreb index," *MATCH - Communications in Mathematical and in Computer Chemistry*, vol. 72, no. 3, pp. 733–740, 2014.

[33] Y. Huang, B. Liu, and L. Gan, "Augmented Zagreb index of connected graphs," *MATCH - Communications in Mathematical and in Computer Chemistry*, vol. 67, no. 2, pp. 483–494, 2012.

[34] W. Nazeer, A. Farooq, M. Younas, M. Munir, and S. Kang, "On Molecular Descriptors of Carbon Nanocones," *Biomolecules*, vol. 8, no. 3, p. 92, 2018.

[35] Y. Kwun, A. Virk, W. Nazeer, M. Rehman, and S. Kang, "On the Multiplicative Degree-Based Topological Indices of Silicon-Carbon Si2C3-I [p, q] and Si2C3-II [p, q]," *Symmetry*, vol. 10, no. 8, p. 320, 2018.

[36] S. Kang, Z. Iqbal, M. Ishaq, R. Sarfraz, A. Aslam, and W. Nazeer, "On Eccentricity-Based Topological Indices and Polynomials of Phosphorus-Containing Dendrimers," *Symmetry*, vol. 10, no. 7, p. 237, 2018.

[37] Y. C. Kwun, A. Ali, W. Nazeer, M. Ahmad Chaudhary, and S. M. Kang, "M-Polynomials and Degree-Based Topological Indices of Triangular, Hourglass, and Jagged-Rectangle Benzenoid Systems," *Journal of Chemistry*, vol. 2018, Article ID 8213950, 8 pages, 2018.

[38] J.-B. Liu, X.-F. Pan, L. Yu, and D. Li, "Complete characterization of bicyclic graphs with minimal Kirchhoff index," *Discrete Applied Mathematics*, vol. 200, pp. 95–107, 2016.

[39] J.-B. Liu, W.-R. Wang, Y.-M. Zhang, and X.-F. Pan, "On degree resistance distance of cacti," *Discrete Applied Mathematics: The Journal of Combinatorial Algorithms, Informatics and Computational Sciences*, vol. 203, pp. 217–225, 2016.

[40] J.-B. Liu and X.-F. Pan, "Minimizing Kirchhoff index among graphs with a given vertex bipartiteness," *Applied Mathematics and Computation*, vol. 291, pp. 84–88, 2016.

[41] W. Gao, M. K. Jamil, A. Javed, M. R. Farahani, S. Wang, and J.-B. Liu, "Sharp Bounds of the Hyper-Zagreb Index on Acyclic, Unicylic, and Bicyclic Graphs," *Discrete Dynamics in Nature and Society*, vol. 2017, Article ID 6079450, 5 pages, 2017.

[42] J.-B. Liu, W. Gao, M. K. Siddiqui, and M. R. Farahani, "Computing three topological indices for Titania nanotubes TiO2[m,n]," *AKCE International Journal of Graphs and Combinatorics*, vol. 13, no. 3, pp. 255–260, 2016.

[43] Q. Liu, J.-B. Liu, and J. Cao, "The Laplacian polynomial and Kirchhoff index of graphs based on R-graphs," *Neurocomputing*, vol. 177, pp. 441–446, 2016.

[44] J.-B. Liu, S. Wang, C. Wang, and S. Hayat, "Further results on computation of topological indices of certain networks," *IET Control Theory & Applications*, vol. 11, no. 13, pp. 2065–2071, 2017.

[45] Y. Huo, J.-B. Liu, A. Q. Baig, W. Sajjad, and M. R. Farahani, "Connective eccentric index of NAnm manotube," *Journal of Computational and Theoretical Nanoscience*, vol. 14, no. 4, pp. 1832–1836, 2017.

7

Preparation of Ampicillin Surface Molecularly Imprinted Polymers for its Selective Recognition of Ampicillin in Eggs Samples

Yang Tian, Yue Wang, Shanshan Wu, Zhian Sun, and Bolin Gong

College of Chemistry and Chemical Engineering, North Minzu University, Yinchuan, 750021, China

Correspondence should be addressed to Bolin Gong; gongbolin@163.com

Academic Editor: Neil D. Danielson

Surface-imprinted polymers (MIPs) microspheres with the ability to specifically recognize water-soluble molecules were prepared using self-made monodisperse porous poly(chloromethylstyrene-co-divinylbenzene) beads as the solid-phase matrix and ampicillin (AMP) as the template molecule. MIPs were synthesized using different template molecule: monomer: crosslinker ratios and the optimum preparation ratio were obtained by measuring adsorption. The maximum equilibrium amount of adsorption by the MIPs reached 115.62 mg/g. Scatchard analysis indicated that the MIPs contained two types of recognition sites: specific and nonspecific. Based on the adsorption kinetics, adsorption equilibrium was reached after 30 minutes. Penicillin G, amoxicillin, and sulbactam acid were used as competitive molecules to research the selective adsorption capacity of the MIPs. The imprinted material was found to have good selectivity with selectivity coefficients for penicillin G, amoxicillin, and sulbactam acid of 5.74, 6.83, and 7.25, respectively. The MIPs were used as solid-phase extraction filler, resulting in successful enrichment and separation of ampicillin residue from egg samples. Standard addition recovery experiments revealed that recovery was good with recoveries from the spiked samples ranging from 91.5 to 94.9% and relative standard deviations from 3.6 to 4.2%. The solid-phase extraction MIPs microcolumn was reused 10 times, where it maintained a recovery rate of over 80%. This work presents a sensitive, fast, and convenient method for the determination of trace ampicillin in food samples.

1. Introduction

Ampicillin (AMP) is a broad-spectrum penicillin used to treat infectious diseases, such as respiratory system and intestinal infections and endocarditis [1, 2]. However, excessive use of AMP can result in varying levels of residues being present in animal-based foods, which can be detrimental to consumers [3]. Current methods of detecting AMP involve high-performance liquid chromatography (HPLC) [4–6], gas chromatography-mass spectrometry (GC-MS) [7, 8], liquid chromatography-tandem mass spectrometry (LC-MS/MS) [9], and surface-enhanced Raman spectroscopy (SERS) [10–12]. However, the majority of these detection methods are used for qualitative analysis and there are few methods available for efficient quantitative analysis of trace residues [13].

Compared with other methods, molecular imprinting technology (MIPs) can produce a polymer with specific recognition ability to quickly and efficiently separate the target substance. Molecularly imprinted polymers microspheres can simulate antigen recognition by antibodies using molecular recognition holes that match the corresponding template molecules in shape, size, and functional groups [11–14]. MIPs generated using molecular imprinting technology have high selectivity and affinity for specific analytes or a group of structurally relevant compounds [15]. Conventional MIPs are synthesized by bulk polymerization, suspension polymerization, and precipitation polymerization [16–18]. However, in the preparation process, due to the use of a large amount of organic solvent, the polymerization reaction is slow (usually the preparation cycle is long), and the formed polymer has a narrow particle size distribution, which is not conducive to mass transfer. In addition, the recognition sites of the MIPs prepared by the conventional method are mostly located inside the polymer, and when the template molecules are eluted, they cannot be completely eluted, and some residues remain. As a result, the binding site is reduced and the adsorption capacity is lowered [19–22]. The emergence of surface molecular imprinting technology is expected to solve the above problems.

The surface molecular imprinting technique can be used to prepare MIPs with recognition sites on the surface. This method can effectively overcome the "embedding" defects that occur in traditional MIPs and increase the mass transfer rate [23, 24]. Surface-initiated atom transfer radical polymerization (SI-ATRP) is a free radical-polymerization technology based on molecular self-assembly that is used to prepare high-density controllable polymer brushes on the surfaces of solid matrixes [25]. This method can effectively improve imprinting efficiency, molecular recognition by the imprinted polymer, specific recognition ability, and mass transfer rate [26–28]. Concurrently, due to the involvement of water-soluble substances, MIPs preparation has always been difficult. It is important to be able prepare MIPs for water-soluble drugs and while the traditional methods of preparing MIPs have issues overcoming water phase recognition, SI-ATRP technology can effectively overcome this problem. Christian et al. [29] prepared AMP-specific recognition material using a complex process because the grinding involved in previously described techniques will destroy the recognition sites and affect imprinting. Mao et al. [30] prepared a magnetic carbon microsphere surface molecularly imprinted adsorbent by solvothermal method to identify and selectively adsorb ampicillin in milk, but the preparation process was complicated. Li et al. [31] prepared magnetic surface-imprinted material with good selectivity towards cephalexin using SI-ATRP, which was applied to detect trace residues in tap water and milk samples and displayed good enrichment.

Chloromethyl styrene resin is a functional material with a special purpose. These resin particles have good single dispersibility, are chemically stable, are easy to render hydrophilic, and have a large specific surface area [32–34]. Monodisperse chlorinated methyl styrene is a porous, highly reactive benzyl chloride that can be used as a trigger in the SI-ATRP reaction, simplifying surface modification of the matrix [35–37]. A combination of surface imprinting of monodisperse porous poly(chloromethylstyrene-co-divinylbenzene) ($P_{VBC\text{-}DVB}$) microspheres and solid phase extraction (SPE) can yield good specificity and recognition in a small column and, thus, high selectivity and separation efficiency, making it one of the most promising current technologies.

In this study, surface-imprinted MIPs that can identify water-soluble AMP molecules were prepared using SI-ATRP. MIPs was characterized by elemental analysis, thermogravimetric analysis (TGA), Fourier transform infrared spectroscopy (FT-IR), and scanning electron microscopy (SEM). The MIPs were subsequently used as an SPE adsorbent in combination with HPLC to detect trace amounts of AMP in egg samples.

2. Materials and Methods

2.1. Reagent and Instrument. AMP (99% purity), styrene (chemically pure), 4-vinylbenzyl chloride (VBC, 90% purity), azodiisobutyronitrile (analytical grade), polyvinyl alcohol (analytical grade), sodium dodecyl sulfate (SDS, analytical grade), polyvinylpyrrolidone (analytical grade), acrylamide (AM), ethylene glycol dimethacrylate (EDMA, 99% purity), α-bromoisobutyryl bromide, cuprous bromide (CuBr), and 2,2′-dipyridyl (Bpy) were purchased from the Aladdin reagent company (Shanghai, China). Toluene, methanol, glacial acetic acid, and tetrahydrofuran were obtained from the Tianjin Damao Chemical Reagent Factory (Tianjin, China). Penicillin G (99% purity), amoxicillin (98% purity), and sulbactam (99% purity) were purchased from Dalian Meilun reagent. Egg samples were purchased from local supermarkets.

EDMA was extracted using aqueous sodium hydroxide and distilled water and then dried with anhydrous calcium chloride. Toluene and tetrahydrofuran were distilled using sodium to remove any water. All other reagents were analytical grade.

All chromatographic tests were performed using an LC-20AT chromatographic system (Shimadzu, Japan) that included two LC-20AT pumps and an SPD-20A UV–VIS detector. Thermogravimetric analysis (TGA) was carried out using a Setsys Evolution (SETARAM, France). A scanning electron microscope was purchased from the JEOL company (JSM-7500F, Japan). FT-IR was performed on FTIR-8400S (Shimadzu, Japan). SEM was performed on JSM-7500F (JEOL, Japan). BET was performed on Brunner-Emmet-Teller measurements (NDVA-2000e, USA) An elemental analyzer was purchased from the YiLe Man element analysis system company (VarioEL III, Germany). A constant temperature water bath oscillator was purchased from the Shanghai pudong physical optics instrument plant (SHZ-C, Shanghai, China). A TG16-WS high-speed centrifuge (Centrifuge Factory, China) was used in this study and a TU-1810-type ultraviolet spectrophotometer was purchased from the Beijing general instrument Co., Ltd. (Beijing, China).

2.2. Preparation of Ampicillin Molecularly Imprinted Polymers

2.2.1. Preparation of Monodisperse $P_{VBC\text{-}DVB}$ Microspheres. Monodisperse $P_{VBC\text{-}DVB}$ microspheres were prepared by one-step seed swelling [38, 39]. The polystyrene seed was placed in a three-neck flask and an appropriate amount of SDS solution was added. This solution was stirred while incubating in a water bath. In a dry beaker containing azodiisobutyronitrile, VBC, and DVB, the ultrasound process was performed with dibutyl phthalate, toluene, 5% polyvinyl alcohol, 0.2% SDS, and distilled water. Following ultrasonic emulsification, the solution was rapidly transferred to a three-neck bottle, and swelling occurred at 25°C for 24 hours and then at 70°C in a nitrogen atmosphere for 24 hours. The product was extracted with a Soxhlet and the $P_{VBC\text{-}DVB}$ microspheres were obtained by drying with a vacuum.

2.2.2. Preparation of Ampicillin Molecularly Imprinted Polymer Microspheres. Template AMP (1 mmol) was placed in a 100 mL round-bottom flask with 40 mL water:acetonitrile (3:5, V/V) as solvent, 4 mmol AM as functional monomer, and 15 mmol EDMA as crosslinker and prepolymerized for 4 hours at 40°C to form a prepolymerized complex solution. $P_{VBC\text{-}DVB}$ microspheres (2 g) were placed in another 100 mL round-bottom flask and 0.1952 g Bpy and 0.0284 g CuBr were added to a self-made N_2 circulation device. After 30 minutes, the above prepolymerized composite solution was quickly transferred to a round-bottom flask containing $P_{VBC\text{-}DVB}$ microspheres. The polymerization reaction was performed at room temperature for 20 hours. After the reaction, the solution was washed with a large amount of deionized water and EDTA solution and then dried. The dried polymer was extracted with methanol: glacial acetic acid (9:1, V/V) and a Soxhlet for 24 hours followed by pure methanol for 8 hours and then dried at 50°C in a vacuum to obtain AMP MIPs microspheres.

The preparation of nonimprinted polymers (NIPs) was the same as above except that no template molecule was added.

2.3. Adsorption Experiment. The adsorption isotherm was determined by adding 10 mg of MIPs to 10 mL of acetonitrile 0.1 mol/L aqueous NaOH (9:1, V / V), wherein the AMP concentration was 1-10 mmol/L. The mixture was shaken at room temperature for 10 hours. The mixture was separated by centrifugation and the solution was filtered through a 0.45μm membrane. The concentration of the filtered solution was measured with a UV spectrophotometer. The same procedure was performed for the NIPs and all tests were performed in triplicate.

The adsorption capacity (Q) is calculated according to

$$Q = \frac{(C_0 - C_t)\,Vm}{M} \quad (1)$$

where C_0 (mmol/L) is the initial concentration of AMP, C_t (mmol/L) is the concentration of equilibrium AMP, Vm (mL) is the total volume of the adsorbed mixture, and M(g) is the mass of the MIPs.

The kinetics of AMP on MIPs was investigated at room temperature by the addition of 10 mg of MIPs in 10 mL of 10 mmol/L AMP solution. The mixture was shaken on a shaker for various times and centrifuged at 4000 rpm for 2 minutes and then filtered to perform UV analysis at 254 nm. Perform the same experimental steps for NIPs.

Penicillin G, amoxicillin, and sulbactam acid were chosen as competitors to estimate the selectivity of MIPs for AMP. Disperse 10 mg of MIPs in 10 mL of 0.1 mol/L NaOH-acetonitrile (1:9, V/V) solution containing Penicillin G, amoxicillin, and sulbactam acid at an initial concentration of 10 mmol/L. After adsorption, UV spectrometry A photometer measures the equilibrium concentration of each analyte.

The values of K_D and IF are the basic measures of imprinted polymers [40], which are calculated by (2) and (3). Larger K_D and IF values mean that the polymer has excellent adsorption affinity.

$$K_D = \frac{Q}{C_t} \quad (2)$$

$$IF = \frac{k_{MIPs}}{k_{NIPs}} \quad (3)$$

where Q (mmol / g) is the binding amount of MIPs and NIPs; Ct (mmol / L) is the equilibrium concentration of AMP; K_D (mL / g) is the partition coefficient; k_{MIPss} and K_{NIPs} are the partition coefficients of MIPss and NIPs. IF stands for the imprinting factor of MIPs.

2.4. Actual Sample Determination. Egg sample (50 g) was mixed with 250 mL of ethanol:water (6:4, V/V), incubated on an oscillator for extraction for 30 min, and then centrifuged. The resulting supernatant was mixed with ethanol:water (1:1, V/V) to a final volume of 100 mL and filtered with a 0.22 μm filter membrane. The filtrate was sealed and refrigerated at 3°C. The final concentrations were 5, 20, and 50 ng·g^{-1}.

For the MIPs-SPE [41], NIPs-SPE, and C_{18}-SPE, 1 mL of the above solution was used. AMP adsorbed on the cartridge was eluted using methanol:glacial acetic acid (8:2, v/v). The extract was blown dry with N_2 and the mobile phase was brought to a constant volume. The concentration of AMP was measured by HPLC and each addition level was measured 3 times.

The chromatographic conditions consisted of a Diamonsil C_{18} column (150mm × 4.6mm, 5 μm), a mobile phase of 40:60 (V/V) acetonitrile-water with 0.1% acetic acid, an injection volume of 10 μL, and a detection wavelength of 230 nm.

3. Results and Discussion

3.1. Preparation of MIPs. $P_{VBC\text{-}DVB}$ microspheres were prepared using the "one-step seed swelling method." The first step was to obtain micron-sized monodisperse styrene microspheres with a lower molecular weight by dispersing polymerization in an organic medium and then using the microspheres as seed liquid. Swelling polymerization was performed directly in the aqueous phase to obtain monodisperse $P_{VBC\text{-}DVB}$ microspheres. Microspheres have the advantages of good hydrophilicity and easy surface modification. When preparing MIPs using SI-ATRP technology, modification of the surface initiator group can be omitted and, thus, preparation can be simplified.

Preparation of AMP MIPs microspheres used CuBr/Bpy as a catalytic system, free radical-initiated polymerization in a water:acetonitrile mixed solution, and a free radical-initiated reaction. The prepolymerized complex formed by the functional monomer AM and imprinted molecule AMP, and EDMA crosslinking agent were grafted onto the surface of the $P_{VBC\text{-}DVB}$ microspheres. Then the template molecules were removed and AMP MIPs microspheres were obtained. The graft density and adsorption properties of

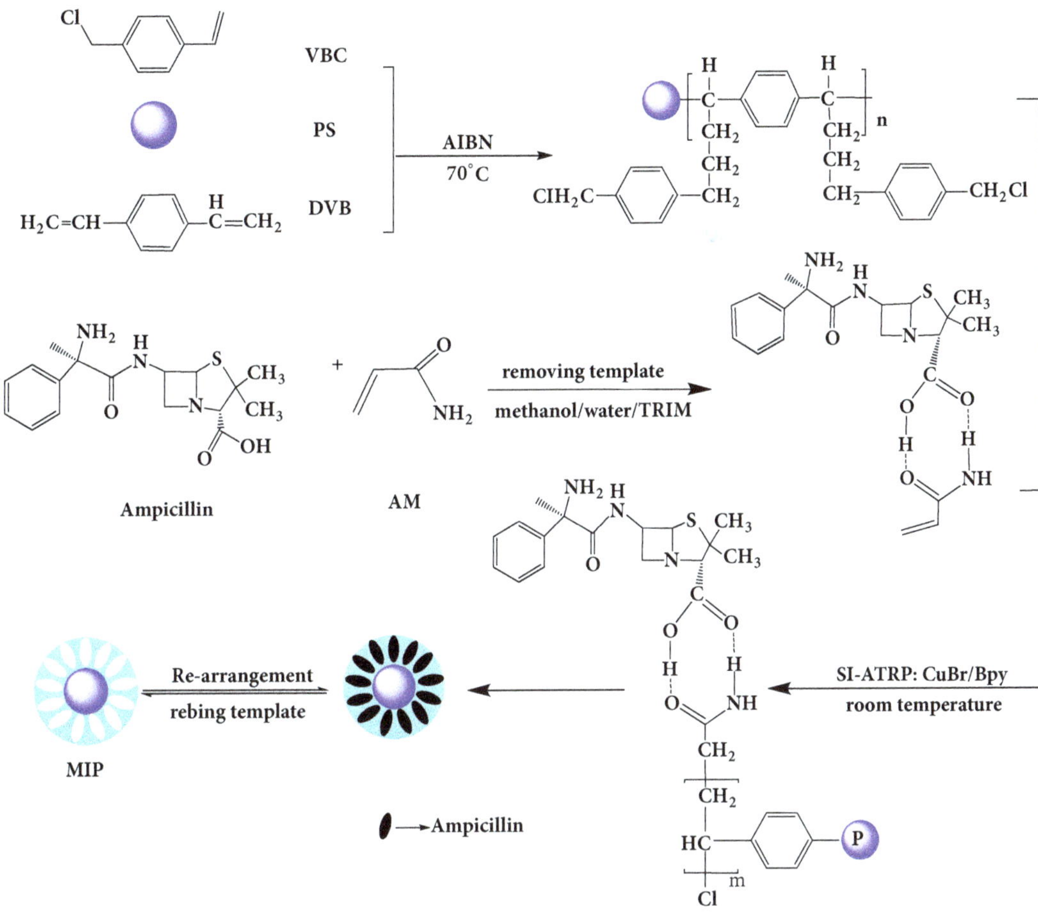

FIGURE 1: Synthetic route of surface-imprinted polymers (MIPs) microspheres.

TABLE 1: Elemental analysis of imprinted materials.

Material	Elemental composition (%,w/w)		
	C	N	H
$P_{VBC\text{-}DVB}$	77.96	0.543	6.951
MIPs	80.98	0.649	6.940

the imprinted materials were controlled by altering the monomer:template:crosslinker ratio. (Figure 1)

3.2. Characterization of MIPs

3.2.1. Elemental Analysis. The polymers of $P_{VBC\text{-}DVB}$ and MIPs were characterized by using elemental analysis. The elemental analysis data are listed in (Table 1). Compared with $P_{VBC\text{-}DVB}$, the contents of the C and H elements were significantly increased. This evidence indicates that the crosslinker EDMA has been successfully grafted to the $P_{VBC\text{-}DVB}$ surface.

3.2.2. Thermogravimetric Analysis. Thermogravimetric analysis results for MIPs and NIPs are shown in (Figure 2) The two polymers lose about 2.8% at 25°C to 110°C, and the main loss component is water. The polymer decomposes rapidly from

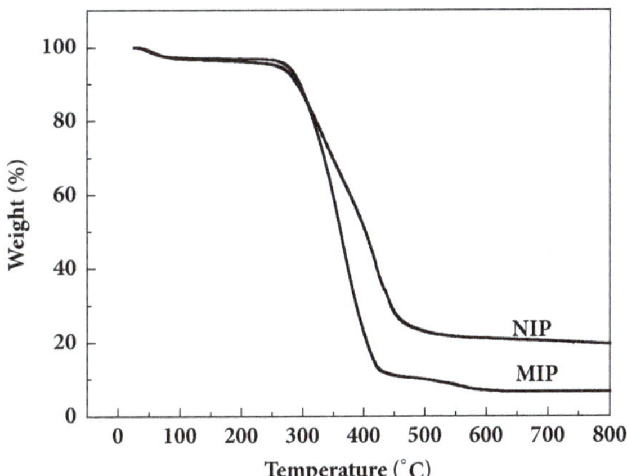

FIGURE 2: Thermogravimetric curves of $P_{VBC\text{-}DVB}$, MIP, and nonimprinted polymers (NIPs).

300°C to 500°C, and the MIPs decompose rapidly at 237°C to 470°C, and the weight loss is 77.8%. The NIPs decompose the fastest at 224°C to 494°C, and the weight loss is 76.88%. The difference may be due to the interaction of template molecules and functional monomers in MIPs affecting the weight loss characteristics of polymers.

Table 2: Comparison of MIPs and NIPs from nitrogen adsorption-desorption analysis.

Sample	Surface Area /($m^2 \cdot g^{-1}$)	Pore Volume /($cm^3 \cdot g^{-1}$)	Average Pore Size /(nm)
MIPs	95.26	0.247	20.76
NIPs	68.39	0.105	12.32

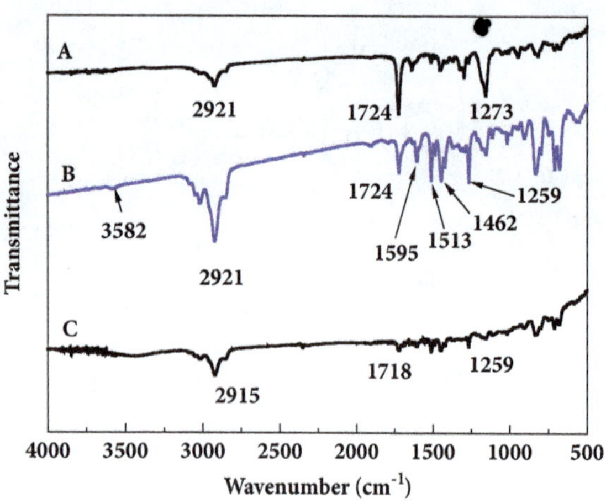

Figure 3: FT-IR spectra of (A) $P_{VBC-DVB}$ microspheres, (B) uneluted MIPs, and (C) eluted MIPs.

3.2.3. FT-IR Characterization of $P_{VBC-DVB}$ Microspheres and MIPs. It can be seen from the figure that the strong absorption peak at 1724 cm^{-1} and 1259 cm^{-1} is the stretching vibration of -C=O in the carboxyl group and the symmetric vibration of -C-O in the EDMA ester group. It is indicated that AM and EDMA successfully polymerize on the surface of polystyrene microspheres in the presence of initiator AIBA [42] (Figure 3(A)). The carbonyl -C=O vibration peak of the β-lactam ring of the AMP molecular structure appeared at 1770 cm^{-1} in (Figure 3(B)), and new peak positions appeared at 1595 cm^{-1}, 1513 cm^{-1}, and 1462 cm^{-1}. These peaks are a characteristic peak of the benzene ring in the AMP molecular structure, indicating the presence of AMP in the imprinted material [43]. These peaks are not seen in (Figure 3(C)), which is the elution of the imprinted material leaving only holes similar in structure to the template molecule and without the template molecule AMP.

3.2.4. Electron Microscopy Analysis and Physisorption Measurements. SEM was used to observe the morphological structure of $P_{VBC-DVB}$ microspheres (a), MIPs (b), and NIPs (c). As shown in (Figure 4(a)), the $P_{VBC-DVB}$ microspheres prepared by the "one-step seed swelling method" were uniform in size and uniform in dispersion and had good monodispersity. After imprinting, the surface of MIPs (Figure 4(b)) is rough and uneven, with morphological features and uniformly distributed pores, which facilitates the adsorption of molecules on mass transfer. The morphological characteristics of NIPs (Figure 4(c)) are not as obvious as MIPs, and the surface pore size is almost absent. The comparison results show that the roughness of the surface of the imprinted polymer increases.

Table 2 shows the specific surface area, pore volume, and average pore size of MIPs and NIPs. As can be seen from Table 2, MIPs have larger specific surface area, pore volume, and average pore size than NIPs. The results showed that the different adsorption properties of MIPs and NIPs could not be completely attributed to the difference in morphology but also related to the imprinting process that produced specific recognition sites [44].

3.3. Binding Properties of the MIPs and NIPs

3.3.1. Equilibrium Adsorption Curve of Ampicillin with Different MIPs. Three groups of imprinted polymers were prepared with fixed template:functional monomer:crosslinker ratios of 1:3:20, 1:4:20, and 1:5:20 and their equilibrium adsorption curves were generated (Figure 5). As can be seen from (Figure 5), as the monomer concentration increased, the amount of imprinted and nonimprinted polymer adsorbed also increased. The difference in the adsorption amounts between MIP3 and NIP3 were small because nonspecific adsorption strengthened as the monomer concentration increased, which reduced the specific adsorption performance of the MIPs. In Figure 5, the adsorptions of MIP1 and NIP1 were relatively low and the difference between them very small because the monomer concentration was too low, rendering it difficult to form effective recognition sites and spatial structures in the polymer and resulting in a low adsorption and poor selectivity. As the AMP concentration increased, as in MIP2, the adsorption capacity of the MIPs also increased and was much larger than that of NIP2. It can be seen that MIP achieved the best recognition when the template molecule:monomer:crosslinker ratio was 1:4:20. The maximum equilibrium adsorption capacity of MIP2 was about 115.62 mg/g. The maximum equilibrium adsorption capacity of NIP2 was only 33.40 mg/g, which was far less than that of MIP2. This is because for the monodisperse imprinted matrix $P_{VBC-DVB}$ microspheres, the surface recognition sites were more evenly distributed and the surface recognition sites formed were more conducive to the enrichment of template molecules, which are jointly affected by both specific and non-specific adsorption. MIP2 exhibited the strongest adsorption performance.

Scatchard plots [45, 46] were created for Q/c and Q. As shown in the MIP graph in Figure 6, there were two distinctly good linear relationships for AMP adsorption on the MIPs, indicating that there were two different types of binding sites. After fitting them separately and calculating the performance of the two types of binding, the maximum adsorption capacities were determined to be 359.08 and 9.04 mg/g. The adsorption capacity of the MIP was higher than for other materials. K_{Ds} were 3.04 and 0.072 mg/L.

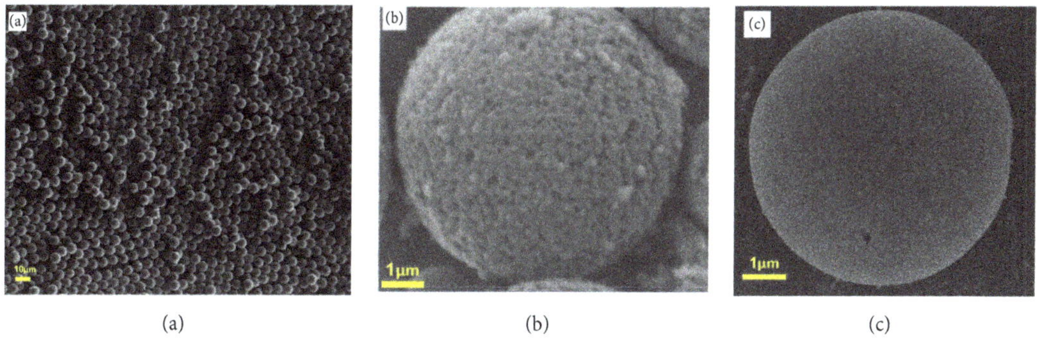

(a) (b) (c)

FIGURE 4: Scanning electron micrographs of (a) $P_{VBC\text{-}DVB}$ microspheres, (b) MIPs, and (c) NIPs.

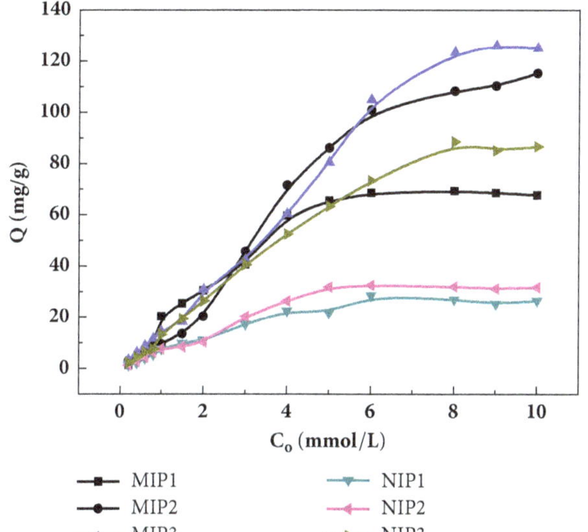

FIGURE 5: Equilibrium adsorption curves of ampicillin with different MIPs and NIPs.

3.3.2. Kinetic Adsorption. Figure 7 shows the kinetics of MIPs and NIPs adsorption. It can be seen from the figure that the adsorption rates of MIPs and NIPs were very fast during the first 20 minutes and the adsorption equilibrium was reached in about 30 minutes. Traditional bulk MIPs has a slow adsorption rate of 10-24 h. This is because the surface imprinting method forms a recognition hole with recognition sites on the surface of the polymer, where uniform distribution facilitates entry of the imprinting molecule into the recognition hole and quick spreading on the inside of the imprinting hole to achieve adsorption equilibrium.

3.3.3. Adsorption Selectivity. Three penicillin antibiotics with structures similar to AMP (Figure 8) were used to study the specific recognition ability of the MIPs [47].

From Figure 9, analysis of the adsorption of the four antibiotics by $P_{VBC\text{-}DVB}$ microspheres matrix, MIPs, and NIPs revealed that MIPs had the largest adsorption capacity and best selectivity for AMP. The differences in the adsorption capacity of NIPs for the four antibiotics were very small because recognition of the molecules by NIPs was controlled by nonspecific adsorption and determined based on the strength of the interactions between the monomer and antibiotic molecules. Recognition of antibiotic molecules by MIPs was influenced by the functional groups and size and spatial structure of the cavities imprinted on the surface.

It can be seen that the relative selectivity coefficient of the MIP for the imprinting molecule AMP was 4.6, showing that the AMP MIPs microspheres had good molecular recognition. Compared to the $P_{VBC\text{-}DVB}$ matrix and NIPs, MIPs displayed corresponding recognition. The MIP had recognition pores the same as the template molecule in terms of size and spatial structure and exhibited specific adsorption of the template molecule. The selectivity coefficients of MIPs for the three structural analogs were 5.74, 6.83, and 7.25, respectively, while the selectivity coefficients of NIPs and $P_{VBC\text{-}DVB}$ microsphere substrates to the structural analogs were between 0.73 and 1.63. Therefore, the MIPs exhibited better recognition selectivity and AMP MIPs microspheres had good molecular affinity and molecular recognition capabilities. (Table 3)

3.4. Reuse Performance of MIPs-SPE. The ability for MIP to be used repeatedly is also an important criterion when evaluating the performance of MIPs. To examine the reusability of MIPs-SPE, 10 recovery experiments were performed on the same cartridge. As shown in (Figure 10), when the MIPs-SPE cartridge was repeatedly used 10 times, it had a recovery rate of over 80%, indicating the repeated use performance was good. Therefore, the MIP was a good reusable SPE sorbent, which greatly reduced experimental cost.

3.5. Molecularly Imprinted Solid-Phase Extraction of AMP from Spiked Samples. Figure 11(d) shows the chromatograms of mixed standard solutions of AMP and penicillin G. The peaks corresponding to AMP and penicillin G occurred at 4.91 and 6.13 min, respectively. The residual amount of AMP in the blank egg sample was approximately 18.87 $\mu g \cdot g^{-1}$, which did not exceed the national minimum limit. The linear equation of the AMP standard solution in the range of 5-10000 ng/g was Y=10.494 X-758.08 and the correlation coefficient was R=0.9995. C_{18}, MIPs, and NIPs were used as SPE matrixes under optimized experimental conditions and the content of AMP in eggs was measured by HPLC.

TABLE 3: Selective coefficients of MIPs, NIPs, and $P_{VBC-DVB}$.

Analyte	C_e (mmol/L)			Q (mmol/g)			K_D (L/g×10^{-3})			IF			β
	$P_{VBC-DVB}$	MIPs	NIPs	$P_{VBC-DVB}$	MIPs	NIPs	$P_{VBC-DVB}$	MIPs	NIPs	$P_{VBC-DVB}$	MIPs	NIPs	
Ampicillin	9.59	6.80	9.07	0.041	0.320	0.093	0.0043	0.0471	0.0103	—	—	—	4.6
peillinG	9.60	9.25	9.28	0.040	0.076	0.072	0.0042	0.0082	0.0078	1.02	5.74	1.32	1.1
amoxicillin	9.88	9.56	9.60	0.035	0.066	0.062	0.0035	0.0069	0.0065	1.23	6.83	1.58	1.1
Salbactam	9.48	9.43	9.44	0.056	0.061	0.059	0.0059	0.0065	0.0063	0.73	7.25	1.63	1.0

∗C_o=10 mmol/L.

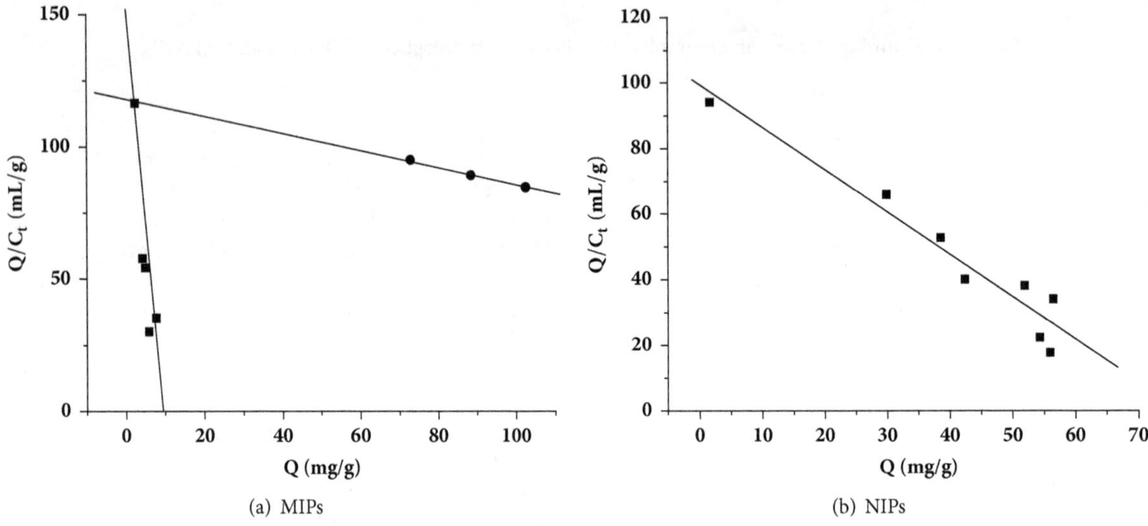

(a) MIPs

(b) NIPs

FIGURE 6: Scatchard fitting curves for (a) MIPs and (b) NIPs microspheres.

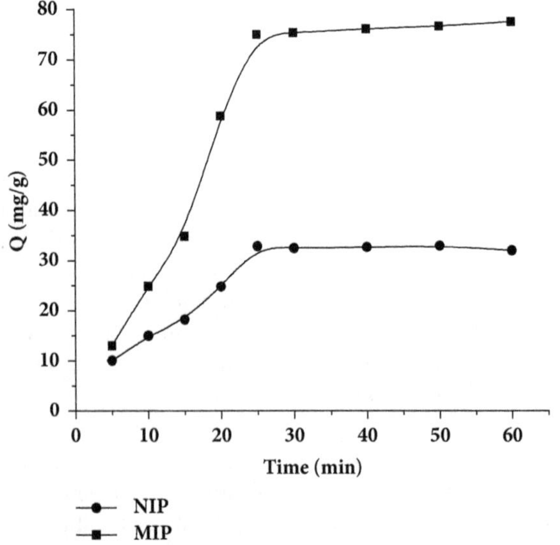

FIGURE 7: Dynamic adsorption curves.

For the MIP-SPE, the average recoveries were 91.5, 92.4, and 94.9% and the relative standard deviations were 3.6, 4.1, and 4.2%. The specific values are listed in Table 4. It can be clearly seen that MIPs-SPE treatment had an enrichment and purification effect on the samples, resulting in a higher extraction recovery rate than when using NIP-SPE or C_{18}-SPE. The results show that MIPs-SPE facilitates the detection of trace AMP in complex samples.

4. Conclusions

In this study, MIPs microspheres with the ability to specifically recognize water-soluble molecules were prepared using SI-ATRP technology and the specificity of recognition by these MIPs was studied. In adsorption tests, the MIPs displayed excellent recognition and AMP selectivity. The MIPs displayed a lower binding capacity to the AMP structural analogues of penicillin G, nafcillin, and sulbactam. The generated material can be successfully used as SPE adsorbent for HPLC separation, enrichment, and detection of trace AMP in food samples. When MIPs was used as SPE filler to measure the remaining trace amounts of AMP in egg, the recovery rate was found to be good, ranging between 91.5 and 94.9% with relative standard deviations of 3.6 to 4.2%. Repeated use of the MIPs as SPE material 10 times yielded a recovery rate above 85%, indicating the MIP had good stability and repeatability with a good rich set and high recovery. Overall, this study shows that MIPs is an ideal material for enrichment and detection of trace AMP residue and provides means for quantitatively and qualitatively analyzing AMP in other complex substrate samples. In addition, this research method

TABLE 4: Standard addition of ampicillin and relative standard deviations of ampicillin recovery from egg samples.

Adsorbent	Amount added (ng·g-1)	Recovery (R/%) 1	2	3	Mean recovery ($\overline{R}$/ %)	RSD/%
	5	90.6	99.1	95.1	94.9	4.2
MIP-SPE	20	92.7	87.4	94.3	91.5	3.6
	50	91.1	97.0	89.2	92.4	4.1
	5	12.3	22.5	25.6	20.2	6.9
NIP-SPE	20	17.8	16.1	8.3	14.1	5.1
	50	26.7	26.7	27.6	27.0	0.5
	5	77.3	78.3	73.9	76.5	2.3
C18-SPE	20	91.5	87.7	93.8	91.0	3.1
	50	92.5	97.1	92.7	94.1	2.6

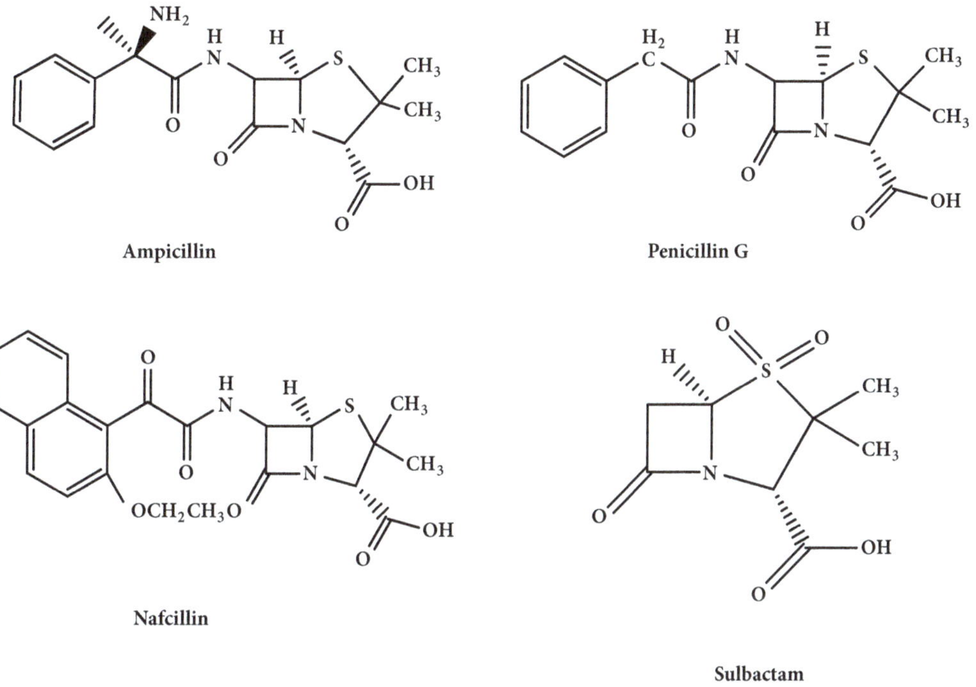

Ampicillin

Penicillin G

Nafcillin

Sulbactam

FIGURE 8: The structures of ampicillin, penicillin G, nafcillin, and sulbactam.

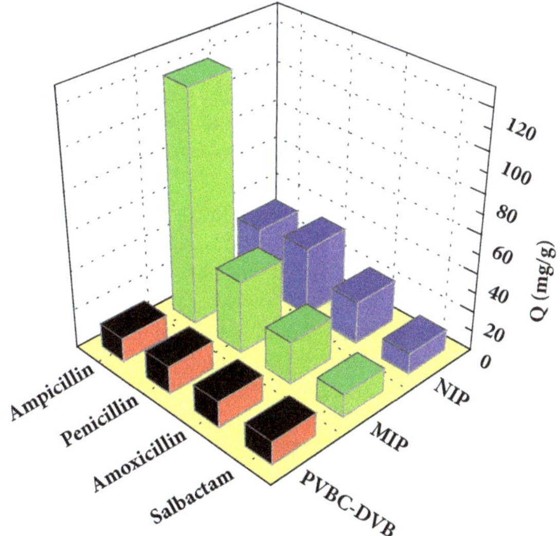

FIGURE 9: Selective adsorptions.

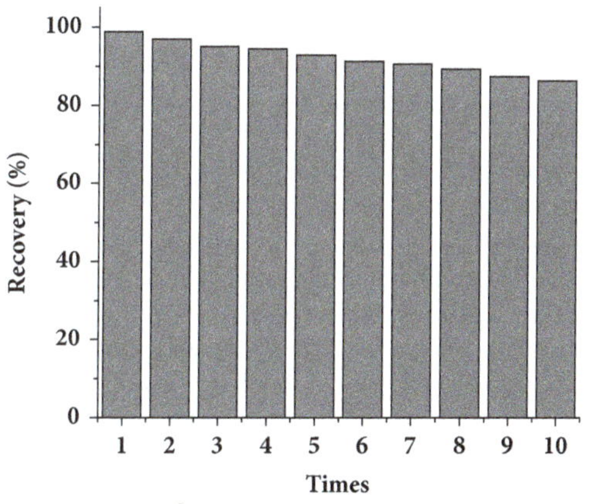

FIGURE 10: MIPs-SPE cycle performance.

provides an important reference for further monitoring and research to improve food safety.

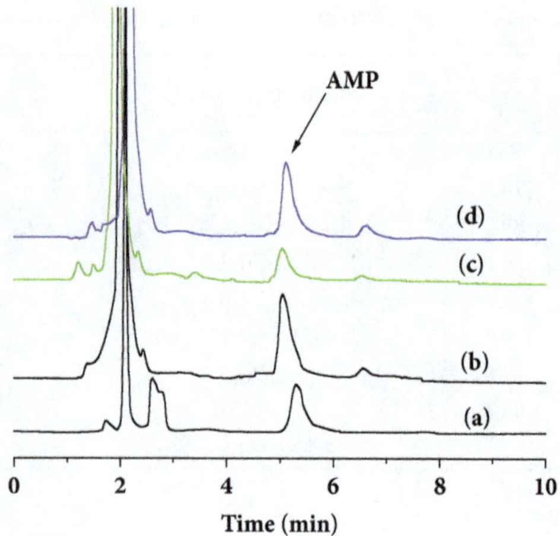

FIGURE 11: Chromatograms of egg samples processed on different solid phase extraction (SPE) columns: (a) C_{18}-SPE cartridge, (b) MIPs-SPE cartridge, (c) NIPs-SPE cartridge, and (d) standard solution.

Authors' Contributions

Yang Tian and Yue Wang contributed equally to this work.

Acknowledgments

This work was financially supported by the National Natural Science Foundation of China (nos. 21565001 and 31271868) and Key Project of North Minzu University (no. 2015KJ30).

References

[1] *National Essential Drugs Clinical Application Guide and Prescription Set Editorial Committee of National Essential Drugs Prescription Set (Chemicals and Biological Products)*, People's Medical Publishing House, Beijing, China, 2012.

[2] S. Liu, "Efficacy analysis of ampicillin sodium and sulbactam sodium for the treatment of bacterial infection in the lower respiratory tract in the elderly," *Journal of China Prescription Drugs*, vol. 14, no. 09, pp. 56-57, 2016.

[3] X.-F. Zhang, M.-Q. Zou, X.-H. Qi, F. Liu, X.-H. Zhu, and B.-H. Zhao, "Detection of melamine in liquid milk using surface-enhanced Raman scattering spectroscopy," *Journal of Raman Spectroscopy*, vol. 41, no. 12, pp. 1655–1660, 2010.

[4] S. Zhou, D. Wang, Y. Zhao, and Y. Wu, "A rapid HPLC method for indirect quantification of β-lactamase activity in milk," *Journal of Dairy Science*, vol. 98, no. 4, pp. 2197–2204, 2015.

[5] C. Nebot, P. Regal, J. M. Miranda, C. Fente, and A. Cepeda, "Rapid method for quantification of nine sulfonamides in bovine milk using HPLC/MS/MS and without using SPE," *Food Chemistry*, vol. 141, no. 3, pp. 2294–2299, 2013.

[6] Y.-K. Lv, L. Yang, X.-H. Liu, Z.-Y. Guo, and H.-W. Sun, "Preparation and evaluation of a novel molecularly imprinted hybrid composite monolithic column for on-line solid-phase extraction coupled with HPLC to detect trace fluoroquinolone residues in milk," *Analytical Methods*, vol. 5, no. 7, pp. 1848–1855, 2013.

[7] R. Santos, E. Limas, M. Sousa, M. da Conceição Castilho, F. Ramos, and M. I. N. da Silveira, "Optimization of analytical procedures for GC-MS determination of phytosterols and phytostanols in enriched milk and yoghurt," *Food Chemistry*, vol. 102, no. 1, pp. 113–117, 2007.

[8] T. Liu, J. Xie, J. Zhao, G. Song, and Y. Hu, "Magnetic Chitosan Nanocomposite Used as Cleanup Material to Detect Chloramphenicol in Milk by GC-MS," *Food Analytical Methods*, vol. 7, no. 4, pp. 814–819, 2014.

[9] T. Delatour, M. Savoy, A. Tarres, T. Bessaire, P. Mottier, and A. Desmarchelier, "Low false response rates in screening a hundred veterinary drug residues in foodstuffs by LC-MS/MS with analyte-specific correction of the matrix effect," *Food Control*, vol. 94, pp. 353–360, 2018.

[10] K. Hamamoto and Y. Mizuno, "Lc-ms/ms measurement of ampicillin residue in swine tissues at 5 days after in-feed administration," *Journal of Veterinary Medical Science*, vol. 77, no. 11, pp. 1527–1529, 2015.

[11] C. Andreou, R. Mirsafavi, M. Moskovits, and C. D. Meinhart, "Detection of low concentrations of ampicillin in milk," *Analyst*, vol. 140, no. 15, pp. 5003–5005, 2015.

[12] P. Wang, S. Pang, H. Zhang, M. Fan, and L. He, "Characterization of Lactococcus lactis response to ampicillin and ciprofloxacin using surface-enhanced Raman spectroscopy," *Analytical and Bioanalytical Chemistry*, vol. 408, no. 3, pp. 933–941, 2016.

[13] M. Zhang, W. Hasi, X. Lin et al., "Rapid and simple detection of pethidine hydrochloride injection using surface-enhanced Raman spectroscopy based on silver aggregates," *Analytical Methods*, vol. 7, no. 19, pp. 8241–8247, 2015.

[14] J. Jiao, Y. Xue, H. Yang, and T. Sun, "Application and Prospect of Molecular Imprinting Technology in Drug Extraction," *Chinese Remedies and Clinics*, vol. 18, no. 02, pp. 214–218, 2018.

[15] H. Yang, H. Zhang, X. Y. Zhu, S. Da Chen, L. Liu, and D. Pan, "Determination of tributyltin in seafood based on magnetic molecularly imprinted polymers coupled with high-performance liquid chromatography-inductively coupled plasma mass spectrometry," *Journal of Food Quality*, vol. 2017, 2017.

[16] J. Wackerlig and R. Schirhagl, "Applications of Molecularly Imprinted Polymer Nanoparticles and Their Advances toward Industrial Use: A Review," *Analytical Chemistry*, vol. 88, no. 1, pp. 250–261, 2016.

[17] S. Daniel, P. Prabhakara Rao, and T. Prasada Rao, "Investigation of different polymerization methods on the analytical performance of palladium(II) ion imprinted polymer materials," *Analytica Chimica Acta*, vol. 536, no. 1-2, pp. 197–206, 2005.

[18] P. Li, F. Rong, and C. Yuan, "Morphologies and binding characteristics of molecularly imprinted polymers prepared by precipitation polymerization," *Polymer International*, vol. 52, no. 12, pp. 1799–1806, 2003.

[19] Y.-M. Yin, Y.-P. Chen, X.-F. Wang, Y. Liu, H.-L. Liu, and M.-X. Xie, "Dummy molecularly imprinted polymers on silica particles for selective solid-phase extraction of tetrabromobisphenol A from water samples," *Journal of Chromatography A*, vol. 1220, pp. 7–13, 2012.

[20] W. Du, H. Zhou, Z. Luo et al., "Selective determination of penicillin G from tap water and milk samples using surface molecularly imprinted polymers as solid-phase extraction sorbent," *Molecular Imprinting*, vol. 2, no. 1, 2014.

[21] J. L. Urraca, M. C. Moreno-Bondi, G. Orellana, B. Sellergren, and A. J. Hall, "Molecularly imprinted polymers as antibody mimics in automated on-line fluorescent competitive assays," *Analytical Chemistry*, vol. 79, no. 13, pp. 4915–4923, 2007.

[22] C.-Y. Huang, M.-J. Syu, Y.-S. Chang, C.-H. Chang, T.-C. Chou, and B.-D. Liu, "A portable potentiostat for the bilirubin-specific sensor prepared from molecular imprinting," *Biosensors and Bioelectronics*, vol. 22, no. 8, pp. 1694–1699, 2007.

[23] A. L. Joke Chow and S. A. Bhawani, "Synthesis and Characterization of Molecular Imprinting Polymer Microspheres of Cinnamic Acid: Extraction of Cinnamic Acid from Spiked Blood Plasma," *International Journal of Polymer Science*, vol. 2016, 2016.

[24] J. Gauczinski, Z. Liu, X. Zhang, and M. Schönhoff, "Surface molecular imprinting in layer-by-layer films on silica particles," *Langmuir*, vol. 28, no. 9, pp. 4267–4273, 2012.

[25] M. J. Shin, Y. J. Shin, S. W. Hwang, and J. S. Shin, "Recognizing amino acid chirality with surface-imprinted polymers prepared in W/O emulsions," *International Journal of Polymer Science*, vol. 2013, 2013.

[26] D. Braun, "Origins and Development of Initiation of Free Radical Polymerization Processes," *International Journal of Polymer Science*, vol. 2009, Article ID 893234, 10 pages, 2009.

[27] Y. Chen, J. Zhang, and X. Yang, "Atom transfer radical polymerization technology and its application in the preparation of separation materials," *Synthetic Materials Aging and Application*, vol. 45, no. 04, pp. 90–95, 2016.

[28] S. Shahabuddin, F. Hamime Ismail, S. Mohamad, and N. Muhamad Sarih, "Synthesis of well-defined three-arm star-branched polystyrene through arm-first coupling approach by atom transfer radical polymerization," *International Journal of Polymer Science*, vol. 2015, 2015.

[29] C. Lubke, M. Lubke, M. J. Whitcombe, and E. N. Vulfson, "Imprinted polymers prepared with stoichiometric template-monomer complexes: efficient binding of ampicillin from aqueous solutions," *Macromolecules* , vol. 33, no. 14, pp. 5098–5105, 2000.

[30] Y. Mao, H. Kang, X. Wang et al., "Selective adsorption of ampicillin by molecularly imprinted polymers based on magnetic carbon microspheres," *Huanjing Kexue Xuebao/Acta Scientiae Circumstantiae*, vol. 36, no. 7, pp. 2451–2459, 2016.

[31] X. Li, J. Pan, J. Dai et al., "Surface molecular imprinting onto magnetic yeast composites via atom transfer radical polymerization for selective recognition of cefalexin," *Chemical Engineering Journal*, vol. 198-199, pp. 503–511, 2012.

[32] H. A. Ezzeldin, A. Apblett, and G. L. Foutch, "Synthesis and properties of anion exchangers derived from chloromethyl styrene codivinylbenzene and their use in water treatment," *International Journal of Polymer Science*, vol. 2010, 2010.

[33] X. Dai, Y. He, Y. Wei, and B. Gong, "Preparation of hydrophilic polymer-grafted polystyrene beads for hydrophilic interaction chromatography via surface-initiated atom transfer radical polymerization," *Journal of Separation Science*, vol. 34, no. 22, pp. 3115–3122, 2011.

[34] J. Hu, H. Lü, H. Liu, W. Liu, and H. Zhang, "Synthesis of Magnetic Polystyrene-Chloromethylstyrene Materials and Their Adsorption Properties for Antimony," *Journal of Instrumental Analysis*, vol. 36, no. 04, pp. 464–470, 2017.

[35] N. Rattanathamwat, J. Wootthikanokkhan, N. Nimitsiriwat, C. Thanachayanont, and U. Asawapirom, "Kinetic Studies of Atom Transfer Radical Polymerisations of Styrene and Chloromethylstyrene with Poly(3-hexyl thiophene) Macroinitiator," *Advances in Materials Science and Engineering*, vol. 2015, 2015.

[36] R. Zhang, Q. Li, D. Ji et al., "Preparation of a novel polymer monolith with high loading capacity by grafting block poly(PEGA-mPEGA) for high-efficiency solid phase synthesis," *Reactive and Functional Polymers*, vol. 94, pp. 63–69, 2015.

[37] R. Zhang, Q. Li, Y. Huang et al., "Preparation of PEGA grafted poly(chloromethylstyrene-co-ethylene glycol dimethacrylate) monolith for high-efficiency solid phase peptide synthesis under continuous flow techniques," *Polymer (United Kingdom)*, vol. 61, pp. 115–122, 2015.

[38] Wang. Dongsha, "Yanjun Liu .Preparation of Monodispersed and Hypercrosslinked, Polystyrene Microspheres by Seed Swelling Method," *Chinese Journal of Applied Chemistry*, vol. 11, pp. 1289–1294, 2007.

[39] X. Zhang, J. Yang, X. Ma, X. Chen, and Q. Wu, "Preparation of Monodisperse Crosslinked Polystyrene Microspheres by Seed Swelling Method," *China Plastics Industry*, vol. 37, no. 03, pp. 37–39, 2009.

[40] A. Ersöz, R. Say, and A. Denizli, "Ni(II) ion-imprinted solid-phase extraction and preconcentration in aqueous solutions by packed-bed columns," *Analytica Chimica Acta*, vol. 502, no. 1, pp. 91–97, 2004.

[41] Z. Zhao, Y. Zhang, Y. Xuan et al., "Ion-exchange solid-phase extraction combined with liquid chromatography-tandem mass spectrometry for the determination of veterinary drugs in organic fertilizers," *Journal of Chromatography B*, vol. 1022, pp. 281–289, 2016.

[42] X. Zou, J. Pan, H. Ou et al., "Adsorptive removal of Cr(III) and Fe(III) from aqueous solution by chitosan/attapulgite composites: Equilibrium, thermodynamics and kinetics," *Chemical Engineering Journal*, vol. 167, no. 1, pp. 112–121, 2011.

[43] H. Zhu, *Organic Molecular Structure Spectrum Analysis*, Chemical Industry Press, Beijing, China, 2005.

[44] Y. Li, X. Li, Y. Li, J. Qi, J. Bian, and Y. Yuan, "Selective removal of 2,4-dichlorophenol from contaminated water using non-covalent imprinted microspheres," *Environmental Pollution*, vol. 157, no. 6, pp. 1879–1885, 2009.

[45] B. Sellergren, "Molecular imprinting by noncovalent interactions. Enantioselectivity and binding capacity of polymers Prepared under conditions favoring the formation of template Complexes," *Die Makromolekulare Chemie*, vol. 190, no. 11, pp. 2703–2711, 1989.

[46] J. Mathew and O. Buchardt, "Molecular Imprinting Approach for the Recognition of Adenine in Aqueous Medium and Hydrolysis of Adenosine 5'-Triphosphate," *Bioconjugate Chemistry*, vol. 6, no. 5, pp. 524–528, 1995.

[47] M. Díaz-Bao, R. Barreiro, J. M. Miranda, A. Cepeda, and P. Regal, "Fast HPLC-MS/MS method for determining penicillin antibiotics in infant formulas using molecularly imprinted solid-phase extraction," *Journal of Analytical Methods in Chemistry*, vol. 2015, 2015.

8

Development of Isocratic RP-HPLC Method for Separation and Quantification of L-Citrulline and L-Arginine in Watermelons

Rasdin Ridwan,[1] Hairil Rashmizal Abdul Razak,[2] Mohd Ilham Adenan,[3,4] and Wan Mazlina Md Saad[1]

[1]Centre of Medical Laboratory Technology, Faculty of Health Sciences, Universiti Teknologi MARA, Puncak Alam Campus, 42300 Bandar Puncak Alam, Selangor, Malaysia
[2]Centre of Medical Imaging, Faculty of Health Sciences, Universiti Teknologi MARA, Puncak Alam Campus, 42300 Bandar Puncak Alam, Selangor, Malaysia
[3]Faculty of Applied Sciences, Universiti Teknologi MARA, 40450 Shah Alam, Selangor, Malaysia
[4]Atta-ur-Rahman Institute for Natural Product Discovery, Level 9, Bangunan FF3, Universiti Teknologi MARA, Puncak Alam Campus, 42300 Bandar Puncak Alam, Selangor, Malaysia

Correspondence should be addressed to Wan Mazlina Md Saad; mazlinasaad14@gmail.com

Academic Editor: Gunther K. Bonn

Watermelons *(Citrullus lanatus)* are known to have sufficient amino acid content. In this study, watermelons grown and consumed in Malaysia were investigated for their amino acid content, L-citrulline and L-arginine, by the isocratic RP-HPLC method. Flesh and rind watermelons were juiced, and freeze-dried samples were used for separation and quantification of L-citrulline and L-arginine. Three different mobile phases, 0.7% H_3P0_4, 0.1% H_3P0_4, and 0.7% H_3P0_4 : ACN (90 : 10), were tested on two different columns using Zorbax Eclipse XDB-C_{18} and Gemini C_{18} with a flow rate of 0.5 mL/min and a detection wavelength at 195 nm. Efficient separation with reproducible resolution of L-citrulline and L-arginine was achieved using 0.1% H_3P0_4 on the Gemini C_{18} column. The method was validated and good linearity of L-citrulline and L-arginine was obtained with $R^2 = 0.9956$, $y = 0.1664x + 2.4142$ and $R^2 = 0.9912$, $y = 0.4100x + 3.4850$, respectively. L-citrulline content showed the highest concentration in red watermelon of flesh and rind juice extract (43.81 mg/g and 45.02 mg/g), whereas L-arginine concentration was lower than L-citrulline, ranging from 3.39 to 11.14 mg/g. The isocratic RP-HPLC method with 0.1% H_3P0_4 on the Gemini C_{18} column proved to be efficient for separation and quantification of L-citrulline and L-arginine in watermelons.

1. Introduction

Citrullus lanatus (Thunb.) Matsum. and Nakai, commonly known as watermelon, is a nonseasonal fruit which is cultivated abundantly in Malaysia and other tropical regions [1]. It belongs to the Cucurbitaceae plant family, which originated from the African Kalahari Desert [1]. Watermelons have high content of phytonutrients and are rich in dietary antioxidants such as carotenoids (lycopene and β-carotene), polyphenolics, ascorbic acid, and significant amino acids [2]. Watermelons are usually consumed by juicing the flesh, beneficial in the prevention and improvement of health problems, such as cardiovascular diseases, erectile dysfunction, hypertension, and cancers [3]. Figueroa et al. [4] demonstrated that watermelon juice supplementation improves aortic hemodynamics by reducing the reflected wave amplitude in prehypertensive individuals. A study by Poduri et al. [5] reported that watermelon attenuated hypercholesterolemia-induced atherosclerosis in mice. Commercial watermelon juices provide enormous marketing potential and nutritious drinks for individuals to maintain a healthy lifestyle.

Amino acids, particularly L-citrulline and L-arginine, are regarded as major types of phytonutrients present in watermelons which may contribute to their reputed and diversified health benefits [6]. L-citrulline, $C_6H_{13}N_3O_3$ (IUPAC name: 2-amino-5-(carbamoylamino)pentanoic acid) (Figure 1), is a nonessential amino acid firstly identified from watermelon, *Citrullus vulgaris* Schrad. [7, 8]. L-citrulline is a physiological

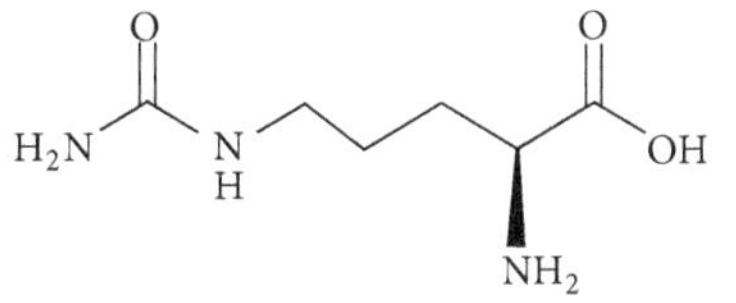

FIGURE 1: Molecular structure of L-citrulline (175.2 g/mol).

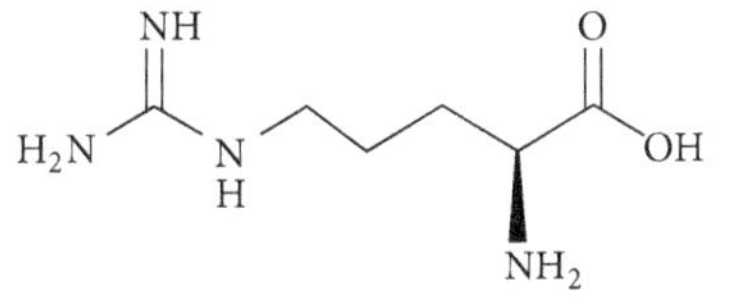

FIGURE 2: Molecular structure of L-arginine (174.2 g/mol).

endogenous amino acid to most living systems involved in protein metabolism and removal of excess metabolic ammonia [9]. It serves as a precursor for L-arginine and product of nitric oxide (NO) cycle [10]. L-arginine, $C_6H_{14}N_4O_2$ (IUPAC name: (*S*)-2-amino-5-guanidinopentanoic acid) (Figure 2), is a semiessential and free form physiological amino acid that functions as one of 20 building block proteins for biological processes such as cell division, ammonia removal, wound healing, and hormone release [7, 8]. Wu et al. [11] demonstrated that supplementation of L-citrulline and L-arginine from watermelon juice improved serum levels of NO metabolites and aortic endothelial-mediated vasodilation in diabetic rats.

L-citrulline and L-arginine are present in all parts of watermelon fruits including flesh, rind, and seed [7]. A study done by Rimando and Perkins-Veazie [12] reported that the rind of red watermelon and yellow watermelon contains more L-citrulline at a concentration ranging from 15.6 to 29.4 mg/g than flesh, 7.9–28.5 mg/g. Similar to the above finding, Jayaprakasha et al. [13] reported that rinds of *Citrullus vulgaris* varieties such as petite treat and jamboree watermelon and also yellow crimson watermelon contained slightly higher L-citrulline ranging from 13.95 to 28.46 mg/g than flesh, 11.25–16.73 mg/g. These findings suggested that watermelon rind has an abundance of L-citrulline content in comparison to its content in flesh.

Analyses of L-citrulline and L-arginine were routinely conducted using capillary electrophoresis and quantification by a spectrophotometric method; however, the method is less sensitive, leading to discrepancies in the outcomes [6]. L-citrulline and L-arginine are polar, nonvolatile, and devoid of chromophores; thus analysis by reverse-phase high performance liquid chromatography (RP-HPLC) commonly employed a derivatization method using pre- or postcolumn derivatization [14–17]. Jayaprakasha et al. [13] stated that precolumn derivatization such as orthophthalaldehyde (OPA), naphthalene-2,3-dicarboxaldehyde, or 4-dimethylaminoazobenzene-4′-sulfonyl chloride (dabsyl chloride) was able to provide accurate and stable chromatography baseline, but the reactions were unstable and affected by the sample matrix [18]. Postcolumn derivatization by ninhydrin is tedious due to long analysis time up to 72 hours and instability of derivatization reagents that may cause poor compound recovery [19]. Analysis of underivatized L-citrulline and L-arginine is warranted for rapid and effective quantification of these compounds. Given that no amino acids content of L-citrulline and L-arginine in Malaysia watermelons has been reported so far, we have developed an isocratic RP-HPLC method for separation and quantification of L-citrulline and L-arginine in watermelons.

2. Materials and Methods

2.1. Chemicals and Reagents. L-citrulline (purity ≥ 99%) and L-arginine (purity ≥ 98%) standard were purchased from Sigma-Aldrich (St. Louis, MO, USA). Methanol and acetonitrile of HPLC grade were purchased from Merck (Germany). Phosphoric acid (purity ≥ 85%) was purchased from Sigma-Aldrich (St. Louis, MO, USA). Deionized water was prepared using ultrapure water purifier system (Elgastat, Bucks, UK).

2.2. Instrumentation. The isocratic RP-HPLC method was carried out using Thermo Scientific™ Dionex-UltiMate™ 3000 HPLC system equipped with solvent reservoirs, LPG-3400SD pump, WPS-3000 autosampler injector, TCC-3000 column oven, and DAD-3000 ultraviolet-visible (UV-Vis) diode array detector module operated at four wavelengths per analysis. Chromeleon data software (Version 7) was used for data analysis.

2.3. Sample Preparation. Citrullus lanatus (Thunb.) Matsum. & Nakai of red watermelon and yellow crimson watermelon was obtained from Selangor Fruit Valley, Selangor. Seeds were removed manually and the edible part was cut into cubes. Watermelon flesh and rind were juiced and frozen at −80°C for at least 2 days. The frozen juices were put in a freeze-drier (Labconco, USA) for 4 days until completely dried. The dried juice powders were kept at −20°C. For the analysis, samples were prepared in the form of juice extract and methanol extract. Juice extract was prepared directly by dissolving the dried juice powder in dH_2O. For methanol extract, a known quantity of dried juice powders was extracted with 30 mL of MeOH and 1 mL of 1 N HCl, vortexed, and sonicated for 30 minutes. The samples were macerated by cold maceration for a period of 72 hours in an orbital shaker. Methanol extracts were then filtered using Whatman filter paper. The residues were reextracted twice using fresh solvent and the three methanol extracts were pooled. The obtained methanol extracts were evaporated to dryness using a rotary vacuum evaporator at 60°C and stored at 4°C until analysis.

2.4. Isocratic RP-HPLC Analysis

2.4.1. Standard Preparation Procedure. A stock solution of L-citrulline and L-arginine was prepared individually in dH_2O at 1 mg/mL and filtered through a 0.45 μm syringe filter (Bioflow). A mixed standard solution was prepared by mixing an equal volume of each standard stock solution. A series of working standard solutions was prepared by diluting the stock solution with dH_2O in the range of 0.1–1000 μg/mL.

TABLE 1: Selection of mobile phases for separation of mixed standard, L-citrulline, and L-arginine by isocratic RP-HPLC.

Mobile phase	Ratio	Solution mixture (%)
0.7% H_3P0_4	100	0.7% H_3P0_4 + 99.3% dH_2O
0.1% H_3P0_4	100	0.1% H_3P0_4 + 99.9% dH_2O
0.7% H_3P0_4 : ACN	90 : 10	(0.7% H_3P0_4 + 99.3% dH_2O) + 100% ACN

2.4.2. Sample Preparation Procedure. Juice extracts were prepared directly by dissolving the dried juices powder in dH_2O at 5 mg/mL. Crude methanol extracts were also dissolved in dH_2O at 5 mg/mL and vortexed for 15 minutes. All extracts were filtered through 0.45 μm filters and injected to isocratic RP-HPLC.

2.4.3. Chromatographic Analysis. Preliminarily, three different concentrations of ion-pair reagents, phosphoric acid (H_3P0_4), or addition of acetonitrile (ACN) as mobile phases, 0.7% H_3P0_4, 0.1% H_3P0_4, and 0.7% H_3P0_4 : ACN (90 : 10) (Table 1), were tested for separation and determination of L-citrulline and L-arginine standard. The column temperature was fixed at room temperature and UV-Vis detection was performed at 195 nm. The RP-HPLC columns [i.e., Zorbax Eclipse XDB-C_{18}, 250 mm × 4.6 mm, 80 Å, 5 μm (Phenomenex, Torrance, CA), and Gemini C_{18}, 250 × 4.6 mm, 110 Å, 3 μm (Phenomenex, Torrance, CA)] were used. The analysis proceeded for quantification of both compounds, L-citrulline and L-arginine in watermelons juice extracts and methanol extracts using the chosen column and mobile phase: Gemini C_{18} eluted by 0.1% H_3P0_4 with a flow rate of 0.5 mL/min at 195 nm. Chromeleon software was used for quantification of L-citrulline and L-arginine. The concentration of L-citrulline and L-arginine content was quantified based on the linear curve of standards. The content of compounds was expressed as milligrams per gram (mg/g) of sample extracts.

2.4.4. Method Validation. The validation of the isocratic RP-HPLC method was performed for linearity of calibration curve, limit of detection (LOD), limit of quantification (LOQ), accuracy, and precision. The linearity of the isocratic RP-HPLC method for quantification of compounds was constructed using the concentration range of 0.1–1000 μg/mL for L-citrulline and 0.1–500 μg/mL for L-arginine. The regression equation was calculated in the form of $y = ax + b$, where x is the concentration and y is the peak area of compounds. Linearity was established by the coefficient of determination (R^2). LOD and LOQ were measured based on signal-to-noise ratio (S/N) method. LOD is the lowest concentration of analyte that can be detected with signal-to-noise ratio of 3 : 1 and LOQ is the lowest concentration that can be quantified with acceptable precision and accuracy with signal-to-noise ratio of 10 : 1. S/N of 3 is considered acceptable for LOD, while LOQ is established at S/N of 10. Precision of the method was determined as percentage relative standard deviation (%RSD) of peak area of intraday and interday analysis data. Intraday (three times in a day operation under the same conditions) and interday (three different days) studies were performed at three different concentrations (Level 1: 20 μg/mL; Level 2: 60 μg/mL; Level 3: 150 μg/mL). The resulting peak area was used to calculate SD and the relative standard deviation (%RSD). Accuracy of the method by recovery study was done by adding a known amount of reference standard solution (three concentrations) to test samples. The spiked extract solutions were injected three times, and the recovery was calculated with the value of detected versus added amounts.

3. Results and Discussion

3.1. Separation of L-Citrulline and L-Arginine by Isocratic RP-HPLC Method. The initial isocratic RP-HPLC method for separation of mixed standard, L-citrulline, and L-arginine was performed using selected mobile phases according to previous literatures with slight modifications [8, 13, 20]. Interaction between mobile phase and stationary phases in isocratic RP-HPLC is important for the determination of solutes' retention time [21].

In this study, separation for determination of mixed standard, L-citrulline, and L-arginine was performed using a hydrophilic anionic ion-pairing reagent with different concentrations of phosphoric acid (H_3P0_4) or addition of acetonitrile (ACN) as mobile phases: 0.7% H_3P0_4, 0.1% H_3P0_4, and 0.7% H_3P0_4 : ACN (90 : 10). The mobile phase at the concentration of 0.7% H_3P0_4 : ACN (90 : 10) resulted in L-citrulline and L-arginine were unretained and coeluted (k value close to 0) as shown in Figure 3(a). The mixture of 0.7% H_3P0_4 : ACN (90 : 10) is highly hydrophilic, leading to rapid elution of L-citrulline and L-arginine with poor separation. Peaks of L-citrulline and L-arginine were slightly retained and partially separated using 0.7% H_3P0_4 (Figure 3(b)). However, optimum resolution was not achieved by 0.7% H_3P0_4 as k value between L-citrulline and L-arginine is close to 1. The mobile phase of 0.1% H_3P0_4 resulted in efficient separation with reproducible peaks of L-citrulline and L-arginine although all chromatograms showed stable baseline (Figure 3(c)). This finding is in agreement with Fekete et al. [22] who noted that 0.1% H_3P0_4 acts as a good separation agent by increasing the polarity and improving the retention time of zwitterionic molecules including amino acids. Dolan [23] supported the notion that 0.1% H_3P0_4 adequately provides reasonable buffering for amino acids separation by RP-HPLC. This showed that a concentration less than 1.0% H_3P0_4 as mobile phase provides efficient separation of amino acids, peptides, or proteins as demonstrated by Shibue et al. [24]. Thus, the mobile phase 0.1% H_3P0_4 is proven to provide efficient separation and the best resolution of mixed standard, L-citrulline, and L-arginine.

The study also evaluated separation of mixed standard in two different columns, Zorbax Eclipse XDB-C_{18} and Gemini C_{18} using 0.1% H_3P0_4. Zorbax Eclipse XDB-C_{18} did not provide good separation and resolution of L-citrulline and L-arginine as shown in Figure 4(a). A study by Barber and Joseph [25] showed that polar compounds were less separated and not well resolved using Zorbax Eclipse XDB-C_{18} with a longer analysis time of 54 minutes. Efficient separation and resolution of L-citrulline and L-arginine from

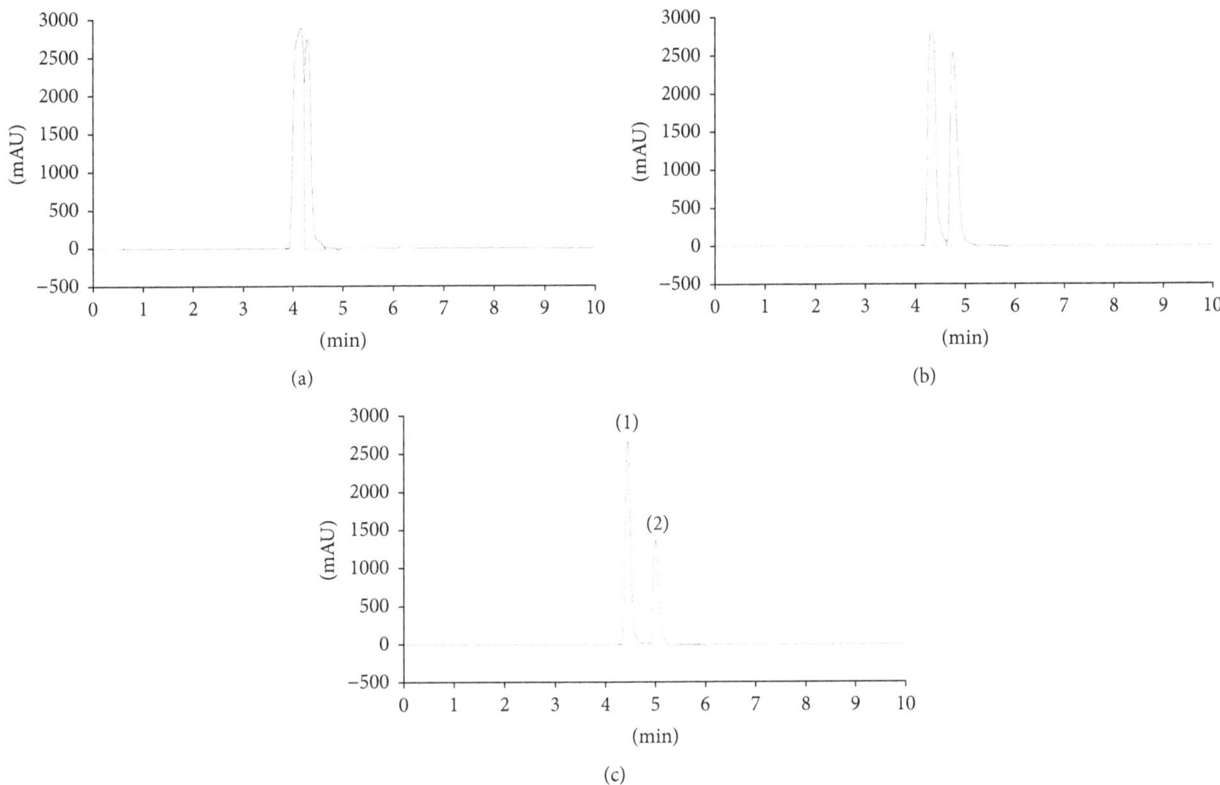

FIGURE 3: Comparative chromatograms showing isocratic RP-HPLC separation of mixed standard, L-citrulline, and L-arginine in different mobile phases: (a) 0.7% H_3P0_4 : ACN (90 : 10); L-citrulline and L-arginine were unretained and coeluted at k value close to zero; (b) 0.7% H_3P0_4; L-citrulline and L-arginine were slightly retained and partially separated; (c) 0.1% H_3P0_4; L-citrulline and L-arginine were efficiently separated with reproducible peaks. The peaks marked represent (1) L-arginine and (2) L-citrulline.

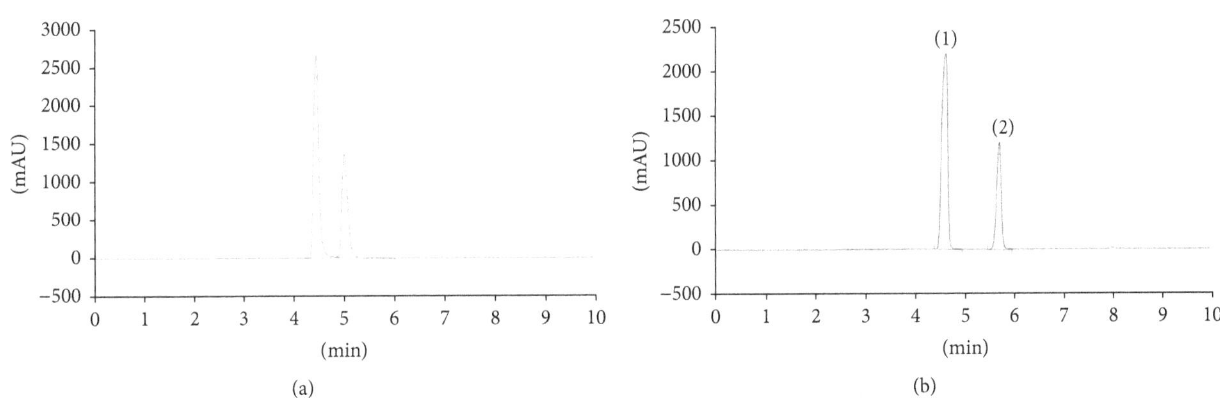

FIGURE 4: Comparative chromatograms showing isocratic RP-HPLC separation of mixed standard, L-citrulline, and L-arginine from 2 different columns: (a) Zorbax Eclipse XDB-C_{18}, 5 μm, and (b) Gemini C_{18}, 3 μm; efficient separation and the best resolution were achieved by the Gemini C_{18} column which showed that compounds are well separated. The peaks marked represent (1) L-arginine and (2) L-citrulline.

mixed standard were achieved using Gemini C_{18} as shown in Figure 4(b). L-citrulline and L-arginine are eluted at a short retention time with L-arginine, 4.773 min, followed by L-citrulline at 5.787 min (Figure 5). Efficient separation of L-citrulline with a retention time of about 4 min was achieved on the Gemini C_{18} column due to the high degree similarity of column with polar compounds [13]. Gemini C_{18} is a new generation hybrid column end-capped with porous silica as base core and polymer media coated on top of the silica core which exhibit silica-like mechanical properties of base material while similarly decreasing the number of residual silanols [26]. This result demonstrated that Gemini C_{18} is the most suited column for efficient separation of mixed standard, L-citrulline and L-arginine.

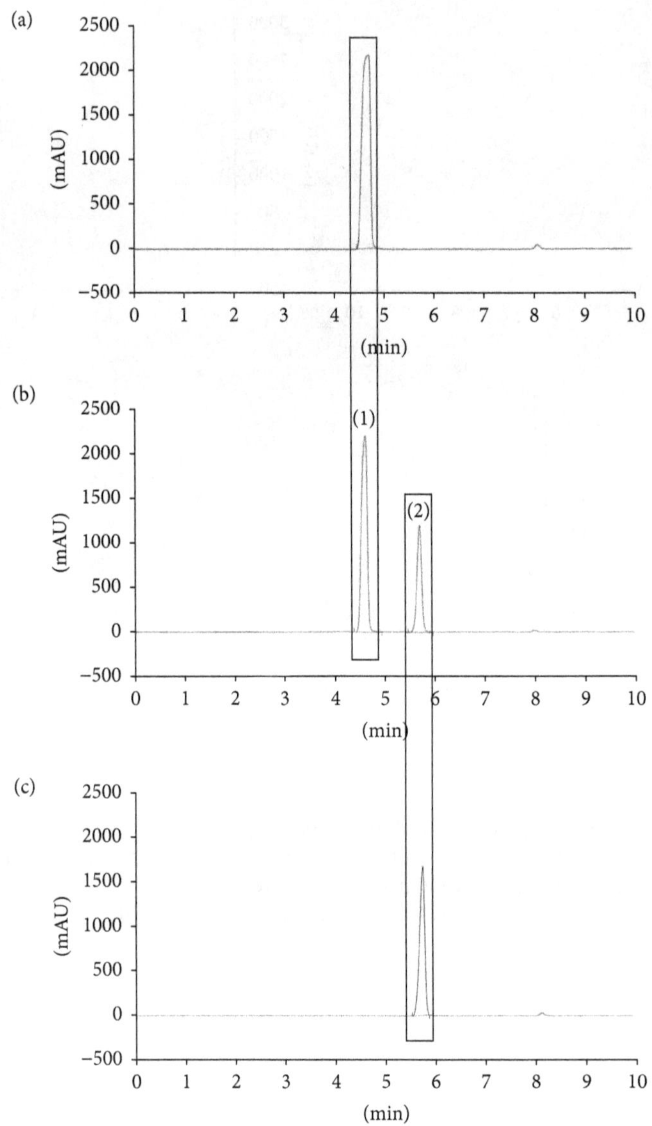

FIGURE 5: Comparative chromatograms showing isocratic RP-HPLC separation of individual and mixed standard, L-citrulline, and L-arginine using Gemini C_{18}: (a) L-arginine, (b) mixed standard, and (c) L-citrulline. The peaks marked represent (1) L-arginine and (2) L-citrulline.

The result from chromatography separation of L-citrulline and L-arginine shown in Figure 5(b) demonstrated that the reverse-phase (RP) mode provided efficient separation and substantial retention achieved on both polar compounds without the need for derivatization. RP mode can efficiently be applied in this study although Brown et al. [27] proposed the use of hydrophilic interaction liquid chromatography (HILIC) mode after cyano- and pentafluorophenylpropyl stationary phases failed to retain target compounds including L-citrulline and L-arginine. HILIC mode is primarily used when separation of very polar compounds is needed or for incomplete chromatographic separation in RP mode [28]. However, HILIC mode required an expensive and robust system equipped with tandem mass spectrometric (MS) detection for L-citrulline and L-arginine separation in high biological matrix samples such as serum and plasma [16, 27]. Complete chromatographic separation of L-citrulline and L-arginine by RP mode in this study eliminates the use of HILIC mode.

3.2. Validation of Isocratic RP-HPLC Method. Method validation of isocratic RP-HPLC method was performed by determination of linearity, LOD, LOQ, recovery, and intraday and interday analysis. The results of linearity, LOD, and LOQ were summarized in Table 2. Good linear regression equations of L-citrulline and L-arginine standard are displayed between corresponding peak areas versus concentrations of compounds based on the correlation coefficients (R^2 = 0.9956, $y = 0.1664x + 2.4142$ and $R^2 = 0.9912$, $y = 0.4100x + 3.4850$, resp.). The LOD and LOQ for L-citrulline were 0.42 μg/mL and 1.28 μg/mL while those for L-arginine were 0.88 μg/mL and 2.66 μg/mL, which demonstrated that isocratic RP-HPLC method was efficiently sensitive. The method had good accuracy that showed efficient recoveries for both

TABLE 2: Calibration data of L-citrulline and L-arginine standard reported from isocratic RP-HPLC method.

Standard	Concentration range (μg/mL)	Regression equation	Correlation coefficient (R^2)	Limit of detection (μg/mL)	Limit of quantification (μg/mL)
L-citrulline	0.01–1000	$y = 0.1664x + 2.4142$	0.9956	0.42	1.28
L-arginine	0.01–500	$y = 0.4100x + 3.4850$	0.9912	0.88	2.66

TABLE 3: Recovery of L-citrulline and L-arginine standard reported from isocratic RP-HPLC method.

Compounds	Added concentration (μg/mL)	Measured concentration (μg/mL)	Recovery (%)	RSD (%)
L-citrulline	100	101.94	101.94	1.70
	60	61.24	102.07	1.46
	30	31.02	103.38	1.00
L-arginine	100	99.87	99.87	1.96
	60	59.33	98.88	1.76
	30	31.02	103.41	1.17

TABLE 4: Intraday and interday analysis of L-citrulline and L-arginine standard reported from isocratic RP-HPLC method.

Compounds	Concentration (μg/mL)	Intraday (n = 3) (%)		Interday (n = 3) (%)	
		Mean	RSD	Mean	RSD
L-citrulline	150	101.25	1.23	100.42	1.09
	60	102.07	1.46	103.44	1.01
	20	116.44	2.03	116.22	0.37
L-arginine	150	102.01	0.56	103.83	2.05
	60	96.26	1.04	97.36	1.13
	20	95.41	0.68	95.94	0.33

compounds ranging from 98.88% to 103.41% (Table 3). The RSD (%) for intraday and interday precision ranged from 0.37% to 1.09% for L-citrulline and 0.33% to 2.05% for L-arginine, in which both RSD ≤ 2% (Table 4). These validation results confirmed that the isocratic RP-HPLC method is precise, accurate, and sensitive for simultaneous quantification of L-citrulline and L-arginine.

3.3. Quantification of L-Citrulline and L-Arginine Contents in Two Different Watermelon Extracts. Consumption of watermelon extracts rich in L-citrulline and L-arginine is proven to be beneficial for diseases prevention. Thus, a rapid, reliable, and efficient isocratic RP-HPLC method is essential for simultaneous quantification of these amino acids in juice extracts and methanol extracts. The chromatographic profiles of both extracts in red watermelon and yellow crimson watermelon are presented in Figure 6. Quantification of L-citrulline and L-arginine was performed using Chromeleon software. The content was calculated based on the calibration curve of L-citrulline and L-arginine standard achieved with good correlation coefficients and linear regression equations, $R^2 = 0.9956$, $y = 0.1664x + 2.4142$ and $R^2 = 0.9912$, $y = 0.4100x + 3.4850$, respectively. The results are tabulated in Table 5.

Red watermelon juice extract showed slightly high yield of L-citrulline in rind, 45.02 mg/g compared to flesh, 43.81 mg/g. Similar trends were shown in L-citrulline content in rind and flesh of yellow crimson juice extract, 16.61 mg/g and 15.77 mg/g, respectively. This finding is in accordance with the study by Jayaprakasha et al. [13] which found that rinds from *C. vulgaris* watermelon varieties of petite treat and jamboree watermelon and yellow crimson watermelon contained significantly high L-citrulline, 13.95 mg/g, 20.84 mg/g, and 28.46 mg/g, respectively, compared to flesh, 11.25 mg/g, 16.73 mg/g, and 14.74 mg/g, respectively, using RP-HPLC method. L-arginine content in red watermelon juice extract was higher in flesh, 11.10 mg/g, compared to rind, 3.39 mg/g. L-arginine content was approximately 3-fold lower than L-citrulline in red watermelon flesh juice extract. Consumption of watermelon flesh juice extract aided in efficient conversion of significantly high L-citrulline, a potent endogenous precursor to L-arginine in the kidney, which resulted in increased plasma L-arginine concentration. Findings by Collins et al. [29] proved that plasma L-arginine concentration increased by 95.2 ± 3.5 μM and 108.0 ± 4.1 μM compared to normal plasma baseline, 86.4 ± 3.5 μM after 3 weeks of consumption of 780 mL (~1 g L-citrulline/day) to 1560 mL (~2 g L-citrulline/day) of watermelon juices. Recently, a study by Bailey et al. [30] supported the notion of the increased plasma L-arginine concentration by 116 ± 9 μM compared to placebo, 67 ± 13 μM after 2 weeks of 300 mL/day watermelon juice consumption, which contains

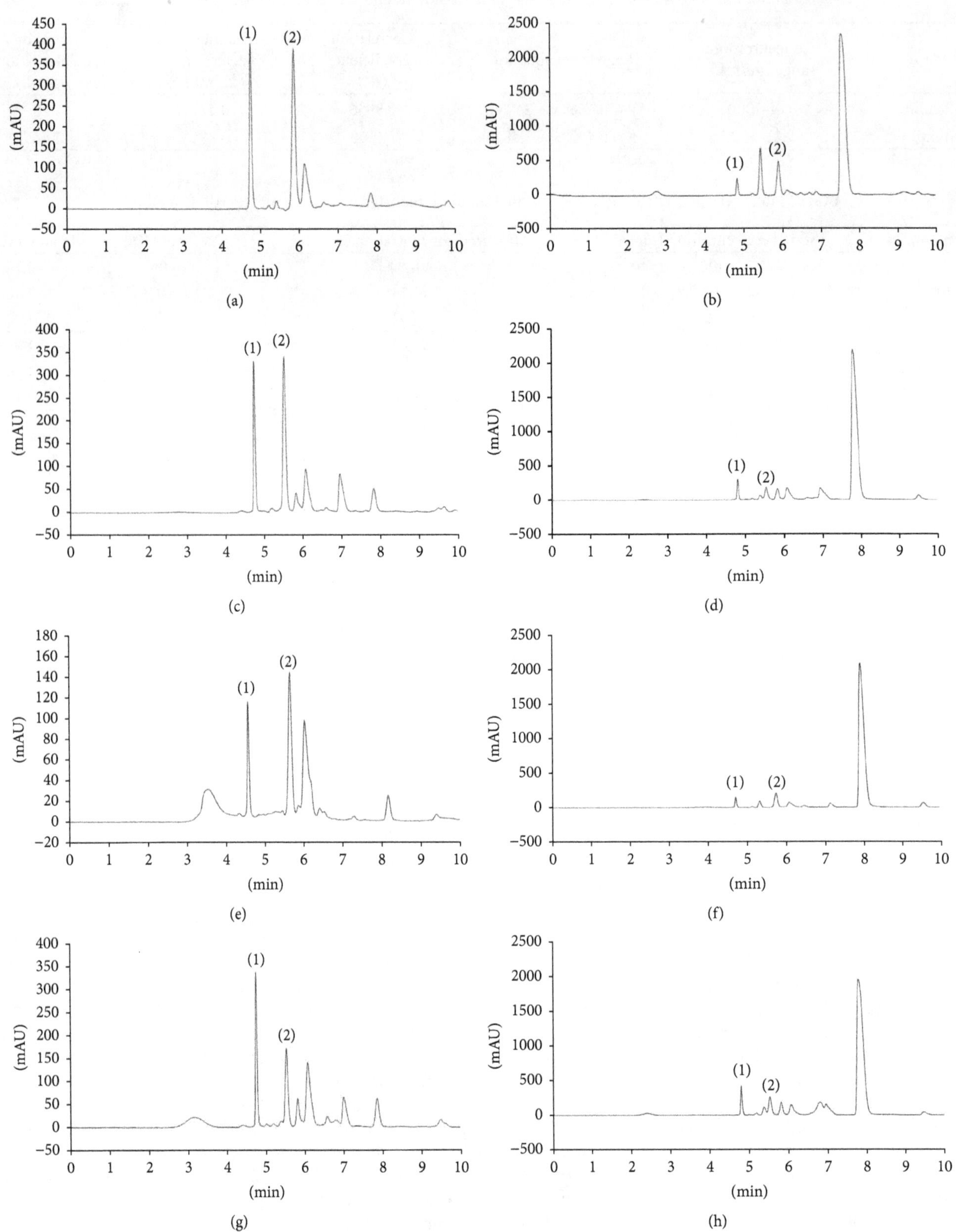

FIGURE 6: Comparative chromatographic profiles showing isocratic RP-HPLC separation of flesh and rind from juice extracts and methanol extracts of red watermelon and yellow crimson watermelon using Gemini C_{18}. Red watermelon: (a) flesh juice extract, (b) rind juice extract, (c) crude flesh extract, and (d) crude rind extract. Yellow crimson watermelon: (e) flesh juice extract, (f) rind juice extract, (g) crude flesh extract, and (h) crude rind extract. The peaks marked represent (1) L-arginine and (2) L-citrulline.

Table 5: L-citrulline and L-arginine contents (mg/g) in juice extract and methanol extract of watermelon flesh and rind juices.

Watermelon	Compounds	Juices	Juice extract (mg/g)	Methanol extract (mg/g)
Red	L-citrulline	Flesh	43.81	16.22
		Rind	45.02	24.99
	L-arginine	Flesh	11.10	6.42
		Rind	3.39	4.08
Yellow crimson	L-citrulline	Flesh	15.77	13.91
		Rind	16.61	16.03
	L-arginine	Flesh	8.23	6.68
		Rind	11.14	8.41

~3.4 g L-citrulline/day. L-citrulline content in crude flesh and rind extract of red watermelon and yellow crimson watermelon varied in the range of 13.91–24.99 mg/g, while L-arginine content was in the range of 4.08–8.41 mg/g. L-citrulline and L-arginine content is much lower in methanol extracts compared to juice extracts. Fish and Bruton [17] stated that methanol extracts may diminish the solubility of amino acids, thus marked reduction in amino acids yield. The quantitative results confirmed that watermelon juice extracts most effectively quantified higher yield of L-citrulline and L-arginine, and this study's outcomes may possibly suggest that juice extraction method is best in optimizing amino acids yield.

4. Conclusion

The isocratic RP-HPLC method has been successfully developed for separation and quantification of L-citrulline and L-arginine content in both watermelon extracts of flesh and rind using the selected mobile phase (0.1% H_3P0_4) in Gemini C_{18}. The established isocratic RP-HPLC method provides evidence that L-citrulline and L-arginine are best retained using Gemini C_{18} column. The validated method is robust, sensitive, accurate, and precise with good linearity ($R^2 \geq 0.99$), low values of LOD and LOQ, recoveries within 98.88%–103.41%, and %RSD precision less than 2%. Juice extract effectively yielded higher L-citrulline and L-arginine content by juice extraction method; thus it is potentially used for quantitative amino acids analysis. The present study procedure may provide a basis for separation and quantification of L-citrulline and L-arginine in local watermelons. The high content of L-citrulline and L-arginine suggested watermelons as a good source of nutraceutical and health benefits ingredients. However, further researches are necessary to explore biological activities such as aphrodisiac properties of these active constituents in watermelons to support their potential in human diet and prevention of health related diseases.

Acknowledgments

The authors gratefully acknowledge (1) the Ministry of Higher Education, Malaysia, through Fundamental Research Grant Scheme [FRGS Grant no. FRGS/1/2016/WAB01/UITM/02/3] and Institute of Research Management & Innovation (IRMI), Universiti Teknologi MARA (UiTM), for funding the study; (2) Dr. Maizatul Hasyima Omar from Phytochemistry Unit, Institute for Medical Research, Kuala Lumpur, for providing analytical training; and (3) Atta-ur-Rahman Institute (AuRIns), UiTM Puncak Alam, Centre of Medical Laboratory Technology, and Centre of Postgraduate Study, Faculty of Health Sciences, UiTM, Selangor, Puncak Alam Campus, for providing facilities throughout this study.

References

[1] M. K. A. Mohammad, M. I. Mohamed, A. M. Zakaria, H. R. Abdul Razak, and W. M. M. Saad, "Watermelon (*Citrullus lanatus* (Thunb.) Matsum. and Nakai) juice modulates oxidative damage induced by low dose X-ray in mice," *BioMed Research International*, vol. 2014, Article ID 512834, 6 pages, 2014.

[2] G. K. Jayaprakasha and B. S. Patil, "A metabolomics approach to identify and quantify the phytochemicals in watermelons by quantitative ^{1}HNMR," *Talanta*, vol. 153, pp. 268–277, 2016.

[3] A. Naz, M. S. Butt, M. T. Sultan, M. M. N. Qayyum, and R. S. Niaz, "Watermelon lycopene and allied health claims," *Experimental and Clinical Sciences Journal*, vol. 13, pp. 650–666, 2014.

[4] A. Figueroa, M. A. Sanchez-Gonzalez, P. M. Perkins-Veazie, and B. H. Arjmandi, "Effects of watermelon supplementation on aortic blood pressure and wave reflection in individuals with prehypertension: A pilot study," *American Journal of Hypertension*, vol. 24, no. 1, pp. 40–44, 2011.

[5] A. Poduri, D. L. Rateri, S. K. Saha, S. Saha, and A. Daugherty, "*Citrullus lanatus* 'sentinel' (watermelon) extract reduces atherosclerosis in LDL receptor-deficient mice," *The Journal of Nutritional Biochemistry*, vol. 24, no. 5, pp. 882–886, 2013.

[6] W. Liu, S. Zhao, Z. Cheng, X. Wan, Z. Yan, and S. R. King, "Lycopene and citrulline contents in watermelon (*Citrullus lanatus*) fruit with different ploidy and changes during fruit development," *Acta Horticulturae*, vol. 871, pp. 543–550, 2010.

[7] A. R. Davis, C. L. Webber III, W. W. Fish, T. C. Wehner, S. King, and P. Perkins-Veazie, "L-citrulline levels in watermelon cultigens tested in two environments," *HortScience*, vol. 46, no. 12, pp. 1572–1575, 2011.

[8] A. Shafaei, A. F. Aisha, M. J. Siddiqui, and Z. Ismail, "Analysis of L-citrulline and L-arginine in *Ficus deltoidea* leaf extracts by reverse phase high performance liquid chromatography," *Pharmacognosy Research*, vol. 7, no. 1, pp. 32–37, 2015.

[9] E. Curis, I. Nicolis, C. Moinard et al., "Almost all about citrulline in mammals," *Amino Acids*, vol. 29, no. 3, pp. 177–205, 2005.

[10] M. J. Romero, D. H. Platt, R. B. Caldwell, and R. W. Caldwell, "Therapeutic use of citrulline in cardiovascular disease," *Cardiovascular Drug Reviews*, vol. 24, no. 3-4, pp. 275–290, 2006.

[11] G. Wu, J. K. Collins, P. Perkins-Veazie et al., "Dietary supplementation with watermelon pomace juice enhances arginine availability and ameliorates the metabolic syndrome in zucker diabetic fatty rats," *Journal of Nutrition*, vol. 137, no. 12, pp. 2680–2685, 2007.

[12] A. M. Rimando and P. M. Perkins-Veazie, "Determination of citrulline in watermelon rind," *Journal of Chromatography A*, vol. 1078, no. 1-2, pp. 196–200, 2005.

[13] G. K. Jayaprakasha, K. N. Chidambara Murthy, and B. S. Patil, "Rapid HPLC-UV method for quantification of L-citrulline in watermelon and its potential role on smooth muscle relaxation markers," *Food Chemistry*, vol. 127, no. 1, pp. 240–248, 2011.

[14] G. Wu and C. J. Meininger, "Analysis of Citrulline, Arginine, and Methylarginines Using High-Performance Liquid Chromatography," *Methods in Enzymology*, vol. 440, pp. 177–189, 2008.

[15] L. Wang, R. Xu, B. Hu et al., "Analysis of free amino acids in Chinese teas and flower of tea plant by high performance liquid chromatography combined with solid-phase extraction," *Food Chemistry*, vol. 123, no. 4, pp. 1259–1266, 2010.

[16] S. Shin, S.-M. Fung, S. Mohan, and H.-L. Fung, "Simultaneous bioanalysis of L-arginine, L-citrulline, and dimethylarginines by LC-MS/MS," *Journal of Chromatography B*, vol. 879, no. 7-8, pp. 467–474, 2011.

[17] W. W. Fish and B. D. Bruton, "Quantification of L-citrulline and other physiologic amino acids in watermelon and various cucurbits," *Cucurbitaceae*, vol. 2010, pp. 152–154, 2010.

[18] P. Markowski, I. Baranowska, and J. Baranowski, "Simultaneous determination of L-arginine and 12 molecules participating in its metabolic cycle by gradient RP-HPLC method. Application to human urine samples," *Analytica Chimica Acta*, vol. 605, no. 2, pp. 205–217, 2007.

[19] D. J. Dietzen, A. L. Weindel, M. O. Carayannopoulos et al., "Rapid comprehensive amino acid analysis by liquid chromatography/tandem mass spectrometry: Comparison to cation exchange with post-column ninhydrin detection," *Rapid Communications in Mass Spectrometry*, vol. 22, no. 22, pp. 3481–3488, 2008.

[20] M. Sadji, P. M. Perkins-Veazie, N. F. Ndiaye et al., "Enhanced L-citrulline in parboiled paddy rice with watermelon (*Citrullus lanatus*) juice for preventing sarcopenia: A preliminary study," *African Journal of Food Science*, vol. 9, no. 10, pp. 508–513, 2015.

[21] M. W. Dong, *Modern HPLC for Practicing Scientists*, John Wiley and Sons, 2006.

[22] S. Fekete, J.-L. Veuthey, and D. Guillarme, "New trends in reversed-phase liquid chromatographic separations of therapeutic peptides and proteins: Theory and applications," *Journal of Pharmaceutical and Biomedical Analysis*, vol. 69, pp. 9–27, 2012.

[23] J. Dolan, *A Guide to HPLC and LC-MS Buffer Selection*, ACE, 2009.

[24] M. Shibue, C. T. Mant, and R. S. Hodges, "Effect of anionic ion-pairing reagent hydrophobicity on selectivity of peptide separations by reversed-phase liquid chromatography," *Journal of Chromatography A*, vol. 1080, no. 1, pp. 68–75, 2005.

[25] W. E. Barber and M. Joseph, *Eclipse XDB-CN Provides Excellent Selectivity and Resolution for Urea Pesticides*, 2004, https://www.agilent.com/cs/library/applications/5989-0930EN_low.pdf.

[26] E. M. Borges, "Silica, hybrid silica, hydride silica and non-silica stationary phases for liquid chromatography," *Journal of Chromatographic Science (JCS)*, vol. 53, no. 4, pp. 580–597, 2015.

[27] C. M. Brown, J. O. Becker, P. M. Wise, and A. N. Hoofnagle, "Simultaneous determination of 6 L-arginine metabolites in human and mouse plasma by using hydrophilic-interaction chromatography and electrospray tandem mass spectrometry," *Clinical Chemistry*, vol. 57, no. 5, pp. 701–709, 2011.

[28] B. Buszewski and S. Noga, "Hydrophilic interaction liquid chromatography (HILIC)-a powerful separation technique," *Analytical and Bioanalytical Chemistry*, vol. 402, no. 1, pp. 231–247, 2012.

[29] J. K. Collins, G. Wu, P. Perkins-Veazie et al., "Watermelon consumption increases plasma arginine concentrations in adults," *Nutrition Journal*, vol. 23, no. 3, pp. 261–266, 2007.

[30] S. J. Bailey, J. R. Blackwell, E. Williams et al., "Two weeks of watermelon juice supplementation improves nitric oxide bioavailability but not endurance exercise performance in humans," *Nitric Oxide: Biology and Chemistry*, vol. 59, pp. 10–20, 2016.

9

Selective Preconcentration of Gold from Ore Samples

Hurmus Refiker ,[1,2] Melek Merdivan,[3] and Ruveyde Sezer Aygun[1]

[1] *Department of Chemistry, Middle East Technical University, Ankara, Turkey*
[2] *Institute of Applied Sciences, University of Kyrenia, Sehit Bakir Yahya Street, Kyrenia, North Cyprus, Mersin 10, Turkey*
[3] *Department of Chemistry, Dokuz Eylul University, Izmir, Turkey*

Correspondence should be addressed to Hurmus Refiker; hurmus.refiker@kyrenia.edu.tr

Academic Editor: Seyyed E. Moradi

A simple and selective method has been developed for preconcentration of gold in ore samples. The method is based on use of N, N-diethyl-N'-benzoylthiourea (DEBT) as selective chelating agent and Amberlite XAD-16 as solid sorbent. Sorption behavior of gold with DEBT impregnated resin under optimized conditions has been studied in batch process. The gold ion capacity of the impregnated resin is calculated as 33.48 mg g^{-1} resin (0.17 mmol g^{-1} resin). The selective preconcentration of metal was examined using gold chelates prepared in column process under optimized conditions: pH, flow rate, volume of sample solution, nature of eluent, flow rate, and volume of eluent. Under optimum conditions, gold ions at the concentration of 0.015 μg mL^{-1} with a preconcentration factor of 6.7 have been determined by flame atomic absorption spectrometry (FAAS). The accuracy of the proposed method was validated by the analysis of a Cu-ore (semi-certified) supplied by CMC (Cyprus Mining Company, North Cyprus) and a certified reference material, Gold Ore (MA-1b Canmet-MMSL). Satisfactory results were obtained with a RSD of 7.6%. The highly selective proposed method does not require any interference elimination process.

1. Introduction

Gold is one of the precious metals which occurs in very low natural contents such as 4 ng g^{-1} in basic rocks, 1 ng g^{-1} in soils, 0.05 μg L^{-1} in sea water, and 0.2 μg/L in river water. Due to its specific physical and chemical properties, gold is widely used in industry, agriculture, and medicine [1]. Low abundance and heterogeneous distribution of gold in geological samples and various interfering matrices requires the development of accurate and reliable analytical procedures for determination of gold in environmental samples. Therefore, a selective separation and preconcentration method is a critical need for sensitive, accurate, and interference free determination of gold [2].

In literature, various techniques have been recorded for separation and preconcentration of gold, such as liquid-liquid extraction [3], coprecipitation [4], solid-phase extraction [5], cloud point extraction [6], and electrodeposition [7]. Solid-phase extraction (SPE) is preferable over all these techniques due to its advantages like high enrichment factor, high recovery, rapid phase separation, low cost, minimum solvent waste generation, and sorption of the target species on the solid surface in a more stable chemical form [8, 9].

The use of solid sorbents for preconcentration and separation has received great attention from analytical chemists [10]. Among wide range of solid phases such as multiwalled carbon nanotubes [11], surfactant coated alumina [12], styrene-divinylbenzene matrix [5, 13, 14], and silica gel [10, 15] have gained much importance for the metal ion enrichment. Amberlite XAD series have been more preferably used as solid support due to their good physical properties such as high porosity, uniform pore size distribution, high surface area as chemical homogenous non-ionic structure, and good adsorbent properties for great amounts of uncharged compounds [5, 16, 17]. Compared to Amberlite XAD-2 and XAD-4 resins, Amberlite XAD-16 has larger surface area [9], which makes possible to increase the number chelating sites hence increasing the selectivity towards target metal ions. This can be achieved by selecting suitable chelating agents. The chelating groups that are widely used for preconcentration of precious metals are imidazole, thioguanidine, dithizone, mercapto groups, amino groups, and thioureas [18]. In several studies, N, N-diethyl-N'-benzoylthiourea (DEBT) has been recorded as a selective complexing agent for precious metals [19, 20]. Its selectivity is mainly controlled by pH. It has very high resistance to hydrolysis and oxidation. In addition to high pKs values, with resonance effects, DEBT can increase the electron density at sulfur donor atom when suitable acceptors are available. DEBT forms stable complexes only with Class b and border line acceptors. Noble metal ions, due to their specific Class b properties, form chelates with DEBT in low oxidation states in strongly acidic solutions [19, 20].

In the present study, DEBT as a chelating ligand and Amberlite XAD-16 as solid support have been used for selective separation and preconcentration of gold which is determined by flame atomic absorption spectrometry (FAAS). Optimum conditions for batch and column processes have been studied in detail. Then the proposed method has been applied to two real samples: Cu-ore supplied from CMC, North Cyprus, and a certified reference material Gold Ore (MA-1b) supplied by Canmet-MMSL, Ontario.

2. Experimental

2.1. Apparatus and Instrumentation. In order to prevent sorption of gold on silica surfaces, equipment made of polytetrafluoroethylene (PTFE) was used. 100 mL DuPont polyethylene containers for the storage and 5-50 and 100-1000 μL adjustable micropipettes (Transferpette, Treff Lab) with disposable polyethylene tips for preparation of solutions were used. In batch process, optimum conditions such as pH, stirring time, metal ion capacity, and agents suitable for desorption were studied by using 50-mL of Falcon tubes. NÜVE SL 350 horizontal shaker was used during sorption optimizations in batch process.

Columns were prepared from 12-mL syringe barrels (1.5 cm i.d., 7.8 cm height, PTFE, Supelco) where disposable porous frits were placed at the bottom of the barrels. 1.0 g resin (unless otherwise stated) slurred in 50 mL water was uniformly placed in column and was covered with cotton wool to prevent dispersion by the addition of sample solution. Tygon® tubing was used to connect the outlet tip of the syringe barrel to a Gilson Miniplus peristaltic pump. A calibration, flow rate mL min^{-1} versus rpm was carried out. This calibration was repeated for each column before the application. Each time, 15 mL blank solutions at a flow rate of 1 mL min^{-1} were passed before sorption and desorption studies.

Philips PU 9200 Atomic Absorption Spectrometer with Epson FX-850 printer was used for determination of gold ions.

2.2. Chemicals. All the reagents were of analytical reagent grade. Deionized water from a Milli-Q system was used throughout the study unless otherwise stated. Amberlite XAD-16 resin was supplied by Sigma. Gold standard solutions were prepared by diluting of 1000 μg mL^{-1} stock solution (Spectrosol) with 1 mol L^{-1} HCl (J.T. Baker, 36-38% w/w). During batch process, pH-adjustments were done using NaOH (Acros, 50% w/w). For desorption studies, $Na_2S_2O_3$ (extra pure, Bilesik Kimya Mekanik) was used.

2.3. Synthesis of DEBT and Impregnation Process. DEBT was synthesized according to the procedure modified in our laboratory [20] where potassium thiocyanate (Fischer, 0.1mol) was dissolved in anhydrous acetone (Riedel-deHaen, 100 mL) by stirring and heating in a reflux condenser. After cooling to room temperature, benzoyl chloride (Merck, 0.1 mol) was added dropwise and stirred for 30 minutes. Then potassium salt was filtered off. The filtrate in orange was reacted with 0.1 mol of diethylamine (Merck) dropwise. The reaction mixture was crystallized in 250 mL of 1 mol L^{-1} HCl solution. After filtering the mixture, the residue was recrystallized with ethanol.

Since the impregnation process deals with physical interactions between the chelating agent and solid support by either inclusion in the pores of the support material or adhesion process or electrostatic interaction, some parameters controlling the impregnation such as stirring time and chelating agent capacity have been optimized as mentioned elsewhere [21].

2.4. Batch Method. With batch studies, sorption behavior of high concentrations of gold on DEBT impregnated Amberlite XAD-16 was investigated. Some critical parameters such as pH, stirring time, and metal ion capacity of resin capacity have been studied to find out the optimum conditions for recovery of gold.

2.4.1. pH Effect. In order to investigate the pH effect on sorption of gold onto impregnated resin, different sets of 10 mL of 10 mg L^{-1} of Au^{3+} solution in the pH range of 1-5 were stirred with samples of 0.1 g impregnated resin for 50 minutes. After filtration under vacuum, metal ions in the filtrate were determined by FAAS.

2.4.2. Stirring Time. Three different sets of 2 mg L^{-1}, 10 mg L^{-1}, and 100 mg^{-1}L of 10 mL of Au^{3+} solutions in 1 mol L^{-1} HCl were stirred with samples of 0.1 g impregnated resin from the periods of 5 minutes to 1 hour. Then solutions were filtered and filtrates were aspirated into FAAS for metal ion determination.

2.4.3. Gold Ion Capacity of Resin. In order to determine the resin capacity, samples of 0.1 g of impregnated resin (1 mmol DEBT g^{-1} resin) were stirred with 10 mL of gold ions solutions in the concentration range of 2 mg L^{-1} to 600 mg L^{-1} in 1 mol /L^{-1} HCl for 15 minutes. Then the solutions were filtered and metal ion concentrations were determined by FAAS.

2.5. Column Method. Since the kinetic and equilibrium aspects of column process are different than batch process, optimization of column conditions is needed. Effect of flow rate and volume of ligand solution on impregnation and effect of ligand concentration on amount of metal chelate adsorption have been studied in column process.

2.6. Proposed Method for Preconcentration of Gold. 100 mL of gold chelate solutions (0.15 μg mL^{-1} Au^{3+} and 3 mL of 2x10^{-3} mol L^{-1} DEBT) was percolated through the column (1.0 g pure resin) at a flow rate of 0.5 mL min^{-1}. Then metal ions could be eluted with 15 mL of 0.2 mol L^{-1} $Na_2S_2O_3$ in water with a recovery of 97.6 ± 2.3% (N=2).

2.7. Preparation of Ore Samples. An acid digestion procedure was applied to Cu-ore and Gold Ore (MA-1b) samples as suggested elsewhere [22]. Accordingly, two parallel 10.0 g of Cu-ore sample and 1.0 g of Gold Ore (MA-1b) were transferred into Teflon beakers. 20 mL of HCl was added to each where the beakers were covered and placed on a warm hot plate. After 15 minutes of digestion, 15 mL of concentrated nitric acid was added and the contents were digested for 20 minutes. Then 25 mL of concentrated HCl and 25 mL of deionized water were added. The contents were boiled to expel nitric acid digestion gases and to dissolve all soluble salts. After cooling they were filtered through Whatman white band filter paper into 100 ml PTFE flask. Once 3 mL of 2 X 10^{-3} mol L^{-1} DEBT solution was added, final volume was completed to 100 mL with deionized water. Later, the proposed procedure for preconcentration of gold was applied.

3. Results and Discussion

3.1. Characterization Studies. The structure of DEBT was characterized by FTIR and UV-VIS spectrophotometer. The synthesized DEBT exhibited 2 strong broad UV-absorption peaks at 237 and 278 nm which were consistent with those given in literature [20]. The characteristic absorption bands for N-H, C–H, and amide I (C=O), amide II, and amide III at 3276, 3066-2936, 1656, 1537, and 1306 cm^{-1}, respectively, appeared in both of the spectra. FTIR studies for pure resin, DEBT, DEBT impregnated resin, and DEBT-metal chelates have been carried out [20]. The characteristic absorption bands for C-H and amide I (C=O), amide II, and amide III at 3276, 3066-2936, 1656, 1537, and 1306 cm^{-1}, respectively, appeared only in the spectra of DEBT and impregnated resin. While the characteristic IR band of $-N(CH_2CH_3)_2$ group in the ligand at 2875 cm^{-1} remained almost unchanged in the spectrum of the complex showing that this group is not involved in coordination, C-H vibration in the aromatic ring is blue shifted upon metal-ligand bond formation. The position of amide I, amide II, and III bands at 1656, 1537, and 1306 cm^{-1}, respectively, arising from the carbonyl of the benzamide moiety and secondary amide of DEBT at 3276 cm^{-1} disappeared in the complex.

3.2. Parameters Optimized in Batch Method

3.2.1. pH Effect. In literature, it is noted that DEBT forms stable and selective complexes with noble metals only in acidic or strongly acidic media [19]. Moreover, Shuster and coworker reported optimum pH range as 0-5 for liquid-liquid extraction of gold with DEBT was previously reported as 0-5 by Schuster and coworkers [31]. Therefore, the pH effect on chelation of gold ions with DEBT impregnated resin is investigated within the pH range from 1 to 5.

It was shown that the maximum percent sorption is obtained at pH ~ 1 (see Figure 1). Therefore, standard solutions were prepared by diluting AAS standard stock solutions with 1 mol L^{-1} HCl.

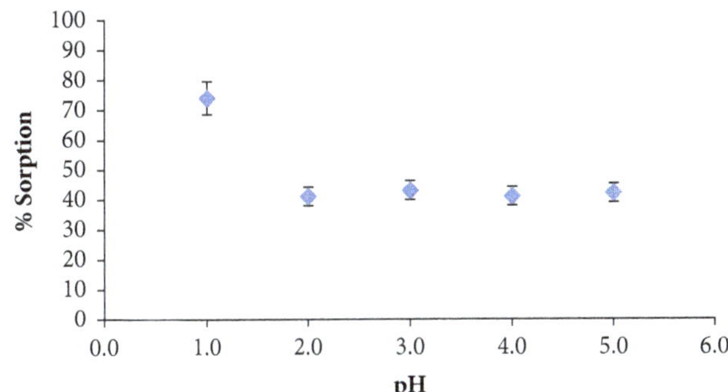

FIGURE 1: Effect of pH on sorption. Amount of resin: 0.1 g resin, amount of DEBT: 1mmol g^{-1} resin, and stirring time: 50 minutes.

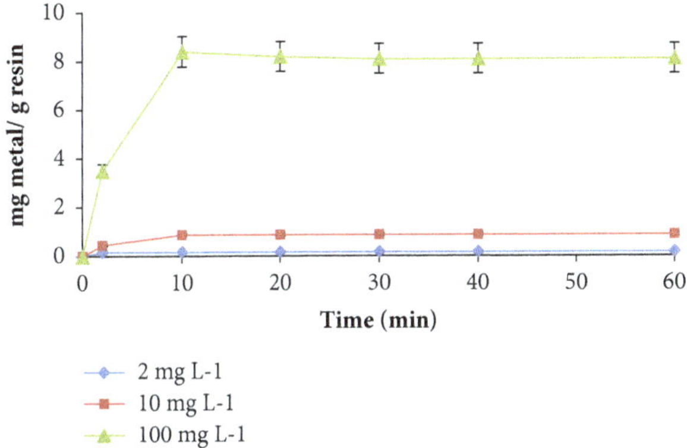

FIGURE 2: Stirring time of gold. Amount of resin: 0.1 g and amount of DEBT: 1 mmol g^{-1} resin.

3.2.2. Effect of Stirring Time. Referring to Figure 2, it can be concluded that 15 minutes of stirring is sufficiently good enough to achieve sorption equilibrium for three different concentrations of gold ion. Higher gold ion concentrations have no effect on the optimum time of sorption. Actually, fast kinetics can be expected in applications of macroporous resins. In addition, high ligand concentration, 1 mmol g^{-1} resin, used in impregnation may increase selectivity of the resin which can also be a reason for fast sorption rate.

As a result, 15 minutes of stirring can be accepted as a suitable stirring time during loading of resin with possible higher concentrations of gold ions to determine gold ion capacity of the resin.

3.2.3. Gold Ion Capacity of Impregnated Resin. Referring to Figure 3, after 500 mg L^{-1} of gold ions solutions, the impregnated resin reaches to saturation. The metal ion capacity of the resin is calculated as 33.48 mg Au^{3+} g^{-1} resin (0.17 mmol Au^{3+} g^{-1} resin) applying the following formula [32].

$$Q = \frac{(C_o - C_A) \times V}{W} \quad (1)$$

where Q is the metal ion capacity (mg/g),

C_o is the initial concentration of metal ion (mg/L),

C_A is the equilibrium concentration of metal ion (mg/L),

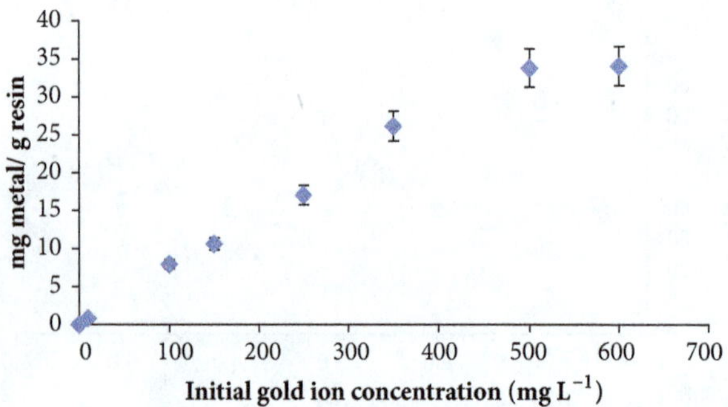

FIGURE 3: Gold ion capacity of resin. Amount of resin: 0.1 g, amount of DEBT: 1 mmol g^{-1} resin, and stirring time: 15 minutes.

V is the volume of the solution (L),

W is the weight of the resin (g).

3.3. Optimized Parameters in Column Method. During the application of the proposed method in column process, it was noticed that excess volume of sample solutions during sorption leached the impregnated DEBT that lead to loss of selectivity and analyte. Moreover, the partial exhaustion of available chelating sites due to leaching of impregnated ligand caused irreproducible results of sorption percentages of metal ion [5]. Therefore, research was continued with preparation of metal chelates before transferring to column and certain limited volume of chelate solution would be percolated through the column containing pure resin under the optimized conditions.

3.3.1. Sample Flow Rate of Gold Chelates. During batch studies, it was recognized that DEBT showed similar kinetics during chelation with gold as that of silver which was reported in the previous study [5], as long as DEBT concentration is kept the same or close optimum pH for sorption is maintained [21]. In the previous study, considering application of larger volume of sample solutions for preconcentration, to be on safe 0.5 mL min^{-1} had been accepted as optimum sample flow rate [5]. The same was also found to be optimum sample flow rate for gold studies.

3.3.2. Effect of Ligand Volume on Impregnation of Gold Chelates onto Resin. Maximum applicable ligand volume and concentration on analyte sorption are important. Therefore, maximum applicable ligand volume on retention is studied before further application of metal chelates for solid-phase extraction. For this reason, 3 mL of $3.75\text{x}10^{-4}$ mol L^{-1} DEBT solution was percolated through column for 4 times and DEBT concentration in the effluent determined by UV spectrometry. In Figure 4, it can be seen that maximum amount of DEBT retained on Amberlite XAD-16 was achieved with the first 3 mL of DEBT solution. Following additions of 3 mL of $3.75\text{x}10^{-4}$ mol L^{-1} DEBT solution showed a decrease in amount of DEBT retained on resin. This may be because of the leaching effect of ethanol on DEBT.

3.3.3. Effect of Optimized Ligand Concentration on Amount of Gold Chelates. Ligand concentration is also important

TABLE 1: Effect of ligand concentration of retention of metal chelates.

Amount of DEBT	Amount of Au^{3+} in sample solution (μg)	% Sorption of gold chelates on resin
3mL of $2\text{x}10^{-3}$ mol L^{-1}	15	100 ± 2
3mL of $2\text{x}10^{-3}$ mol L^{-1}	100	100 ± 2
3mL of $2\text{x}10^{-3}$ mol L^{-1}	500	94 ± 3

Amount of resin: 1.0 g, sample volume: 10 mL, and sample flow rate: 0.5 mL min^{-1}.

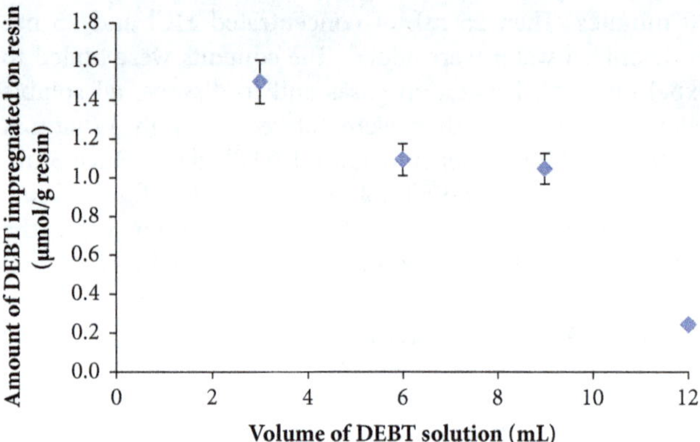

FIGURE 4: Effect of ligand volume on impregnation. Initial DEBT-ethanol concentration: 3.75 x 10^{-4} mol L^{-1} and flow rate: 0.5 mL min^{-1}.

because if it is not excessively present, the chelate formation may not be complete so metal ions may not be selectively retained on resin. However, excess DEBT (in case of inadequate amount of resin) may prevent retention of metal chelates because of the competition for sorption on resin between excess DEBT and metal chelates.

Considering the further applications of the proposed method to a real sample and limitations related to ligand mentioned above, it was decided to use 1.0 g of resin and amount of DEBT as 3 mL of $2\text{x}10^{-3}$ mol L^{-1}. This amount of DEBT is always in excess considering the amounts of analyte metal that is our concern (15 μg gold ions).

As indicated in Table 1, up to 100 μg gold ions in 10 mL sample solution can be safely retained on 1.0 g of resin as metal chelates. When 10 mL of 50 μg mL^{-1} metal chelate solution was passed through the column, a decrease in percent retention was observed.

During this study, we dealt with quite low amounts of metal ions and carried out the optimizations accordingly. Referring to results in Table 1, any researchers interested in higher amounts of gold up to 100 μg can study safely with the proposed method under the same conditions (such as amount of resin, sample flow rate, pH of sorption media, and ligand volume) as long as only the concentration of eluent and its volume are reoptimized according to the interested amount of metal ions.

3.3.4. Choice of Eluent: Its Nature, Concentration, Volume, and Flow Rate. In the previous study, we used sodium thiosulfate

Table 2: Analytical figures of merit.

Initial concentration of solution[a]	Regression equation[b]	R^2	LOD[c] ($\mu g\ mL^{-1}$)	LOQ[d] ($\mu g\ mL^{-1}$)	%RSD[e]	PF[f]
0.15 $\mu g\ mL^{-1}$	A = 0.0125C + 0.0003	0.9998	0.025	0.085	7.56	6.7

[a]sample volume = 100 mL, [b]A(absorbance) = slope x C(concentration $\mu g\ mL^{-1}$) + intercept, [c]limit of detection (2.5 $\mu g\ g^{-1}$ ore), [d]limit of quantitation (8.5 $\mu g\ g^{-1}$ ore), [e]percentage relative standard deviation, and [f]preconcentration factor.

Table 3: Determination of Au in CMC ore sample and Gold Ore (MA-1b) CRM.

Samples	Au concentration ($mgkg^{-1}$)	Corrected values according to 97% desorption
CMC sample		
Au (found) *	< LOD	< LOD
Au (claimed)	1.34	1.34
Au (spiked found) *	13.6 ± 0.6	14.1 ± 0.6
Au (spiked)	15	15
Gold Ore (MA-1b)		
Au (found) *	15.0 ± 1.0	15.5 ± 1.0
Au (certified value)	17.0 ± 0.3	17.0 ± 0.3

* Values are given as mean ± SD, N = 3 (number of replicates).

which was found as the most suitable eluent for desorbing silver ions from Amberlite XAD-16 [5]. The sorption of silver ions was governed by the chelation mechanism in that silver ions (belong to class of soft acids) have affinity for (S-O) chelating group of DEBT. During desorption, (S-S) chelating groups of $Na_2S_2O_3$ provide a stronger complex formation as silver has a higher affinity for (S-S) than (S-O) chelating group [21]. Since gold metal also belongs to class of soft acid and same discussion could be valid as well, as a result, $Na_2S_2O_3$ was selected as suitable eluent for desorbing gold ions.

Series of experiments were conducted to optimize the concentration and the volume of eluent [21]. When 0.1 mol L^{-1} $Na_2S_2O_3$ in water was used as an eluent, at a flow rate of 0.3 mL min^{-1}, only 65% desorption was obtained. Later, it was found that the highest recovery with 97.6 ± 2.3% desorption was achieved when 15 mL of 0.2 mol L^{-1} $Na_2S_2O_3$ in water was percolated through the column at a flow rate of 0.3 mL min^{-1}.

3.4. Effect of Electrolytes and Competing Ions. In geological samples like ores, some metals in higher concentrations such as Na^+, K^+, Cu^{2+}, Ni^{2+}, Pb^{2+}, Mn^{2+}, Fe^{3+}, Zn^{2+}, Al^{3+}, and Cr^{3+} can coexist with gold. The anions Cl^-, NO_3^-, SO_4^{2-}, PO_4^{3-}, and ClO_4^- are the anions that are capable of forming complexes with several metal ions.

Considering the real sample amount weighed according to gold digestion procedure (at least 10.0 g), 50 mL of 300 $\mu g\ mL^{-1}$ of copper standard solution was prepared in 1 mol L^{-1} HCl. Initial metal ion concentration was determined by FAAS. Then the proposed method was applied. The metal ion concentration in the effluent was determined by FAAS. Initial metal ion concentration was found to be 292 $\mu g\ mL^{-1}$ and the metal ion concentration in the effluent was found as 288 $\mu g\ mL^{-1}$. As a result, only 1.37% of copper ions was adsorbed on resin. Although it is known that copper ion forms complex with DEBT in pH range of 0-7 [20], this result showed that the formation of Cu-DEBT complex in 1 mol L^{-1} HCl is quite lower to compete with gold ions. N-benzoylthioureas are bidentate chelating ligands with S and O as donor atoms. The possibility of increasing the electron density at the sulfur atom by means of resonance effect leads to selective complex behavior of DEBT which can be influenced by the adjustment of pH, where competing metal ions could be eliminated.

The effect of various electrolytes like $NaNO_3$, Na_2SO_4, Na_3PO_4, and Na_2CO_3 on the sorption of gold (1 mg L^{-1}) as Au-DEBT chelate on Amberlite XAD-16 resin was studied as well. Na_2SO_4 was tolerable up to 0.04 mol L^{-1}, Na_3PO_4 up to 0.1 mol L^{-1}, and $NaNO_3$ and Na_2CO_3 up to 0.15 mol L^{-1}.

3.5. Analytical Figures of Merit. The calibration graph for the determination of gold was plotted according to the proposed procedure under the optimum conditions. The equation of the line was derived as A = 0.0125C + 0.0003 with the regression coefficient 0.9998 where A is the absorbance and C is concentration of the metal ion ($\mu g\ mL^{-1}$).

The limit of detection (LOD) and limit of quantitation (LOQ) for gold ions were determined employing the standard solutions giving absorbance signal slightly recognizable than blank. The LOD and LOQ were calculated based on 3s/slope and 10s/slope of 10 measurements of the blank, respectively, where s is the standard deviation of the sample solution. The results of the LOD, LOQ, and precision (RSD %) for gold and its concentration are shown in Table 2.

3.6. Analysis of Real Samples. In order to demonstrate the accuracy of the proposed method, analyses of two real samples, one of which is the Cu-ore supplied by Cyprus Mining Company (CMC), North Cyprus, and the other one certified Gold Ore (MA-1b) by Canmet, Ontario, were carried out and the results were compared with the values reported. The results of spiked CMC samples and the results that were corrected for 97% desorption recovery value which was found for 100 mL sample solution were also tabulated in Table 3. Student's t-test was performed to statistically evaluate

TABLE 4: Comparison of the proposed method with some studies based on SPE and determination of gold reported in literature.

Adsorbent	Medium	Eluent	D. M.	LOD	Matrix	Ref.
Octadecyl silica membrane discs modified with pentathia-15-crown-5	pH 4.5-7.00	0.5 mol L^{-1} Sodium thiosulphate	FAAS	1.0 $\mu g\ L^{-1}$	Pharmaceutical and water samples	[23]
Diethyldithiocarbamate complex on Amberlite XAD-2000	0.5-2.5 mol L^{-1} HNO_3	1 mol L^{-1} HNO_3 in acetone	FAAS	16.6 μg	Environmental samples	[14]
1-phenyl-1,2-propanedione-2-complex on oximethiosemicarbazone SP Sephadex C25	pH 3	--	ICP-MS	$1.6X10^{-8}$-$141X10^{-8}$ mol L^{-1}	Minerals and natural water samples	[24]
Poly(N-(hydroxymethyl)methacylamide 0-1-allyl-thiourea) hydrogels	pH 0.5	0.8 mol L^{-1} thioura in 3 mol L^{-1} HCl	GFAAF	3 ng L^{-1}	Anode slime and geological samples∗	[25]
Dowex M 4195 chelating resin	pH 4	2 mol L^{-1} H_2SO_4 + 2 mol L^{-1} NH_3	FAAS	1.61 $\mu g\ L^{-1}$	Water, soil and sediment samples	[26]
Multi-walled carbon nanotubes	pH 1-6	3% thiourea in 1 mol L^{-1} HCl	FAAS	0.15 $\mu g\ L^{-1}$	Geological and water samples	[11]
2-pyridine-5-(4-tolyl)-1,3,4-oxadiazole complex on Amberlite XAD-4	0.5 mol/L HNO_3	1 mol/L HCl in acetone	FAAS	1.03 $\mu g\ L^{-1}$	Environmental Samples	[27]
Polyethylenimine coated on Al_2O_3	pH 5.7	0.5 mol L^{-1} thiourea then 1.0 mol/L HCl	FAAS	26.2 ng L^{-1}	Water Samples	[28]
Rubeanic acid complex on silica gel	pH 3.5	0.5 mol L^{-1} thiourea then 1.0 mol L^{-1} HCl	FAAS	0.80 ng mL^{-1}	Water Samples	[29]
Silica gel (SG-CIPrNTf$_2$)	pH 2	--	ICP-OES	--	Water Samples	[30]
DEBT complex on Amberlite XAD-16	pH~1	0.2 mol L^{-1} sodium thiosulphate	FAAS	0.025 μg mL^{-1}	Cu and Au Ores	This study

D.M.: detection method, Ref.: references. ∗Matrix elimination method is used.

the found and certified values. The found values were in good agreement with the certified ones and the difference was found to be statistically insignificant (at 95% confidence interval level).

4. Conclusions

Highly selective, reliable, and low cost method has been proposed for preconcentration of gold ions from highly interfering matrices, namely ores. The validity of the proposed method was demonstrated by the analyses of two geological samples: Cu-ore (supplied by CMC) and Gold Ore (MA-1b) as a certified reference material. The results are in good agreement with the given values. Comparison of the proposed method with some similar studies in literature is summarized in Table 4. Although there are more sensitive methods applied to similar samples, such as ICP-MS and GFAAS, these are much more expensive and sophisticated. The gold ion at such a low concentration of 0.15 $\mu g\ mL^{-1}$ could be preconcentrated selectively and determined by the proposed method without any matrix elimination processes.

Acknowledgments

This study is the unpublished part of Hurmus Refiker's thesis [21] and the authors thank Middle East Technical University-Scientific Research Fund, Project No.: 2002-07-02-00-39 for financial assistance.

References

[1] H. Fazelirad and M. A. Taher, "Ligandless, ion pair-based and ultrasound assisted emulsification solidified floating organic drop microextraction for simultaneous preconcentration of ultra-trace amounts of gold and thallium and determination by GFAAS," *Talanta*, vol. 103, pp. 375–383, 2013.

[2] R. Dobrowolski, M. Kuryło, M. Otto, and A. Mróz, "Determination of gold in geological materials by carbon slurry sampling graphite furnace atomic absorption spectrometry," *Talanta*, vol. 99, pp. 750–757, 2012.

[3] S. Rastegarzadeh, N. Pourreza, and A. Larki, "Determination of trace silver in water, wastewater and ore samples using dispersive liquid-liquid microextraction coupled with flame

atomic absorption spectrometry," *Journal of Industrial and Engineering Chemistry*, vol. 24, pp. 297–301, 2015.

[4] A. Iraji, D. Afzali, and A. Mostafavi, "Separation for trace amounts of gold (III) ion using ion-pair dispersive liquid-liquid microextraction prior to flame atomic absorption spectrometry determination," *International Journal of Environmental Analytical Chemistry*, vol. 93, no. 3, pp. 315–324, 2013.

[5] H. Refiker, M. Merdivan, and R. S. Aygün, "Solid-phase extraction of silver in geological samples and its determination by FAAS," *Separation Science and Technology*, vol. 43, no. 1, pp. 179–191, 2008.

[6] G. Hartmann and M. Schuster, "Species selective preconcentration and quantification of gold nanoparticles using cloud point extraction and electrothermal atomic absorption spectrometry," *Analytica Chimica Acta*, vol. 761, pp. 27–33, 2013.

[7] E. A. Moawed and M. F. El-Shahat, "Synthesis, characterization of low density polyhydroxy polyurethane foam and its application for separation and determination of gold in water and ores samples," *Analytica Chimica Acta*, vol. 788, pp. 200–207, 2013.

[8] J. S. Fritz, *Analytical Solid-Phase Extraction*, Wiley-VCH, New York, NY, USA, 1999.

[9] R. K. Sharma and P. Pant, "Preconcentration and determination of trace metal ions from aqueous samples by newly developed gallic acid modified Amberlite XAD-16 chelating resin," *Journal of Hazardous Materials*, vol. 163, no. 1, pp. 295–301, 2009.

[10] M. Ghaedi, M. Rezakhani, S. Khodadoust, K. Niknam, and M. Soylak, "The solid phase extraction of some metal ions using palladium nanoparticles attached to silica gel chemically bonded by silica-bonded N-propylmorpholine as new sorbent prior to their determination by flame atomic absorption spectroscopy," *The Scientific World Journal*, vol. 2012, Article ID 764195, 9 pages, 2012.

[11] P. Liang, E. Zhao, Q. Ding, and D. Du, "Multiwalled carbon nanotubes microcolumn preconcentration and determination of gold in geological and water samples by flame atomic absorption spectrometry," *Spectrochimica Acta Part B: Atomic Spectroscopy*, vol. 63, no. 6, pp. 714–717, 2008.

[12] Ş. Tokalıoğlu, A. Papak, and Ş. Kartal, "Separation/preconcentration of trace Pb(II) and Cd(II) with 2-mercaptobenzothiazole impregnated Amberlite XAD-1180 resin and their determination by flame atomic absorption spectrometry," *Arabian Journal of Chemistry*, vol. 10, no. 1, pp. 19–23, 2017.

[13] S. Sivrikaya, B. Karslı, and M. Imamoglu, "On-line Preconcentration of Pd(II) Using Polyamine Silica Gel Filled Mini Column for Flame Atomic Absorption Spectrometric Determination," *International Journal of Environmental Research*, vol. 11, no. 5-6, pp. 579–590, 2017.

[14] H. B. Senturk, A. Gundogdu, V. N. Bulut et al., "Separation and enrichment of gold(III) from environmental samples prior to its flame atomic absorption spectrometric determination," *Journal of Hazardous Materials*, vol. 149, no. 2, pp. 317–323, 2007.

[15] R. Liu and P. Liang, "Determination of gold by nanometer titanium dioxide immobilized on silica gel packed microcolumn and flame atomic absorption spectrometry in geological and water samples," *Analytica Chimica Acta*, vol. 604, no. 2, pp. 114–118, 2007.

[16] A. Ahmad, J. A. Siddique, M. A. Laskar et al., "New generation Amberlite XAD resin for the removal of metal ions: A review," *Journal of Environmental Sciences*, vol. 31, pp. 104–123, 2015.

[17] D. A. Chowdhury, M. I. Hoque, and Z. Fardous, "Solid Phase Extraction of Copper, Cadmium and Lead Using Amberlite XAD-4 Resin Functionalized with 2-Hydroxybenzaldehyde Thiosemicarbazone and its Application on Green Tea Leaves," *Jordan Journal of Chemistry*, vol. 8, no. 2, pp. 90–102, 2013.

[18] V. Losev, E. Elsufiev, O. Buyko, A. Trofimchuk, R. Horda, and O. Legenchuk, "Extraction of precious metals from industrial solutions by the pine (Pinus sylvestris) sawdust-based biosorbent modified with thiourea groups," *Hydrometallurgy*, vol. 176, pp. 118–128, 2018.

[19] M. Schuster and M. Schwarzer, "Selective determination of palladium by on-line column preconcentration and graphite furnace atomic absorption spectrometry," *Analytica Chimica Acta*, vol. 328, no. 1, pp. 1–11, 1996.

[20] M. Merdivan, "The Analysis of Platinum Metals in Platinum Catalysts by Thin Layer Chromatography," *PhD Thesis*, Middel East Technical University, Ankara, Turkey, 1997.

[21] H. Refiker, "Preconcentration of Some Precious Metals Using Debt Impregnated Resin," MSc Thesis, Middle East Technical University, Ankara, Turkey, 2005.

[22] Perkin Elmer, "Analytical methods for Absorption spectrometry, manual, United States of America," 1996.

[23] M. Bagheri, M. H. Mashhadizadeh, and S. Razee, "Solid phase extraction of gold by sorption on octadecyl silica membrane disks modified with pentathia-15-crown-5 and determination by AAS," *Talanta*, vol. 60, no. 4, pp. 839–844, 2003.

[24] L. Morales and M. I. Toral, "Simultaneous determination of Au(III) and Cu(II) with 1-phenyl-1,2-propanedione-2-oximethiosemicarbazone (PPDOT) on solid phase," *Minerals Engineering*, vol. 20, no. 8, pp. 802–806, 2007.

[25] B. Salih, Ö. Çelikbıçak, S. Döker, and M. Doğan, "Matrix elimination method for the determination of precious metals in ores using electrothermal atomic absorption spectrometry," *Analytica Chimica Acta*, vol. 587, no. 2, pp. 272–280, 2007.

[26] M. Tuzen, K. O. Saygi, and M. Soylak, "Novel solid phase extraction procedure for gold(III) on Dowex M 4195 prior to its flame atomic absorption spectrometric determination," *Journal of Hazardous Materials*, vol. 156, no. 1-3, pp. 591–595, 2008.

[27] H. Elvan, D. Ozdes, C. Duran, V. N. Bulut, N. Gümrükçüoglu, and M. Soylak, "Development of a new solid phase extraction procedure for selective separation and enrichment of Au(III) ions in environmental samples," *Journal of the Brazilian Chemical Society*, vol. 24, no. 10, pp. 1701–1706, 2013.

[28] D. Afzali, Z. Daliri, and M. A. Taher, "Flame atomic absorption spectrometry determination of trace amount of gold after separation and preconcentration onto ion-exchange polyethylenimine coated on Al2O3," *Arabian Journal of Chemistry*, vol. 7, no. 5, pp. 770–774, 2014.

[29] F. Sabermahani, M. A. Taher, and H. Bahrami, "Separation and preconcentration of trace amounts of gold from water samples prior to determination by flame atomic absorption spectrometry," *Arabian Journal of Chemistry*, vol. 9, pp. S1700–S1705, 2016.

[30] H. M. Marwani, A. E. Alsafrani, H. A. Al-Turaif, A. M. Asiri, and S. B. Khan, "Selective extraction and detection of noble metal Based on ionic liquid immobilized silica gel surface using ICP-OES," *Bulletin of Materials Science*, vol. 39, no. 4, pp. 1011–1019, 2016.

[31] M. Schuster, B. Kugler, and K. König, "The chromatography of metal chelates Fresen," *Journal of Analytical Chemistry*, vol. 338, no. 6, pp. 717–720, 1990.

[32] E. A. Moawed, I. Ishaq, A. Abdul-Rahman, and M. F. El-Shahat, "Synthesis, characterization of carbon polyurethane powder and its application for separation and spectrophotometric determination of platinum in pharmaceutical and ore samples," *Talanta*, vol. 121, pp. 113–121, 2014.

10

Selective Recognition of Myoglobin in Biological Samples using Molecularly Imprinted Polymer-Based Affinity Traps

Rüstem Keçili

Anadolu University, Yunus Emre Vocational School of Health Services, Department of Medical Services and Techniques, 26470 Eskisehir, Turkey

Correspondence should be addressed to Rüstem Keçili; rkecili@anadolu.edu.tr

Academic Editor: Jan Åke Jönsson

The current work demonstrates the design, characterization, and preparation of molecularly imprinted microspheres for the selective detection of myoglobin in serum samples. The suspension polymerization approach was applied for the preparation of myoglobin imprinted microspheres. For this purpose, N-methacryloylamino folic acid-Nd^{3+} (MAFol- Nd^{3+}) was chosen as the complex functional monomer. The optimization studies were performed changing the medium pH, temperature, and myoglobin concentration. pH 7.0 was determined as the optimum value where the prepared imprinted microspheres displayed maximum binding for myoglobin. The maximum binding capacity was achieved as 623 mgg^{-1}. In addition, the selectivity studies were conducted. The results confirmed that the imprinted microspheres showed great selectivity towards myoglobin in the existence of hemoglobin, cytochrome c, and lysozyme which were chosen as potentially competing proteins.

1. Introduction

Myoglobin (Mb) is a hemeprotein which is primarily found in muscle tissues which binds oxygen through its heme functional group. Mb is also a crucial biomarker for the early recognition of various diseases such as acute myocardial infarction also called "heart attack" [1–3]. Because of its small size, Mb in muscle tissues is released to the bloodstream when muscle tissue is damaged. The increase in the level of Mb may lead to kidney failure because it may be converted to the undesired toxic molecules [4, 5]. After the acute myocardial infarction, Mb level in serum starts to increase in a short time and its level reaches the maximum value in 6-9 h. Therefore, the early determination of Mb level is vital for the detection of acute myocardial infarction [6–13].

Several conventional approaches such as enzyme-linked immunosorbent assay (ELISA) [14, 15], chromatography [16–19], electrochemical sensors [20–24], and surface plasmon resonance (SPR) sensors [25–27] were applied for the detection of myoglobin. Most of these approaches show high sensitivity towards myoglobin. However, they have some disadvantages such as high production costs (especially for antibody-based assays), high process time, low stability, and need of special and expensive equipment. Thus, innovative approaches that have higher selectivity are required.

Molecularly imprinted polymers (MIPs) also known as "artificial antibodies" are artificial materials that possess specific binding regions towards a desired molecule. During the preparation of selective MIPs, appropriate functional monomers are chosen and polymerized with a cross-linker in the presence of the desired molecule (template). Since MIPs display excellent affinity and selectivity towards the target compound, they can efficiently be used in different applications such as biosensor platforms, catalysis, and extraction [28–30]. In addition, MIPs are low-cost and robust materials at harsh process conditions such as high pressure, high temperature, and low and high pH values [31–37].

In our previous study [38], we developed a molecularly imprinted cryogel column using the functional monomer N-methacryloylamino antipyrine for the recognition of myoglobin. Unlike this reported study, molecularly imprinted microspheres which show excellent selectivity and binding capacity towards myoglobin in serum samples compared to other reported studies were prepared in the present work. For this purpose, the complex functional monomer N-methacryloylamino folic acid-Nd^{3+} (MAFol-Nd^{3+}) was polymerized with the target protein myoglobin (template) in the presence of ethylene glycol dimethacrylate which is the cross-linker. To the best of our knowledge, this is the first report in which the complex functional monomer MAFol-Nd^{3+} was used for the design and preparation of selective imprinted microspheres towards proteins. The prepared myoglobin imprinted microspheres were characterized and their binding behaviour towards target protein myoglobin was evaluated.

2. Experimental Section

2.1. Chemicals. Myoglobin, hemoglobin, lysozyme, cytochrome c, folic acid, 2,2′-azobisisobutyronitrile, poly (vinyl alcohol) (molecular weight: 27.000), ethylene glycol dimethacrylate, and solvents were provided by Sigma-Aldrich (St. Louis, MO, USA).

2.2. Analytical Instruments. FT-IR spectroscopic studies were conducted using Perkin-Elmer 400 FT-IR spectrometer. The morphological characterization of the imprinted microspheres was done using scanning electron microscope (SEM) (FEI-Quanta-FEG 250 model) and Shimadzu-UV-3600 model spectrophotometer was used for the UV spectroscopic studies. Circular dichroism (CD) spectroscopic studies of the extracted and commercial myoglobin were carried out by using Applied Photophysics Chirascan CD Spectropolarimeter.

2.3. The Synthesis of the Functional Monomer MAFol. The MAFol was successfully synthesized by applying the following recipe which has been already reported [39].

Firstly, folic acid (1 eq) in H_2O was prepared and solution pH was then adjusted to 9.0 with 1 M NaOH. Then, 15 mL of methacryloyl benzotriazole in dioxane was mixed with this solution. The final mixture was allowed to stir for 60 min. Once the reaction was finished, evaporation of dioxane was carried out and excess of the benzotriazole was removed by extraction using EtOAc. Then, the pH of the aqueous phase was adjusted to pH 6. Finally, the product MAFol was obtained after aqueous phase was removed from the reaction medium.

The MAFol synthesis is schematically demonstrated in Figure 1.

2.4. Preparation of Complex Monomer MAFol- Nd^{3+}. For the preparation of MAFol- Nd^{3+}, 1.0 mmol Nd $(NO_3)_3.6H_2O$ was added to 2.0 mmol MAFol solution in chloroform and stirred for 18 h. After filtration, the obtained complex was washed with H_2O and EtOH and allowed to dry at 55°C for 18 h.

2.5. Preparation of Imprinted Microspheres for Myoglobin. Suspension polymerization was used for the preparation of the myoglobin imprinted microspheres (MIPs). The following protocol was applied.

0.2 g poly (vinyl alcohol) in 50 mL H_2O was prepared for the dispersion phase. Then, 100 mg MAFol-Nd^{3+}-myoglobin preorganized complex monomer in 5 mL DMSO was added to the mixture of ethylene glycol dimethacrylate: toluene (1.0 mL:5.0 mL). The prepared final solution was then mixed with the aqueous dispersion phase. Then, initiator 2,2′-azobisisobutyronitrile (ca. 30 mg) was added and the solution was stirred at 70°C for 8 h and 90°C for the next 4 h. Once the polymerization was completed, the prepared myoglobin imprinted microspheres (MIPs) were washed with EtOH and deionized H_2O and dried at 55°C for 18 h. Template removal from the imprinted microspheres was carried out by extraction with 1.0 M NaCl for 24 h. The same procedure was used for the preparation of nonimprinted microspheres (NIPs) without myoglobin.

2.6. Characterization Studies. SEM, FT-IR analyses, and swelling tests were performed for the characterization experiments of the prepared MIPs and NIPs. To perform the FT-IR studies, 10 mg of the dry particles was placed on the ATR crystal surface and the FT-IR spectrum was recorded.

For the SEM analyses, thin gold layer (approximately 20 nm) was deposited on the surface of the microspheres to provide conductivity. The SEM images were then recorded.

The swelling tests of the prepared MIPs and NIPs were also carried out. In these experiments, the dried MIPs were put into the distilled water in a NMR tube for 3 h. Then, the volume of the MIPs in swollen state was determined.

Equation (1) was used for the calculation of the % swelling ratio of the prepared polymers.

$$\%\ \text{Swelling ratio} = \left[\frac{V\,\text{swollen} - V\,\text{dry}}{V\text{dry}}\right] \times 100 \qquad (1)$$

where ***Vdry*** is volume of the MIPs in dry state and ***Vswollen*** is volume of the MIPs in swollen state.

For the CD spectroscopic measurements, 500 ppm myoglobin in pH 7.0 phosphate buffer was prepared. Analyses of the samples were carried out using a quartz cuvette. The CD spectra were obtained at 180-340 nm wavelength.

2.7. Binding of Myoglobin to the Imprinted Microspheres. In the binding studies for myoglobin in batch mode, 20 mg of each MIP and NIP was put into glass vials. Then, 2 mL of 0.5 $mgmL^{-1}$ myoglobin in pH 7.0 phosphate buffer was added and these mixtures were allowed to stir for 1 h and then aliquots of 1 mL were taken and analyzed by UV-VIS spectrophotometer at 410 nm wavelength.

To evaluate the pH effect on the binding of myoglobin to the MIP/NIP, 2 mL of 0.5 $mgmL^{-1}$ myoglobin in different buffer solutions pH 4 to 9 was added to 20 mg of the polymer. Then, the solutions were allowed to stir for 1 h and 1 mL aliquots of each solution were spectrophotometrically analyzed at 410 nm.

On the other hand, 2 mL of 0.5 $mgmL^{-1}$ myoglobin in pH 7.0 phosphate buffer was added to 20 mg of MIP and NIP to test the time effect on the binding of myoglobin. The solutions were stirred and aliquots of 1 mL of each solution were taken at different time (10, 20, 30, 40, 50, and 60 min) and spectrophotometrically analyzed at 410 nm.

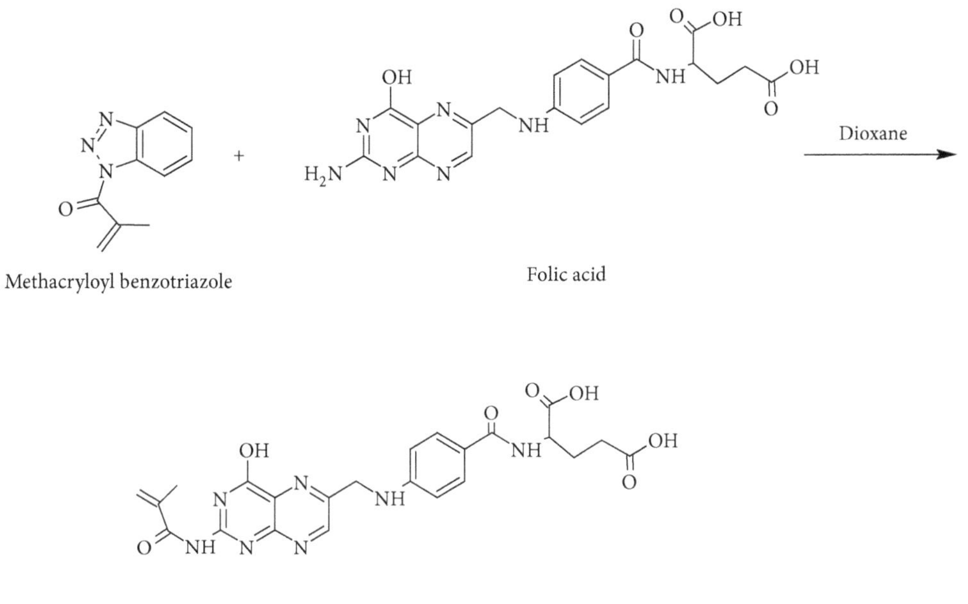

Dioxane
Methacryloyl benzotriazole
Folic acid
N-Methacryloylamino folic acid (MAFol)

FIGURE 1: MAFol synthesis.

MIP
NIP
% T
1643 cm-1
1558 cm-1
1162 cm-1
1651 cm-1
1555 cm-1
4000
3500
3000
2500
2000
1500
1000
500
Wavenumber (cm-1)

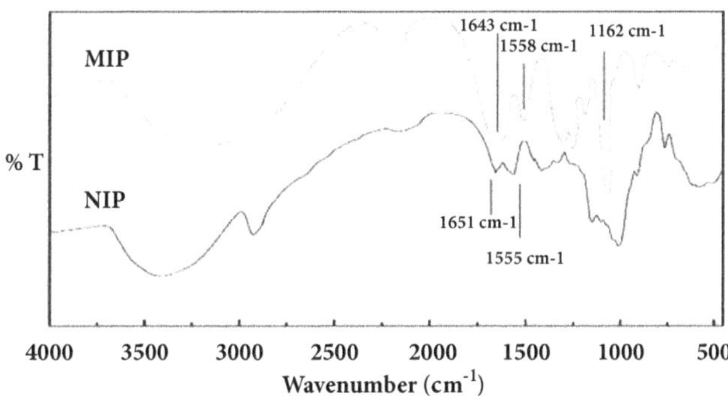

FIGURE 2: FT-IR spectra of imprinted microspheres (MIP) and nonimprinted microspheres (NIP).

The effect of myoglobin concentration on binding efficiency of the polymer was also studied. 2 mL solution of various concentrations of myoglobin in pH 7.0 phosphate buffer was interacted with 20 mg of MIP and NIP. The solutions were then allowed to stir for 1 h and 1 mL aliquots of each solution were spectrophotometrically analyzed at 410 nm.

2.8. Reusability of the Imprinted Microspheres and Selectivity Studies for Myoglobin. To evaluate the reusability of the imprinted microspheres towards myoglobin, myoglobin binding cycles were repeated 10 times using the same material. Then, microspheres were washed with 1.0 M NaCl after each cycle for the regeneration.

The studies for the selectivity of the prepared imprinted microspheres for myoglobin were conducted in the presence of hemoglobin, cytochrome c, and lysozyme which were chosen as competitor proteins. For this purpose, 20 mg imprinted microspheres were put into the 2 mL of 0.5 $mgmL^{-1}$ solution composed of myoglobin, hemoglobin, cytochrome c, and lysozyme in pH 7.0 phosphate buffer. The samples were allowed to stir for 1 h. Then, 1 mL of the collected aliquots was spectrophotometrically analyzed at 410 nm.

2.9. Binding of Myoglobin in Human Serum. In these experiments, 20 mg of MIP and NIP was put into glass vials and 0.1 μgmL^{-1} myoglobin was spiked in the 2 mL of serum sample and interacted with 20 mg of MIP/NIP. Then, the solutions were allowed to stir for 1 h. 1 mL of the solutions was then spectrophotometrically analyzed at 410 nm.

3. Results and Discussion

3.1. Characterization Experiments of the Imprinted Microspheres towards Myoglobin. The results obtained from the FT-IR experiments are shown in Figure 2. As seen, the obtained FT-IR spectra of MIP and NIP in Figure 2 were similar which confirms that the prepared MIPs and NIPs have

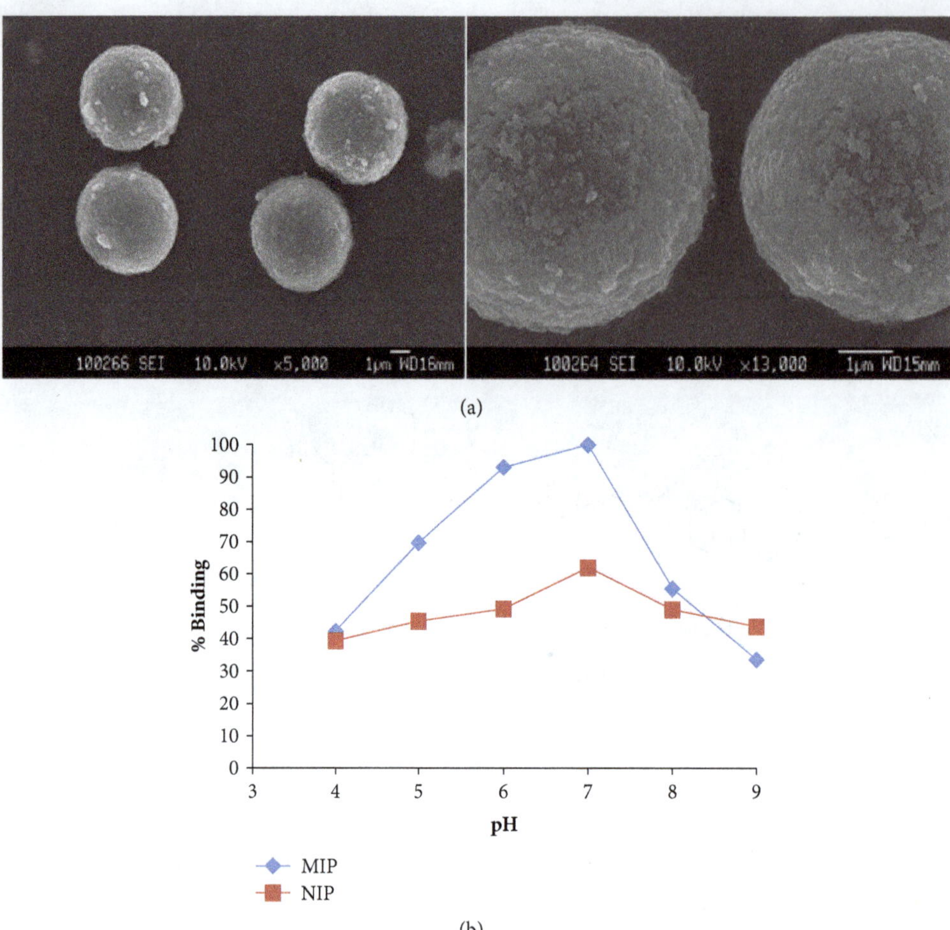

FIGURE 3: (a) SEM images of the myoglobin imprinted microspheres. (b) The pH effect on myoglobin binding [C_{Mb}: 0,5 $mgmL^{-1}$, $m_{polymer}$: 20 mg, t: 1 h].

similar backbone. In the FT-IR spectra, absorption peaks observed at 1645 cm^{-1} (COO- stretching) and at 2900 cm^{-1} (C-H stretching) belong to the functional monomer and the peak at 1160 cm^{-1} (C-O stretching) belongs to the cross-linker.

The obtained SEM images of the myoglobin imprinted microspheres are shown in Figure 3(a). The images indicated that the prepared imprinted microspheres are porous and spherical. The swelling behaviours of the microspheres in H_2O were also investigated. The swelling ratios of the MIP and NIP were obtained as 57.4% and 36.3%, respectively.

3.2. pH, Time, Ionic Strength, and Concentration Effects on the Binding of Myoglobin to the Imprinted Microspheres. The pH effect on the binding of myoglobin was evaluated in the pH range between 4.0 and 9.0. The obtained experimental outcomes are shown in Figure 3(b). The outcomes confirmed that the maximum binding of myoglobin was achieved at pH 7.0.

The reason of this result can be explained by the strong coordination and electrostatic interaction of myoglobin and the complex functional monomer MAFol-Nd^{3+} through Nd^{3+} ions at pH 7.0. Ionization states of various groups on the amino acid residues of the myoglobin may also lead to high binding of myoglobin at this pH value. The binding behaviour of the myoglobin imprinted microspheres considerably decreased at higher and lower than pH 7.0. The repulsive electrostatic forces between Nd^{3+} ions and myoglobin cause the decrease in the myoglobin binding

The binding kinetics were also carried out to evaluate how the binding capability of the imprinted microspheres for myoglobin changes over time. For this purpose, the experiments were conducted using 0.5 $mgmL^{-1}$ myoglobin. The obtained results are shown in Figure 4(a). The binding of myoglobin to the imprinted microspheres gradually increased within 40 min. Then, the myoglobin binding reached an equilibrium. Myoglobin binding to the imprinted microspheres is fast at the beginning of binding process. Then, it becomes much more difficult due to the difficulties on the penetration of myoglobin molecules into the MIP particles. The rapid binding of myoglobin until 40 min can be explained by the strong affinity interactions and complexation between the complex functional monomer MAFol-Nd^{3} and the target protein myoglobin.

The myoglobin concentration effect on the binding is shown in Figure 4(b). As seen, the imprinted microspheres

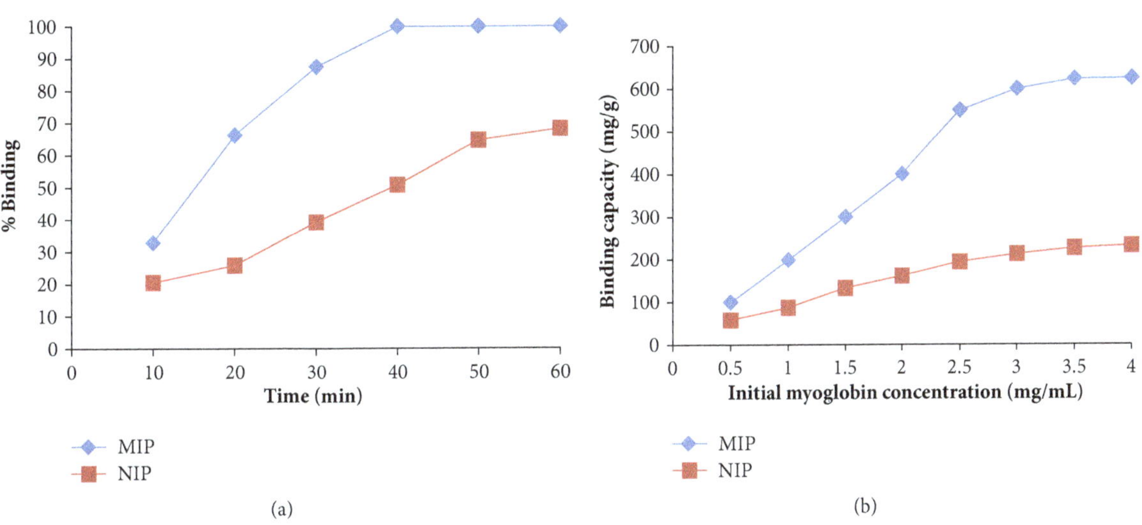

FIGURE 4: (a) The time effect on the binding of myoglobin [C_{Mb}:0,5 $mgmL^{-1}$, $m_{microbeads}$: 20 mg, pH: 7.0, t: 1 h]. (b) Effect of initial myoglobin concentration on binding [$m_{microbeads}$: 20 mg, pH: 6.0, t: 1 h, T: 25°C].

displayed high binding behaviour towards myoglobin at higher myoglobin concentrations in the range from 0.5 to 3.5 $mgmL^{-1}$. After saturation value which is 3.5 $mgmL^{-1}$, the myoglobin binding to the imprinted microbeads reached an equilibrium since myoglobin molecules occupied all binding regions of the imprinted microspheres. The highest binding of myoglobin to the imprinted microspheres was achieved as 623 mgg^{-1}.

Ertürk and her colleagues developed cryogel-based column systems for the detection of myoglobin in human plasma. In their study, N-methacryloyl-(L)-tryptophan (MATrp) was chosen as the functional monomer [40]. The highest binding myoglobin was obtained as 35.9 6 mgg^{-1}.

Turan et al. prepared imprinted hydrogels towards myoglobin using N-isopropylacrylamide and 2-acrylamido-2-methyl-propanosulfonic acid. The maximum myoglobin binding was determined as 97.40 mgg^{-1} [41].

The ionic strength effect on the binding of myoglobin was also investigated (Figure 5). The binding behaviour of the imprinted microspheres for myoglobin significantly decreased at higher concentrations of NaCl that was changed from 0 to 1.0 M. The reason of this can be the ionic interactions of the counter NaCI ions with the myoglobin molecules. This ionic interaction may lead to masking of the binding regions of the microspheres towards myoglobin. In addition, the repulsive electrostatic interactions between the microspheres and myoglobin molecules at higher NaCl concentrations may also cause the decrease of myoglobin binding to polymer. The nonspecific binding behaviour of nonimprinted microspheres towards myoglobin can be explained by the cooperative effects of various binding mechanisms such as ion-exchange or hydrophobic interactions.

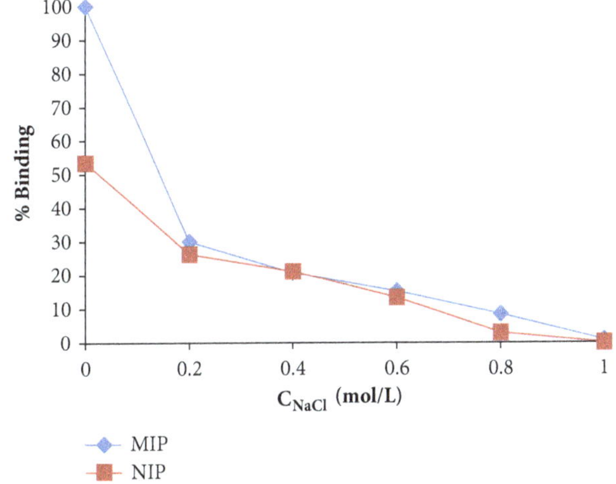

FIGURE 5: The ionic strength effect on myoglobin binding [$m_{microbeads}$: 20 mg, pH: 7.0, t: 1 h, T: 25°C].

3.3. Binding Isotherms. Binding isotherms were used for the characterization of the interactions between myoglobin and the prepared imprinted microspheres.

In the Langmuir binding isotherm [42], binding data can be obtained using the following:

$$\frac{1}{Q} = \left[\frac{1}{Qmax \times b}\right]\left[\frac{1}{Ceq}\right] + \frac{1}{Qmax} \tag{2}$$

In this equation, $\boldsymbol{Q}$ represent the myoglobin amount bound to the MIP (mgg^{-1}), $\boldsymbol{Qmax}$ is the highest myoglobin binding to the imprinted microspheres (mgg^{-1}), $\boldsymbol{Ceq}$ is the myoglobin concentration at equilibrium (mgL^{-1}), and $\boldsymbol{b}$ is the Langmuir constant that shows the affinity interactions between myoglobin and imprinted microspheres.

Another binding isotherm is Freundlich isotherm [43], which is defined in the following:

$$\log Q = \log kF + \frac{1}{n}\log Ceq \tag{3}$$

TABLE 1: The data calculated from the Freundlich and Langmuir isotherm equations.

Experimental	Langmuir			Freundlich		
Q (mgg^{-1})	Q (mgg^{-1})	b ($mLmg^{-1}$)	R^2	K_F(mgg^{-1})	n	R^2
623	615.15	5.34	0.98	583.45	3.35	0.93

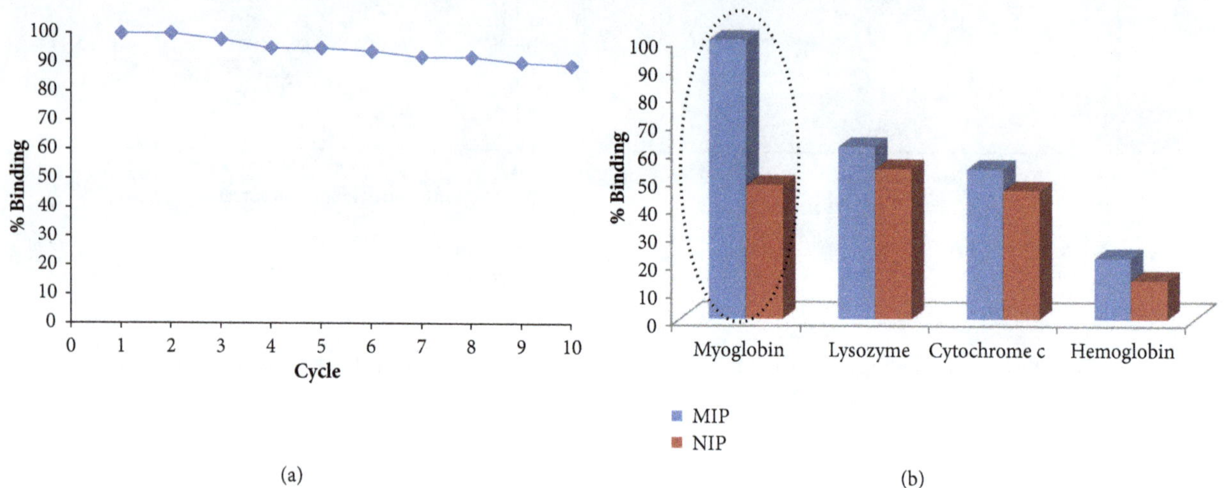

FIGURE 6: (a) Reusability of the myoglobin imprinted microbeads [C_{Mb}: 0,5 $mgmL^{-1}$, $m_{microbeads}$: 20 mg, pH: 7.0, t: 1 h, T: 25°C]. (b) Selectivity of the MIPs and NIPs for myoglobin.

where$\boldsymbol{Ceq}$ is the concentration of myoglobin at equilibrium (mgL^{-1}),$\boldsymbol{kF}$ is the Freundlich constant, and $\boldsymbol{n}$ is the Freundlich exponent. **1/n** is a measure of heterogeneity of the myoglobin binding regions of the imprinted microspheres changes from 0 to 1. When **1/n** value gets closer to zero heterogeneity increases.

The obtained results from the equations above for myoglobin binding to the polymers are shown in Table 1. The obtained results confirmed that the Langmuir binding isotherm well demonstrates the myoglobin binding to the prepared imprinted microspheres. The determined Langmuir constant **b** for myoglobin was 5.34 $mLmg^{-1}$.

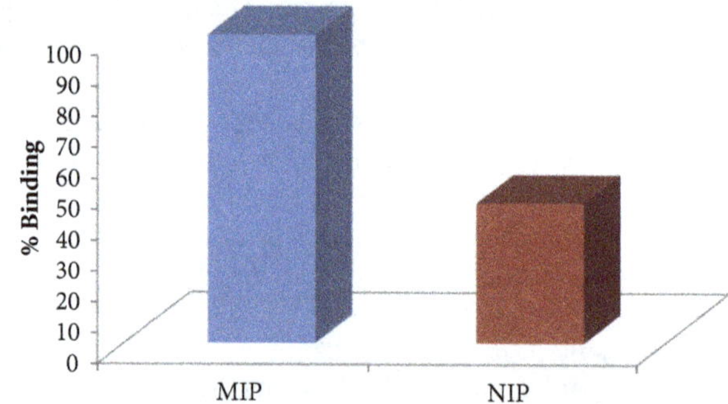

FIGURE 7: Extraction of myoglobin from human serum [C_{Mb}: 0.1 μgmL^{-1}, $m_{polymer}$: 20 mg, pH: 7.0, t: 1 h, T: 25°C].

3.4. Reusability of the Imprinted Microspheres and Selectivity Studies for Myoglobin. In this part of the study, same microspheres were used 10 times after each binding cycle. The obtained outcomes are shown in Figure 6(a). The prepared imprinted microspheres for myoglobin preserved their robustness even after 10 cycles.

The selectivity of the imprinted microspheres towards myoglobin was also tested. For this purpose, cytochrome c, lysozyme, and hemoglobin were chosen as competitive proteins. The imprinted microspheres towards myoglobin displayed excellent selectivity for myoglobin in the existence of hemoglobin, cytochrome c, and lysozyme (Figure 6(b)). On the other hand, nonimprinted microspheres showed lower binding behaviour for myoglobin. Although their isoelectric points (pIs) are very close to each other (pI of myoglobin is 6.9 and pI of hemoglobin is 7.2), the molecular weights of myoglobin and hemoglobin are 17 and 65 kDa, respectively. In addition, the shapes of these two proteins are completely different. Myoglobin has ellipsoidal shape while the hemoglobin has biconcave shape. Because the 3D cavities in the polymeric network of the imprinted microspheres are matched to the shape of the myoglobin, the entrance and binding of different molecules having different shapes and molecular weights to the 3D cavities are difficult.

3.5. Myoglobin Extraction from Real Samples. The results from myoglobin extraction from human serum are given in Figure 7. The imprinted microspheres (MIP) showed 100% binding towards myoglobin while the nonimprinted microspheres (NIP) exhibited 59% binding.

CD analyses of the commercial and extracted myoglobin from human serum were also performed to obtain detailed information about the secondary structure of the myoglobin

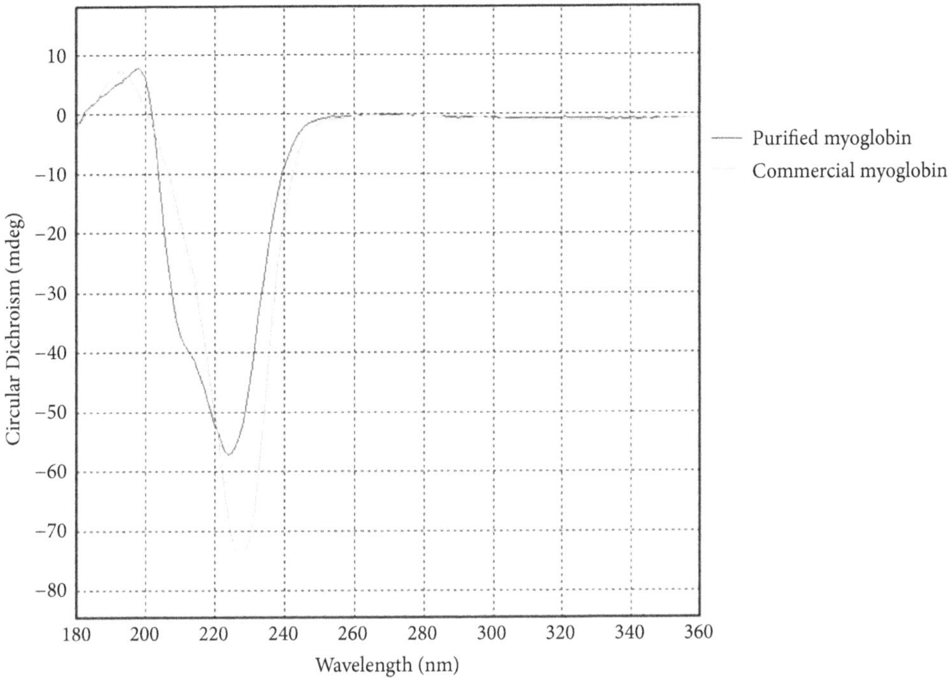

Figure 8: CD spectra of purified myoglobin and commercial myoglobin.

backbone. Figure 8 shows the CD spectra of commercial and extracted myoglobin from human serum. As can be seen from the figure, bands between 190 and 230 nm show the α-helical type of secondary structure of 500 mg/mL myoglobin (blue spectrum) [44, 45]. The red spectrum showed that α-helical type of secondary structure in 208-222 nm interval remained for isolated myoglobin. Peaks between 208 and 222 nm interval showed this data. This means that the isolated protein has preserved the secondary structure without any deformation in its structure during isolation processes.

4. Conclusions

In the current study, molecular imprinting-based microspheres were prepared for the selective detection and extraction of myoglobin from human serum. The outcomes confirmed that the prepared imprinted microspheres can be efficiently employed for the selective detection of target protein myoglobin in the existence of various competing proteins such as hemoglobin, cytochrome c, and lysozyme. The achieved maximum binding capacity was 623 mgg^{-1} at pH 7.0

In conclusion, the current study demonstrated the design and development of low-cost, reusable, robust, and selective materials using molecular imprinting technique for the sensitive recognition of a crucial protein myoglobin in real samples.

Acknowledgments

The author gratefully acknowledges the Anadolu University and Dicle University for providing laboratory facilities throughout this study.

References

[1] A. H. Wu, "Markers for early detection of cardiac diseases," *Scandinavian Journal of Clinical and Laboratory Investigation, Supplementum*, vol. 240, pp. 112–121, 2005.

[2] Y. Rozenman and M. S. Gotsman, "The earliest diagnosis of acute myocardial infarction," *Annual Review of Medicine*, vol. 45, pp. 31–44, 1994.

[3] M. J. Murphy and C. B. Berding, "Use of measurements of myoglobin and cardiac troponins in the diagnosis of acute myocardial infarction," *Critical Care Nurse*, vol. 19, no. 1, pp. 58–66, 1999.

[4] I. D. Laios, R. Caruk, and A. H. Wu, "Myoglobin clearance as an early indicator for rhabdomyolysis-induced acute renal failure," *Annals of Clinical and Laboratory Science*, vol. 25, pp. 179–184, 1995.

[5] S. N. Heyman, S. Rosen, S. Fuchs, F. H. Epstein, and M. Brezis, "Myoglobinuric acute renal failure in the rat: A role for medullary hypoperfusion, hypoxia, and tubular obstruction," *Journal of the American Society of Nephrology*, vol. 7, no. 7, pp. 1066–1074, 1996.

[6] S. S. Wong, "Strategic utilization of cardiac markers for the diagnosis of acute myocardial infarction," *Annals of Clinical and Laboratory Science*, vol. 26, pp. 301–312, 1996.

[7] J. E. Adams and V. A. Miracle, "Cardiac biomarkers: past, present, and future," *American Journal of Critical Care*, vol. 7, pp. 418–423, 1998.

[8] K. Penttila, H. Koukkunen, A. Kemppainen et al., "Myoglobin, creatine kinase MB, troponin T, and troponin I - rapid bedside assays in patients with acute chest pain," *International Journal of Clinical and Laboratory Research*, vol. 29, pp. 93–101, 1999.

[9] D. J. Karras and D. L. Kane, "Serum markers in the emergency department diagnosis of acute myocardial infarction," *Emer-*

gency Medicine Clinics of North America*, vol. 19, no. 2, pp. 321–337, 2001.

[10] C. Montague and T. Kircher, "Myoglobin in the early evaluation of acute chest pain," *American Journal of Clinical Pathology*, vol. 104, no. 4, pp. 472–476, 1995.

[11] R. J. De Winter, J. G. Lijmer, R. W. Koster, F. J. Hoek, and G. T. Sanders, "Diagnostic accuracy of myoglobin concentration for the early diagnosis of acute myocardial infarction," *Annals of Emergency Medicine*, vol. 35, no. 2, pp. 113–120, 2000.

[12] K. T. Moe and P. Wong, "Current trends in diagnostic biomarkers of acute coronary syndrome," *Annals Academy of Medicine Singapore*, vol. 39, pp. 210–215, 2010.

[13] S. M. Sallach, R. Nowak, M. P. Hudson et al., "A change in serum myoglobin to detect acute myocardial infarction in patients with normal troponin I levels," *American Journal of Cardiology*, vol. 94, no. 7, pp. 864–867, 2004.

[14] M. Sarkar and C. Mandal, "Immobilization of antibodies on a new solid phase for use in ELISA," *Journal of Immunological Methods*, vol. 83, no. 1, pp. 55–60, 1985.

[15] A. W. Hodson, A. W. Skillen, and N. B. Argent, "An ELISA method to measure human myoglobin in urine," *Clinica Chimica Acta*, vol. 209, no. 3, pp. 197–207, 1992.

[16] B. M. Mayr, O. Kohlbacher, K. Reinert et al., "Absolute myoglobin quantitation in serum by combining two-dimensional liquid chromatography-electrospray ionization mass spectrometry and novel data analysis algorithms," *Journal of Proteome Research*, vol. 5, no. 2, pp. 414–421, 2006.

[17] H.-Y. Lin, J. Rick, and T.-C. Chou, "Optimizing the formulation of a myoglobin molecularly imprinted thin-film polymer-formed using a micro-contact imprinting method," *Biosensors and Bioelectronics*, vol. 22, no. 12, pp. 3293–3301, 2007.

[18] S. C. Powell, E. R. Friedlander, and Z. K. Shihabi, "Myoglobin determination by high-performance liquid chromatography," *Journal of Chromatography A*, vol. 317, no. C, pp. 87–92, 1984.

[19] D. Han, K. W. McMillin, and J. S. Godber, "Hemoglobin, myoglobin, and total pigments in beef and chicken muscles: Chromatographic determination," *Journal of Food Science*, vol. 59, pp. 1279–1282, 1994.

[20] I. Lee, X. Luo, X. T. Cui, and M. Yun, "Highly sensitive single polyaniline nanowire biosensor for the detection of immunoglobulin G and myoglobin," *Biosensors and Bioelectronics*, vol. 26, no. 7, pp. 3297–3302, 2011.

[21] S. S. Mandal, K. K. Narayan, and A. J. Bhattacharyya, "Employing denaturation for rapid electrochemical detection of myoglobin using TiO2 nanotubes," *Journal of Materials Chemistry B*, vol. 1, no. 24, pp. 3051–3056, 2013.

[22] S. K. Mishra, D. Kumar, A. M. Biradar, and Rajesh, "Electrochemical impedance spectroscopy characterization of mercaptopropionic acid capped ZnS nanocrystal based bioelectrode for the detection of the cardiac biomarker-myoglobin," *Bioelectrochemistry*, vol. 88, pp. 118–126, 2012.

[23] F. T. C. Moreira, R. A. F. Dutra, J. P. C. Noronha, and M. G. F. Sales, "Electrochemical biosensor based on biomimetic material for myoglobin detection," *Electrochimica Acta*, vol. 107, pp. 481–487, 2013.

[24] F. T. C. Moreira, S. Sharma, R. A. F. Dutra, J. P. C. Noronha, A. E. G. Cass, and M. G. F. Sales, "Smart plastic antibody material (SPAM) tailored on disposable screen printed electrodes for protein recognition: Application to myoglobin detection," *Biosensors and Bioelectronics*, vol. 45, no. 1, pp. 237–244, 2013.

[25] O. V. Gnedenko, Y. V. Mezentsev, A. A. Molnar, A. V. Lisitsa, A. S. Ivanov, and A. I. Archakov, "Highly sensitive detection of human cardiac myoglobin using a reverse sandwich immunoassay with a gold nanoparticle-enhanced surface plasmon resonance biosensor," *Analytica Chimica Acta*, vol. 759, pp. 105–109, 2013.

[26] J.-F. Masson, T. M. Battaglia, P. Khairallah, S. Beaudoin, and K. S. Booksh, "Quantitative measurement of cardiac markers in undiluted serum," *Analytical Chemistry*, vol. 79, no. 2, pp. 612–619, 2007.

[27] E. Matveeva, Z. Gryczynski, I. Gryczynski, J. Malicka, and J. R. Lakowicz, "Myoglobin immunoassay utilizing directional surface plasmon-coupled emission," *Analytical Chemistry*, vol. 76, no. 21, pp. 6287–6292, 2004.

[28] R. Gui, H. Jin, H. Guo, and Z. Wang, "Recent advances and future prospects in molecularly imprinted polymers-based electrochemical biosensors," *Biosensors and Bioelectronics*, vol. 100, pp. 56–70, 2018.

[29] R. Kecili, R. Say, A. Ersöz, D. Hür, and A. Denizli, "Investigation of synthetic lipase and its use in transesterification reactions," *Polymer Journal*, vol. 53, no. 10, pp. 1981–1984, 2012.

[30] O. A. Attallah, M. A. Al-Ghobashy, A. T. Ayoub, and M. Nebsen, "Magnetic molecularly imprinted polymer nanoparticles for simultaneous extraction and determination of 6-mercaptopurine and its active metabolite thioguanine in human plasma," *Journal of Chromatography A*, vol. 1561, pp. 28–38, 2018.

[31] B. Sellergren, *Molecularly IMprinted Polymers: Man-Made Mimics of Antibodies and their Application in Analytical Chemistry: Techniques and Instrumentation in Analytical Chemistry*, Elsevier Science, Amsterdam, The Netherlands, 2001.

[32] D. R. Kryscio and N. A. Peppas, "Critical review and perspective of macromolecularly imprinted polymers," *Acta Biomaterialia*, vol. 8, no. 2, pp. 461–473, 2012.

[33] W. J. Cheong, S. H. Yang, and F. Ali, "Molecular imprinted polymers for separation science: a review of reviews," *Journal of Separation Science*, vol. 36, no. 3, pp. 609–628, 2013.

[34] G. Vasapollo, R. D. Sole, L. Mergola et al., "Molecularly imprinted polymers: present and future prospective," *International Journal of Molecular Sciences*, vol. 12, no. 9, pp. 5908–5945, 2011.

[35] R. Keçili, A. Atilir Özcan, A. Ersöz, D. Hür, A. Denizli, and R. Say, "Superparamagnetic nanotraps containing MIP based mimic lipase for biotransformations uses," *Journal of Nanoparticle Research*, vol. 13, no. 5, pp. 2073–2079, 2011.

[36] S. E. Diltemiz, R. Keçili, A. Ersöz, and R. Say, "Molecular imprinting technology in Quartz Crystal Microbalance (QCM) sensors," *Sensors*, vol. 17, no. 3, p. 454, 2017.

[37] J. Kupai, M. Razali, S. Buyuktiryaki, R. Kecili, and G. Szekely, "Long-term stability and reusability of molecularly imprinted polymers," *Polymer Chemistry*, vol. 8, no. 4, pp. 666–673, 2017.

[38] İ. Dolak, R. Keçili, R. Onat, B. Ziyadanoğulları, A. Ersöz, and R. Say, "Molecularly imprinted affinity cryogels for the selective recognition of myoglobin in blood serum," *Journal of Molecular Structure*, 2018.

[39] I. Dolak, R. Keçili, D. Hür, A. Ersöz, and R. Say, "Ion-imprinted polymers for selective recognition of neodymium(III) in environmental samples," *Industrial & Engineering Chemistry Research*, vol. 54, no. 19, pp. 5328–5335, 2015.

[40] G. Ertürk, N. Bereli, P. W. Ramteke, and A. Denizli, "Molecularly imprinted supermacroporous cryogels for myoglobin recogni-

tion," *Applied Biochemistry and Biotechnology*, vol. 173, no. 5, pp. 1250–1262, 2014.

[41] E. Turan, G. Özçetin, and T. Caykara, "Dependence of protein recognition of temperature-sensitive imprinted hydrogels on preparation temperature," *Macromolecular Bioscience*, vol. 9, no. 5, pp. 421–428, 2009.

[42] I. Langmuir, "The adsorption of gases on plane surfaces of glass, mica and platinum," *Journal of the American Chemical Society*, vol. 40, no. 9, pp. 1361–1403, 1918.

[43] H. M. F. Freundlich, "Über die Adsorption in Lösungen," *Zeitschrift für Physikalische Chemie*, vol. 57, no. A, pp. 385–470, 1906.

[44] A. F. Mehl, M. A. Crawford, and L. Zhang, "Determination of myoglobin stability by circular dichroism spectroscopy: Classic and modern data analysis," *Journal of Chemical Education*, vol. 86, no. 5, pp. 600–602, 2009.

[45] M. Nagai, Y. Nagai, K. Imai, and S. Neya, "Circular dichroism of hemoglobin and myoglobin," *Chirality*, vol. 26, no. 9, pp. 438–442, 2014.

11

A Green and Efficient Method for the Preconcentration and Determination of Gallic Acid, Bergenin, Quercitrin, and Embelin from *Ardisia japonica* using Nononic Surfactant Genapol X-080 as the Extraction Solvent

Ying Chen,[1,2] **Kunze Du,**[1,2] **Jin Li,**[1,2] **Yun Bai,**[1,2] **Mingrui An,**[3] **Zhijing Tan,**[3] **and Yan-xu Chang**[1,2]

[1]*Tianjin State Key Laboratory of Modern Chinese Medicine, Tianjin University of Traditional Chinese Medicine, Tianjin 300193, China*
[2]*Key Laboratory of Formula of Traditional Chinese Medicine, Tianjin University of Traditional Chinese Medicine, Ministry of Education, Tianjin 300193, China*
[3]*Department of Surgery, University of Michigan, Ann Arbor, MI 48109, USA*

Correspondence should be addressed to Yan-xu Chang; tcmcyx@126.com

Academic Editor: Barbara Bojko

A simple cloud point preconcentration method was developed and validated for the determination of gallic acid, bergenin, quercitrin, and embelin in *Ardisia japonica* by high-performance liquid chromatography (HPLC) using ultrasonic assisted micellar extraction. Nonionic surfactant Genapol X-080 was selected as the extraction solvent. The effects of various experimental conditions such as the type and concentration of surfactant and salt, temperature, and solution pH on the extraction of these components were studied to optimize the conditions of *Ardisia japonica*. The solution was incubated in a thermostatic water bath at 60°C for 10 min, and 35% NaH_2PO_4 (w/v) was added to the solution to promote the phase separation and increase the preconcentration factor. The intraday and interday precision (RSD) were both below 5.0% and the limits of detection (LOD) for the analytes were between 10 and 20 ng·mL^{-1}. The proposed method provides a simple, efficient, and organic solvent-free method to analyze gallic acid, bergenin, quercitrin, and embelin for the quality control of *Ardisia japonica*.

1. Introduction

Ardisia japonica, one of common traditional Chinese medicines, belongs to the family of Ardisia [1]. *Ardisia japonica* has many medicinal and ornamental values and has drawn a global attention [2, 3]. Recent studies have reported a myriad of chemical compounds, antibacterial activity, and pharmacological properties. Many compounds including gallic acid [4], bergenin [5], quercitrin, and embelin have been elucidated (Figure 1). *Ardisia japonica* has been used to cure pancreatic, pneumonia, bronchitis conjunctivitis, and trauma and other types of cancer [1]. It has also been proven to possess anticancer activity [6] and anti-HIV activity [7]. Therefore, it has become imperative to develop a method of separation and purification of the compounds presenting in *Ardisia japonica.*

Many analytical methods for the extraction of bioactive compounds are employed in sample preparation [8, 9], but it is very challenging to determine trace bioactive compounds at low concentrations in the matrixes. It is therefore necessary to separate and concentrate the targeted analytes from the matrix sample to improve the detection sensitivity [10]. Many papers have reported enrichment methods including free air carbon dioxide enrichment (FACE) [11], cloud point extraction (CPE) [12], ionic liquid/ionic liquid liquid-liquid microextraction [13], solid-phase extraction (SPE) [14], and aqueous two-phase system (ATPS) [15]. The micelle-mediated extraction and cloud point preconcentration method provide a suitable alternative to the common extraction methods [16]. The surfactant-rich phase is a small volume of with a high concentration of surfactant and the

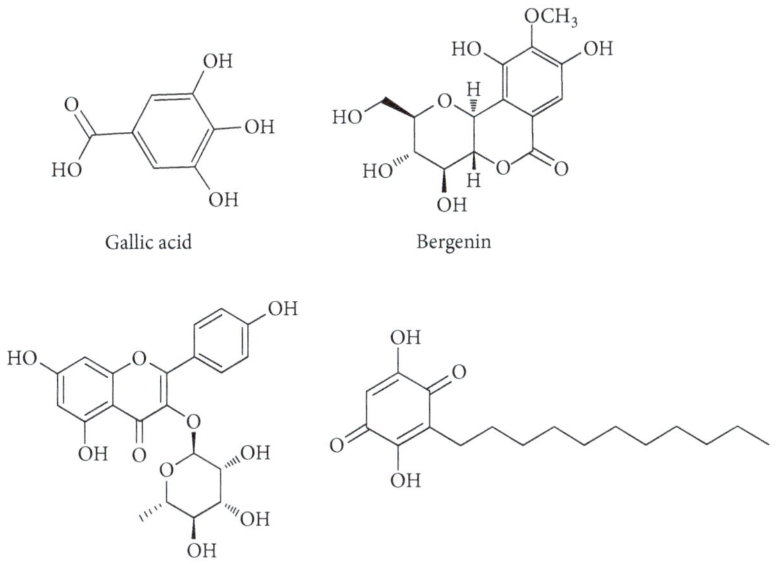

Gallic acid Bergenin

Quercitrin Embelin

FIGURE 1: Chemical structures of gallic acid, bergenin, quercitrin, and embelin.

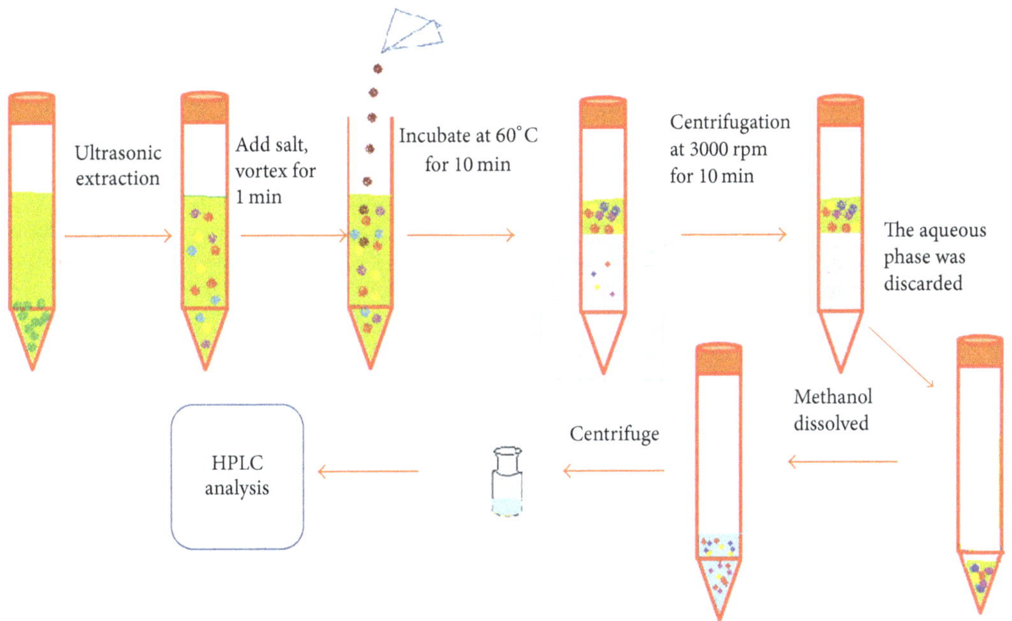

FIGURE 2: Schematic diagram for micellar extraction and cloud point preconcentration.

aqueous phase is a large volume with a low concentration of surfactant. The aqueous phase is discarded while the supernatant (surfactant-rich phase) was leaved and dissolved by methanol. Compared with common extraction methods, cloud point extraction (CPE) bears many advantages such as less cost, smaller aliquot of organic solutions, less toxicity, and higher precision. Thus, it is regarded as a rapid eco-friendly technique with the high precision due to the small volume of surfactant-rich phase and the reduction in the use of toxic organic solvents. At present, CPE has been used in different areas of extraction and preconcentration, including blood [17, 18], urine [19], water [18], food [20], metal ions [21], rare earth elements [22], and Chinese herbs [23]. Considering the complex matrixes of Chinese herbs, a micelle-mediated extraction and cloud point preconcentration method is fast becoming important for separation and preconcentration of analytes in Chinese herbs.

The aim of this study was to develop CPE method based on surfactants for extraction and preconcentration of gallic acid, bergenin, quercitrin, and embelin from *Ardisia japonica* samples (Figure 2). Through a series of optimum conditions, including the effects of ionic strength, pH of extraction solvent, bath temperature, and time which were all studied, the best conditions for *Ardisia japonica* samples were obtained. It was successfully employed to the quantification of gallic acid, bergenin, quercitrin, and embelin from *Ardisia japonica* samples.

2. Experimental

2.1. Plant Materials. The dried *Ardisia japonica* was obtained from a local pharmacy (China). The dried *Ardisia japonica* was pulverized using a grinder and sieved with a 100 mesh sieve to produce sample. The sample solutions were stored under controlled moisture and temperature.

2.2. Chemicals and Reagents. The standard of gallic acid, bergenin, quercitrin, and embelin (purity > 98%) were obtained from Chinese Academy of Sciences (Chengdu, China). Nonionic surfactant Genapol X-080 was purchased from Sigma (USA). Various concentrations (v/v) of Genapol X-080 solutions were prepared by dissolving appropriate amounts of Genapol X-080 in deionized water. Sodium dihydrogen phosphate, hydrochloric acid, and sodium hydroxide were of analytical grade and obtained from Beijing Chemical Factory (Beijing, China); acetonitrile was HPLC-grade and obtained from Fisher (Leicestershire, UK). All other reagents used in this work were of analytical grade. Deionized water (Milli-Q) was used throughout this study.

2.3. Preparation of Standard Solutions. The stock standard solutions of gallic acid, bergenin, quercitrin, and embelin (1.0 mg mL^{-1}) were made by dissolving appropriate amounts of these compounds in methanol. Then, the working standard solution was obtained by mixing certain concentration of the four stock solutions together and diluted with 4% Genapol X-080.

2.4. Instrumentation. All analysis was performed on Waters HPLC system (2695 series). The HPLC system consisted of a quaternary pump, a photodiode array detector and a column thermostat. The analytical column was an Ultimate XB-C18 (250 mm × 4.6 mm i.d, 5 μm) connected to an Agilent Zorbax Extend-C18 guard column (12.5 mm × 2.1 mm i.d., 5 μm). The column temperature was 35°C. Chromatographic data processing was aided by Empower software in HPLC systems.

Samples were pulverized and sieved (Zhejiang, China). A vortex mixer (WH-861, Taicang, China), a pH meter (Crison pH 2000, Shanghai, China), and a thermostatic bath (HH-2, Guangzhou, China) were used to performance cloud point extraction. High-speed centrifugation (Sigma) was applied to increase the phase separation process.

2.5. Experimental Procedures

2.5.1. Micelle-Mediated Extraction Procedure. The dried *Ardisia japonica* powder (0.1000 g) was accurately weighed and put it into a 50 mL centrifuge tube. 20 mL Genapol X-080 solution (4%) (v/v) was then added. The powders and Genapol X-080 solution (4%) (v/v) were thoroughly mixed and then were extracted (400 W) for 40 min at room temperature (22°C) in the ultrasonic cleaning bath. The extract solution was centrifuged at 5000 rpm for 10 min and the supernatant was filtered through a 0.45 μm membrane.

2.5.2. Cloud Point Preconcentration. 10 mL supernatant of extract was accurately measured and placed in a 15 mL centrifuge tube. 35% (w/v) sodium dihydrogen phosphate was diluted by the supernatant and the solution pH was adjusted to 5 by 10% of HCL and 0.1 M NaOH solutions in a test tube. The sodium dihydrogen phosphate was added to the sample solution and dissolved thoroughly with a Vortex Genie Mixture for 1 min and then incubated in the thermostatic bath at 60°C for 10 min. The phase separation was accelerated by centrifugation at 3000 rpm for 10 min. The aqueous phase was then discarded, leaving the surfactant-rich phase of volume 600 μL. The surfactant-rich phase was dissolved in methanol and topped up to 2 mL using a glass volumetric flask. After ultrasonic processing for 3 min and centrifuging at 14000 r min^{-1} for 10 min, a 0.45 μm nylon membrane was applied to filter the samples. 10 μL of the solution was injected into the HPLC system for analysis.

2.6. HPLC Analysis. 0.1% formic acid (A) and acetonitrile (B) was selected as mobile phase. The linear gradient was 5–9% B over 0–4 min, 9–14% B over 4–7 min, 14–30% B over 7–20 min, 30–60% B over 20–25 min, 60–86% B over 25–30 min, 86–100% B over 30–40 min, and 100% B over 40–45 min and then returned to 5% B at 46 min immediately. The flow rate was 1.0 mL min^{-1}. The detector was at 275 nm and the column temperature was maintained at 35°C. In order to avoid the influence of salt and Genapol X-080 to the separation of analytes and to protect the column, acetonitrile and water were used to sufficiently elute.

3. Results and Discussion

3.1. Optimization of Micellar Extraction. In order to optimize the CPE of gallic acid, bergenin, quercitrin, and embelin from the herbal sample, a series of experiments including the type of surfactant, surfactant concentration, ultrasonic extracting time, ultrasonic extraction power, solid-liquid rate, ionic strength, pH of sample solution, bath temperature, and time under different conditions were studied. The peak area of the analyte was used to evaluate the effect of these factors.

3.1.1. Selection of the Surfactant in Micellar Extraction. The type of surfactant plays an important role on micellar extraction. Seven surfactants (Triton X-100, Triton X-114, Triton X-305, Triton X-405, Triton X-45, and Genapol X-080) were optimized as extraction solvents (Figure 3(a)). It was found that highest peak areas of analytes were obtained when Genapol X-080 and Triton X-100 were selected as extraction solvents. Compared with Triton X series surfactant, the Genapol X-080 does not absorb at 275 nm by HPLC-UV method. In addition, Genapol X-080 is the inexpensive and eco-friendly surfactant [24]. Therefore, Genapol X-080 was chosen as the CPE surfactant for next step in this research.

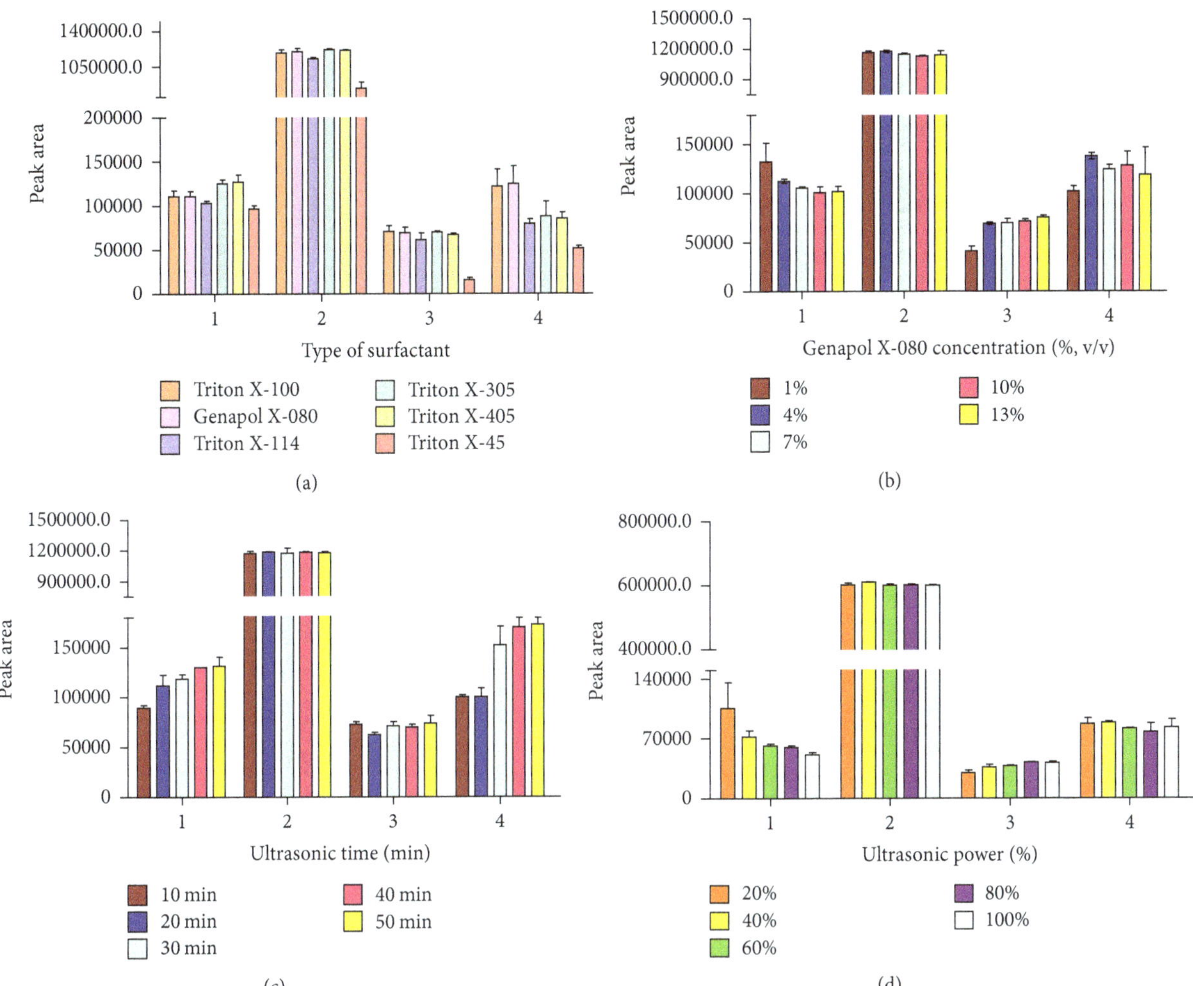

FIGURE 3: Effect of some parameters of ultrasonic assisted micellar extraction on the extraction effect of 4 components: (a) type of surfactant; (b) Genapol X-080 concentration (%, v/v); (c) ultrasonic time (min); (d) ultrasonic power (%). 1, gallic acid; 2, bergenin; 3, quercitrin; 4, embelin.

3.1.2. Genapol X-080 Concentration in Micellar Extraction. Genapol X-080 concentration was studied in the range of 1–13% (v/v). Other conditions were kept constant, including 20 mL Genapol X-080 as extraction solvent and 40 min as ultrasound time. The critical micellar concentration (CMC) of Genapol X-080 is 0.05 mM (2.9%, v/v) [25]. When the concentration of surfactant is at 1% (below CMC), it is difficult to form two phases. When the Genapol X-080 concentration increases to 4%, the peak areas of analytes were highest. When the Genapol X-080 concentration continues to rise, the solution becomes too sticky to be handled so that the peak area of analytes would not increase. The character of the aqueous nonionic Genapol X-080 solution could increase the solubility of analytes via the surfactant micelles. According to above-mentioned results (Figure 3(b)), 4% Genapol X-080 was selected for obtaining highest peak areas.

3.1.3. Ultrasonic Extraction Time Effect in Micellar Extraction. The influence of ultrasonic time on extraction effect of gallic acid, bergenin, quercitrin, and embelin were studied by changing the ultrasonic time from 10 to 50 min (Figure 3(c)). The results indicated that the amount of extracted gallic acid, bergenin, quercitrin, and embelin increased with the increase of extraction time and the extraction yield of gallic acid, quercitrin, and embelin reached the highest value at 50 min. There was no significant difference between the peak areas of the four analytes extracted by ultrasonic extraction for 50 min and those extracted by ultrasonic extraction for 40 min ($P > 0.05$). Hence, 40 min was chosen for the ultrasonic extraction of gallic acid, bergenin, quercitrin, and embelin.

3.1.4. Ultrasonic Power Effect in Micellar Extraction. The effect of ultrasonic power on extraction ability of gallic acid, bergenin, quercitrin, and embelin were studied by varying the ultrasonic power from 20% to 100% at the optimal condition (Figure 3(d)). For bergenin and embelin, the peak area reached the highest under the ultrasonic power of 40%. For gallic acid and quercitrin, the peak area reached the highest under the ultrasonic power of 20% and 80%, respectively. The result of this method is consistent with the reported research

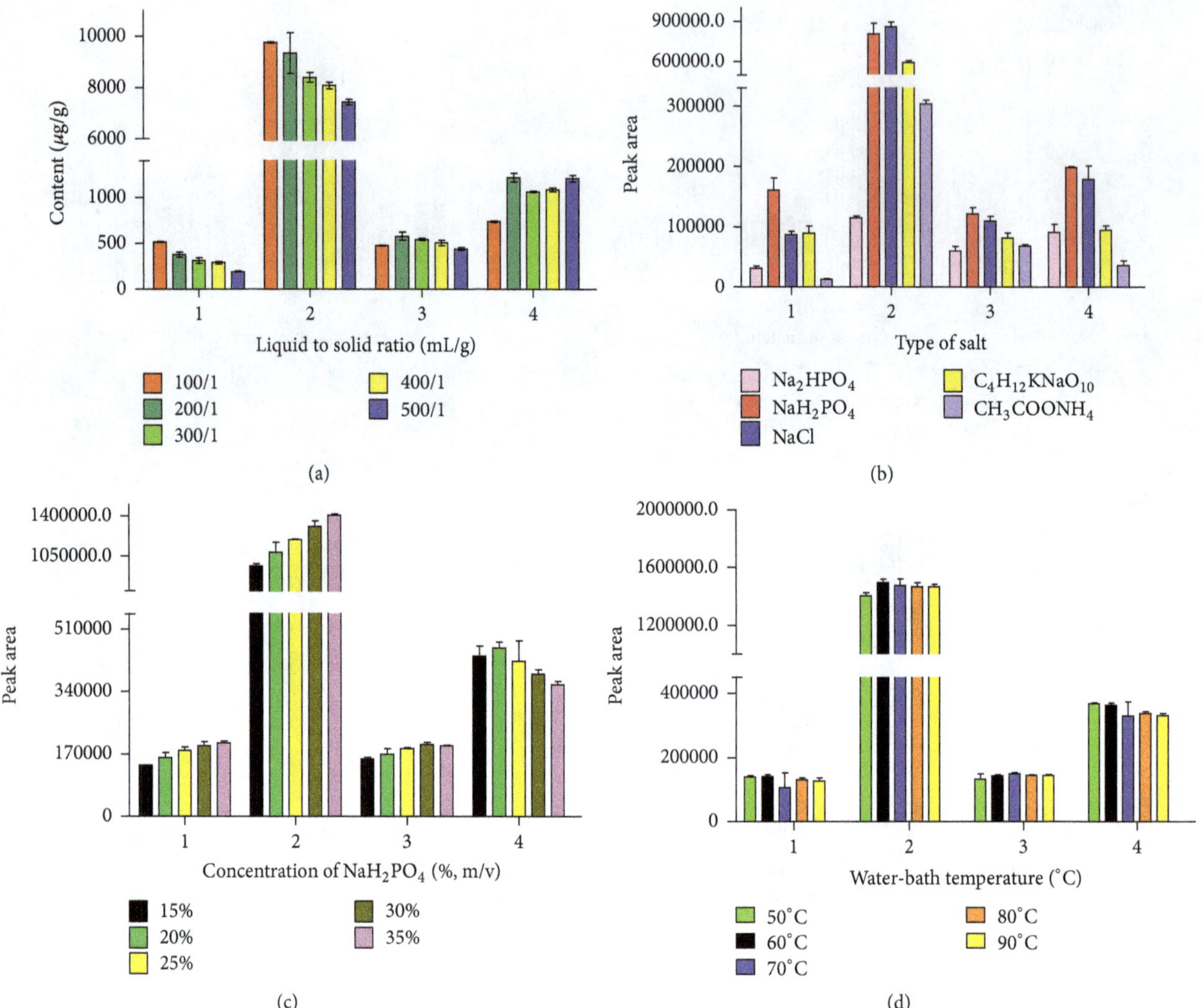

FIGURE 4: Effect of some parameters of cloud point extraction on the extraction effect of 4 components: (a) liquid to solid ratio (mL/g); (b) type of salt; (c) concentration of NaH2PO4 (%, m/v); (d) water-bath temperature (°C). 1, gallic acid; 2, bergenin; 3, quercitrin; 4, embelin.

that ultrasound can improve diffusion and then increases the extract yield of analytes [26]. There is no difference between the peak area of gallic acid under the ultrasonic power 20% and that under the ultrasonic power 40% ($P > 0.05$). Consequently, 40% of the ultrasonic power was chosen for the extraction of the sample.

3.1.5. Liquid/Solid Ratio Effect in Micellar Extraction. Liquid/solid ratio is the ratio between the volume of solvent and the amount of crude material. The effects of liquid/solid ratio on extraction ability of gallic acid, bergenin, quercitrin, and embelin were studied by varying the liquid/solid ratio from 100 : 1 to 500 : 1 ($mL\,g^{-1}$) at the optimal condition above (Figure 4(a)). The extraction yield of quercitrin and embelin was the highest under the liquid/solid ratio at 200 : 1 (20 mL/0.1 g). The extraction yield of gallic acid and bergenin was the highest under the liquid/solid ratio at 100 : 1 (10 mL/0.1 g). There was no significant difference between the peak areas the extraction yield of the bergenin at the liquid/solid ratio of 100 : 1 and those extracted by at the liquid/solid ratio of 200 : 1 ($P > 0.05$). Considering the extraction efficiency of embelin and bergenin, the liquid/solid ratio of 200 : 1 ($mL\,g^{-1}$) was used to perform the following experiments.

The optimal condition was obtained from micellar extraction by serious of experiment above: 0.1 g of the powders and 20 mL Genapol X-080 solution (4%) (v/v) were thoroughly mixed and then put through ultrasonic extraction (400 W) for 40 min.

3.2. Optimization of Cloud Point Preconcentration

3.2.1. The Effect of the Type of Salt. The type of salt is a key factor in CPE. The addition of salt could promote the separation of surfactant from water [25]. It is one of the factors influencing the cloud point extraction efficiency of gallic acid, bergenin, quercitrin, and embelin. Supernatant was obtained from the optimal condition of micellar extraction. Five additions (NaH_2PO_4, Na_2HPO_4, NaCl, CH_3COONH_4, and $C_4H_{12}KNaO_{10}$) were chosen to optimize the conditions. As shown in Figure 4(b), the highest extract efficiency of gallic

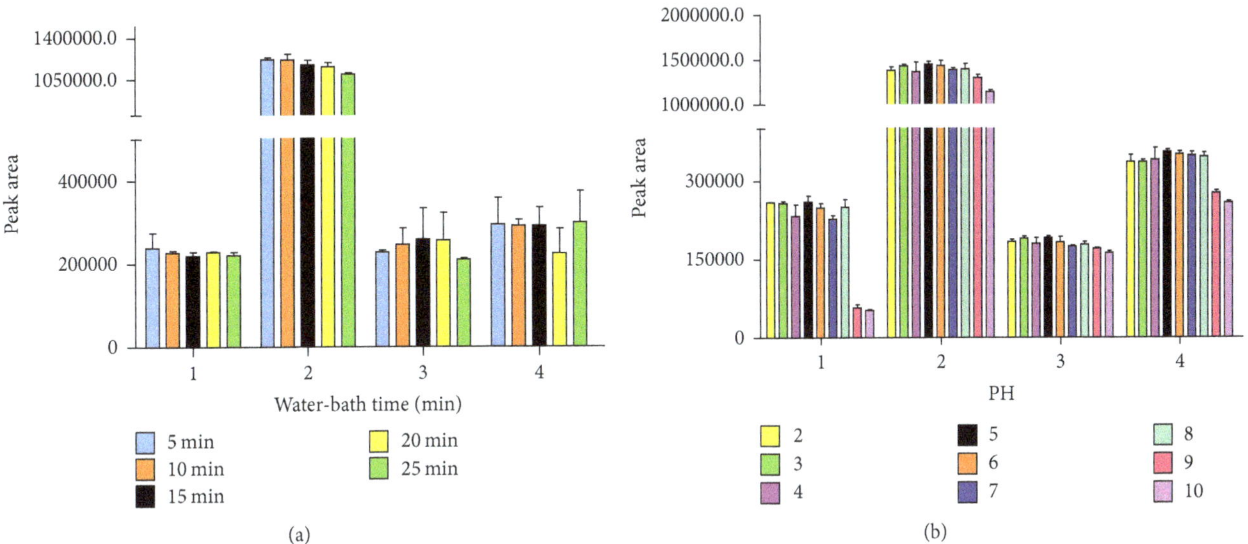

FIGURE 5: Effect of water-bath time (a) and pH of cloud point extraction (b) on the extraction effect of 4 components. 1, gallic acid; 2, bergenin; 3, quercitrin; 4, embelin.

acid, quercitrin, and embelin was obtained when NaH_2PO_4 was selected as addition salt. In addition, it was found that there is no significant difference ($P > 0.05$) for the extract of bergenin between NaH_2PO_4 and NaCL (Figure 4(b)). Thus, NaH_2PO_4 was employed in the following experiments.

3.2.2. The Effect of Concentration of NaH_2PO_4. The concentration of sodium dihydrogen phosphate is also a key factor in CPE [27]. The addition of salt to the sample solution can influence the extraction resulting from a changing in density of the aqueous phase [28]. The influence of the ionic strength on the extraction efficiency was carried out by changing the concentration of sodium dihydrogen phosphate between 15% and 35% (w/v). The results showed that the addition of sodium dihydrogen phosphate promoted the separation between the surfactant-rich phase and the aqueous phase. With the salt concentration increased, the micelle size and the aggregation number were also increasing and analytes might become more soluble in the surfactant-rich phase, while the critical micellar concentration stayed constant [29]. Therefore, the addition of salt not only promoted the transform hydrophilic compounds into the surfactant-rich phase, but also reduced the volume of the surfactant-rich phase, most likely via some type of dehydration mechanism [30–32]. In this method, 35% sodium dihydrogen phosphate was applied. The results in Figure 4(c) indicated that the CPE at a salt concentration of 35% (w/v) obtained the optimum extraction recovery of gallic acid and bergenin. Higher peak area of quercitrin was obtained at the salt concentration of 30% (w/v). Although the highest peak area of embelin was obtained at the salt concentration of 20% (w/v), 35% sodium dihydrogen phosphate was selected as addition to rich the four analytes according to other three compounds.

3.2.3. The Effect of Incubation Temperature and Time. Incubation temperature and time are two important factors of the cloud point extraction. The aqueous solution of surfactant divides into two phases when at certain temperature, called cloud point temperature (CPT) of the surfactant [33]. The optimal incubation temperature for the extraction was obtained by 15–20°C, which was higher than the cloud point of the surfactant (42°C in pure water [25]). The influence of temperature on the extraction effect was studied in the range of 50–90°C (Figure 4(d)). The results showed that the highest peak areas of gallic acid, bergenin, and embelin were obtained at 60°C and the highest peak area of quercitrin was obtained at 50°C. It was found that there no significant different between peak area of quercitrin at 60°C and that at 50°C. Therefore, 60°C was employed as optimized incubation temperature. The effect of different incubation time on the extraction efficiency was studied by varying the incubation time from 5 to 25 min (Figure 5(a)). The results indicated that higher peak areas of gallic acid and embelin were obtained at the incubation time of 5 min while higher peak areas of bergenin and quercitrin were obtained at the incubation time of 10 min. A slight decrease of peak area of three analytes (gallic acid, bergenin, and quercitrin) was also found when the extraction time incubation was longer than 20 min. There was no significance between peak areas of gallic acid and embelin at 60°C for 10 min and those at 60°C for 5 min ($P < 0.05$). Therefore, the CPE was performed at 60°C for 10 min.

3.2.4. The Effect of pH. It was well-known that partition of some ionizable organic analytes in two immiscible phases depends on solution pH. The effect of pH on the analytes extraction efficiency was studied over the range of 2–10 and adjusted by 10% HCL and 0.01 M NaOH solution (Figure 5(b)). The results showed that the solution pH in acidic environment had no significant effect on the extraction of analytes while the pH of solution has the significant effect on the extraction of gallic acid and embelin in alkaline environment. This phenomenon might be due to the formation

TABLE 1: Calibration curve, linear range, LOD, and LOQ of the investigated analytes.

Analytes	Calibration curve	Linear range (μg mL^{-1})	r^2	LOD (μg mL^{-1})	LOQ (μg mL^{-1})
Gallic acid	$y = 19752x + 5015.7$	0.4–50	0.999	0.013	0.04
Bergenin	$y = 5179.4x + 29596$	4–500	0.9988	0.012	0.04
Quercitrin	$y = 12748x - 6495.3$	1–50	0.9961	0.010	0.03
Embelin	$y = 15290x + 4114.9$	4–250	0.9964	0.016	0.04

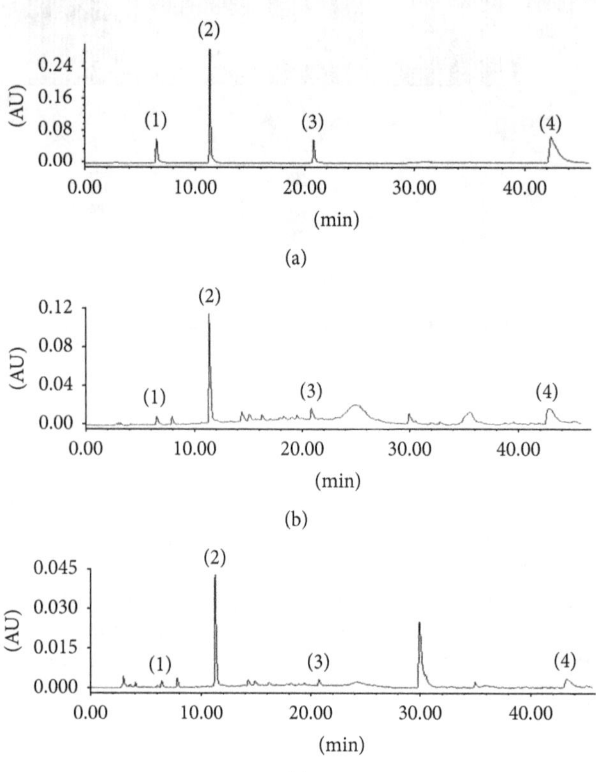

FIGURE 6: The typical HPLC chromatograms of (a) the mixed standard solutions: (1) gallic acid, 50 μg mL^{-1}; (2) bergenin, 500 μg mL^{-1}; (3) quercitrin, 50 μg mL^{-1}; (4) embelin, 250 μg mL^{-1}; (b) sample extracted by ultrasonic assisted micellar extraction and then preconcentrated by cloud point extraction: (1) gallic acid, 5.58 μg mL^{-1}; (2) bergenin, 216.65 μg mL^{-1}; (3) quercitrin, 12.72 μg mL^{-1}; (4) embelin, 48.45 μg mL^{-1}; (c) sample extracted by ultrasonic assisted micellar extraction. (1) gallic acid, 1.09 μg mL^{-1}; (2) bergenin, 39.93 μg mL^{-1}; (3) quercitrin, 2.59 μg mL^{-1}; (4) embelin, 10.46 μg mL^{-1}.

of ion-pairs of the analytes. It was found that the maximum extraction efficiency of four analytes was achieved at pH 5. Therefore, the CPE was performed at pH 5.

3.3. HPLC Profiles of the Extracted Four Compounds. Based on the experiments result above, the optimum PCE conditions were as follows: 4% (v/v) Genapol X-080, 35% (w/v) NaH_2PO_4 in 10 mL solution (pH 5), and 10 min incubation at 60°C. Under the optimized conditions, enrichment factor of all the four compounds was greater. The results showed that the extraction efficiency of the analytes increased significantly. Figure 6 showed the chromatogram of four compounds extracted from *Ardisia japonica* samples after mixed cloud point extraction. It could be observed that the chromatographic condition allowed a good separation of analytes.

3.4. Analytical Method Validation. At least six described concentrations of the standard solution extracted by ultrasonic assisted micellar extraction and then preconcentrated by cloud point extraction method were analyzed. Then the calibration curves were constructed by plotting the peak area versus the concentrations of compounds. Correlation coefficients of the calibration curves were found to be higher than 0.99. The regression data obtained showed good linear relationship. Limits of detection (LOD) for the four compounds were less than 16.0 ng mL^{-1} ($S/N = 3$), and limits of quantification (LOQ) were less than 40.0 ng mL^{-1} ($S/N = 10$), which showed a high sensitivity under these experimental conditions (Table 1).

The intraday precision and interday precision were studied by analyzing three different concentrations of standard solutions. The intraday variance was determined by analyzing the spiked samples six times at one day. The interday variance was studied for three continuous days. Variations were indicated by the RSD. The RSDs were all less than 5% for intraday and interday. The accuracies were determined by comparing the mean calculated concentration with the spiked target concentration of the three different concentrations samples. The intraday and interday accuracies for the four analytes were found to be within 95.9% and 100% (Table 2). The recovery of the method was determined using the standard addition method. The standard analytes were added to approximately 0.025 g of the *Ardisia japonica* extract sample and then extracted and analyzed using the method described above. The total content of each analyte was calculated by the corresponding calibration curve. The formula was used to calculate the recoveries: recovery (%) = (amount found − original amount)/amount spiked × 100%. The recoveries of the four analytes were between 96.3% and 100% (Table 3), which showed good analytical characteristics of this method.

3.5. Application of Method. To validate the method, three batches of *Ardisia japonica* extract samples were extracted using the optimized method. Table 4 showed the content of all of the four analyses. For the four compounds, the average content of bergenin was the highest, between 3.07 mg g^{-1} and 11.77 mg g^{-1}. The content of gallic acid ranged from 0.16 mg g^{-1} to 0.37 mg g^{-1}, the content of quercitrin was between 0.21 mg g^{-1} and 0.88 mg g^{-1}, and the content of embelin was changed from 0.66 mg g^{-1} to 2.08 mg g^{-1},

Table 2: Precision and accuracy of the method ($n = 6$).

Analytes	Assay	Concentration ($\mu g\ mL^{-1}$)	Accuracy (%)	Precision (RSD%)
Gallic acid	Intra-assay	0.4	96.5	4.96
		2	97.9	0.79
		10	98.6	0.13
	Inter-assay	0.4	97.1	3.65
		2	97.7	0.93
		10	97.4	0.11
Bergenin	Intra-assay	4	97.7	0.47
		20	98.7	0.03
		100	98.7	0.03
	Inter-assay	4	97.6	0.44
		20	98.4	0.01
		100	97.0	0.02
Quercitrin	Intra-assay	0.4	98.6	4.10
		2	95.9	1.19
		10	100.0	0.25
	Inter-assay	0.4	98.4	4.23
		2	98.5	1.30
		10	97.8	0.17
Embelin	Intra-assay	1	98.2	2.53
		10	97.8	0.21
		50	97.9	0.03
	Inter-assay	1	98.4	0.46
		10	96.8	0.12
		50	97.3	0.03

Table 3: Recovery ($n = 6$) for four analytes in samples.

	Sample[1] (μg)	Spiked (μg)	Determined[2] (μg)	Recovery (%)	RSD (%)
Gallic acid	0.82	0.80	1.59	96.3	4.22
Bergenin	21.07	21.00	41.66	98.0	3.21
Quercitrin	1.39	1.40	2.77	98.6	1.28
Embelin	2.22	2.24	4.46	100	3.51

[1]The amounts of analytes in plant samples extracted by ultrasonic assisted micellar extraction and then preconcentrated by cloud point extraction. [2]The amounts of analytes in the spiked plant samples extracted by ultrasonic assisted micellar extraction and then preconcentrated by cloud point extraction.

respectively. The average of contents of the four analyses in the *Ardisia japonica* extract samples was different due to the difference of growth place of the raw materials. The average of the four analyses content was the highest of the herb grown in Hubei.

3.6. Comparison with Conventional Methanol Method. The chromatograms using different extraction methods for the same sample are compared. It was found that the chromatographic signals obtained by the CPE method were higher than those by the conventional methanol extraction method (Pharmacopeia of China 2015) because of its good enrichment capacity (Table 4). At the same time, peak areas of four components increased nearly 3 times by using cloud point preconcentration than those by using single micellar extraction (the enrichment ratio increased nearly three times) (Figure 6).

4. Conclusion

An environmentally friendly micelle-mediated extraction and cloud point preconcentration method was developed and validated for analysis of the four bioactive compounds in *Ardisia japonica* samples by using Genapol X-080 nonionic as extraction solvent. The optimal concentration of Genapol X-080, solution pH, liquid/solid ratio, ultrasonic time, concentration of NaH_2PO_4, temperature, and time of water bath were 4% (v/v), pH 5.0, 200 : 1 (mL/g), 40 min, 35% (w/v), 60°C, and 10 min, respectively. Under these conditions, high yield of the four compounds was obtained, which was higher

Table 4: The concentrations (mg g^{-1}) of four analytes in the three batches of *Ardisiajaponica* from different area (mg g^{-1}) (n = 3).

Analytes	Hubei	Hunan	Guizhou
Gallic acid	0.29 ± 0.01	0.16 ± 0.00	0.37 ± 0.01
	0.18 ± 0.00	0.11 ± 0.01	0.21 ± 0.02
Bergenin	11.77 ± 0.40	3.07 ± 0.05	10.28 ± 0.28
	10.07 ± 0.11	2.44 ± 0.17	7.04 ± 0.50
Quercitrin	0.88 ± 0.01	0.21 ± 0.00	0.56 ± 0.01
	0.85 ± 0.02	0.16 ± 0.02	0.51 ± 0.04
Embelin	1.18 ± 0.01	0.66 ± 0.01	0.92 ± 0.00
	1.07 ± 0.03	0.66 ± 0.10	0.97 ± 0.02

Note. Upper line: micelle-mediated extraction and cloud point preconcentration method. Lower line: conventional methanol extraction method.

than those with methanol method. It was proved that the method not only is simple, green, and reliable but also can obtain high enhancement factors and low limits of detection for the studied of four compounds of *Ardisia japonica*. The inexpensive Genapol X-080 solvent was used to extract and preconcentrate the trace bioactive compounds at low concentrations in the matrixes. It was concluded that micelle-mediated extraction and cloud point preconcentration will maintain a promising role in extraction and purification of active compounds from herbs medicines.

Authors' Contributions

Ying chen and Jin Li were involved in detailed experimental design and carried out the main part of the study responsible for data analysis, interpretation, and manuscript writing; Yun Bai provided some of the experimental materials; Mingrui An and Zhijing Tan were involved in some part of the experimental study; Kunze Du was involved in manuscript writing: Yan-xu Chang was responsible conceptually for the project design, data analysis, manuscript writing, and final approval of the manuscript. All authors read and approved the final manuscript.

Acknowledgments

This research was supported by National Natural Science Foundation of China (81374050 and 81503213) and Special Program of Talents Development for Excellent Youth Scholars in Tianjin of China and PCSIRT (IRT-14R41).

References

[1] H. Kobayashi and E. De Mejía, "The genus Ardisia: A novel source of health-promoting compounds and phytopharmaceuticals," *Journal of Ethnopharmacology*, vol. 96, no. 3, pp. 347–354, 2005.

[2] A. M. B. Newell, G. G. Yousef, M. A. Lila, M. V. Ramírez-Mares, and E. G. de Mejia, "Comparative in vitro bioactivities of tea extracts from six species of Ardisia and their effect on growth inhibition of HepG2 cells," *Journal of Ethnopharmacology*, vol. 13, no. 3, pp. 536–544, 2010.

[3] E. G. De Mejía and M. V. Ramírez-Mares, "Ardisia: Health-promoting properties and toxicity of phytochemicals and extracts," *Toxicology Mechanisms and Methods*, vol. 21, no. 9, pp. 667–674, 2011.

[4] C. Y. Xu, L. Y. Zhang, Y. Wang et al., "Study on HPLC fingerprint of miao medicine Ardisia japonica," *Journal of Chinese Medicinal Materials*, vol. 37, no. 9, pp. 1570–1573, 2014.

[5] W. Liu, "Quantiative analysis of bergenin in herta ardisia japonica BI by TLC-densitttometric method," *China Journal of Chinese Materia Medica*, 1991.

[6] X. Chang, W. Li, Z. Jia, T. Satou, S. Fushiya, and K. Koike, "Biologically active triterpenoid saponins from Ardisia japonica," *Journal of Natural Products*, vol. 70, no. 2, pp. 179–187, 2007.

[7] S. Piacente, C. Pizza, and N. De Tommasi, "Constituents of *Ardisia japonica* and their in vitro anti-HIV activity," *Journal of Natural Products*, vol. 59, no. 6, pp. 565–569, 1996.

[8] K. Madej, "Microwave-assisted and cloud-point extraction in determination of drugs and other bioactive compounds," *TrAC - Trends in Analytical Chemistry*, vol. 28, no. 4, pp. 436–446, 2009.

[9] T. Y. See, S. I. Tee, T. N. Ang, C.-H. Chan, R. Yusoff, and G. C. Ngoh, "Assessment of various pretreatment and extraction methods for the extraction of bioactive compounds from orthosiphon stamineus leaf via microstructures analysis," *International Journal of Food Engineering*, vol. 12, no. 7, pp. 711–717, 2016.

[10] S. Z. Mohammadi, T. Shamspur, D. Afzali, M. A. Taher, and Y. M. Baghelani, "Applicability of cloud point extraction for the separation trace amount of lead ion in environmental and biological samples prior to determination by flame atomic absorption spectrometry," *Arabian Journal of Chemistry*, vol. 26, no. 5, pp. 1916–1934, 2011.

[11] A. Fangmeier, V. Torres-Toledo, J. Franzaring, and W. Damsohn, "Design and performance of a new FACE (free air carbon dioxide enrichment) system for crop and short vegetation exposure," *Environmental and Experimental Botany*, vol. 130, pp. 151–161, 2016.

[12] J. Zhou, J. B. Sun, X. Y. Xu et al., "Application of mixed cloud point extraction for the analysis of six flavonoids in Apocynum venetum leaf samples by high performance liquid chromatography," *Journal of Pharmaceutical and Biomedical Analysis*, vol. 107, pp. 273–279, 2015.

[13] Z. Liu, W. Yu, H. Zhang, F. Gu, and X. Jin, "Salting-out homogenous extraction followed by ionic liquid/ionic liquid liquid–liquid micro-extraction for determination of sulfonamides in blood by high performance liquid chromatography," *Talanta*, vol. 161, pp. 748–754, 2016.

[14] K. Xu, Y. Wang, Y. Li, Y. Lin, H. Zhang, and Y. Zhou, "A novel poly(deep eutectic solvent)-based magnetic silica composite for solid-phase extraction of trypsin," *Analytica Chimica Acta*, vol. 946, pp. 64–72, 2016.

[15] W. Zhang, D. Zhu, H. Fan et al., "Simultaneous extraction and purification of alkaloids from Sophora flavescens Ait. by microwave-assisted aqueous two-phase extraction with ethanol/ammonia sulfate system," *Separation and Purification Technology*, vol. 141, pp. 113–123, 2015.

[16] Z. Shi, X. Zhu, and H. Zhang, "Micelle-mediated extraction and cloud point preconcentration for the analysis of aesculin and

aesculetin in Cortex fraxini by HPLC," *Journal of Pharmaceutical and Biomedical Analysis*, vol. 44, no. 4, pp. 867–873, 2007.

[17] H. Shirkhanloo, M. Ghazaghi, and M. M. Eskandari, "Cloud point assisted dispersive ionic liquid -liquid microextraction for chromium speciation in human blood samples based on isopropyl 2-[(isopropoxycarbothiolyl)disulfanyl] ethane thioate," *Analytical Chemistry Research*, vol. 10, pp. 18–27, 2016.

[18] H. Shirkhanloo, A. Khaligh, H. Z. Mousavi, and A. Rashidi, "Ultrasound assisted-dispersive-ionic liquid-micro-solid phase extraction based on carboxyl-functionalized nanoporous graphene for speciation and determination of trace inorganic and organic mercury species in water and caprine blood samples," *Microchemical Journal*, vol. 130, pp. 245–254, 2017.

[19] M. Rahimi, P. Hashemi, and F. Nazari, "Cold column trapping-cloud point extraction coupled to high performance liquid chromatography for preconcentration and determination of curcumin in human urine," *Analytica Chimica Acta*, vol. 826, no. 1, pp. 35–42, 2014.

[20] R. Gürkan, S. Korkmaz, and N. Altunay, "Preconcentration and determination of vanadium and molybdenum in milk, vegetables and foodstuffs by ultrasonic-thermostatic-assisted cloud point extraction coupled to flame atomic absorption spectrometry," *Talanta*, vol. 155, pp. 38–46, 2016.

[21] R. Galbeiro, S. Garcia, and I. Gaubeur, "A green and efficient procedure for the preconcentration and determination of cadmium, nickel and zinc from freshwater, hemodialysis solutions and tuna fish samples by cloud point extraction and flame atomic absorption spectrometry," *Journal of Trace Elements in Medicine and Biology*, vol. 28, no. 2, pp. 160–165, 2014.

[22] M. M. Hassanien, I. M. M. Kenawy, M. E. Khalifa, and M. M. Elnagar, "Mixed micelle-mediated extraction approach for matrix elimination and separation of some rare earth elements," *Microchemical Journal*, vol. 127, pp. 125–132, 2016.

[23] M. P. K. Choi, K. K. C. Chan, H. W. Leung, and C. W. Huie, "Pressurized liquid extraction of active ingredients (ginsenosides) from medicinal plants using non-ionic surfactant solutions," *Journal of Chromatography A*, vol. 983, no. 1-2, pp. 153–162, 2003.

[24] C. Sun and H. Liu, "Application of non-ionic surfactant in the microwave-assisted extraction of alkaloids from Rhizoma Coptidis," *Analytica Chimica Acta*, vol. 612, no. 2, pp. 160–164, 2008.

[25] Z. Shi, J. Yan, Y. Ma, and H. Zhang, "Cloud point extraction-HPLC determination of polycyclic aromatic hydrocarbons residues in traditional Chinese medicinal herbs," in *Proceedings of the 2011 3rd International Conference on Environmental Science and Information Application Technology (ESIAT '11)*, vol. 10, pp. 1216–1221, China, August 2011.

[26] L. C. Klein-Júnior, Y. Vander Heyden, and A. T. Henriques, "Enlarging the bottleneck in the analysis of alkaloids: A review on sample preparation in herbal matrices," *TrAC - Trends in Analytical Chemistry*, vol. 80, pp. 66–82, 2016.

[27] S. Li, M. Wang, Y. Zhong, Z. Zhang, and B. Yang, "Cloud point extraction for trace inorganic arsenic speciation analysis in water samples by hydride generation atomic fluorescence spectrometry," *Spectrochimica Acta Part B: Atomic Spectroscopy*, vol. 111, pp. 74–79, 2015.

[28] H. Wang, M. Gao, M. Wang et al., "Integration of phase separation with ultrasound-assisted salt-induced liquid-liquid microextraction for analyzing the fluoroquinones in human body fluids by liquid chromatography," *Journal of Chromatography B*, vol. 985, pp. 62–70, 2015.

[29] M. Poša, S. Bjedov, D. Škorić, and M. Sakač, "Micellization parameters (number average, aggregation number and critical micellar concentration) of bile salt 3 and 7 ethylidene derivatives: Role of the steroidal skeleton II," *Biochimica et Biophysica Acta (BBA) - General Subjects*, vol. 1850, no. 7, pp. 1345–1353, 2015.

[30] W. J. Horvath and C. W. Huie, "Screening of urinary coproporphyrin using cloud point extraction and chemiluminescence detection," *Talanta*, vol. 40, no. 9, pp. 1385–1390, 1993.

[31] C. R. Parish, B. J. Classon, J. Tsagaratos, I. D. Walker, L. Kirszbaum, and I. F. C. McKenzie, "Fractionation of detergent lysates of cells by ammonium sulphate-induced phase separation," *Analytical Biochemistry*, vol. 156, no. 2, pp. 495–502, 1986.

[32] W. J. Horvath and C. W. Huie, "Salting-out surfactant extraction of porphyrins and metalloporphyrin from aqueous non-ionic surfactant solutions," *Talanta*, vol. 39, no. 5, pp. 487–492, 1992.

[33] N. Pourreza, S. Rastegarzadeh, and A. Larki, "Micelle-mediated cloud point extraction and spectrophotometric determination of rhodamine B using Triton X-100," *Talanta*, vol. 77, no. 2, pp. 733–736, 2008.

12

Mid and Near-Infrared Reflection Spectral Database of Natural Organic Materials in the Cultural Heritage Field

Claudia Invernizzi,[1,2] **Tommaso Rovetta,**[1,3] **Maurizio Licchelli,**[1,4] **and Marco Malagodi**[1,5]

[1]*Arvedi Laboratory of Non-Invasive Diagnostics, CISRiC, University of Pavia, Via Bell'Aspa 3, 26100 Cremona, Italy*
[2]*Department of Mathematical, Physical and Computer Sciences, University of Parma, Parco Area delle Scienze, 7/A, 43124 Parma, Italy*
[3]*Department of Physics, University of Pavia, Via Bassi 6, 27100 Pavia, Italy*
[4]*Department of Chemistry, University of Pavia, via Taramelli 12, 27100 Pavia, Italy*
[5]*Department of Musicology and Cultural Heritage, University of Pavia, Corso Garibaldi 178, 26100 Cremona, Italy*

Correspondence should be addressed to Marco Malagodi; marco.malagodi@unipv.it

Academic Editor: David Touboul

This study presents mid and near-infrared (7500-375 cm^{-1}) total reflection mode spectra of several natural organic materials used in artworks as binding media, consolidants, adhesives, or protective coatings. A novel approach to describe and interpret reflectance bands as well as calculated absorbance after Kramers-Kronig transformation (KKT) is proposed. Transflection mode spectra have represented a valuable support both to study the distorted reflectance bands and to validate the applicability and usefulness of the KK correction. The aim of this paper is to make available to scientists and conservators a comprehensive infrared reflection spectral database, together with its detailed interpretation, as a tool for the noninvasive identification of proteins, lipids, polysaccharides, and resins by means of portable noncontact FTIR spectrometers.

1. Introduction

For about ten years, modern heritage science community has been moving toward the development of nondestructive and noninvasive diagnostic methodologies that do not require any removal of small fragments from precious or unique works of art and preferably enable in situ investigations [1–4]. Furthermore, the omission of invasive sampling makes the measurements repeatable at any time, thus allowing the potential monitoring of materials conservation state over the years and the overcoming of the limit of poor representativity typical of microdestructive analyses [5].

Fourier transform infrared spectroscopy (FTIR) is one of the most important analytical techniques for the identification of organic and inorganic chemical compounds on the basis of the functional group characterization. Until recently, infrared spectroscopy applied to the study of Cultural Heritage artefacts has been mainly used in either transmission, attenuated total reflection (ATR) or diffuse reflection (DRIFT) modes [6–8]. These techniques are very accurate and precise, giving a truthful characterization of the single specimen, at the expense of the need for sampling or the contact with the surface. In order to gradually reduce the use of sampling methods to the benefit of the artwork integrity, the last decades have seen an increasing use of noninvasive infrared reflection techniques thanks to the development in the infrared instrumentation technology. Portable noninvasive FTIR spectrophotometers work in total reflection mode by means of fiber optics [9–11] or, just more recently, reflection mode accessories [12–14]. Yet the main drawback is related to the interpretation of the reflectance spectral data, which can present large distortions of band shape, absorption frequency, and intensity compared to transmittance spectra. These anomalies are due to several factors, such as absorption (k) and refraction (n) indices and surface roughness, which affect the extent of specular (or surface) and diffuse (or volume) reflection contributions. In particular, the specular component is ruled by Fresnel's law giving rise to first-derivative-like spectral bands from materials with $k < 1$ and/or to Reststrahlen (or inverted) bands when $k \gg 1$, while the diffuse component appears as features that are very similar to those collected in transmission mode except for some differences in terms of relative band intensities [6, 15]. It is worth reporting that the application of specific data-processing algorithms, such as the Kramers-Kronig transformation (KKT), can give accurate and reliable results on condition that specific required conditions are correctly fulfilled [16–18].

As regards natural organic materials composing the art objects, existing mid-infrared (MIR) reflection studies are limited to their interaction, as binding media, with inorganic pigments [9, 19, 20] or to very specific fields of interest [21]. A valuable reflectance spectral database in the near-infrared (NIR) region has been set up [1]. In order to correctly interpret infrared reflection spectra of natural organic components in the expanded region of MIR and a portion of NIR, we considered the building of a reflectance spectral database as mandatory. This necessity is mainly due to the fact that reflection spectra cannot be generally identified using the available archives of spectral signatures acquired in transmission or ATR modes.

The paper here presented aims to fill this gap, by providing a detailed interpretation of mid (4000-375 cm^{-1}) and near (7500-4000 cm^{-1}) infrared total reflection mode spectra of sixteen pure, nonaged natural organic compounds used in artworks as binding media, consolidants, adhesives or protective coatings. Firstly, a description of mid-IR absorption bands acquired in transflection mode (TR) was given. In TR spectrometry, the beam is transmitted through a surface film, whose thickness must be on the order of one-half wavelength or more, and reflected from a metallic surface. Following these conditions, the collected spectrum is quite similar to the transmission spectrum of the material and the bands are considered absorption ones. These data represented a crucial starting point for the study and the interpretation of the distorted total reflectance spectral behavior in the MIR region, as well as having made available an accurate, direct visual comparison in frequency and lineshape between absorption and reflection bands. TR spectra were then used to assess the applicability and usefulness of the Kramers-Kronig correction for an easier interpretation of reflectance bands. For the sake of completeness, the interpretation of reflection spectral signals in the restricted portion of NIR region was given, as added value for the identification of the substances. Final output of this work is to provide a comprehensive database of FTIR reflectance spectra to scientists and conservators as a tool for the noninvasive identification of natural organic materials belonging to the classes of proteins, lipids, polysaccharides, and resins. In conclusion, it is worth emphasizing that the paper represents a fundamental, preliminary study for tackling the subsequent interpretation of the more complex mixtures, layered systems or aged materials occurring in real cases.

2. Materials and Methods

2.1. Reference Materials. Sixteen commercially available materials (Kremer Pigmente GmbH & Co., Aichstetten, Germany) were selected according to their widespread use in art. The reference compounds belong to the classes of polysaccharide (cellulose 63600; Arabic gum 63300), lipid (carnauba wax 62300; beeswax 62200; shellac wax 60550), proteinaceous (hide glue 63020; bone glue 63000; casein 63200) and natural resinous (sandarac 60100; manila copal 60150; colophony 60310; Venetian turpentine 62010; dammar 60001; mastic 60050; shellac 60480; dragon's blood 37000) materials.

2.2. Sample Preparation. In order to prepare the samples for the infrared analyses, each substance was dissolved in a proper solvent (20% w/v): distilled water for animal glues, cellulose and Arabic gum, ammonium hydroxide for casein, and absolute ethanol for all resins with the exception of Venetian turpentine for which 1-propanol was required. The solutions were then purified using a polyethylene filter funnel (stitch size of 60 μm). Waxes were simply heated in water bath until reaching the melting point (60-90°C). With the aim of collecting mainly the specular component and minimizing the diffuse one in the total reflection analysis, bulk samples with flat surface were prepared using the procedure described in a preliminary paper [23]. To compare and prove the accuracy of the method, transflection mode spectra were acquired from a thin film laid on a smooth reflective substrate (i.e., aluminum foil).

2.3. Portable FTIR Reflectance Spectroscopy. Both total reflection and transflection mode FTIR spectra were recorded using the Alpha portable spectrometer (Bruker Optics, Germany/USA-MA) equipped with a SiC globar source, a permanently aligned RockSolid interferometer (with gold mirrors) and a DLaTGS detector. Measurements were performed at a working distance of 15 mm by an external reflectance module with an optical layout of 23°/23°. Pseudoabsorbance spectra [log(1/R); R = reflectance] were acquired between 7500 and 375 cm^{-1} from areas of about 5 mm in diameter, at a resolution of 4 cm^{-1} and with an acquisition time of 1 min. Spectra from a gold flat mirror were used as background. An average of three spectra for each sample was carried out. Mid-infrared total reflection mode spectra were transformed to absorbance spectra by applying the Kramers-Kronig algorithm (included in the OPUS 7.2 software package). A large negative band can appear in the corrected reflection spectra around 3600 cm^{-1}, as an artefact of KK transformation [13].

2.4. Spectral Analysis. The extent of specular reflection contribution turned out to be predominant over the mid-IR region in all acquired total reflection mode spectra resulting in a derivative-like band profile. The specular reflection bands have been discussed by the functional group giving rise to characteristic slope, maximum position and shape. These

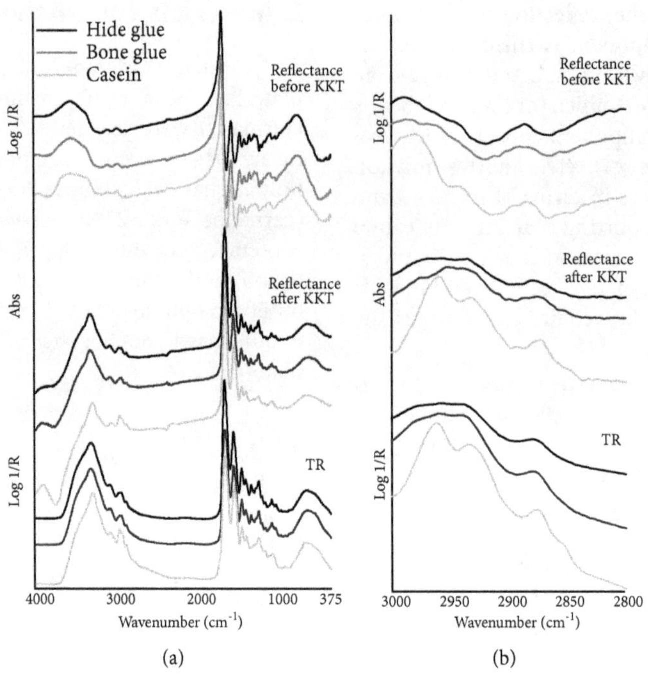

FIGURE 1: FTIR spectra in the mid-infrared region of the studied proteinaceous materials (black line: hide glue, gray line: bone glue, light gray line: casein) using different analysis modes. From top to bottom: total reflection mode before the KKT correction, total reflection mode after the KKT correction, and transflection mode. Spectral regions of (a) 4000-375 cm^{-1} and (b) CH stretching bands.

derivative-shaped bands are correlated to their corresponding KK transformed bands, which are superimposable, in turn, on the absorption signals acquired in transflection mode and whose frequency attribution is widely known by literature [24–26]. The fact that the KK corrected and TR spectra are comparable means that the required conditions for a successful Kramers-Kronig transformation have been correctly fulfilled, and confirms the specular reflection as the main collected contribute of the total reflection mode spectra in the mid-IR region. From a comparative evaluation of the collected spectra, we can see that the steeper the slope of a derivative-shaped band is the sharper the lineshape of the corresponding absorption band gets, whereas slope changes mean the presence of multiple bands. Instead, the inflection point (i.e., the point on a curve at which the concavity changes) of specular reflectance bands corresponds to the band maximum of the absorption ones. However, the difficulty often detected in determining the precise point of inflection has made it more relevant to refer to the maximum of specular reflectance bands, the wavenumber value resulting up-shifted compared to that one of absorbance bands. On the other hand, specular distortions due to the Reststrahlen effect have not appeared in the analyzed spectra, as expected by low absorption coefficient materials. As regards near-IR region, combination and overtone bands were not subjected to spectral anomalies; because of the small absorption coefficients, the diffuse contribution results, indeed the dominant factor, in determining the behavior of the NIR reflection bands [6, 27].

3. Results and Discussion

In this section, the detailed analysis of the spectra recorded in transflection and total reflection modes from the natural organic materials, grouped into four classes, will be proposed and discussed.

3.1. Proteinaceous Materials. Proteins are macromolecules made up of one or more unbranched chains of amino acids, which are joined together by peptide bonds between the carboxyl and amino groups of adjacent amino acids residues. Concerning the analyzed materials, casein is a phosphoprotein complex that is precipitated by the acidification of skimmed milk whereas animal glues are obtained by dissolving collagen, the fibrous protein of connective tissues in animals which contains three polypeptide α-chains in a triple helix conformation. In art and art conservation, they have been mainly used as binding media, consolidants and adhesives [28] and, less frequently, in protective treatments on different material typologies (e.g., stones, fossils) [29].

In TR mode, the mid-IR spectral range of the selected proteinaceous materials is characterized by consistent recognizable absorption peaks (Figure 1(a)). A typical stair-step pattern is formed by the amide I band in the region of 1650 cm^{-1}, due to the strong carbonyl stretching vibration (C=O), then the amide II band near 1550 cm^{-1} as a combination of C-N stretching and N-H bending vibrations and finally the CH bending vibration occurring near 1450 cm^{-1}, occasionally referred to as an amide III. The asymmetrical

N-H stretching vibration occurring near 3300 cm^{-1} and the first overtone of amide II band positioned at 3080 cm^{-1} both appear as peaks on the broader O-H stretching band that overlaps this region. Additionally, the methyl (CH_3) and methylene (CH_2) groups produce asymmetric stretching vibrations respectively at 2960 and 2935 cm^{-1} and small symmetric stretching bands respectively at 2875 and 2850 cm^{-1}. It should be noticed that these CH stretching bands appear better resolved in the spectrum of casein, with CH_3 groups of greater intensity than CH_2 groups, whereas both bone and hide glues show CH asymmetric stretches as broad overlapping bands with the CH_2 symmetric stretching very difficult to be discerned because of its weakness (Figure 1(b)). Other small CH bending vibrations are found in the 1400-1300 cm^{-1} region as well as several C-O vibrations occur from 1250 to 1000 cm^{-1}.

In total reflection mode, the mid-IR collected spectra (Figure 1(a)) display an intense and narrow maximum of the amide I band at 1698 cm^{-1}, appearing predominant over the mid-infrared range, followed with decreasing intensity by the amide II and amide III bands whose sharp maxima are positioned at 1578 and 1470 cm^{-1}, respectively. In the 3800-3200 cm^{-1} region, the lineshape of the ν_{as}N-H band is characterized by a steep slope with a wide-ranging maximum near 3370 cm^{-1} while the O-H vibration produces a broad band having a slight slope and an unstructured maximum near 3570 cm^{-1}. The region comprising the overtone of the amide II band and the CH stretches shows low-intensity bands with the maxima placed respectively near 3095, 2985, 2945, 2885 and 2855 cm^{-1}. As observed also in TR mode, casein slightly differs from animal glues in the νCH spectral range here resulting in a major slope and more structured maxima of the band lineshape (Figure 1(b)). Concerning the more pronounced intensity difference between the CH_3 and CH_2 bands in the spectrum of casein compared to animal glues, this could be ascribable to the presence in the former of a higher percentage amount of amino acids containing two methyl groups each (i.e., valine, leucine and isoleucine). Some other small differences can be detected in the low-wavenumber region from 1350 to 1000 cm^{-1}, where less-structured and broader C-O and δCH band maxima characterize casein compared to animal glues. The experimental wavenumbers corresponding to the maxima of the mid-IR total reflection mode bands of proteinaceous materials and their assignment are summarized in Table 1. After applying the KK correction, an accurate match in position and shape between corrected reflectance and transflectance bands can be observed (Figure 1).

Regarding the near-IR region, all three total reflection mode spectra exhibit a similar absorption pattern (Figure 2(a)). The first overtones of the asymmetric and symmetric CH_2 stretching modes occur respectively at 5900 and 5775 cm^{-1}, whereas the combination bands ν(OH) + δ(OH) and 1st overtone ν(C-O) amide I + amide II are visible near 5160 and 4600 cm^{-1} respectively [1]. As expected after observing the fundamental vibrations in the mid-infrared range, casein shows slightly more defined shapes of the $\nu_a(CH_2)+\delta(CH_2)$ and $\nu_s(CH_2)+\delta(CH_2)$ combination bands

TABLE 1: Experimental wavenumber values corresponding to the maxima of the mid-IR total reflection mode bands of proteinaceous materials and their tentative assignment.

Experimental wavenumbers (cm^{-1})	Band assignment
Casein and animal glues	
3570	νOH
3370	ν_{as}NH
3095	Overtone of combination band δNH and νCN (amide II)
2985	$\nu_{as}CH_3$
2945	$\nu_{as}CH_2$
2885	$\nu_s CH_3$
2855	$\nu_s CH_2$
1698	νC=O (amide I)
1578	Combination band δNH and νCN (amide II)
1470	δCH (amide III)
1400-1300	δCH
1300-1000	C-O (COO^-)

respectively at 4375 and 4260 cm^{-1}, if compared to animal glues. Moreover, the dairy protein exhibits more resolved bands at 4865 cm^{-1}, due to the ν(NH)+δ(NH) combination band, and near 4055 cm^{-1} which could be attributed to the CH functional group as a combination or overtone band [30].

3.2. Lipid Materials. The use of lipids in art objects is essentially limited to waxes and drying oils, respectively made up of nonglyceryl and glyceryl esters. However, only the physical properties of the waxes fulfilled the required sample preparation conditions for reflection analysis. Associated with long-chain nonglyceryl esters are one or more of the following: free fats or wax acids, alcohols, sterols, ketones and aliphatic hydrocarbons. These components vary greatly according to the origin of the wax, which can be vegetal (e.g., carnauba wax), animal (e.g., beeswax, shellac wax) or mineral (e.g., ceresin). Since ancient times, they have been employed in art conservation as adhesives and surface coatings, and particularly as waterproofing agents thanks to their hydrophobic properties [28].

In TR mode, the MIR spectra collected from the three waxes (Figure 3) display the characteristic sharp bands produced by the many CH_2 groups: the predominant asymmetric and symmetric stretches near 2920 and 2850 cm^{-1}, and the peak splitting of scissoring and rocking vibrations into doublets respectively at 1472/1462 cm^{-1} and 730/720 cm^{-1} which indicates the semicrystalline structure of the waxes [21, and references therein]. The CH_3 methyl end groups of pure, long-chain hydrocarbons produce stretching bands near 2955 and 2905 cm^{-1}. Ester and acid/ketone groups account for the C=O stretching bands respectively at 1735 and 1715 cm^{-1}, appearing as an overlapping band in the shellac wax spectrum, and for the C-O bands in the 1172 cm^{-1} region.

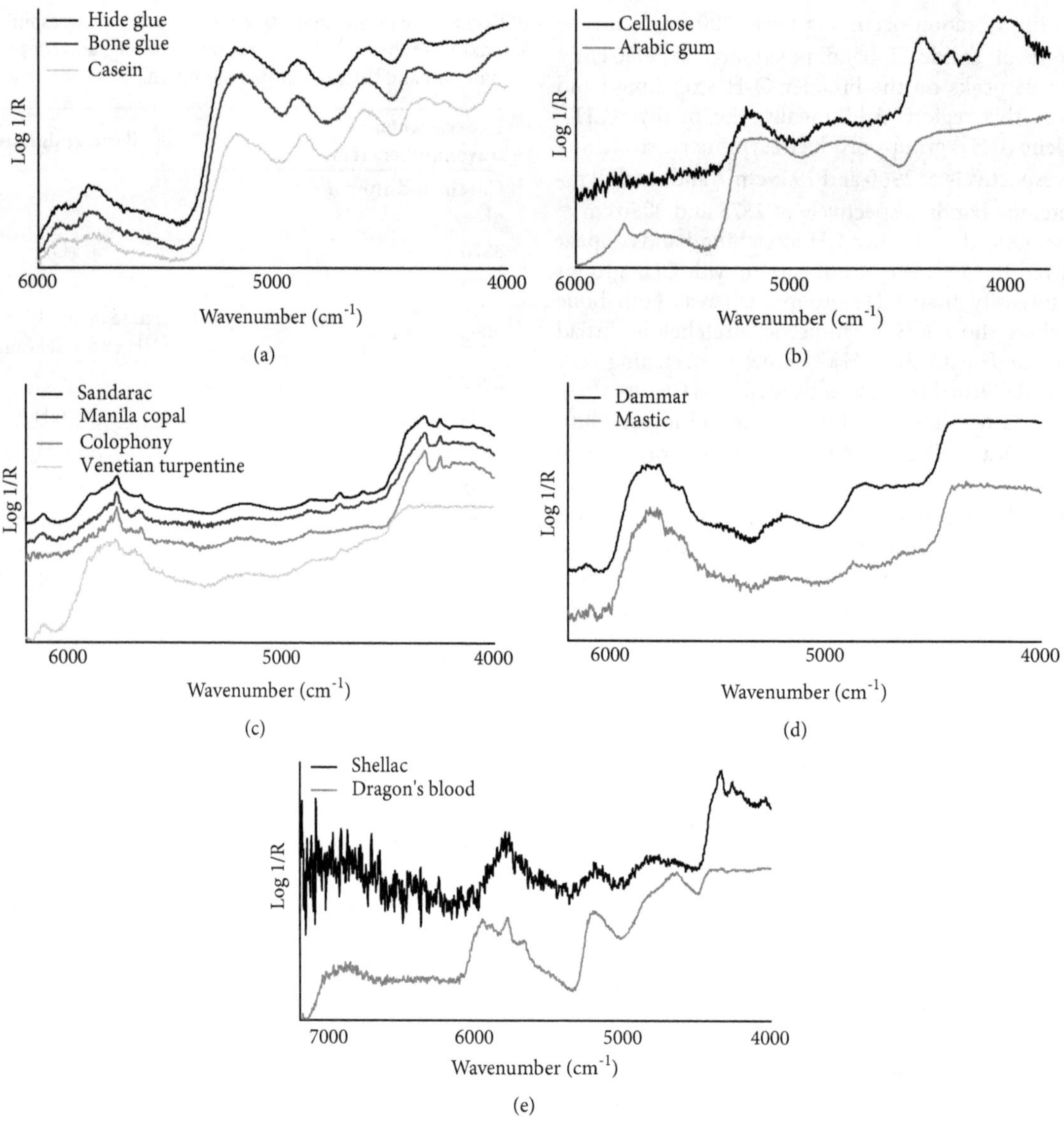

FIGURE 2: FTIR total reflection mode spectra in the near-infrared region of the studied (a) proteinaceous materials, (b) polysaccharide materials, (c) diterpenoid resins, (d) triterpenoid resins, and (e) not (exclusively) terpenoid-based resins.

Additional characteristic sharp peaks of the carnauba wax are found at 1635, 1605, 1515, 832 and 520 cm^{-1} whereas the shellac wax spectrum shows a further well-resolved band in the region of C-O vibrations at 1240 cm^{-1} and a broad band near 1060 cm^{-1}.

In total reflection mode, all acquired mid-IR spectra (Figure 3) exhibit sharp maxima of the CH_2 stretching bands at 2935 and 2858 cm^{-1}. These bands are characterized by a very steep slope of the lineshape and appear predominant over the mid-infrared range because of their intensity. The asymmetric CH_3 stretching mode produces a wide and low-intensity band maximum near 2965 cm^{-1} while the symmetric methyl band is difficult to be discerned in these spectra. This νCH region is clearly shown in Figure 4(a). Beeswax and shellac wax spectra show two close narrow maxima both of CH_2 scissoring and rocking bands respectively at 1478/1468 and 735/725 cm^{-1}, corresponding to the previously described absorption doublets (Figures 4(c) and 4(d)). Conversely, the peak splitting of these bands does not appear in the spectrum of the carnauba wax and this is confirmed observing the corresponding KK corrected spectrum. The absence of these doublets is indicative of a loss of the crystal structure of hydrocarbon chains during the bulk sample preparation procedure [31]. The C=O stretching vibrations produce a high slope ester band having the sharp maximum placed at 1745 cm^{-1} in the beeswax and carnauba wax spectra and at 1752 cm^{-1} in the spectrum of shellac wax, and a weaker acid/ketone band with the maximum respectively at 1716 and 1718 cm^{-1} (Figure 4(b)). Concerning this spectral region, shellac wax shows an additional well-defined narrow C=O band (acids/ketones) with the maximum being positioned at 1735 cm^{-1}. Another band characterized by the steep slope of

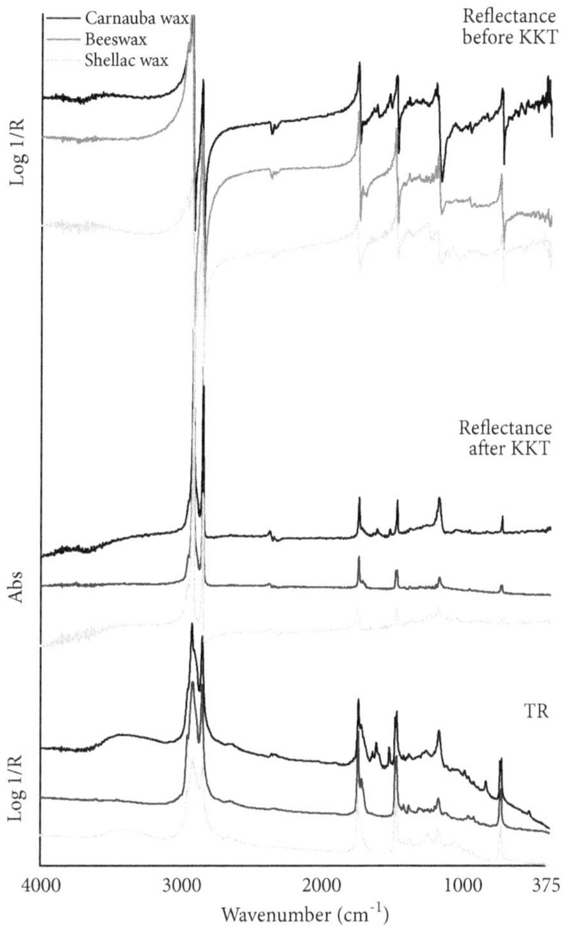

FIGURE 3: FTIR spectra in the mid-infrared region of the studied lipid materials (black line: carnauba wax, gray line: beeswax, light gray line: shellac wax) using different analysis modes. From top to bottom: total reflection mode before the KKT correction, total reflection mode after the KKT correction, and transflection mode.

the lineshape, with the maximum positioned at 1183 cm^{-1}, is due to C-O vibration. Those additional features detected in TR mode show the reflection maxima at 1612, 1522 and 832 cm^{-1} in carnauba wax and near 1250 and 1075 cm^{-1} in shellac wax. The experimental wavenumber values corresponding to the maxima of the mid-IR total reflection mode bands of lipid materials and their suggested assignment are summarized in Table 2.

In the NIR region, only overtones from methylenic stretching at 5770 and 5660 cm^{-1} can be clearly detected in total reflection mode spectra whereas combination bands from ν and δCH_2 do not occur except as broad and unresolved features. It is worth noting that NIR spectra, here not shown in figure, do have a prominent noise that could make it difficult to identify the characteristic bands.

3.3. Polysaccharide Materials. Polysaccharides are polymers made up of many monosaccharide units joined together by glycoside bonds, and include cellulose, starch, honey and plant gums. As regards the analyzed materials, cellulose consists of high molecular weight polymer of D-glucose with β(1-4)-glycosidic bonds, whereas the long-chain polymers of plant gums, exudates from several species of plants or extracted from the endosperm of some seeds, are made of aldopentoses, aldohexoses and uronic acids which condense through glycosidic bonds. Arabic gum (exuded by *Acacia senegal* and *Acacia seyal*) is not only one of the most important representatives of the plant gums' family in the field of artistic and historic works, it is also of broader economic relevance and therefore produced in rather large quantities [32, 33]. Over the time, polysaccharides have widely been used in art and art conservation as painting media and sizing agents as well as adhesives [22] and, to a lesser extent, in consolidation and protective treatments [34].

The mid-infrared TR spectra of cellulose and Arabic gum (Figure 5(a)) display the characteristic IR pattern for the polysaccharides produced by the high proportion of O-H groups bound to the carbons. Indeed, both spectra show two strong, broad stretching bands: one in the 3400-3300 cm^{-1} region due to O-H groups and the other at about 1080 cm^{-1} due to C-O groups. A moderately strong band found at 1650 cm^{-1} in cellulose and at 1610 cm^{-1} in Arabic gum (with asymmetric lineshape) is partially associated with intramolecular bound water and partially due to the presence

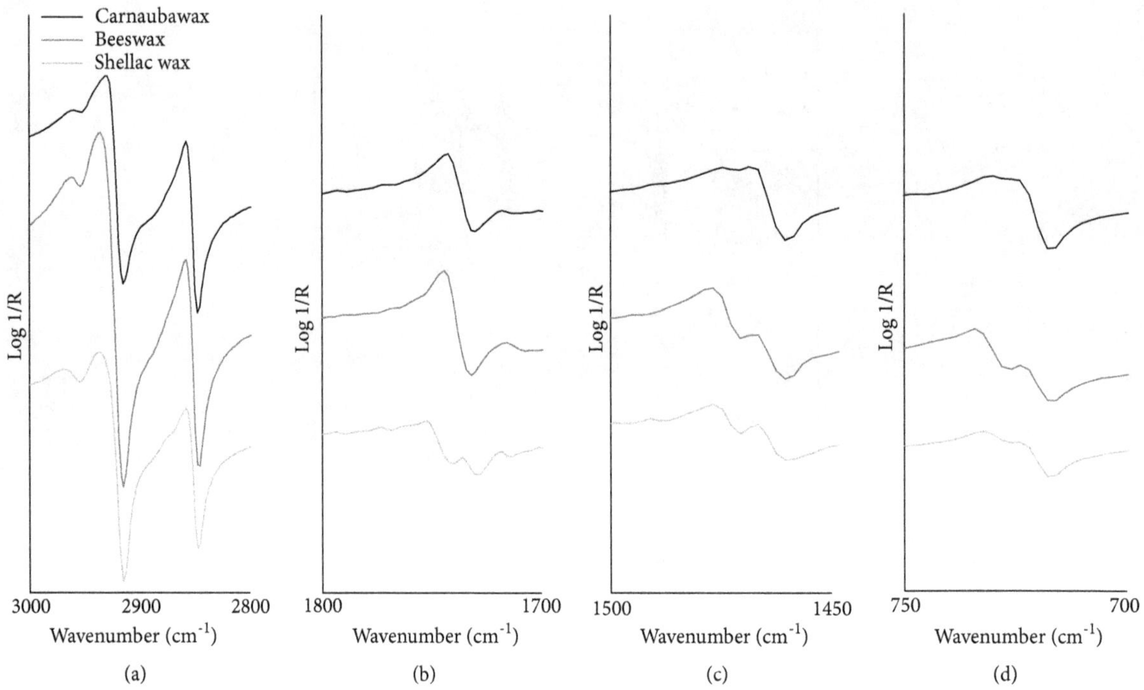

FIGURE 4: FTIR total reflection mode spectra (before the application of the KKT correction) of the studied lipid materials (black line: carnauba wax, gray line: beeswax, light gray line: shellac wax) in the regions of (a) CH stretching, (b) C=O stretching, (c) CH_2 scissoring, and (d) CH_2 rocking.

TABLE 2: Experimental wavenumber values corresponding to the maxima of the mid-IR total reflection mode bands of lipid materials and their tentative assignment.

Experimental wavenumbers (cm^{-1})			Band assignment
Beeswax	Carnauba wax	Shellac wax	
2965	2965	2965	$\nu_{as}CH_3$
2935	2935	2935	$\nu_{as}CH_2$
2858	2858	2858	$\nu_s CH_2$
1745	1745	1752	νC=O (esters)
		1735	νC=O (acids/ketones)
1716	1716	1718	νC=O (acids/ketones)
	1612		
	1522		
1478/1468	1468	1478/1468	$\delta_{sciss}CH_2$
		1250	C-O
1183	1183	1183	C-O
		1075	
	832		
735/725	725	735/725	$\delta_{rock}CH_2$

of a carboxyl group, whereas the only gum spectrum presents an additional very weak, sharp peak at 1720 cm^{-1} which is attributed to the C=O stretching of an ester-containing component. The lineshape of CH stretching and bending absorptions, respectively, found in the 3000-2800 and 1480-1300 cm^{-1} regions, appears sharper and more resolved in cellulose compared to Arabic gum, even though this is generally true for all spectral features. Moreover, a further well-defined peak at 950 cm^{-1} can be ascribed to the CH rocking vibration which occurs in cellulose.

In total reflection mode, the most significant region of the MIR spectra (Figure 5(a)) corresponds to the C-O vibrations: both the polysaccharides display indeed an intense and narrow band maximum at 1171 cm^{-1}, followed with decreasing intensity by well-defined peak maxima positioned at 1144 and 1100 cm^{-1} in cellulose and at 1108 in Arabic gum. The lineshape of all these combined bands is characterized by a very steep slope, thus resulting in sharp absorptions after the application of the KK logarithm. This region is more clearly shown in Figure 5(b). As regards the lower wavenumber range, the well-resolved CH vibration band occurring in cellulose at 958 cm^{-1}, already highlighted at 950 cm^{-1} in the TR spectrum, can be considered a diagnostic feature for the distinction of the analyzed polysaccharides. The O-H groups produce a moderately strong band which appears more marked in the gum than in cellulose: it is characterized by a broad band maximum near 3500 cm^{-1} and a rather slight slope of the lineshape. As already pointed out by comparison with TR spectra, the CH stretching and bending regions of cellulose turn out to have more resolved features compared to those of Arabic gum (for details, see Table 3), and this is also confirmed by their corresponding KK transformed spectra. On the contrary, the combined C=O and H-O-H (intramolecular water) vibrations give a response which is

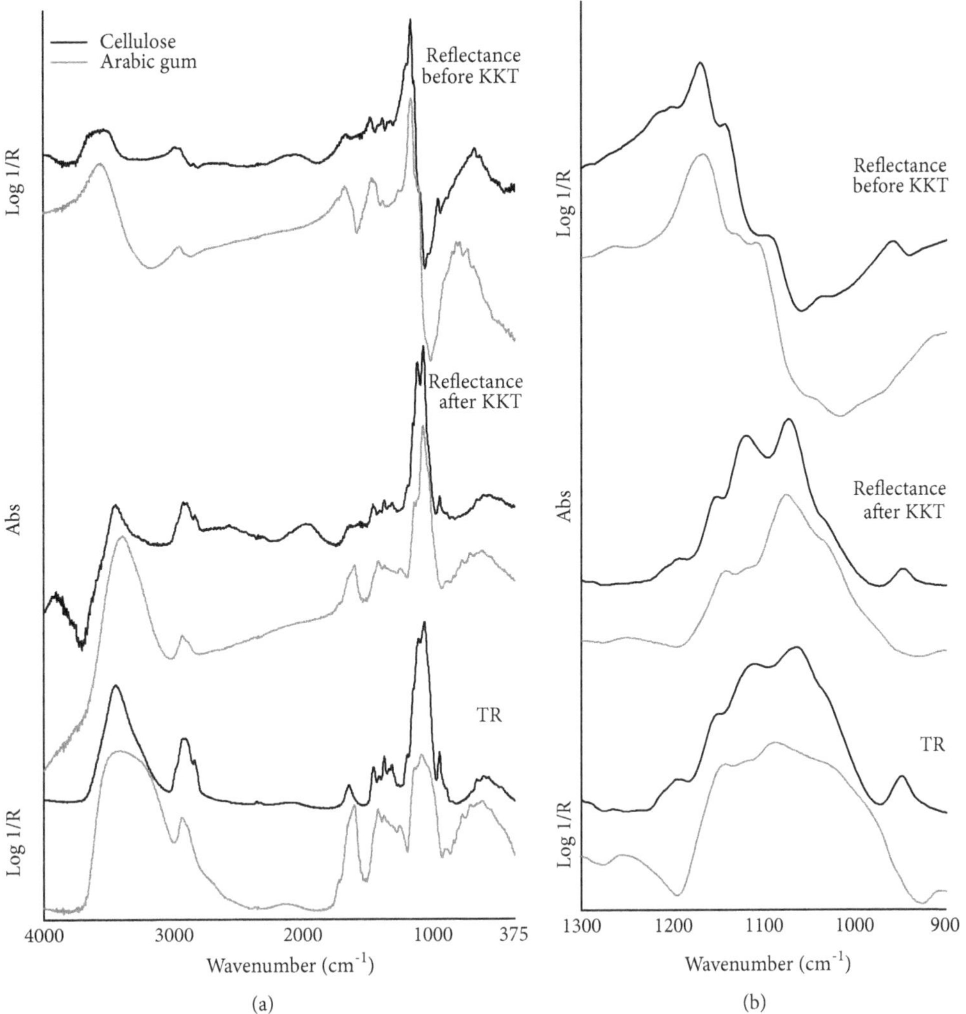

FIGURE 5: FTIR spectra in the mid-infrared region of the studied polysaccharide materials (black line: cellulose, gray line: Arabic gum) using different analysis modes. From top to bottom: total reflection mode before the KKT correction, total reflection mode after the KKT correction, and transflection mode. Spectral regions of (a) 4000-375 cm^{-1} and (b) C-O stretching bands.

more intense in the gum than in cellulose. Concerning this region (1700-1550 cm^{-1}), the reflectance spectrum of Arabic gum displays an overlapping band characterized by two different slopes which give rise to a sharp peak at 1605 cm^{-1} and a shoulder near 1645 cm^{-1} in the corresponding KK spectrum. Furthermore, it is possible to notice that only a weak and poorly resolved feature near 1730 cm^{-1} (C=O vibration) occurs in the reflectance spectrum of Arabic gum, whereas a small sharp peak has been well distinguished using the TR technique. The detailed experimental wavenumber values corresponding to the maxima of the mid-IR total reflection mode bands, and the relative suggested assignment, are reported in Table 3.

As regards NIR region (Figure 2(b)), the investigated total reflection mode spectra are characterized by different features except for the common broad bands respectively found near 5190 cm^{-1}, due to the combination of O-H stretching and H-O-H bending, and near 4760 cm^{-1} which is attributed both to the overtone of C-O stretching (including C=O and C-O) and to the combination of O-H bending and C-O stretching. In addition to those ones, the spectrum of cellulose displays the O-H and C-O stretching combination band at 4400 cm^{-1}, the overtone band of CH bending vibration at 4250 cm^{-1} with a possible further contribute of CH_2 stretching and bending combination band and, finally, the combination band of CH and C-O-C stretches and C-C vibration positioned at 4010 cm^{-1}. On the other hand, Arabic gum exhibits the overtones bands of CH_2 stretching vibrations at 5775 (asymmetric) and 5660 (symmetric) cm^{-1} whereas the combination bands produced by the same CH_2 vibrations, which are expected to be found in the 4300-4200 cm^{-1} region [30], do not clearly appear.

3.4. Natural Resinous Materials. Natural resins are polymers exuded from plants or secreted by insects: although their chemistry is diverse, most are mainly composed of terpenoids, which are formally considered compounds made up of units of isoprene. Since ancient times natural resins,

TABLE 3: Experimental wavenumber values corresponding to the maxima of the mid-IR total reflection mode bands of polysaccharide materials and their tentative assignment.

Experimental wavenumbers (cm^{-1})		Band assignment
Cellulose	**Arabic gum**	
3510	3560	OH
2985		$\nu_{as}CH_3$
2950	2950	$\nu_{as}CH_2$
2918	2900	$\nu_s CH_3$
2845		$\nu_s CH_2$
	1730	νC=O (esters)
1670	1665	C=O and HOH (intramolecular water)
	1620	C=O and HOH (intramolecular water)
1470	1445	δCH, νC-O and δCOH
1385	1380	δCH, νC-O and δCOH
1345		δCH, νC-O and δCOH
1320		δCH, νC-O and δCOH
	1265	δCH, νC-O and δCOH
1200		νC-O
1171	1171	νC-O
1144		νC-O
1100	1108	νC-O
958		δ_{rock}CH

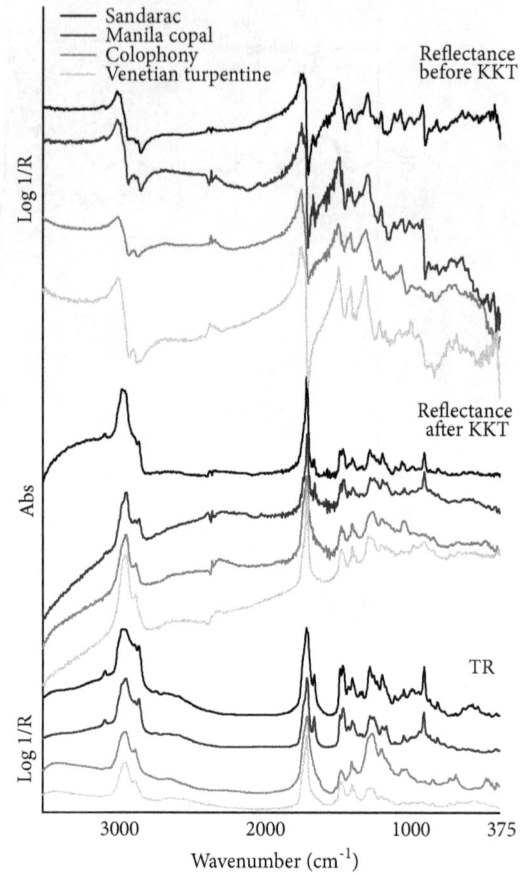

FIGURE 6: FTIR spectra in the mid-infrared region of the studied diterpenoid resins (black line: sandarac, dark gray line: Manila copal, gray line: colophony, light gray line: Venetian turpentine) using different analysis modes. From top to bottom: total reflection mode before the KKT correction, total reflection mode after the KKT correction, and transflection mode.

alone, or in mixture with other substances, have been used extensively by artists and conservators as varnishes, consolidants, adhesives, hydrorepellents and sealing agents thanks to their intrinsic properties [22, 28]. According to their chemical composition, this section is structured into three subdivisions: diterpenoid, triterpenoid and not (exclusively) terpenoid-based resins.

Diterpenoid resins Diterpenoid resins are primarily composed of mixtures of tricyclic (abietanes and pimaranes) and dicyclic (labdanes) resin acids. The botanical origin and the chemical composition of the analyzed diterpenoid resins (i.e., colophony, Venetian turpentine, sandarac and Manila copal) are listed in Table 4 [22]. For the spectral interpretation, it is worth noting that the composition of Venetian turpentine and colophony is similar, with the former containing a remarkable amount of labdane alcohols in addition to resinous acids. As far as their composition is concerned, also sandarac and Manila copal are very similar one to the other both consisting of free diterpenoids and a highly polymerized fraction of polycommunic acid.

In TR mode, the spectra of the four diterpenoid resins show an overall similar pattern in the MIR region (Figure 6). In particular, an accurate wavenumber correspondence is observed between Venetian turpentine and colophony, on one hand, and between sandarac and Manila copal, on the other, reflecting the respective alike compositions as reported by Daher et al. [24]. All spectra are characterized by intense signals in the CH stretching region, with an overlapping band at about 2945 cm^{-1} due to the asymmetric CH_3/CH_2 modes and a defined peak at 2870 cm^{-1} due to the symmetric CH_3 mode. In addition, sandarac and Manila copal exhibit a sharp band produced by symmetric CH_2 vibrations at 2850 cm^{-1}, which suggests a considerable number of CH_2 groups occurring in these resins. Broad O-H bands are found near 3420 and 2650 cm^{-1}, respectively due to the hydroxyl stretching absorption and to the O-H vibration of dimerized carboxyl groups. The C=O stretching vibration produces a characteristic sharp, strong band with the maximum falling from 1698 to 1692 cm^{-1}. CH bending vibrations are found at about 1460 cm^{-1}, with well-resolved features at 1468 (CH_2 groups) and 1450 (CH_3 groups) cm^{-1} occurring only in sandarac and Manila copal, and at 1385 cm^{-1} due to the symmetric CH_3 mode. Moreover, several bands due both to C-OH and C-O groups of esters and acids and to CH_2 vibrations are visible from 1300 to 900 cm^{-1}, resulting in a more defined and well separated band lineshape for sandarac and Manila copal if compared to Venetian turpentine and colophony. Concerning this spectral

TABLE 4: Botanical origin and chemical composition of the analyzed plant resins [22].

Order	Family	Genus (type of resin)	Main composition
Coniferales	Pinaceae	*Pinus* (colophony)	Abietadienic acids, pimaradienic acids
		Larix (Venetian turpentine)	Abietadienic acids, pimaradienic acids, epimanool, larixol, larixyl acetate
	Araucariaceae	*Agathis* (Manila copal)	Sandaracopimaric acid, communic acid, agathic acid, abietic acid
	Cupressaceae	*Tetraclinis articulata* (sandarac)	Pimaradienic acids (sandaracopimaric acid), communic acid, totarol
Guttiferales	Dipterocarpaceae	*Hopea* (dammar)	Dammaranes (hydroxydammarenone, dammaradienol), ursanes (ursonic acid, ursonaldehyde)
Sapindales	Anacardiaceae	*Pistacia* (mastic)	Euphanes (masticadienonic acid, isomasticadienonic acid), dammaranes, oleananes (oleanonic acid, moronic acid)
Arecales	Arecaceae	*Daemonorops* (dragon's blood)	Dracoresinotannol, dracorubin, dracorhodin, abietic acid

region, colophony presents the most intense feature with respect to the other resins, falling at about 1250 cm^{-1}. In addition to all these described common frequencies, it is important to notice that sandarac and Manila copal show a recognizable, characteristic system of peaks mostly related to the conjugated double bonds occurring in the side chain of communic acid, which predominantly composed the highly cross-linked fraction of these resins. These sharp, narrow and moderately strong bands are positioned at 3080 (νCH from C=C bonds, appearing very weak in Venetian turpentine and colophony), 1645 (νC=C) and 890 (out-of-plane bending of the exomethylene groups) cm^{-1}. Furthermore, these two resins exhibit other well resolved, weak bands at 2730, 1410, 1330, 1315, 1030, 795 cm^{-1}.

In total reflection mode spectra (Figure 6), the most intense band over the MIR of all these substances is produced by C=O groups of resinous acids, with the maximum falling from approximately 1730 to about 1715 cm^{-1} and the lineshape being characterized by a rapid, steep slope. As can be seen in the same figure, these groups give rise to the strongest and sharpest peak of the KK transformed spectra. Another very striking region corresponds to the νCH vibrations where the asymmetric CH_3/CH_2 modes produce an intense band maximum near 2985 cm^{-1} followed by a steep band slope. The weak symmetric CH_3 stretching band, with the maximum being placed near 2880 cm^{-1}, appears more resolved in Venetian turpentine and colophony with respect to sandarac and Manila copal which present the additional, sharp symmetric CH_2 stretching band maximum at 2857 cm^{-1}. Moderately strong δCH bands, all characterized by a rapid slope of the lineshape, present sharp maxima at around 1475 cm^{-1} and 1395 cm^{-1}, in addition to which the 1455 cm^{-1} band maximum (due to CH_3 groups) occurs in all resins with the exception of Venetian turpentine. This last further feature contributes to form the well-resolved doublet that is visible in the 1490-1420 cm^{-1} region of the KKT spectra. The C-O vibrations produce a significant medium-intensity band with the maximum falling from 1290 to 1280 cm^{-1} and gradually weaker bands with the maxima placed near 1190 and 1050 cm^{-1}. On the other hand, the O-H bands cannot be easily discerned in reflection mode. As previously reported, sandarac and Manila copal exhibit some characteristic additional features related to the presence of conjugated double bonds in their composition. In detail, the weak and well-defined band with the maximum at 3085 cm^{-1}, then the weak-to-moderate band with the sharp maximum at 1650 cm^{-1}, and finally the moderately strong and well-resolved band having the maximum positioned at 900 cm^{-1} as well as a very steep slope of the lineshape. The experimental wavenumber values corresponding to the maxima of the mid-IR total reflection mode bands, and the relative suggested assignment, are reported in Table 5.

As regards NIR range, all reflectance spectra (Figure 2(c)) are characterized by similar pattern except for Venetian turpentine which displays no features over the 4400-4000 cm^{-1} region (a flat, steady line is here visible). In this region, indeed, the other three resins exhibit sharp, well resolved $\nu_a(CH_2)+\delta(CH_2)$ and $\nu_s(CH_2)+\delta(CH_2)$ combination bands respectively found at 4325 and 4252 cm^{-1}. Common to all resins are the first overtone bands from methylenic stretching at 5775 (asymmetric) and 5665 (symmetric) cm^{-1}, then the weak and broad combination band of ν(OH)+δ(OH) which is centered at around 5170 cm^{-1}, and finally the poorly defined band near 4865 cm^{-1}. Weak-to-moderate and well-defined bands are also found at 6120 and 4725 cm^{-1}, respectively due to the first overtone of $\nu(CH_2)$ cyclic and the combination of C-O and OH stretching. However, these two features do not occur in colophony. Moreover, sandarac and Manila copal display an additional weak peak at 4615 which can be attributed to the combination of ν(C-O)+ $\nu(CH_2)$ [1].

Triterpenoid resins Triterpenoid resins consist of mixtures of triterpenoid molecules with mainly pentacyclic (ursanes, oleananes, lupanes and hopanes) and tetracyclic (dammaranes and lanostanes) skeletons. Table 4 reports the botanical origin and the kind of terpenoid compounds of the analyzed triterpenoid resins (i.e., mastic, dammar).

The mid-IR TR spectra of the two analyzed resins are characterized by the same band frequencies, with some slight differences appearing in the low-wavenumber peaks intensity (Figure 7). Moreover, the main fundamental frequencies are in common with the diterpenoid resins, resulting in extremely similar IR profiles which make it difficult to identify the different plant resins. The acquired spectra exhibit the most intense absorptions at 2945 (a saturated, less defined

TABLE 5: Experimental wavenumber values corresponding to the maxima of the mid-IR total reflection mode bands of diterpenoid resins and their tentative assignment.

Experimental wavenumbers (cm^{-1})				Band assignment
Sandarac	Manila copal	Colophony	Venetian turpentine	
3085	3085			νHC=C
2985	2985	2985	2985	$\nu_{as}CH_3/CH_2$
2880	2880	2880	2880	ν_sCH_3
2857	2857			ν_sCH_2
1720	1730	1730	1718	νC=O
1650	1650			νC=C
1475	1475	1475	1475	δCH_2
1455	1455	1455		$\delta_{as}CH_3$
1395	1395	1395	1395	δ_sCH_3
1280	1280	1290	1290	C-O
1190	1190	1190	1190	C-O
1050	1050	1050	1050	C-O
900	900			$\delta H_2C=C$

signal occurs in dammar) and 2872 cm^{-1}, respectively due to asymmetric CH_3/CH_2 and symmetric CH_3 stretches, and at 1705 cm^{-1}, due to the C=O stretching of the resinous acids. Further characteristic sharp, moderately strong peaks are produced by the CH bending vibrations at 1455 (CH_3/CH_2 groups) and 1380 (CH_3 group) cm^{-1}. On the other hand, the intensity of all the C-O vibrations falling in the 1300-900 cm^{-1} region appears low to very low. Another weak band, characterized by a quite broad shape, is due to the O-H stretching vibration and is centered at around 3440 cm^{-1}. Moreover, the presence of carbon-carbon double bonds in the cycling ring structure or in its side chains produces the weak bands positioned at 3070 (νCH from C=C bonds) and 1645 (νC=C) cm^{-1}, and the peak at 890 cm^{-1} (out-of-plane bending of the exomethylene groups) which appears sharp and weak-to-moderate in dammar resin. Differently from diterpenoid resins, mastic and dammar do not clearly show the O-H vibrations of dimerized carboxyl groups in the region of 2700-2500 cm^{-1}.

The MIR total reflection mode spectra of mastic and dammar (Figure 7) exhibit four recognizable, strong bands each of which is characterized by very steep slope and sharp, well-resolved maximum. Their band maxima fall at about 2980 ($\nu_{as}CH_3/CH_2$), 1715 (νC=O), 1475 ($\delta CH_3/CH_2$) and 1392 (δCH_3) cm^{-1}. It should be noticed that the carbonyl stretching vibration produces another well-defined, weak band maximum at about 1735 cm^{-1} that can be ascribed to oxidation products (ketones, esters, lactones) [32] whose formation could have occurred during the bulk sample preparation. This feature gives rise to a slight asymmetry toward higher wavenumbers of the C=O band in the corresponding KK transformed spectra (Figure 7), which is not observed in TR mode. Other weak and narrow band maxima are found at 2880 cm^{-1} (ν_sCH_3) and over the C-O vibrations region 1300 to 900 cm^{-1}, whereas the weak-to-moderate $\delta H_2C=C$ band with the lineshape having a rapid slope and the maximum being positioned at 896 cm^{-1} clearly occurs only in dammar resin. However, the other bands due to the carbon-carbon double bonds vibrations do not appear in reflection mode. The only broad band over the mid-IR range is produced by the νO-H vibrations and is characterized by a slight slope of the lineshape and a maximum falling near 3500 cm^{-1}; this band does not clearly occur though after the KK correction. The experimental wavenumber values corresponding to the maxima of the mid-IR total reflection mode bands, and the relative suggested assignment, are given in Table 6.

TABLE 6: Experimental wavenumber values corresponding to the maxima of the mid-IR total reflection mode bands of triterpenoid resins and their tentative assignment.

Experimental wavenumbers (cm^{-1})		Band assignment
Mastic	Dammar	
3500	3500	νOH
2988	2978	$\nu_{as}CH_3/CH_2$
2880	2880	ν_sCH_3
1737	1731	νC=O (ketones, esters, lactones)
1715	1715	νC=O (acids)
1475	1475	$\delta CH_3/CH_2$
1392	1392	δCH_3
1270	1270	C-O
1207	1207	C-O
1116	1116	C-O
1085	1094	C-O
1050	1055	C-O
	896	$\delta H_2C=C$

The NIR total reflection mode spectra display the first overtone bands of asymmetric and symmetric CH_3/CH_2 stretches, respectively near 5800 and 5760 cm^{-1}, as predominant over this spectral region. Moreover, the weak first

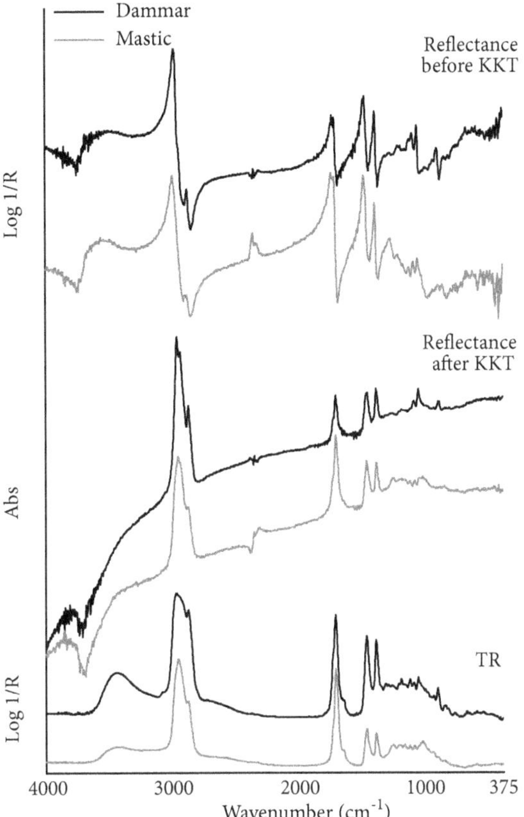

FIGURE 7: FTIR spectra in the mid-infrared region of the studied triterpenoid resins (black line: dammar, gray line: mastic) using different analysis modes. From top to bottom: total reflection mode before the KKT correction, total reflection mode after the KKT correction, and transflection mode.

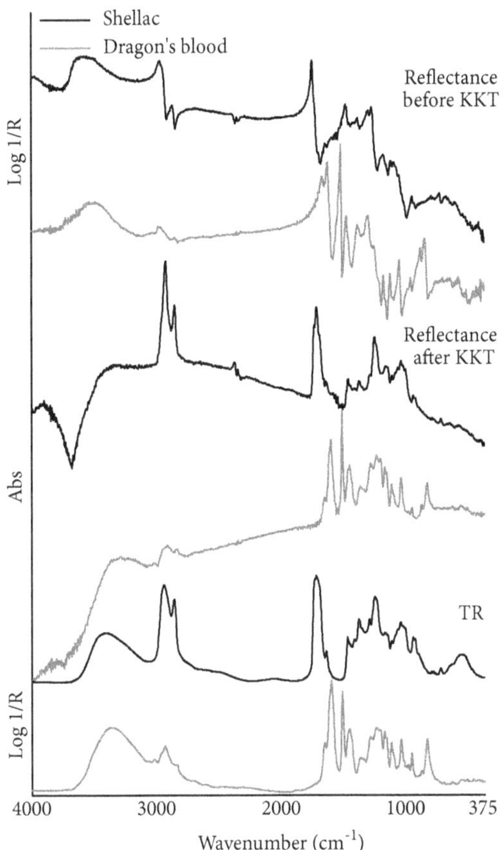

FIGURE 8: FTIR spectra in the mid-infrared region of the studied not (exclusively) terpenoid-based resins (black line: shellac, gray line: dragon's blood) using different analysis modes. From top to bottom: total reflection mode before the KKT correction, total reflection mode after the KKT correction, and transflection mode.

overtone band of the νCH_2 cyclic is found at 6100 cm^{-1} in dammar resin, while it is not clearly visible in mastic because of its marked spectral noise. In addition, less pronounced combination bands occur near 5180 (νOH+δOH) and 4850 (νC-O+νOH) cm^{-1}. It is worth noting that the 4400-4000 cm^{-1} region, where the combination bands of CH stretching and bending vibrations typically occur [1], does not show any features (a flat, steady line is here visible, as found in Venetian turpentine). The NIR total reflection mode spectra are reported in Figure 2(d).

Not (Exclusively) Terpenoid-Based Resins Among those resins that are not composed only of terpenes, shellac and dragon's blood have been playing a prominent role in the field of Cultural Heritage. Shellac, derived from secretions of the lac beetle and widely used in protective coatings for wooden surfaces, paintings, metal artworks and many other objects, is a complex mixture made of mono- and polyesters of hydroxy-aliphatic and sesquiterpene acids whereas dragon's blood (botanical origin and chemical composition are reported in Table 4) has been mainly used as a coloring matter in paint, enhancing the color of precious stones and glass, marble and the wood for violins [35].

In TR mode (Figure 8), the MIR spectrum of shellac shows the intense, sharp CH stretching bands at 2945 (asymmetric) and 2860 (symmetric) cm^{-1} and the strong, poorly resolved doublets at 1730-1715 and 1245-1235 cm^{-1} due to the stretching of the carbonyl (from acids and esters) and the C-O groups, respectively. Medium-intensity bands are produced by CH bending vibrations lying at 1375 (CH_3 asymmetric mode) and 1465 cm^{-1} (CH_2 in-plane bending or scissoring), with the weak CH_3 asymmetric mode shoulder occurring at 1448 cm^{-1}, and by C-O stretching vibrations from approximately 1200 to about 1000 cm^{-1}. A similar intensity characterizes the broad O-H stretching band centered at 3420 cm^{-1} and the characteristic doublet at 945-930 cm^{-1}. Moreover, the carbon-carbon stretching vibrations produce a weak-to-moderate and sharp olefinic band at 1636 cm^{-1} as well as a weak and well-defined C-C band at 725 cm^{-1} from partially crystalline long-chain hydrocarbons. On the other hand, the spectrum of dragon's blood is characterized by many sharp, well-resolved bands in the fingerprint region. The strongest peaks over this region are produced by the stretching vibration of aromatic carbon-carbon double bonds at 1596 and 1507 cm^{-1}, with the band at 1450 cm^{-1} appearing less pronounced [25]. Bands of medium intensity are found at 1650 cm^{-1}, due to the C=C stretching mode, in the region

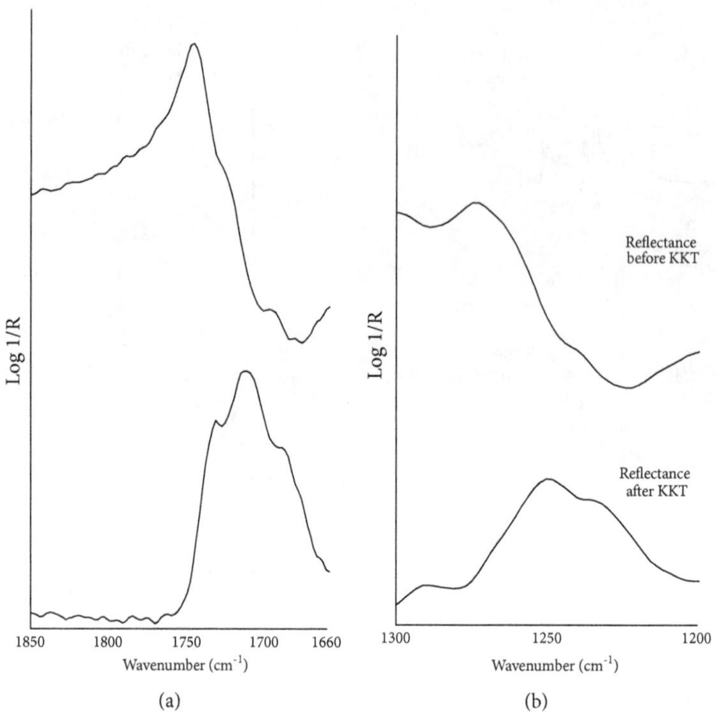

FIGURE 9: FTIR total reflection mode spectra of shellac before and after the KKT correction in the regions of (a) C=O stretching bands and (b) C-O stretching bands.

of the C-O stretching vibrations at 1235-1210, 1172-1160 and 1115 cm^{-1}, and, finally, from approximately 1000 to about 800 cm^{-1} where in-plane aromatic CH bending vibrations occur at 1040 cm^{-1} and out-of-plane aromatic and olefinic CH bending vibrations appear at 955 and 830 cm^{-1} [36]. Toward the high-wavenumber region, a moderately strong, broad O-H stretching band lies near 3370 cm^{-1} while CH stretching vibrations produce several weak bands at 3025 (aromatic and olefinic groups), 2865 (CH_3 groups) and 2840 (CH_2 groups) cm^{-1} and a weak-to-moderate band at 2935 cm^{-1} (CH_2 groups) with the shoulder of CH_3 groups positioning at 2965 cm^{-1}.

In the total reflection mode spectrum of shellac (Figure 8), the most intense features over the MIR range appear with a steep slope of the lineshape and sharp band maxima that fall at 2970 and 2873 (νCH), 1747 (νC=O) and 1273 (νC-O) cm^{-1}. It is worth noting that both C=O and C-O stretching bands present a slight slope variation along the lineshape, which is responsible for the formation of the 1731-1712 and 1250-1235 cm^{-1} doublets after the KK transformation, respectively (Figure 9). Weak-to-moderate bands are all characterized by a slighter lineshape slope with the less resolved maxima falling near 1470, 1385, 1180, 950-930 and 733 cm^{-1}. Their corresponding peak assignments are reported in Table 7. The νC-O bands in the 1200-1000 cm^{-1} region are partially well resolved and partially defined only by slope changes of the lineshape, as confirmed observing the KKT spectrum. The νO-H band differs from all others because of its broad maximum, centered at around 3570 cm^{-1}, and its slightly pronounced slope which extends from approximately 3500 to about 3200 cm^{-1}. However, in correspondence of this high wavenumbers region the KK algorithm produces a flat line which is preceded by a deep falling as artefact. The total reflection mode spectrum of dragon's blood (Figure 8) presents a peculiar, distinctive fingerprint region where most of the bands have a well-resolved lineshape with steep to very steep slopes and sharp maxima (see Table 7 for the wavenumber values). As expected, the KK spectrum has the same profile of the TR one whose bands and relative assignment has been already accurately described in the paragraph above. Observing the mid-IR region over the 2000 cm^{-1}, the medium-intensity O-H stretching band exhibits a wide maximum near 3500 cm^{-1} and a characteristic slightly pronounced slope, while the CH stretching vibrations produce weak-to-moderate bands with the lineshape being characterized by a few slope variations which give rise to CH_3/CH_2 overlapping bands after the application of the KK transformations. The detailed experimental wavenumber values of both shellac and dragon's blood, and the relative suggested assignment, are reported in Table 7.

As regards NIR region, the reflectance spectra of shellac and dragon's blood (Figure 2(e)) display a few common bands positioned at 5778 cm^{-1}, due to the first overtone of asymmetric CH_2 stretching vibration, and near 5200, 4800, and 4650 cm^{-1} which are attributed respectively to ν(OH)+δ(OH), ν(C-O)+ν(OH) and ν(C-O)+ν(CH_2) combination bands. It should be noted that the described ν(C-O)+ν(CH_2) band appears more intense and better resolved in dragon's blood compared to shellac, because of the likely additional contribution of the aromatic ν(CH)+ν(CC)

TABLE 7: Experimental wavenumber values corresponding to the maxima of the mid-IR total reflection mode bands of not (exclusively) terpenoid-based resins and their tentative assignment.

Experimental wavenumbers (cm^{-1})	Band assignment
Shellac	
3570	νOH
2970	ν_{as}CH
2873	ν_sCH
1747	νC=O
1470	In-plane δCH_2
1385	$\delta_{as}CH_3$
1273	νC-O
1180	νC-O
1120	νC-O
1095	νC-O
1075	νC-O
1055	νC-O
1015	νC-O
1030	νC-O
950-930	
733	νC-C
Dragon's blood	
3500	νOH
3030	νCH (aromatic, olefinic)
2963	ν_{as}CH
2855	ν_sCH
1665	νC=C, νC=O
1623	νC=C (aromatic)
1520	νC=C (aromatic)
1473	νC=C (aromatic)
1300	
1255	νC-O
1218	νC-O
1178	νC-O
1164	νC-O
1123	νC-O
1052	In-plane δCH (aromatic)
960	Out-of-plane δCH (aromatic, olefinic)
885	Out-of-plane δCH (aromatic, olefinic)
850	Out-of-plane δCH (aromatic, olefinic)

combination band. Other common features are found in the 4500-4000 region, where the $\nu_{as}(CH_2)+\delta(CH_2)$ and $\nu_s(CH_2)+\delta(CH_2)$ combination bands lie respectively at 4342 and 4263 cm^{-1}. These bands are strong, sharp, and well defined in the spectrum of shellac because of its characteristic aliphatic chains while they appear broad, very weak in the spectrum of dragon's blood. Moreover, in this spectral portion shellac shows another signal at 4040 cm^{-1}, which is due to the third overtone of the CC bending vibration. Conversely, additional high-wavenumber bands linked to the presence of aromatic structures characterize the dragon's blood resin at about 5950 and 6900 cm^{-1}, respectively attributed to the first overtone of aromatic CH stretching vibration and to the dual contribution of the first overtone of ν(OH) and the aromatic CH combination. Finally, it can be noticed that the first overtone band of $\nu(CH_2)$ at 5668 cm^{-1} appears well defined only in the spectrum of dragon's blood, whereas the marked noise toward the high wavenumbers does not allow us to discern it in the shellac resin. It is important to point out that the interpretation of the NIR signals of dragon's blood has been based on vibration charts [30], because no band assignment has occurred in literature using the infrared spectroscopy technique.

4. Conclusions

The aim of the present paper is to provide a valuable analytical tool to conservators and scientists for a complete comprehension of infrared spectra acquired in total reflection mode by means of portable noncontact FTIR spectrometers. The knowledge of the reflectance spectral behavior in the mid and near-infrared region of pure standards represents, in fact, the first necessary step for subsequently tackling the more complex interpretation of aged substances, mixtures (e.g., varnishes) or layered systems occurring in real cases. Understanding when specific data-processing algorithms can be applied with accurate results represents another important issue here discussed.

In this work, we present FTIR total reflection mode spectra of sixteen pure, nonaged natural organic materials widely spread in works of art. The spectral analysis of absorption bands acquired in transflection mode has been significant, firstly, to approach the study of distorted reflection bands and, then, to evaluate the applicability of KK correction in the MIR region. Moreover, a visual comparison between reflection and absorption bands is useful to make the different spectral behavior understandable.

The study mainly focuses on the mid-IR region (4000-375 cm^{-1}) where the specular reflection contribution appears predominant giving rise to a derivative-shaped spectral profile. According to low absorption coefficient materials, Reststrahlen bands do not occur. A novel, tentative approach to the discussion of specular reflectance bands compared to absorbance ones is here provided. The good wavenumber and lineshape correspondence between KK corrected and transflection bands allowed us to conclude that an accurate result from KK algorithm has been achieved for the sample references, thus confirming the specular reflection as the main collected contribute of the total reflection mode spectra in the mid-IR region. Then, the work tackles the interpretation of combination and overtone bands in the near-IR range (7500-4000 cm^{-1}), where the diffuse reflection phenomenon is dominant.

Despite the possible limitation represented by artwork superficial conditions, which necessarily have to be evaluated case by case, we consider our results promising for their application in the Cultural Heritage field.

Acknowledgments

The authors wish to thank Prof. Claudio Canevari and Dr. Maduka Weththimuni for the support in the choice and preparation of the reference samples.

References

[1] M. Vagnini, C. Miliani, L. Cartechini, P. Rocchi, B. G. Brunetti, and A. Sgamellotti, "FT-NIR spectroscopy for non-invasive identification of natural polymers and resins in easel paintings," *Analytical and Bioanalytical Chemistry*, vol. 395, no. 7, pp. 2107–2118, 2009.

[2] C. Ricci, C. Miliani, B. G. Brunetti, and A. Sgamellotti, "Non-invasive identification of surface materials on marble artifacts with fiber optic mid-FTIR reflectance spectroscopy," *Talanta*, vol. 69, no. 5, pp. 1221–1226, 2006.

[3] B. Brunetti, C. Miliani, F. Rosi et al., "Non-invasive Investigations of Paintings by Portable Instrumentation: The MOLAB Experience," *Topics in Current Chemistry*, vol. 374, no. 1, 2016.

[4] J. Echard, "In situ multi-element analyses by energy-dispersive X-ray fluorescence on varnishes of historical violins," *Spectrochimica Acta Part B: Atomic Spectroscopy*, vol. 59, no. 10-11, pp. 1663–1667, 2004.

[5] G. Valentina Fichera, M. Albano, G. Fiocco et al., "Innovative Monitoring Plan for the Preventive Conservation of Historical Musical Instruments," *Studies in Conservation*, vol. 63, no. supl, pp. 351–354, 2018.

[6] C. Miliani, F. Rosi, A. Daveri, and B. G. Brunetti, "Reflection infrared spectroscopy for the non-invasive in situ study of artists' pigments," *Applied Physics A: Materials Science & Processing*, vol. 106, no. 2, pp. 295–307, 2012.

[7] F. Casadio and L. Toniolo, "The analysis of polychrome works of art: 40 years of infrared spectroscopic investigations," *Journal of Cultural Heritage*, vol. 2, no. 1, pp. 71–78, 2001.

[8] G. Fiocco, T. Rovetta, M. Gulmini, A. Piccirillo, M. Licchelli, and M. Malagodi, "Spectroscopic Analysis to Characterize Finishing Treatments of Ancient Bowed String Instruments," *Applied Spectroscopy*, vol. 71, no. 11, pp. 2477–2487, 2017.

[9] F. Rosi, A. Daveri, C. Miliani et al., "Non-invasive identification of organic materials in wall paintings by fiber optic reflectance infrared spectroscopy: A statistical multivariate approach," *Analytical and Bioanalytical Chemistry*, vol. 395, no. 7, pp. 2097–2106, 2009.

[10] T. Poli, E. Alice, and O. Chiantore, "Surface Finishes and Materials: Fiber-Optic Reflectance Spectroscopy (FORS) Problems in Cultural Heritage Diagnostics," in *e-Preservation Science*, vol. 6, pp. 174–179, 2009.

[11] T. Poli, O. Chiantore, M. Nervo, and A. Piccirillo, "Mid-IR fiber-optic reflectance spectroscopy for identifying the finish on wooden furniture," *Analytical and Bioanalytical Chemistry*, vol. 400, no. 4, pp. 1161–1171, 2011.

[12] F. Rosi, A. Daveri, P. Moretti, B. G. Brunetti, and C. Miliani, "Interpretation of mid and near-infrared reflection properties of synthetic polymer paints for the non-invasive assessment of binding media in twentieth-century pictorial artworks," *Microchemical Journal*, vol. 124, pp. 898–908, 2016.

[13] C. Daher, V. Pimenta, and L. Bellot-Gurlet, "Towards a non-invasive quantitative analysis of the organic components in museum objects varnishes by vibrational spectroscopies: Methodological approach," *Talanta*, vol. 129, pp. 336–345, 2014.

[14] D. Saviello, L. Toniolo, S. Goidanich, and F. Casadio, "Non-invasive identification of plastic materials in museum collections with portable FTIR reflectance spectroscopy: Reference database and practical applications," *Microchemical Journal*, vol. 124, pp. 868–877, 2016.

[15] M. Fabbri, M. Picollo, S. Porcinai, and M. Bacci, "Mid-Infrared Fiber-Optics Reflectance Spectroscopy: A Noninvasive Technique for Remote Analysis of Painted Layers. Part II: Statistical Analysis of Spectra," *Applied Spectroscopy*, vol. 55, no. 4, pp. 428–433, 2016.

[16] P. R. Griffiths and J. A. De Haseth, "Fourier Transform Infrared Spectrometry: Second Edition," *Fourier Transform Infrared Spectrometry: Second Edition*, pp. 1–529, 2006.

[17] M. Fabbri, M. Picollo, S. Porcinai, and M. Bacci, "Mid-infrared fiber-optics reflectance spectroscopy: A noninvasive technique for remote analysis of painted layers. Part II: Statistical analysis of spectra," *Applied Spectroscopy*, vol. 55, no. 4, pp. 428–433, 2001.

[18] M. Picollo, G. Bartolozzi, C. Cucci, M. Galeotti, V. Marchiafava, and B. Pizzo, "Comparative study of fourier transform infrared spectroscopy in transmission, attenuated total reflection, and total reflection modes for the analysis of plastics in the cultural heritage field," *Applied Spectroscopy*, vol. 68, no. 4, pp. 389–397, 2014.

[19] C. Miliani, F. Rosi, I. Borgia, P. Benedetti, B. G. Brunetti, and A. Sgamellotti, "Fiber-optic Fourier transform mid-infrared reflectance spectroscopy: A suitable technique for in Situ studies of mural paintings," *Applied Spectroscopy*, vol. 61, no. 3, pp. 293–299, 2007.

[20] C. Invernizzi, G. V. Fichera, M. Licchelli, and M. Malagodi, "A non-invasive stratigraphic study by reflection FT-IR spectroscopy and UV-induced fluorescence technique: The case of historical violins," *Microchemical Journal*, vol. 138, pp. 273–281, 2018.

[21] C. Invernizzi, A. Daveri, M. Vagnini, and M. Malagodi, "Non-invasive identification of organic materials in historical stringed musical instruments by reflection infrared spectroscopy: a methodological approach," *Analytical and Bioanalytical Chemistry*, vol. 409, no. 13, pp. 3281–3288, 2017.

[22] M. P. Colombini and F. Modugno, *Organic Mass Spectrometry in Art and Archaeology*, John Wiley & Sons, Ltd, Chichester, UK, 2009.

[23] C. Invernizzi, A. Daveri, T. Rovetta et al., "A multi-analytical non-invasive approach to violin materials: The case of Antonio Stradivari "Hellier" (1679)," *Microchemical Journal*, vol. 124, pp. 743–750, 2016.

[24] C. Daher, C. Paris, A.-S. Le Hô, L. Bellot-Gurlet, and J.-P. Échard, "A joint use of Raman and infrared spectroscopies for the identification ofnatural organic media used in ancient varnishes," *Journal of Raman Spectroscopy*, vol. 41, no. 11, pp. 1494–1499, 2010.

[25] M. Koperska, T. Łojewski, and J. Łojewska, "Vibrational spectroscopy to study degradation of natural dyes. Assessment of

oxygen-free cassette for safe exposition of artefacts," *Analytical and Bioanalytical Chemistry*, vol. 399, no. 9, pp. 3271–3283, 2011.

[26] J. J. Bischoff, M. R. Derrick, D. Stulik, and J. M. Landry, "Infrared Spectroscopy in Conservation," *Journal of the American Institute for Conservation*, vol. 40, no. 3, p. 268, 2001.

[27] R. K. Vincent and G. R. Hunt, "Infrared reflectance from mat surfaces," *Applied Optics*, vol. 7, no. 1, pp. 53–59, 1968.

[28] M. T. Doménech-Carbó, "Novel analytical methods for characterising binding media and protective coatings in artworks," *Analytica Chimica Acta*, vol. 621, no. 2, pp. 109–139, 2008.

[29] M. de Buergo and R. González, "Protective patinas applied on stony façades of historical buildings in the past," *Construction and Building Materials*, vol. 17, no. 2, pp. 83–89, 2003.

[30] J. Workman Jr. and L. Weyer, *Practical Guide to Interpretive Near-Infrared Spectroscopy*, CRC Press, 2007.

[31] S. Merk, A. Blume, and M. Riederer, "Phase behaviour and crystallinity of plant cuticular waxes studied by Fourier transform infrared spectroscopy," *Planta*, vol. 204, no. 1, pp. 44–53, 1998.

[32] D. Scalarone, M. Lazzari, and O. Chiantore, "Ageing behaviour and analytical pyrolysis characterisation of diterpenic resins used as art materials: Manila copal and sandarac," *Journal of Analytical and Applied Pyrolysis*, vol. 68-69, pp. 115–136, 2003.

[33] M. Größl, S. Harrison, I. Kaml, and E. Kenndler, "Characterisation of natural polysaccharides (plant gums) used as binding media for artistic and historic works by capillary zone electrophoresis," *Journal of Chromatography A*, vol. 1077, no. 1, pp. 80–89, 2005.

[34] T. Geiger and F. Michel, "Studies on the polysaccharide Jun-Funori used to consolidate Matt Paint," *Studies in Conservation*, vol. 50, no. 3, pp. 193–204, 2005.

[35] D. Gupta, B. Bleakley, and R. K. Gupta, "Dragon's blood: botany, chemistry and therapeutic uses," *Journal of Ethnopharmacology*, vol. 115, no. 3, pp. 361–380, 2007.

[36] B. Stuart, "Infrared Spectroscopy: Fundamentals and Applications," in *Climate Change 2013 – The Physical Science Basis*, 2014.

13

Structural Study of Europium Doped Gadolinium Polyphosphates $LiGd(PO_3)_4$ and its Effect on their Spectroscopic, Thermal, Magnetic, and Optical Properties

Saoussen Hammami,[1] Nassira Chniba Boudjada,[2] and Adel Megriche[1]

[1] *Université de Tunis El Manar, Faculté des Sciences de Tunis, UR11ES18 Unité de recherche de Chimie Minérale Appliquée. Campus Universitaire Farhat Hached, 2092 Manar 1. Tunis, Tunisia*
[2] *Institut Néel, BP 166, 38042 Grenoble Cedex 9, France*

Correspondence should be addressed to Saoussen Hammami; sosanehammami@gmail.com

Academic Editor: Adil Denizli

Alkali metal-rare earth polyphosphates $LiGd_{(1-x)}Eu_x(PO_3)_4$ (LGP:Eu^{3+}) (where x= 0, 0.02 and 0.04) were synthesized by solid-state reaction. The Rietveld refinement showed the following cell parameters: I 2/a space group, a=9.635(3) Å, b=7.035(3) Å, c=13.191(3) Å, β=90.082°, V= 894.214Å3, and Z=4. The similarity between R_F=4.21% and R_B=4.31% indicated that the realized refinement is reliable. The crystal structure consists of infinite zig-zag chains of $(PO_4)^{3-}$ tetrahedra, linked by bridging oxygen. The acyclic structure of polyphosphates is confirmed by infrared and Raman (IR) spectroscopies. A good thermal stability up to 940°C and paramagnetic behavior of these compounds were also proved by thermal analyses and magnetic susceptibility measurements, respectively. Excitation spectra revealed the charge transfer phenomenon between O^{2-} and Eu^{3+} (CTB), the energy transfer from Gd^{3+} to Eu^{3+}, and the intrinsic 4f-4f transitions of Eu^{3+} where the electronic transitions were also identified. Moreover, LGP:Eu^{3+} can emit intense reddish orange light under excitation at 394 nm. The strongest tow at 578 and 601 nm can be attributed to the transitions from excited state $^5D_0$ to ground states $^7F_1$ and $^7F_2$, respectively.

1. Introduction

Condensed alkali metal-rare earth polyphosphates, with the general formula $M^IRE^{III}(PO_3)_4$ (where M^I= are alkali metal and RE^{III}= are rare earth ions), have been extensively investigated thanks to their structural diversity [1–4] and their interesting magnetic [5], optic [6], and electric [7] proprieties. These polyphosphates are generally stable in normal conditions of temperature and humidity [8], which makes them useful for industrial applications. For example, Yamada et al. have used the $LiNd(PO_3)_4$ polyphosphate as a solid-state laser material [9, 10]. However, Z. Mua et al. have used $LiEu(PO_3)_4$ compound for white light-emitting diodes [11]. They are also used as promising scintillation material such as the Ce^{3+} doped $MGd(PO_3)_4$ compound [12].

In order to enhance the optical properties of polyphosphates, researches were oriented to doping them with metal-rare earths. In recent years, a great interest was accorded to europium earth-rare due to its outstanding photoluminescence feature [13–19]. In fact, the presence of well-defined energy levels in europium allows the emission of monochromatic and coherent radiations in solid laser.

The Gd–Eu couple is well known for its efficient conversion of the absorbed high-energy photons into two visible ones [20–24]. This phenomenon may be followed by a sequence of two steps of energies transfer. The first step is the transition $^6G_J \rightarrow {}^6P_J$ of Gd^{3+}, involving the $^5D_0 \rightarrow {}^7F_1$ transition of the Eu^{3+} ion. The one red photon related to the Eu^{3+} $^5D_0 \rightarrow {}^7F_J$ transition is created. In the second step, the energy of Gd^{3+} is related to $^6P_{7/2} \rightarrow {}^8S$ transition which is transported over the Gd^{3+} sublattice and then transferred to another Eu^{3+} ion. This phenomenon leads to ^5D_J emission (J = 0, 1, 2, or 3) [25].

This work describes the synthesis of $LiGd_{(1-x)}Eu_x(PO_3)_4$ polyphosphate, doped with different low percentages of europium (2 and 4%). The structural study of all obtained compounds is carried out with XRD diffraction. The infrared and Raman spectroscopies and magnetic and thermal analyses were recorded at room temperature. Moreover, the optical study through excitation and emission of Eu^{3+} ions spectra was also undertaken.

2. Experimental

The condensed phosphates $LiGd_{(1-x)}Eu_x(PO_3)_4$ (where x= 0, 0.02 and 0.04) were synthesized by solid-state reaction (methods of Hammami et al. 2017) [26]. A mixture of the reagents, Li_2CO_3, Gd_2O_3, Eu_2O_3, and $NH_4H_2PO_4$, was prepared with the molar ratio (2.1:1:8) of Li:Gd:P, respectively. First, the raw materials were grounding in an agate mortar for one hour at least to homogenize the solid phase and improve the interatomic diffusion. Second, the mixture was introduced into the oven and submitted to the following thermal program. The first level was at 430°C to eliminate H_2O, NH_3, and CO_2, the second one was at 730°C to get $LiGd_{(1-x)}Eu_x(PO_3)_4$ pure phase. Then, the obtained products were cooled with the rate of 2°C/min to ensure a better crystallinity. Finally, the synthesized polyphosphates were washed with boiling water and nitric acid solution (1mol/L) to eliminate the residual raw materials from the final product.

The proposed chemical reaction for polyphosphate synthesis is

$$8NH_4H_2PO_4 + (1-x)\,Gd_2O_3 + xEu_2O_3 + Li_2CO_3 \longrightarrow 2\,LiGd_{(1-x)}Eu_x\,(PO_3)_4 + 8NH_3\uparrow + 12H_2O + CO_2\uparrow \quad (1)$$

Samples were characterized using an INEL XRG 3000 (D5000T) diffractometer with monochromatic Cu K_α radiation. The diffraction pattern was recorded under 300K over the angular range 10-90° (2θ). The luminescence spectra were performed under ambient atmosphere via Xenius (the fluorescence Genius) spectrophotometer, at 591nm and 394nm for excitation and emission, respectively. The infrared spectra were recorded in the range of 250–1500 cm^{-1} with a Thermo Scientific Nicolet N10 MX using sample dispersed in KBr pellets. Raman analysis was carried out at room temperature, with 514.5 nm radiation from an argon ion laser as the excitation beam. A microscope allowed a selection of high optical quality regions in the crystalline sample. Thermal stability of Eu^{3+} doped LGP was measured with differential thermal analysis SETARAM TAG 16 operating from room temperature up to 1000°C with heating rate of 5°C min^{-1}. Magnetic measurements were carried out using Quantum Design MPMSXL magnetometer with detection SQUID (at institute NEEL France).

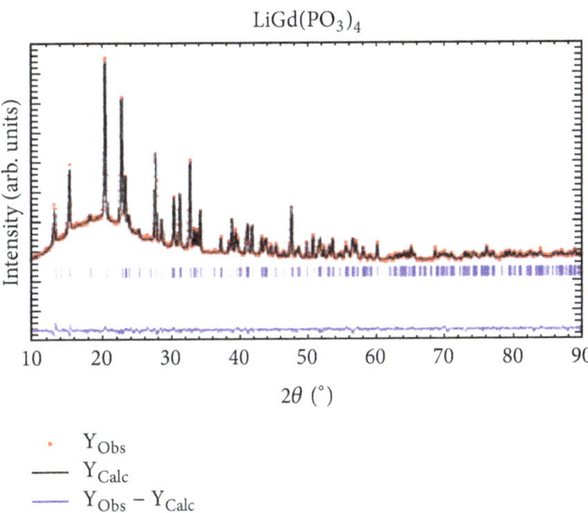

FIGURE 1: The Rietveld analysis of X-ray diffraction patterns for $LiGd(PO_3)_4$.

3. Results and Discussion

3.1. Rietveld Refinement Data Analyses. The Rietveld refinement of X-ray diffraction patterns of synthesized $LiGd(PO_3)_4$ samples is shown in Figure 1. The graph presents the experimental and the calculated data as well as the difference between them. As it is shown, the presence of only single phase was checked by Rietveld fitting quality through the reliability R factors: R_{exp}, R_{brag}, profile R_p, and weighted profile R_{wp}, which should be less than 10%. The final R factors, atomic coordinates, site occupancy, thermal displacement parameters, and their estimated standard deviations in parentheses for LGP are shown in Table 1. Interatomic bond distances and angles are given in Table 2. The new lattice parameters, derived from the Rietveld refinement, are a=9.635(3) Å, b= 7.035(3) Å, c= 13.191(3) Å, and β= 90.082° and with monoclinic space group I 2/a.

The LGP structure can be simply described as three-dimensional framework of GdO_8 polyhedra linked to $(PO_4)^{3-}$ rings by Gd-O-P bridges. This framework delimits interesting tunnels with Li^+ ions which are bonded to four oxygen atoms (LiO_4). Each LiO_4 tetrahedron shares all four O atoms with two LaO_8 polyhedra and four different PO_4 tetrahedra. A view of this structure projected along the b axis is shown in Figure 2.

3.2. X-Ray Powder Diffraction. X-ray diffraction patterns of $LiGd_{(1-x)}Eu_x(PO_3)_4$ (where x= 0, 0.02 and 0.04) are shown in Figure 3. The obtained crystalline phases are isotypes of the mother-phase $LiGd(PO_3)_4$ [27]. Mainly, it is shown that XR diffraction peaks of studied solids, with different percentages of europium, are described in a cell with a super space group I 2/a instead of C 2/c usually used in crystallographic data of the old LGP. The same XRD pattern is obtained for almost of synthesized compounds, even at high Eu^{3+} concentrations. However, a shift of the diffraction peaks to the lower 2θ is observed. This shift can be explained by Bragg's theory "$\mathbf{n\lambda= 2d_{hkl}\sin\theta}$" (where λ is the X-ray wavelength (Cu K_α

TABLE 1: Refined structure parameters from powder X-ray Rietveld analysis for $LiGd(PO_3)_4$ in space group I 2/a.

Atom	Wyck	x	y	z	B	Occ
Gd	4e	0.75000	0.2982(3)	0.00000	1.35(8)	0.50
Li	4e	0.75000	0.79603(4)	0.00000	1.88(16)	0.50
P1	8f	0.48189(3)	0.06204(4)	0.143135(20)	1.88(16)	1.00
P2	8f	0.54601(3)	0.66399(4)	0.15354(2)	1.88(16)	1.00
OL12	8f	0.5665(6)	0.87995(6)	0.1598(14)	2.7(2)	1.00
OL21	8f	0.4057(4)	0.0815(16)	0.2447(3)	2.7(2)	1.00
OE21	8f	0.6520(11)	0.59342(6)	0.0754(8)	2.7(2)	1.00
OE11	8f	0.5743(13)	0.2282(14)	0.1098(11)	2.7(2)	1.00
OE22	8f	0.3953(6)	0.626(2)	0.1220(12)	2.7(2)	1.00
OE12	8f	0.3696(12)	0.0106(3)	0.0652(10)	2.7(2)	1.00
	R_p = 1.237	R_{wp} = 1.598	R_{exp} = 1.134	R_{Bragg} = 4.301	χ^2 = 2	

TABLE 2: Atomic distances(Å) and angles in $LiGd(PO_3)_4$ with standard deviations in parentheses.

Tetrahedra	**aroundP1**				
P1-OL12	1.534(4)	OL12 -OL21	2.381(12)	OL12^{i} —P1—OE11	111.61(38)
P1-OL21	1.535(4)	OL12 -OE21	2.446(11)	OL12^{i} —P1—OL21	101.76(23)
P1-OE11	1.534(11)	OL12 -OL21	2.463(14)	OL12^{i} —P1—OE12	105.86(29)
P1-OE12	1.535(12)	OL12 -OE11	2.538(11)	OL21—P1—OE11	117.46(53)
		OL12 -OE22	2.482(12)	OL21—P1—OE12	105.55(40)
		OL12 -OE12	2.448(15)	OE12—P1—OE11	113.37(47)
Tetrahedra	**around P2**				
P2 -OL12	1.5343(13)	OE22 -OL12	2.482(12)	OE22—P2—OE21	113.09(39)
P2 -OL21	1.534(6)	OE22 -OL21	2.619(12)	OE22—P2—OL12	108.04(41)
P2 -OE21	1.535(10)	OE22 -OE21	2.560	OE22—P2—OL21^{vi}	117.27(36)
P2 -OE22	1.533(7)	OL21 -OL12	2.381(12)	OE21—P2—OL12	105.66(4)
		OL21 -OL12	2.440(11)	OE21—P2—OL21^{vi}	105.25(16)
		OE21 -OE22	2.560(13)	OL12—P2—OL21^{vi}	106.79(38)
Polyhedra	**around Gd**	**Tetrahedra around Li**			
Gd -OE21	2.489(6)	Li -OE21		Gd -Gd^{iv}	5.592(2)
Gd -OE21	2.489(6)	Li -OE21		Li -Gd^{ii}	3.533(2)
Gd -OE11	2.283(13)	Li -OE12		Li -Li^{v}	5.068(0)
Gd -OE11	2.283(13)	Li -OE12		P1 - P2	2.8710(4)
Gd-OE22	2.197(13)			P2 -P1	2.7897(4)
Gd-OE22	2.197(13)			Gd -P1^{iii}	6.256(2)
Gd-OE12	2.605(7)				
Gd -OE12	2.605(7)				
Symmetry code	i: x,1+y,z	ii:x,1+y,z iii:1.5-x,1+y,-z		iv:2-x,1-y,-z v:2-x,2-y,-z	vi:1-x,0.5+y,0.5-z

=1.5406Å), θ is diffraction angle, and d is interplanar distance of corresponding diffraction peaks). Therefore, λ is constant; it can be concluded that this shift is due to the increase of interplanar distance "d". Considering the characteristics of Gd and Eu (ionic radius of Gd^{+3}: 1.05 Å, Eu^{+3}: 1.07 Å and atomic volume Gd: 19.9 cm^3/mol, Eu: 28.9 cm^3/mol), this phenomenon can be attributed to the distortion of the tetrahedra of polyphosphates upon europium insertion [28].

The crystallite size of obtained polyphosphates is calculated using Sherrer's equation below [29] and values are summarized in Table 3. Results show that the range of calculated crystallite size is between 42.49 and 42.79 nm, which prove that the synthesized compounds are nanometric.

$$\text{Sherre'r equation: } D = \frac{0.9\lambda}{\beta \cos\theta} \quad (2)$$

TABLE 3: The size of the crystallite according to percentage of europium.

Percentages (%)	FWHM	Position(2θ)	D(nm)
0	0.189	20.630	42.49
2	0.189	20.628	42.79
4	0.189	20.621	42.79

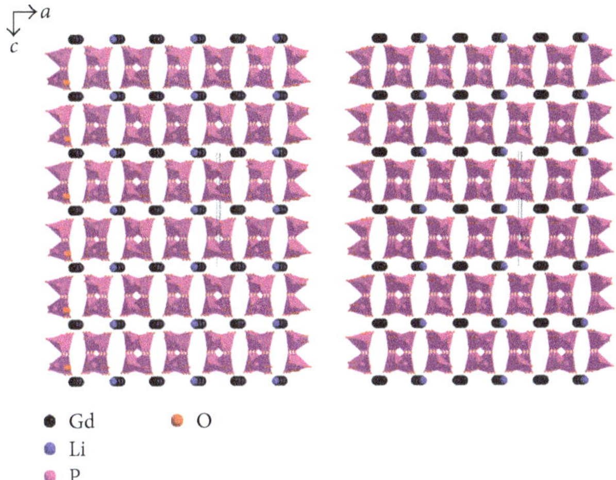

FIGURE 2: The structural arrangement of the $LiGd(PO_3)_4$ viewed in the (0 1 0) plane.

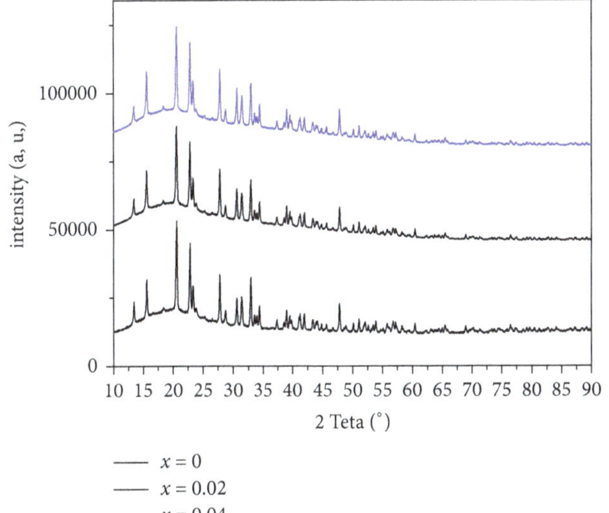

FIGURE 3: XRD patterns of $LiGd_{(1-x)}Eu_x(PO_3)_4$ (x=0, 0.02 and 0.04).

where λ is the X-ray wavelength (Cu K_α =1.5406Å), θ is the Bragg diffraction angle, and βe is the full width at half - maximum (FWHM) in radian of the main peak of each XRD pattern.

3.3. Infrared and Raman Spectroscopy Investigations

3.3.1. Infrared. Figure 4 shows the IR spectra of all studied compounds. The comparison of spectra (Figure 4) and those obtained in previous works in literature for condensed polyphosphates [30, 31] proves that positions of infrared absorption bands are characteristic of phosphates with chain structures.

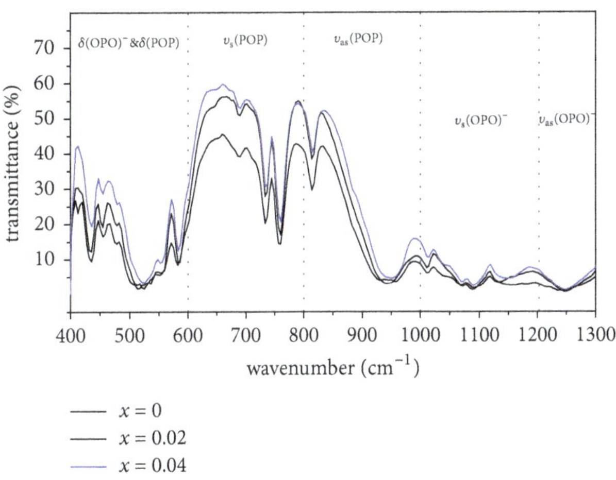

FIGURE 4: IR spectra of $LiGd_{(1-x)}Eu_x(PO_3)_4$ (x=0, 0.02 and 0.04).

IR bands attribution is carried out based on $(O\text{-}P\text{-}O)^-$ groups and P-O-P bridges vibrations [32, 33]. The IR absorption spectra show the presence of two bands around 1249 cm^{-1} which are assigned to the asymmetric stretching vibration (υ_{as}) of O-P-O. The weak band observed between 1071 and 1136 cm^{-1} is attributed to the symmetric stretching vibration υ_s of O-P-O. The large and intense band around 944 cm^{-1} is assigned to the asymmetric vibration υ_{as} of P-O-P. We can also attribute the few bands at 689-818 cm^{-1} to the symmetric vibration υ_s (P-O-P). At low frequencies region, below 600 cm^{-1}, it is very difficult to distinguish the symmetric and antisymmetric bending modes of the (O-P-O) and (P-O-P) groups. The frequencies of the corresponding bands are given in Table 4.

The major difference between the IR spectra of cyclic polyphosphate and polyphosphate is the absence of vibration bands between 750 and 1000 cm^{-1}. In this range, IR spectroscopy confirms the structure as long as polyphosphates chains.

3.3.2. Raman. The Raman spectra of $LiGd_{(1-x)}Eu_x(PO_3)_4$ (where x= 0, 0.02 and 0.04) at room temperature are shown in Figure 5. These spectra show the presence of many bands; the first intense band at 1178 cm^{-1} and the second at 700 cm^{-1} are assigned to antisymmetric stretching vibration mode υ_{as} (O-P-O) and symmetric stretching vibrations mode υ_s (P-O-P), respectively. The υ_{as} (P-O-P) asymmetric and υ_s (O-P-O) symmetric stretching vibration modes, respectively, appear in the 1000-1100 cm^{-1} and 1212-1296 cm^{-1} ranges. The bands under 599 cm^{-1} are attributed to the symmetric and the asymmetric bending vibrations (δ_{as} and δ_s) of $(O\text{-}P\text{-}O)^-$ and (P-O-P). The intense symmetric stretching vibrations bands around 700 and 1178 cm^{-1} are characteristic of phosphoric anions $(PO_3)_4^{4-}$ [34].

Distinguishing characteristic cyclotetraphosphates and polyphosphates compounds exit also in the Raman spectrum.

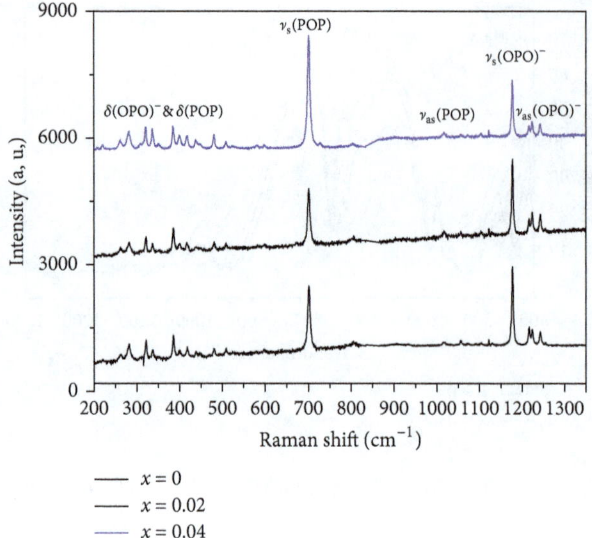

FIGURE 5: Raman spectra of $LiGd(PO_3)_4$ (x=0, 0.02 and 0.04) at 300 K.

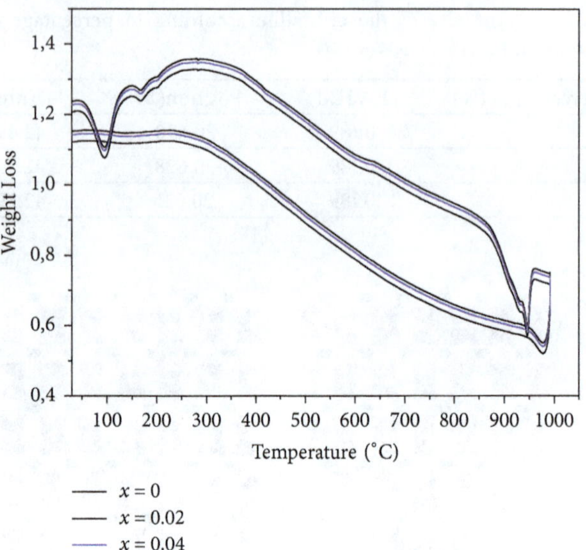

FIGURE 6: DTA of $LiGd_{(1-x)}Eu_x(PO_3)_4$ (x=0, 0.02 and 0.04).

TABLE 4: Attributions of main IR bands (cm^{-1}) of $LiGd_{(1-x)}Eu_x(PO_3)_4$ samples.

Assignment	x=0	x=0.02	x= 0.04
υ_{as} OPO	1257	1241	1250
	1137	1127	1149
υ_s OPO	1089	1089	109
υ_{as} POP	1014	1014	1014
	944	944	947
υ_s POP	689	689	689
	736	736	736
	761	758	761
	818	818	818
δPOP	437-	437	437
δPOP	456	456	457
	519	519	521
	586	585	585

The symmetric stretching vibration of the P-O-P (υ_s (P-O-P)) has a single band at 700 cm^{-1}. This is the strongest of all the Raman vibration bands. That is because of the monoclinic symmetry of LGP doped Eu and the different positions of the lanthanide and alkali ions. The results of Raman spectroscopy can identify the structure of alkali metal lanthanide polyphosphates.

3.4. DTA (Differential Thermal Analysis). The thermal stability of lithium polyphosphate is investigated using DTA. The curves of the Eu: $LiGd(PO_3)_4$ crystal are given in Figure 6. It is clearly observed that the curves present the same shape (evolution). Indeed, a single sharp endothermic peak is observed between 900 and 1000°C for all samples, which exhibits the characteristics of a first-order phase transition. This signal can be attributed to the decomposition of polyphosphates to $GdPO_4$. The stability of lithium gadolinium polyphosphates

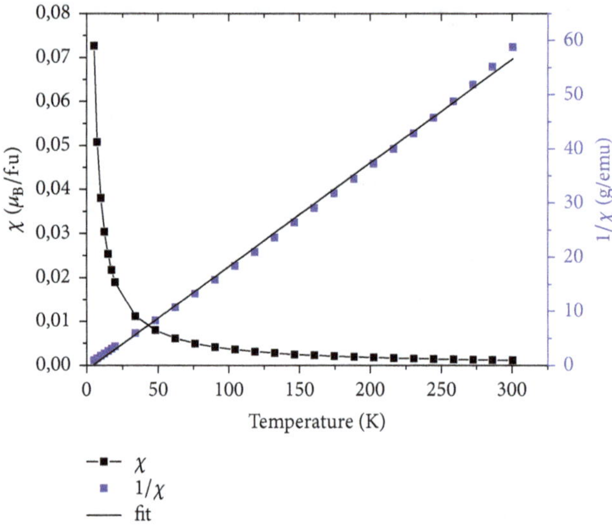

FIGURE 7: The magnetic susceptibility (χ) and inverse magnetic susceptibility (1/χ) measurements as a function of temperature of $LiGd(PO_3)_4$.

can be explained by heavily distorted of PO_4 tetrahedra as are the GdO_8 polyhedra. We thus conclude that all compounds are stable at high temperatures and it is monophasic.

3.5. Magnetic Study. The magnetic susceptibility and inverse magnetic susceptibility versus temperature of $LiGd(PO_3)_4$, $LiGd_{0.98}Eu_{0.02}(PO_3)_4$, and $LiGd_{0.96}Eu_{0.04}(PO_3)_4$ are shown in Figures 7, 8, and 9, respectively. The only other reported type of rare earth polyphosphate structures are those of LGP: Eu^{3+}; and these were chosen because Gd^{3+} has an effective magnetic moment and the $4f^n$ electrons.

These curves prove that all three rare earth polyphosphate compounds exhibit a paramagnetic response. The nondoped LGP is the most paramagnetic one; this is explained by

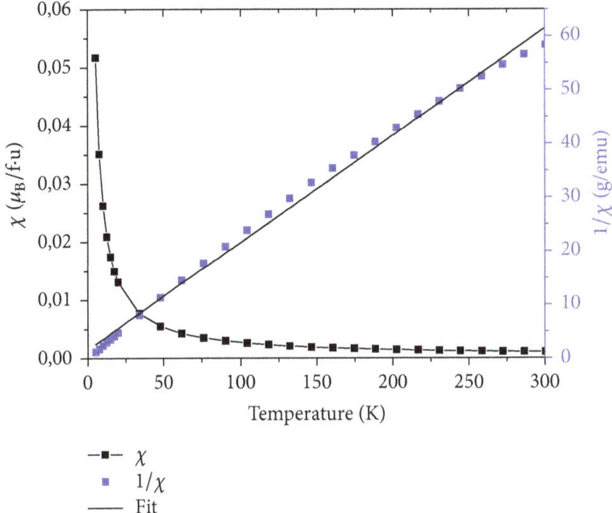

FIGURE 8: The magnetic susceptibility (χ) and inverse magnetic susceptibility ($1/\chi$) measurements as a function temperature of $LiGd_{0.98}Eu_{0.02}(PO_3)_4$.

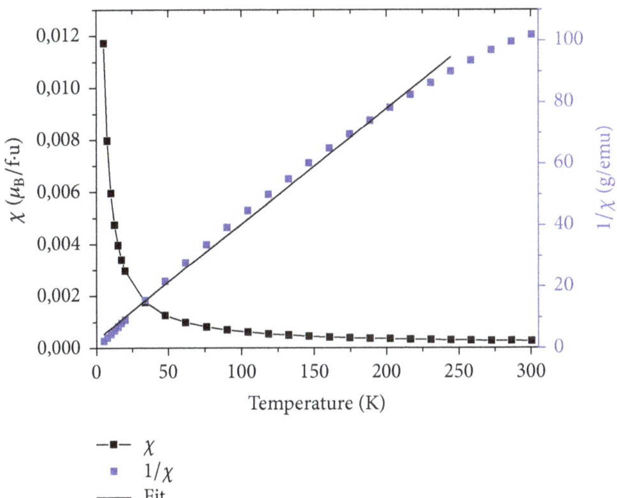

FIGURE 9: The magnetic susceptibility (χ) and inverse magnetic susceptibility ($1/\chi$) measurements as a function temperature of $LiGd_{0.96}Eu_{0.04}(PO_3)_4$.

their structural stability. Indeed, the addition of europium in the host disturbs samples in crystallinity. The response for LGP obeys Curie's Law very well; this is consistent with the ($^8S_{7/2}$) ground state of Gd^{3+}, which has no orbital angular momentum and so is unaffected by crystal field effects. Fitting, the temperature dependence of the inverse of susceptibility χ^{-1} in high temperatures is given by the formula [35]

$$\frac{1}{\chi}=\frac{(T-\theta_p)}{C} \tag{3}$$

TABLE 5: Values of C, μ_{eff}^{the} (μ_B), and μ_{eff}^{exp} (μ_B) for the $LiGd_{(1-x)}Eu_x(PO_3)_4$ (x=0, 0.02 and 0.04) compounds.

	x=0	x=0.02	x=0.04
C (μ_B.KT^{-1})	5.26	5	2.86
μ_{eff}^{the} (μ_B)	7.94	7.86	7.78
μ_{eff}^{exp} (μ_B)	6.50	6.32	4.65

where θ_P is the Weiss temperature and C is the Curie constant given by

$$C \approx \frac{\mu_0 N \mu_{eff}^2}{3K_B} \tag{4}$$

where N is the number of carriers of magnetic moment, μ_0 is the vacuum permeability, K_B is the Boltzmann constant, μ_B is the Bohr magnetron, and μ_{eff} is effective moment of carriers. Samples' structure consists of one magnetic species (i), possessing each a magnetic moment μ_{eff} (i); the magnetic susceptibility is given by the relation:

$$\chi=\mu_0 \frac{n_1\mu_{eff}^2(1)+n_2\mu_{eff}^2(2)+\cdots+n_i\mu_{eff}^2(i)}{3K_B T} \tag{5}$$

Generally, the magnetic moment is determined by

$$\mu_{eff}=g_J\sqrt{J(J+1)} \tag{6}$$

where g_J is the Lande factor and J is the total angular moment. The theoretical effective paramagnetic moment μ_{eff}^{the} for the samples can be calculated by

$$\mu_{eff}^{2\ the} = \left\{x g_{Gd^{3+}}^2 J_{Gd^{3+}}\left(J_{Gd^{3+}}+1\right)\right\}\mu_B^2 \tag{7}$$

Curves of χ^{-1} versus temperature allow deducing μ_{eff}^{exp} values, which are summarized in Table 5. We can notice that the values of μ_{eff}^{the} decrease with the decrease of Gd percentage in the system, due to the important magnetic moment of Gd^{3+} ions (7.94μ_B). The comparison between the theoretical and the experimental effective moment values shows that the former are higher than the latter. This result can be associated with the increase of disorder in the matrix (LGP). On the other side, when the temperature increases to more than 75K, it induces a thermal agitation and causes magnetic moments disorientation of atoms in Eu doped LGP polyphosphates. Consequently, a decrease of paramagnetism is clearly observed (Figure 10).

3.6. Luminescence Properties

3.6.1. Excitation. Excitation spectra of $LiGd(PO_3)_4$, doped with europium (2, 4%) (Figure 11), are measured at 300K under emission with λ_{em}= 591 nm. Figure 11 shows broad band from 254 to 271 nm. These bands are assigned to the charge transfer bands (CTB), resulting from the transfer of an electron from the orbital $2p^6$ of the ligand O^{2-} to the empty

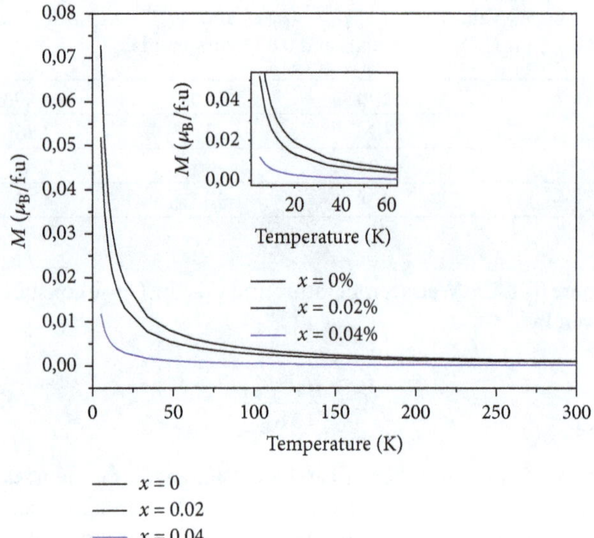

FIGURE 10: Magnetic measurements of $LiGd_{(1-x)}Eu_x(PO_3)_4$ (x=0, 0.02 and 0.04).

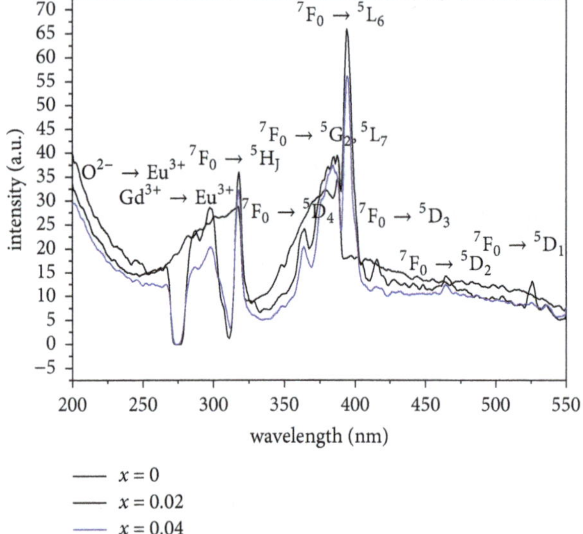

FIGURE 11: Excitation spectra with λ_{em}=591 nm of $LiGd_{(1-x)}Eu_x(PO_3)_4$ (x=0, 0.02 and 0.04) at 300 K.

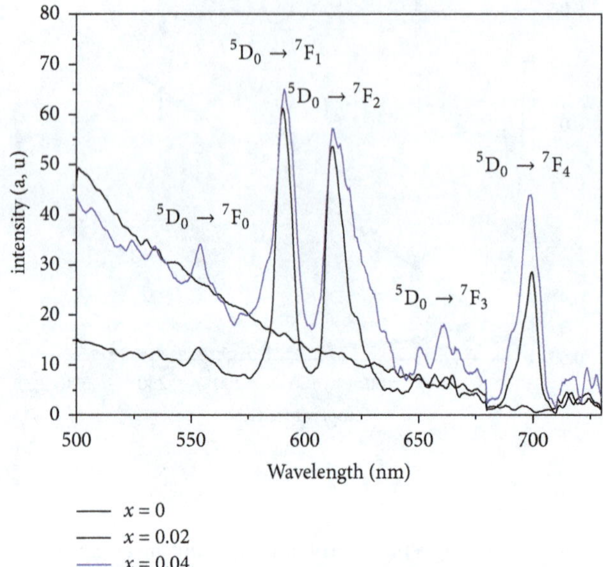

FIGURE 12: Emission spectra with λ_{ex}=394 nm of $LiGd_{(1-x)}Eu_x(PO_3)_4$ (x=0, 0.02 and 0.04) at 300 K.

state of the configuration [Xe]4f^6 of the Eu^{3+} ion (Eu^{3+}-O^{2-} transition). The maximum of the CTB is located at 245 nm. The differences between broadening and positions of the maxima intensities of the CTB in polyphosphate indicate their dependence on the host lattices [36]. This is due to the strong binding of the oxygen ligands in the polyphosphate compound [37].

At low frequencies, several groups of narrow bands in the spectral region 271-310 nm are clearly observed and assigned to $^8S_{7/2}\rightarrow^6H_J$, $^8S_{7/2}\rightarrow^6G_J$, $^8S_{7/2}\rightarrow^6D_J$, $^8S_{7/2}\rightarrow^6I_J$, and $^8S_{7/2}\rightarrow^6P_J$ transitions of the Gd^{3+} ion. $LiYF_4$:Gd^{3+} crystal is used as reference to identify excitation bands, which describe the basis of the detailed energy level scheme proposed for the trivalent gadolinium [38]. The presence of band in the range between 271 and 310 nm indicates the presence of energy transfer between the two rare earths, which occurs from Gd^{3+} to Eu^{3+} in the matrix. However, there is no CTB of Eu^{3+}–O^{2-} or energy transfer band Gd^{3+}-Eu^{3+} above 310 nm. Excitation spectra within the wavelength range of 310–550 nm, show only the intrinsic transitions 4f-4f from the ground state $^7F_0$ to different excited levels (^{5}D, or ^{5}L) of Eu^{3+} ion. These transitions are assigned as follows: $^7F_0\rightarrow$ ^5H_J at 316 nm, $^7F_0\rightarrow^5D_4$ at 362 nm, $^7F_0\rightarrow^5G_2$, $^5L_7$ at 382 nm, $^7F_0\rightarrow^5L_6$ at 393 nm, $^7F_0\rightarrow^5D_3$ at 417 nm, $^7F_0\rightarrow^5D_2$ at 464 nm, and $^7F_0\rightarrow^5D_1$ at 502 nm. All these assignments and wavelengths are given in Table 6. Most of the excitation bands are broadened and some of them overlap together to form a strong band, particularly the band between 369 and 409 nm with FWHM of about 18 nm.

The perfect match of this excitation band with the emission wavelength of NUV In GaN-based LED chips makes these phosphors conveniently useful in white LEDs [39]. Figure 11 shows that the band intensities increase with europium concentration. However, they maintain the same shape and position.

3.6.2. Emission. The emission spectra of condensed phosphates are recorded at ambient temperature (300K) and in the range of 500-750 nm after excitation with λ_{ex}= 394 nm (Figure 12). These spectra present the same shapes, with bands intensity proportional to Eu^{3+} active ion concentration. However, we notice that the undoped $LiGd(PO_3)_4$ polyphosphate does not emit light. The observed emission bands are attributed to the following transitions: $^5D_0\rightarrow^7F_J$ (where J = 0,1,2,3 or 4) of Eu^{3+} ion in the matrix $LiGd(PO_3)_4$ [40, 41].

Table 6: Excitation lines attribution of Eu^{3+} doped $LiGd(PO_3)_4$.

Wavelength (nm)	Attribution
287	$^7F_0 \rightarrow ^5I_6$
294	$^7F_0 \rightarrow ^5F_4$
297	$^7F_0 \rightarrow ^5F_2$
6318	$^7F_0 \rightarrow ^5H_6$
321	$^7F_0 \rightarrow ^5H_4$
328	$^7F_0 \rightarrow ^5H_7$
363	$^7F_0 \rightarrow ^5D_4$
376	$^7F_1 \rightarrow ^5D_4$
373-390	$^7F_0 \rightarrow ^5G_{J(2,4)}$
394	$^7F_0 \rightarrow ^5L_6$
405	$^7F_1 \rightarrow ^5L_6$
416	$^7F_0 \rightarrow ^5D_3$
464	$^7F_0 \rightarrow ^5D_2$
526	$^7F_0 \rightarrow ^5D_1$

Table 7: Emission attribution of Eu^{3+} doped $LiGd(PO_3)_4$.

Transitions	Wavelengths (nm)
$^5D_0 \rightarrow ^7F_0$	554
$^5D_0 \rightarrow ^7F_1$	578-601
$^5D_0 \rightarrow ^7F_2$	604-634
$^5D_0 \rightarrow ^7F_3$	660
$^5D_0 \rightarrow ^7F_4$	686-706

Figure 12 proves the presence of five bands in the emission spectra where the most intense ones are those situated at 578-600 nm ($^5D_0 \rightarrow ^7F_1$) and 604-634 nm ($^5D_0 \rightarrow ^7F_2$). The other emission bands are observed at 554 nm ($^5D_0 \rightarrow ^7F_0$), 660 nm ($^5D_0 \rightarrow ^7F_3$), and 686-706 nm ($^5D_0 \rightarrow ^7F_4$). The corresponding assignments and wavelengths of these emissions are given in Table 7. The relative intensities of the most intense transitions $^5D_0 \rightarrow ^7F_0$, $^7F_1$, $^7F_2$ $^7F_3$ and $^7F_4$ are strongly influenced by the nature of the host and the crystalline environment [42]. Therefore, the dominance of magnetic dipole (MD) transition $^5D_0 \rightarrow ^7F_1$ of Eu^{3+} means that Eu^{3+} occupies a site in the crystal lattice with inversion symmetry. However, in the case of absence of symmetry inversion in the site of Eu^{3+}, the main emission would be the electric dipole (ED) transition $^5D_0 \rightarrow ^7F_2$ [43]. The synthesized polyphosphates showed that orange emission transition ($^5D_0 \rightarrow ^7F_1$) is slightly dominated. This indicates that Eu^{3+} occupies a site in the crystal lattice with symmetry inversion.

4. Conclusion

Polyphosphates of rare earth and alkali metal LGP:Eu^{3+} were successfully synthesized by solid-state reaction at 730°C. XRD patterns proved that the obtained samples crystallize in a monoclinic single phase with space group I 2/a and following cell parameters a= 9.635(3) Å, b= 7.035(3) Å, c= 13.191(3) Å, β= 90.082°, V= 894.214 Å^3, and Z= 4. The synthesized polyphosphates showed a good thermal stability until 940°C. Spectroscopic analyses by IR and Raman spectra confirmed the acyclic zig-zag chain of $(PO_4)^{3-}$ in LGP structure, involving GdO_8 dodecahedra and LiO_4 polyhedra. The magnetic susceptibility carried out on single crystals revealed that the title compounds were paramagnetic between 5 and 300 K. An increase in excitation and emission bands intensities was observed with the increase of europium concentration. The presence of band in the range between 271 and 310 nm in excitation spectra proved the energy transfer process from Gd^{3+} to Eu^{3+}. The dominance of $^5D_0 \rightarrow ^7F_1$ transition in the emission spectra confirms that Eu^{3+} occupies a site in the crystal lattice with symmetry inversion. The change in transition bands intensity proves that LGP phosphates affect europium environment.

References

[1] L. Campayo, F. Audubert, and D. Bernache-Assollant, "Synthesis study of alkaline-bearing rare earth phosphates with rhabdophane structure," *Solid State Ionics*, vol. 176, no. 35-36, pp. 2663–2669, 2005.

[2] J. Zhu, H. Chen, Y. Wang, H. Guan, and X. Xiao, "Structure determination, electronic and optical properties of rubidium holmium polyphosphate $RbHo(PO_3)_4$," *Journal of Molecular Structure*, vol. 1030, pp. 204–208, 2012.

[3] S. Yang and Z. Chen, "The Study on Aging and Degradation Mechanism of Ammonium Polyphosphate in Artificial Accelerated Aging," *Procedia Engineering*, vol. 211, pp. 906–910, 2018.

[4] M. El Masloumi, I. Imaz, J. P. Chaminade et al., "Synthesis, crystal structure and vibrational spectra characterization of $MILa(PO_3)_4$ (MI = Na, Ag)," *Journal of Solid State Chemistry*, vol. 178, pp. 3581–3588, 2005.

[5] J. M. Cole, M. R. Lees, J. A. K. Howard, R. J. Newport, G. A. Saunders, and E. Schönherr, "Crystal Structures and Magnetic Properties of Rare-Earth Ultraphosphates, RP_5O_{14} (R=La, Nd, Sm, Eu, Gd)," *Journal of Solid State Chemistry*, vol. 150, pp. 377–382, 2000.

[6] J. J. Gavaldà, I. Parreu, R. Solé, X. Solans, F. Díaz, and M. Aguiló, "Growth and Structural Characterization of $Rb_3Yb_2(PO_4)_3$: A New Material for Laser and Nonlinear Optical Applications," *J. Chemistry of Materials*, vol. 17, no. 26, pp. 6746–6754, 2005.

[7] M. Ferid and K. Horchani-Naifer, "Structure and ionic conductivity of $NaCeP_2O_7$," *Journal of Solid State Ionics*, vol. 176, pp. 1949–1953, 2005.

[8] K. Jaouadi, N. Zouari, T. Mhiri, and M. Pierrot, "Synthesis and crystal structure of sodium-bismuth polyphosphate $NaBi(PO_3)_4$," *Journal of Crystal Growth*, vol. 273, no. 3-4, pp. 638–645, 2005.

[9] M. Amria, N. Zouaria, T. Mhiri, A. Daoud, and P. Gravereau, "Crystal structure and conductivity investigation of $KDyP_4O_{12}$: a new potassium dysprosium cyclotetraphosphate," *Journal of Molecular Structure*, vol. 782, pp. 16–23, 2006.

[10] D. P. Minh, A. R. Sane, N. Semlal, P. Sharrock, and A. Nzihou, "Alkali polyphosphates as new potential materials for thermal

energy storage," *Journal of Solar Energy*, vol. 157, pp. 277–283, 2017.

[11] Z. Mua, Y. Hub, L. Chen et al., "A reddish orange stoichiometric phosphor $LiEu(PO_3)_4$ for white light-emitting diodes," *Journal of Ceramics International*, vol. 40, pp. 2575–2579, 2014.

[12] J. Zhong, H. Liang, Q. Su, J. Zhou, I. V. Khodyuk, and P. Dorenbos, "Radioluminescence properties of Ce^{3+}-activated $MGd(PO_3)_4$ (M = Li, Na, K, Cs)," *Optical Materials*, vol. 32, no. 2, pp. 378–381, 2009.

[13] J. Gu, J. Zhong, H. Liang, J. Zhang, and Q. Su, "Competitive absorption of Eu^{3+} and Tb^{3+}codoped in $NaGd(PO_3)_4$ phosphors," *Chemical Physics Letters*, vol. 592, pp. 261–264, 2014.

[14] B. Han, H. Liang, H. Ni et al., "Intense red light emission of Eu^{3+}-doped $LiGd(PO_3)_4$ for mercury-free lamps and plasma display panels application," *J. of Optics Express*, vol. 17, pp. 7138–7144, 2009.

[15] T. Shalapska, G. Stryganyuk, P. Demchenko, A. Voloshinovskii, and P. Dorenbos, "Luminescence properties of Ce^{3+}-doped $LiGdP_4O_{12}$ upon vacuum-ultraviolet and x-ray excitation," *Journal of Physics: Condensed Matter*, vol. 21, no. 44, Article ID 445901, 2009.

[16] B. Han, H. Liang, Y. Huang, Y. Tao, and Q. Su, "Vacuum Ultraviolet–Visible Spectroscopic Properties of Tb^{3+} in Li(Y, Gd)$(PO_3)_4$: Tunable Emission, Quantum Cutting, and Energy Transfer," *The Journal of Physical Chemistry*, vol. C114, no. 14, pp. 6770–6777, 2010.

[17] R. Shi, G. Liu, H. Liang, Y. Huang, Y. Tao, and J. Zhang, "Consequences of ET and MMCT on Luminescence of Ce^{3+}, Eu^{3+}, and Tb^{3+}-doped $LiYSiO_4$," *Inorganic Chemistry*, vol. 55, no. 15, pp. 7777–7786, 2016.

[18] A. Krasnikov, T. Shalapska, G. Stryganyuk, A. Voloshinovskii, and S. Zazubovich, "Photoluminescence and energy transfer in Eu^{3+}-doped alkali gadolinium phosphates," *Physica Status Solidi (b) - Basic Solid State Physics*, vol. 250, no. 7, pp. 1418–1425, 2013.

[19] J. Zhong, H. Liang, Q. Su, and J. Zhou, "Luminescence properties of $NaGd(PO_3)_4$:Eu^{3+} and energy transfer from Gd^{3+} to Eu^{3+}," *Applied Physics B: Lasers and Optics*, vol. 98, pp. 139–147, 2010.

[20] W. Xu, J.-G. Dai, Z. Ding, and Y. Wang, "Polyphosphate-modified calcium aluminate cement under normal and elevated temperatures: Phase evolution, microstructure, and mechanical properties," *Ceramics International*, vol. 43, no. 17, pp. 15525–15536, 2017.

[21] B. Liu, Y. Chen, C. Shi, H. Tang, and Y. Tao, "Visible quantum cutting in BaF_2: Gd, Eu downconversion," *Journal of Luminescence*, vol. 101, no. 1-2, pp. 155–159, 2003.

[22] H. Kondo, T. Hirai, and S. Hashimoto, "Dynamical behavior of quantum cutting in alkali gadolinium fluoride phosphors," *Journal of Luminescence*, vol. 108, no. 1-4, pp. 59–63, 2004.

[23] N. Kodama and Y. Watanabe, "Visible quantum cutting through downconversion in Eu^{3+}-doped KGd_3F_{10} and KGd_2F_7 crystals," *Applied Physics Letters*, vol. 84, no. 21, pp. 4141–4143, 2004.

[24] P. Ghosh, S. Tang, and A.-V. Mudring, "Efficient quantum cutting in hexagonal $NaGdF_4$:Eu^{3+}nanorods," *Journal of Materials Chemistry*, vol. 21, no. 24, pp. 8640–8644, 2011.

[25] J. Legendziewicz, M. Guzik, and J. Cybińska, "VUV spectroscopy of double phosphates doped with rare earth ions," *Optical Materials*, vol. 31, no. 3, pp. 567–574, 2009.

[26] S. Hammami, S. Sebai, D. Jegouso, V. Reita, N. C. Boudjada, and A. Megriche, "Magnetic and Photon Cascade Emission of Gd^{3+} of $NaGd(PO_3)_4$ Monocrystal Under Appropriate Synthesis Conditions," *Journal of Physical Chemistry & J Biophysics*, vol. 7, pp. 2161-0398, 2017.

[27] H. Ettis, H. Naïli, and T. Mhiri, "The crystal structure, thermal behaviour and ionic conductivity of a novel lithium gadolinium polyphosphate $LiGd(PO_3)_4$," *Journal of Solid State Chemistry*, vol. 179, no. 10, pp. 3107–3113, 2006.

[28] R. Shannon, "Revised effective ionic radii and systematic studies of interatomic distances in halides and chalcogenides," *Acta Crystallographica Section A: Foundations of Crystallography*, vol. 32, pp. 751–767, 1976.

[29] T. Gholami and M. Salavati-Niasari, "Green facile thermal decomposition synthesis, characterization and electrochemical hydrogen storage characteristics of $ZnAl_2O_4$nanostructure," *International Journal of Hydrogen Energy*, vol. 42, no. 27, pp. 17167–17177, 2017.

[30] S. Sebai, S. Hammami, A. Megriche, D. Zambon, and R. Mahiou, "Synthesis, structural characterization and VUV excited luminescence properties of $Li_xNa_{(1-x)}Sm(PO_3)_4$polyphosphates," *Optical Materials*, vol. 62, pp. 578–583, 2016.

[31] T. Sun, Y. Zhang, P. Shan et al., "Growth, structure, thermal properties and spectroscopic characteristics of Nd^{3+}- doped $KGdP_4O_{12}$ crystal," *PLoS ONE*, vol. 9, no. 6, Article ID e100922, 2014.

[32] W. Rekik, H. Naïli, T. Mhiri, and T. Bataille, "$[NH_3(CH_2)_2NH_3][Co(SO_4)_2(H_2O)_4]$: Chemical preparation, crystal structure, thermal decomposition and magnetic properties," *Materials Research Bulletin*, vol. 43, no. 10, pp. 2709–2718, 2008.

[33] J. Zhang, Z. Zhang, W. Zhang et al., "Polymorphism of $BaTeMo_2O_9$: a new polar polymorph and the phase transformation," *Chemistry of Materials*, vol. 23, no. 16, pp. 3752–3761, 2011.

[34] K. Jaouadi, H. Naili, N. Zouari, T. Mhiri, and A. Daoud, "Synthesis and crystal structure of a new form of potassium–bismuth polyphosphate $KBi(PO_3)_4$," *Alloys Compds*, vol. 354, no. 30, p. 104, 2003.

[35] D. Petrov, B. Angelov, and V. Lovchinov, "Magnetic and XPS studies of lithium lanthanide tetraphosphates $LiLnP_4O_{12}$ (Ln=Nd, Gd, Er)," *Journal of Rare Earths*, vol. 31, no. 5, pp. 485–489, 2013.

[36] S. Hachani, B. Moine, A. El-akrmi, and M. Férid, "Luminescent properties of some ortho- and pentaphosphates doped with Gd^{3+}–Eu^{3+}: Potential phosphors for vacuum ultraviolet excitation," *Optical Materials*, vol. 31, no. 4, pp. 678–684, 2009.

[37] M. Ferhi, K. Horchani-Naifer, and M. Férid, "Spectroscopic properties of Eu^{3+}-doped $KLa(PO_3)_4$ and $LiLa(PO_3)_4$ powders," *Optical Materials*, vol. 34, no. 1, pp. 12–18, 2011.

[38] R. T. Wegh, H. Donker, A. Meijerink, R. J. Lamminmäki, and J. Hölsä, "Vacuum-ultraviolet spectroscopy and quantum cutting for Gd^{3+} in $LiYF_4$," *Physical Review B*, vol. 56, no. 21, pp. 13841–13848, 1997.

[39] M. Abdelhedi, K. Horchani-Naifer, M. Dammak, and M. Ferid, "Structural and spectroscopic properties of pure and doped $LiCe(PO_3)_4$," *Materials Research Bulletin*, vol. 70, pp. 303–308, 2015.

[40] E. E. S. Teotonio, H. F. Brito, M. C. F. C. Felinto, L. C. Thompson, V. G. Young, and O. L. Malta, "Preparation, crystal structure and optical spectroscopy of the rare earth complexes (RE^{3+}=Sm, Eu, Gd and Tb) with 2-thiopheneacetate anion," *Journal of Molecular Structure*, vol. 751, no. 1-3, pp. 85–94, 2005.

[41] A. Mbarek, "Synthesis, structural and optical properties of Eu^{3+}-doped $ALnP_2O_7$(A = Cs, Rb, Tl; Ln = Y, Lu, Tm) pyrophosphates phosphors for solid-state lighting," *Journal of Molecular Structure*, vol. 1138, pp. 149–154, 2017.

[42] D. C. Tuan, R. Olazcuaga, F. Guillen, A. Garcia, B. Moine, and C. Fouassier, "Luminescent properties of Eu^{3+}-doped yttrium or gadolinium phosphates," *J. of Physics*, vol. 123, pp. 259–263, 2005.

[43] M. Gaft, G. Panczer, R. Reisfeld, I. Shinno, B. Champagnon, and G. Boulon, "Laser-induced Eu^{3+} luminescence in zircon $ZrSiO_4$," *Journal of Luminescence*, vol. 87, pp. 1032–1035, 2000.

14

TEMPO-Functionalized Nanoporous Au Nanocomposite for the Electrochemical Detection of H_2O_2

Dongxiao Wen,[1] Qianrui Liu,[2] Ying Cui,[1] Huaixia Yang,[1] and Jinming Kong[2]

[1] *Pharmacy College, Henan University of Chinese Medicine, Zhengzhou 450008, China*
[2] *School of Environmental and Biological Engineering, Nanjing University of Science and Technology, Nanjing 210094, China*

Correspondence should be addressed to Huaixia Yang; yanghuaixia886@163.com and Jinming Kong; j.kong@njust.edu.cn

Academic Editor: Seyyed E. Moradi

A novel nanocomposite of nanoporous gold nanoparticles (np-AuNPs) functionalized with 2,2,6,6-tetramethyl-1-piperidinyloxy radical (TEMPO) was prepared; assembled carboxyl groups on gold nanoporous nanoparticles surface were combined with TEMPO by the "bridge" of carboxylate-zirconium-carboxylate chemistry. SEM images and UV-Vis spectroscopies of np-AuNPs indicated that a safe, sustainable, and simplified one-step dealloying synthesis approach is successful. The TEMPO-np-AuNPs exhibited a good performance for the electrochemical detection of H_2O_2 due to its higher number of electrochemical activity sites and surface area of 7.49 m^2g^{-1} for load bigger amount of TEMPO radicals. The TEMPO-functionalized np-AuNPs have a broad pH range and shorter response time for H_2O_2 catalysis verified by the response of amperometric signal under different pH and time interval. A wide linear range with a detection limit of 7.8×10^{-7} M and a higher sensitivity of 110.403 μA $mM^{-1}cm^{-2}$ were obtained for detecting H_2O_2 at optimal conditions.

1. Introduction

Hydrogen peroxide (H_2O_2) is the smallest and simplest peroxide, which could be generated in many biological processes and cause severe oxidative damage [1–3]. It has been applied in many biosynthetic reactions and also plays an important role in various fields, especially in immune cell activation, vascular remodelling, apoptosis, stomatal closure, root growth, and so on [1, 4]. Because of this, many practical explorations have been done in the detection and monitoring of H_2O_2 in pharmaceutical [1–6], biological [5, 6], clinical [1, 3, 7], chemical, textile [1, 2, 8–10], and food industries [1–6, 10]. The dynamic equilibrium of the production and consumption of H_2O_2 is closely interlinked with the quality of our life; therefore, the development of new materials and techniques for quantitative detection of H_2O_2 at a trace level has presented a significant role in the fundamental studies of diagnostic and monitoring applications. However, most existing techniques suffer from many drawbacks, such as inherent instability, time consuming, poor selectivity, low activity, complicated, and costly immobilization procedures [1–3, 9–11]. Thus, it is necessary to develop new electrochemical sensors overcoming those drawbacks for accurate and sensitive detection of H_2O_2.

A number of new materials such as metals nanomaterials, carbon nanotubes, quantum dots, nanocomposites, and redox substances have been employed for the detection of H_2O_2 [1–9]. In recent studies, nanoscale hollow materials have attracted scientists' attention for their unique porous structural and versatile properties. There have been some reports already about its applications in fuel cell and remarkable catalytic oxidation activities to CO, methanol, hydrazine, and so on [12–15]. Particularly, nanoporous gold nanoparticles (np-AuNPs), one of the most important nanoscale hollow materials, combining their high surface areas with tunable surface plasmon resonance (SPR) features, can serve larger immobilized surface and excellent catalytic sites in chemical reactions and also provide a substrate material to manufacture functionalized nanocomposite [12, 15]. Meanwhile, TEMPO plays a key role in chemistry and biology as an organic redox catalyst for alcohol, aldehydes, and ketones [16–21] and has shown its catalytic oxidation potential to

H_2O_2 due to the electrochemical oxidation of stable nitroxyl radical [22, 23]. Recently, the nitroxyl radical of TEMPO has been reported to attach to solid surfaces by chemical routes to form supramolecular assemblies with high coverage of catalytic radicals [24, 25]. Based on considerations above, in order to take full advantage of their inherent catalytic oxidation activity of the np-AuNPs and the TEMPO, we explore the synergistic effect of TEMPO-functionalized np-AuNPs for the peroxidase-like activity response and use this feature for H_2O_2 detection.

Most nanoporous materials are prepared by using traditional dealloying, template, electrochemical methods, and directed self-assembly [9, 10, 26], and the dealloying approach is mostly adopted to synthesize various nanoporous hollow structures [12–15], using AgCl templates to get the nanoporous gold [12, 26, 27]. However, this traditional dealloying approach is limited by its harsh cumbersome and high-heat conditions [12, 28]. In this work, we prepared the zero-dimensional hollow np-AuNPs via an improved one-step dealloying synthesis in moderate conditions instead of traditional two-step dealloying methods. To improve the peroxidase-like activity of the np-AuNPs and TEMPO, we assembled mercaptoacetic acid (MA) on the surfaces of np-AuNPs/GCE, and then the 4-carboxy-TEMPO was connected to the mercaptoacetic acid through Zr^{4+} as the bridge bond. The np-AuNPs were characterized by scanning electron microscopy (SEM) and UV-Vis spectroscopy, and the TEMPO-contained nanocomposites characterized its electrochemical active area by cyclic voltammetry in different electrolytic buffer. Finally, its amperometric response of TEMPO-np- AuNPs/ GCE to detect H_2O_2 was exploited.

2. Experimental

2.1. Chemical Reagents. Silver nitrate (99%), $HAuCl_4{\cdot}3H_2O$ (>99.9%), hydroquinone (>99%), H_2O_2 (30wt% aqueous), and mercaptoacetic acid (MA) were obtained by Sinopharm Chemical Reagent Co., Ltd. (Beijing). The polyvinylpyrrolidone (PVP, MW 1,300,000) was purchased from J&K Scientific Ltd. Zirconium dichloride oxide octahydrate ($ZrOCl_2{\cdot}8H_2O$) and 4-carboxy-2,2,6,6-tetramethylpiperidine-1-oxyl free radical (4-carboxy-TEMPO) came from Sigma-Aldrich (St. Louis, MO) and TCI (Shanghai) Development Co., Ltd., respectively. 0.1 M PBS supporting electrolyte was prepared by orthophosphoric acid and its salts (0.1 M Na_2HPO_4, 0.1 M NaH_2PO_4), pH=4.0-9.0. All reagents were of analytical grade.

2.2. Apparatus. A scanning electron microscopy (SEM, Hitachi S-4800, Japan) was employed to study the morphology of nanoporous gold nanoparticles and TEMPO-MA-np-AuNPs on the electrode. The UV-Vis absorption peak was carried on a UV-3600 spectrophotometer (SHIMADZU). All data of electrochemical studies were obtained with an electrochemical workstation (CHI 760D, Chenhua, Shanghai) at the room temperature. A platinum wire auxiliary electrode, a saturated calomel reference electrode, and a modified glassy carbon electrode (GCE, Φ=3 mm, as the working electrode) are included in a standard three-electrode cell. All ultrapure water (≥18.25 MΩ) for experiments was obtained from a Millipore Milli-Q water purification system.

2.3. Preparation of Nanoporous Gold Nanoparticles. The nanoporous gold nanoparticles (np-AuNPs) were prepared according to the literature with some modification [12]. Briefly, 160 μL solution of hydroquinone (28 mM) and 10 mM $AgNO_3$ aqueous solution (60 μL) were mixed firstly into 4.5 mL PVP solution (90 mM) in sequence. Then the mixed system came to an equilibrium stirring for 5s, 650 r.m.p. Next, 40 mM $HAuCl_4$ (100 μL) was added dropwise into the system at room temperature under gentle stirring. After 3 min standing, the color of the reaction liquid turned to stable (reddish brown), and then the residual AgCl, which is formed during the reaction, was removed by the additional concentrated NH_4OH (1 mL). Finally, the resulting solution was centrifuged and washed repeatedly with ultrapure water, 7,000 r.p.m. 5 min, to collect the np-AuNPs. Finally, 1.0 mg of np-AuNPs sample was fully suspended in a mixed solution (1.0 mL ethanol, 1.0 mL of 90 mM PVP) by ultrasonication.

2.4. Electrode Fabrication of Tempo-Np-Au NPs/GCE. The bare glassy carbon electrode (GCE) was polished with 0.30μM Al_2O_3 slurries on the chamois leather until a mirror-like surface is obtained. Next, it was ultrasonically cleaned with absolute ethanol and ultrapure water, dried with N_2. Then, the electrode was subjected to cyclic voltammetry (CV) in 0.1 M KCl with the potential of -0.4V and 1.6 V, at a scan rate of 100 mVs^{-1}, until the reproducible cyclic voltammograms were obtained.

A quantity of 1μL of the prepared np-AuNPs suspension was dropped on the surface of freshly pretreated bare GCE above. After the GCE was air dried at room temperature, 20 μL of 0.2 mM MA solution was added onto the np-AuNPs/GCE surface and incubated for 1.0 h at room temperature; by so doing, a MA self-assembled monolayer formed on the surface of np-AuNPs (MA-np-AuNPs/GCE) via chemisorption and the chemistry of formation of MA-SAM on the np-AuNPs surface; a procedure probably involves the oxidative addition to form the S-Au bond by losing the hydrogen as H_2 or H_2O [29, 30]. Subsequently, the TEMPO-np-AuNPs/GCE was prepared by carboxylate- zirconium-carboxylate chemistry [31–33]. After washing the MA-np-AuNPs/GCE with ultrapure water to remove the remaining MA, the electrode was immersed in $ZrOCl_2{\cdot}8H_2O$ 60% ethanol solution (5.0 mM) for 30 min. The modified electrode was taken out and then washed with absolute ethanol, dried with N_2. Next, the electrode was incubated in 20 μL of 0.2 mM 4-carboxy-TEMPO solution for 30 min, followed by rinsing with ultrapure water to remove remaining reactants. TEMPO-functionalized nanoporous Au nanocomposite electrode was prepared. The establishment of this sensor for H_2O_2 detection is depicted in Figure 1.

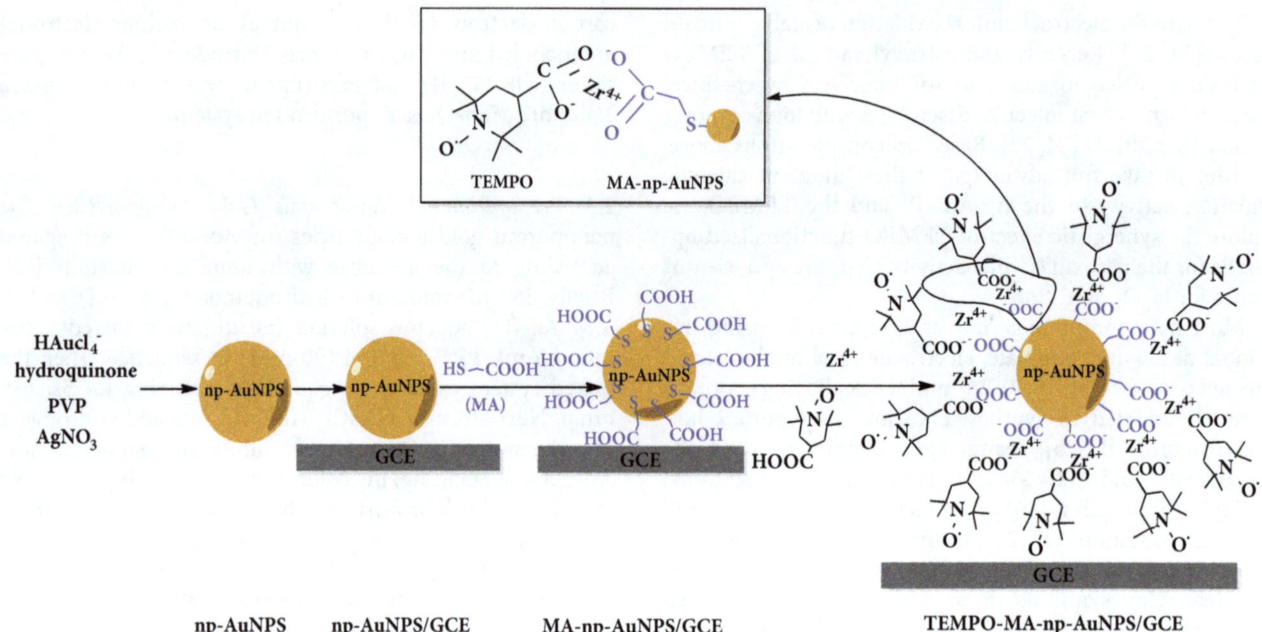

FIGURE 1: Preparation of TEMPO-MA-np-AuNPs/GCE for the electrochemical determination of H_2O_2.

3. Results and Discussion

3.1. Characterization of Modified Electrode by SEM and UV-Vis. By following procedures above, we successfully synthesized the free-standing np-AuNPs under moderate conditions by using a one-step aqueous solution-based approach, which circumvents the limits of stringent and harsh multistep protocols of traditional dealloying approaches. As shown in Figures 2(a) and 2(b), SEM images indicate that the as-synthesized np-AuNPs have a shape of spherical and exhibit an extremely roughened surface, which is consistent with the result of Srikanth's report [12].

As it has been known to all, AgNPs and AuNPs including np-AuNPs are attractive due to their surface plasmon resonance (SPR) properties. Ag^+ AuNPs and np-AuNPs often exhibit different spectral absorptions in the UV-Vis wavelength region [26, 34]. Therefore, we can use absorption spectroscopy to monitor the process of reaction; the UV-Vis spectral absorptions of the as-prepared np-AuNPs are shown In Figure 2(c). A peak appeared at 324 nm, which is the typical UV-Vis absorbance peak of Ag^+ [34]. After 100μL $HAuCl_4$ solution was added in the system, a significant increase of signal appeared at 530 nm, which is caused by the grown of AuNPs on the surface of AgCl templates in the kinetic-controlled process [35]. Subsequently, with NH_4OH added to stop the reaction, the surface plasmon resonance (SPR) peak of 530 nm underwent a red shift to 596 nm-700 nm with longer and broader profile, which belongs to the characteristic peak of porous gold nanoparticles. The typical SPR peak of AuNPs always exhibited around 530 nm; the Ag° broad peak maximum usually occurs at 410 nm, in solution [34]. It was found that, after the centrifuge step to remove residual AgCl, the absorbance peak at 324 nm disappeared, indicating the formation of the np-AuNPs.

3.2. Characterization of Modified Electrode by Cyclic Voltammetry. To characterize the modified electrode TEMPO-np-AuNPs, the typical cyclic voltammetry was performed. The cyclic voltammogram (CV) curves of different modified electrodes in the absence of oxygen in 0.1 M PBS supporting electrolyte (pH = 7.0), at 50 mVs^{-1}, are shown in Figure 3(a). For the bare GCE, there are no redox peaks (A). With the np-AuNPs added on the GCE surface, a broad peak arose at 1.2-1.5V (B) due to the increased effective electroactive area of np-AuNPs. What is more, when MA self-assembled on the np-AuNPs, an obvious peak increase was observed due to the S-Au bond formation (C). By comparing the CVs of C and D, a new pair of well-behaved redox peaks of 0.38 V (cathodic peak) and 0.82 V (anodic oxidation peak) was obtained after 4-carboxy-TEMPO was finally attached to the former-electrode surface. In addition, the decreases of a pair of peaks at -0.1 V/1.2-1.5 V due to the steric hindrance also indicted that 4-carboxy-TEMPO was successfully connected with MA-np-AuNPs via carboxylate-zirconium-carboxylate chemistry.

Further, we characterized the formed np-AuNPs with electroactive activity through CV at different scan rates from 10 mVs^{-1} to 200 mVs^{-1}. As shown in Figure 3(b), the intensities of the electric current increased linearly along with the increase of scan rates (k_1 = 3.934 (± 0.122)), which confirmed that the electroactive np-AuNPs attached on the GCE electrode surface by adsorption according to the theory of homogeneous redox catalysis [22], rather than other interactions. As mentioned above, the CV of TEMPO-np-AuNPs has a new pair of redox peak (0.38 V/0.82 V), while the CV of np-AuNPs did not. As displayed in Figure 2(c), with the scan rates increasing, the current intensity of this redox couple enhanced with a slight redshift; both the I_{pa} (black spots) and I_{pc} (red spots) peak currents were linearly proportional to

FIGURE 2: SEM images (a and b) of nanoporous gold nanoparticles (np-AuNPs) and typical UV-Vis absorption spectra (c) of the reactions during the prepared process of np-AuNPs.

the scan rate. Their slopes, respectively, were k_{pa} = 0.290 (± 0.005) and k_{pc} = -0.456 (± 0.011). Those behaviors illustrate that the modified np-AuNPs (TEMPO-np-AuNPs) had the adsorption with the GCE electrode surface, which are in accordance with the np-AuNPs and literatures [42, 43].

3.3. Electrochemical Measurement of Np-AuNPs Active Surface Areas. The effective areas of different surface modification were estimated by CV method. As shown in Figure 3(d), the anodic oxidation current of the three curves all rose at about 1.2 V and had a typical reduction peak around 0.75 V, which is caused by the reversible redox reaction in 0.5 M H_2SO_4. However, the MA-np-AuNPs (B) exhibit a higher peak at 0.75 V and sustain a large redshift as compared with np-AuNPs (A), indicating that it is hard to be oxidized with H_2SO_4 due to the increased impedance of charge transfer after MA is immobilized on the surface of np-AuNPs. When the TEMPO is chemically modified on MA-np-AuNPs (C), the cathodic peak (around 0.75 V) had an obvious enhancement, which may be caused by the diffusion layer of TEMPO· and is a three-dimensional steady-state. Besides, the rough porous surface of np-AuNPs contributed to generate the multimodal of the curve (e.g., 0.20 V-0.45 V of curve C).

For each experiment, the amount of np-AuNPs used was the same; CV of np-AuNPs in 0.5 M H_2SO_4 (A) were measured to calculate their electroactive surface areas via integrating the area of the gold oxide reduction curve. An electroactive surface area of 7.49 m^2g^{-1} for np-AuNPs is obtained by Randles-Sevcik equation and assuming a specific charge of 450 $\mu C\ cm^{-2}$ for the gold oxide reduction [12, 44]. Notably, the active surface areas of np-AuNPs are higher than that of the commercial Au electrodes, Au nanoparticle, Au

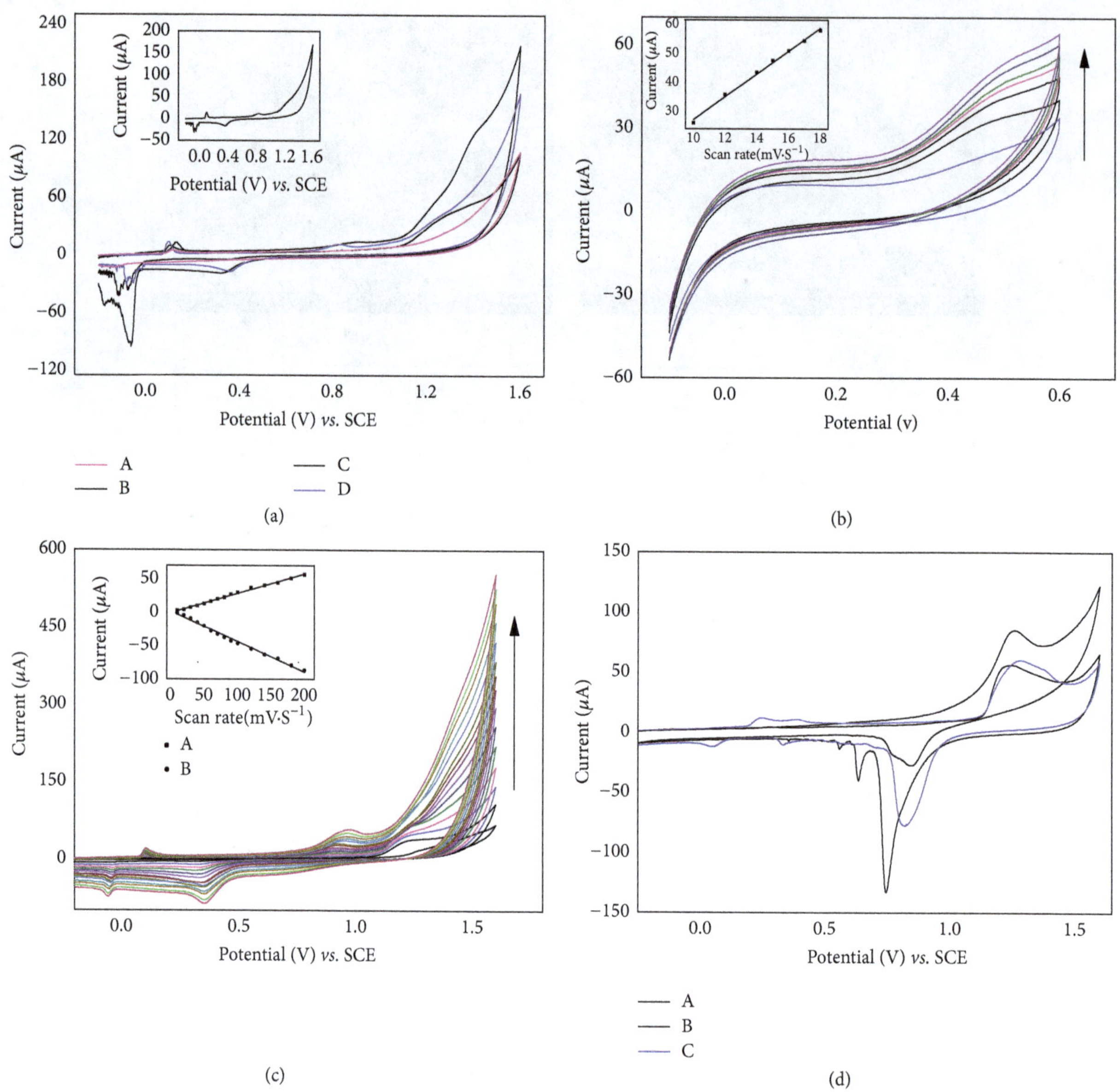

FIGURE 3: **(a)** Different modified GCE in 0.1 M PBS (pH = 7.0) under N_2. Scan rate: $50mVs^{-1}$. (A) Bare GCE, (B) np-AuNPs/GCE, (C) MA-np-AuNPs/GCE, and (D and inset) TEMPO-np-AuNPs/GCE. CVs of np-AuNPs **(b)** and TEMPO-np-AuNPs **(c)** on the GC electrode surface under N_2, 0.1 M PBS (pH = 7.0) at different scan rates from $10mVs^{-1}$ to $200mVs^{-1}$. Inset: **(b)** the linear relationship between the scan rate and the currents at a potential of 0.55V. **(c)** The linear relationship between anodic (black spots, at 0.95 V) and cathodic (red spots, at 0.38 V) peak currents and scan rate. **(d)** CVs of different electrodes in 0.5 M H_2SO_4 under N_2, 0.1 M PBS (pH = 7.0) at a scan rate of 100 mVs^{-1}.(A) np-AuNPs, (B) MA-np-AuNPs, and (C) TEMPO-np-AuNPs.

nanocoral, and other kinds of Au electrodes [45], near 49 times high according to report. The large surface-to-volume ratio of metal nanoparticles is closely related to its high electrical conductivity, catalytic ability, and surface reaction activity such as for the detection of H_2O_2 [2, 45].

3.4. Electrochemical Detection of H_2O_2. As we have known already, the electrochemical behavior of np-AuNPs modified GCE electrode for H_2O_2 was studied by CV. As shown in Figure 4(a), when 3 mM H_2O_2 was added to 0.1 M phosphate buffer saline (pH=7.0) under N_2, compared to the bare GCE, distinct increases of the AuNPs/GCE and np-AuNPs/GCE response currents were observed. Interestingly, the np-AuNPs/GCE had a significant advantage in the difference of the current intensity (ΔI) under the same condition in the presence of H_2O_2. As one of the most famous catalytic materials, Au nanomaterials with different shapes and structures have been suggested good responses for electrocatalytic H_2O_2 [1–3].The current responses toward H_2O_2 concentration over the range of 0.5 μM-100 μM on the np-AuNPs/GCE (Figure 4(b)) were studied. The current intensity had a steady rise with the increasing of H_2O_2 concentration. The HO· radical resulted from H_2O_2 would

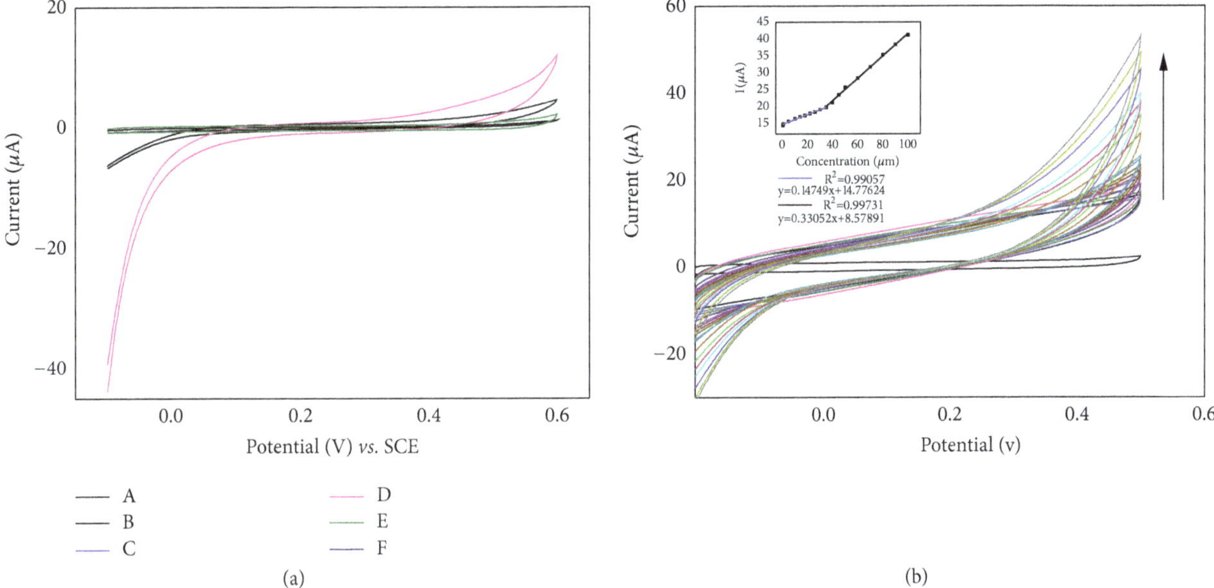

FIGURE 4: **(a)** CVs of electrodes in the absence (A-C) and presence (D-F) of H_2O_2 (3mM) in N_2-saturated 0.1 M PBS (pH = 7.0). Scan rate: $20mVs^{-1}$. (A and E) Bare GCE (curves B and F) AuNPs/GCE (curves C and D) np-AuNPs/GCE. **(b)** CVs of np-AuNPs/GCE in 0.1 M PBS (pH = 7.0) with N_2 toward different concentrations of H_2O_2 over the range of 0.5 μM to 100 μM. Applied potential: 0.55 V. Inset: plot of electrocatalytic current of H_2O_2 versus its concentrations.

be stabilized by the np-AuNPs [3, 46]. The surface property of np-AuNPs may influence the catalytic ability of H_2O_2 and the charge-transfer processes. When the H_2O_2 concentration is below 26 μM, it exhibited a linear correlation of I (μA) = 0.14749 C (μM) +14.77624 (R^2 = 0.99057, RSD = 4.2%) between 0.5 μM and 26 μM. With the increase of stable surface charge transfer for the continuous regeneration of charge-transfer complex [47], it may fill the concave surface of np-AuNPs, and the slope (K_b) of the oxidation catalytic linear response grew to 0.3305 (R^2 = 0.9973) in the range of 26 μM-100 μM. Results of np-AuNPs /GCE show a good catalytic detection activity to H_2O_2.

CV voltammograms of different product on GC electrode in the absence and presence of hydrogen peroxide were shown in Figure 5(a). When H_2O_2 was put into 0.1 M PBS electrolyte buffer, an obvious increase of the peak current at 0.95 V in the CV of TEMPO-np-AuNPs is observed unlike that of smooth curves in other CVs. To ascertain the synergistic peroxidase-like activity between the TEMPO and np-AuNPs, we compared the net current strengths (ΔI) of TEMPO-np-AuNPs and np-AuNPs, where the ΔI refers to the difference strength value within H_2O_2 in and without it. The nitroxide mediator can improve the free diffuse state, which is adjacent to the electrode surface [22, 23]. The ΔI of np-AuNPs sharply rose to 403 μA after it was modified with the TEMPO. All of these manifest that TEMPO-np-AuNPs have a higher potential peroxidase-like activity to H_2O_2. Usually, the H_2O_2 biosensor is constructed via the transfers of the two consecutive single electron transfers (Scheme 1) [22]. First of all, the TEMPO-np-AuNPs undergo a stable reversible one-electron oxidation to produce the intermediate at the TEMPO-functionalized electrode. Finally, the intermediate provides an electrocatalytic electron transfer way to detect H_2O_2.

3.5. Factors Influencing Detection. We all know that time and pH can directly affect the stability of the reaction proceeding and catalytic of enzymes. The analytical performance of the sensor is usually closely linked with the stability of the materials on electrode, which was partially influenced by the time and pH of electrolyte solution. So the pH and time were tested for their electrical catalytic activity to an optimal condition in this work. Firstly, the effect of pH on the potential of electrocatalytic activity on the TEMPO-np-AuNPs was tested at the range of 2.0 to 9.0 under the presence of 5 mM H_2O_2.

As can be seen in Figures 5(b) and 5(c), there is a slight negative shift in the potential catalytic site due to the reversible anodic oxidation of nitroxide derivatives and the inherent peroxidase-like activity response to H_2O_2 of np-AuNPs. Besides, the current intensity at 0.95V has an obvious enhancement under acidic environment and a distinct reduction followed when it was substituted with alkaline buffer (pH > 7.0). Briefly, TEMPO-np-AuNPs can reach the highest catalytic activity to H_2O_2, which is consistent with the result of the one-electron behavior of TEMPO/$TEMPO^+$ [22], under similar physiological conditions (0.1 M PBS, pH 7.0). Results show that pH can inappreciably influence the catalytic activity of TEMPO-np-AuNPs at the range of 5.0 to 8.5. Compared with the pH decided enzyme-modified electrode detection methods, this sensor can advance the application of TEMPO-np-AuNPs to detect H_2O_2 *in vivo* measurements [1].

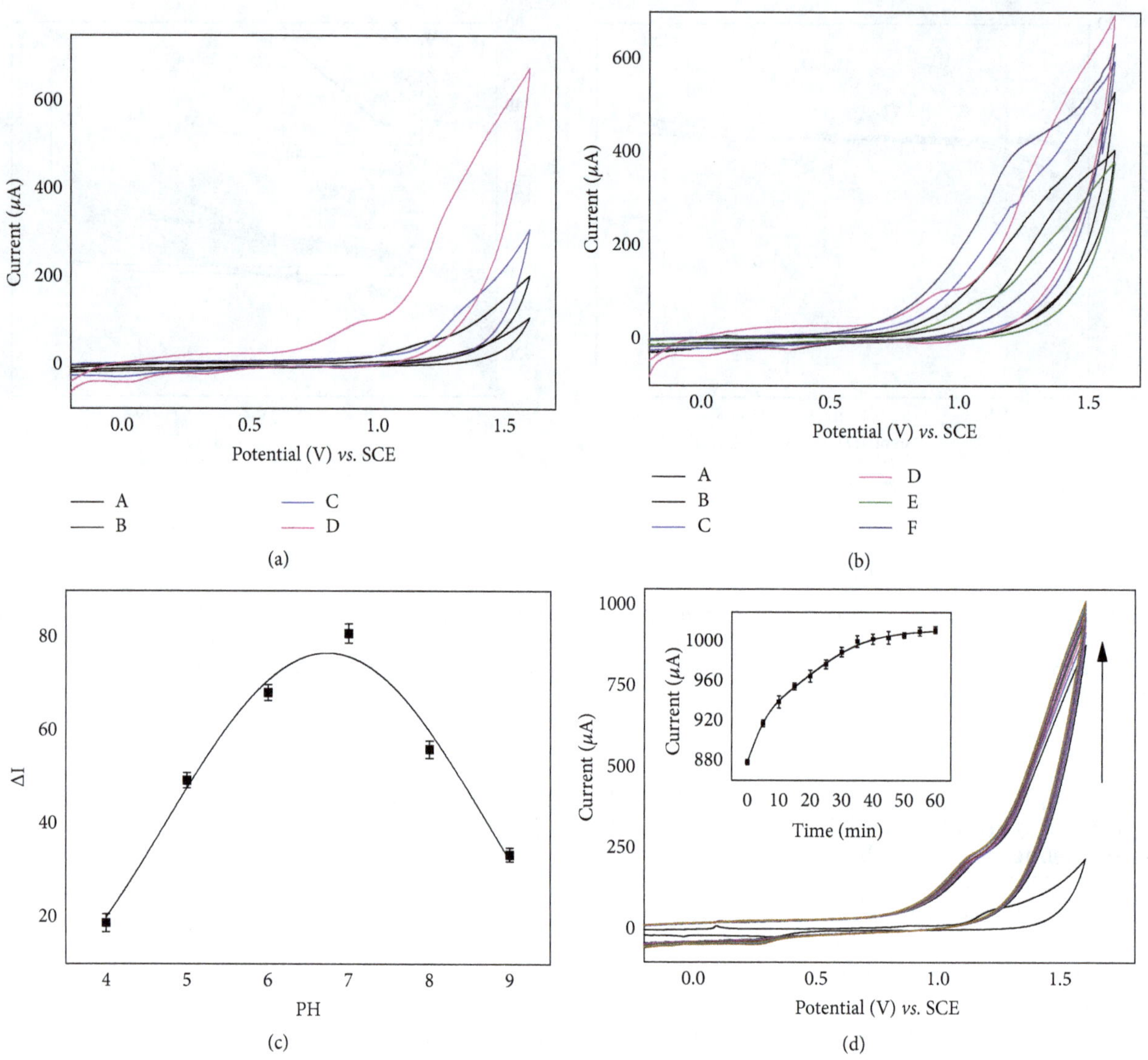

FIGURE 5: **(a)** CVs of different GCE under the absence (A and B) and presence (C and D) of 5mM H_2O_2 in N_2-saturated 0.1 M PBS (pH = 7.0) at a scan rate of $50mVs^{-1}$. (A) Bare GCE (B and C) np-AuNPs/GCE, and (D) 4-carboxy-TEMPO-np-AuNPs/GCE. Optimization of experimental conditions: **(b)** CVs of the TEMPO -np-AuNPs/GCE in different pH values at the range of 2.0~9.0 and **(c)** the relationship between the net current at 0.95V with different pH. **(d)** CVs of TEMPO-np-AuNPs/GCE in 0.1 M PBS (pH = 7.0), in the presence 5 mM H_2O_2, in different times from 0 min to 60 min, N_2-saturated, $50mVs^{-1}$. Inset: the changes of the electric current (1.6 V) with the time increasing.

Further, we investigated the time factor in the presence of 5 mM H_2O_2. Figure 5(d) shows TEMPO-np-AuNPs can catalyze H_2O_2 to produce O_2, immediately, which then comes to an equilibrium state with a maximum current strength in 40 min. By comparing the current strength at 1.6 V, we found that the floating electric potential of 40 min only takes 13.64% in the whole catalytic oxidation process. It is possible that this remaining increase is closely related to the irregular gaps and surface of the np-AuNPs, which may lead to the time retardation to the reversible one-electron behavior of the redox couple TEMPO/$TEMPO^+$ [46]. So we selected 40 min stirring constantly after H_2O_2 was added to the 0.1 M PBS with the buffer system of pH=7.0 to ensure the sufficient current response in the following measurements.

3.6. Steady-State Amperometric Response of H_2O_2. To evaluate the applied potential of TEMPO-MA-np-AuNPs/GCE as peroxidase-like. We investigated the response of the amperometric signal under the optimal condition with the 0.95 V as the applied potential (E_{app}). As shown in Figure 6, the TEMPO-np-AuNPs/GCE not only can achieve a quick steady-state current within 10 s, but also has a stronger response (C) compared to the np-AuNPs/GCE (B).

Moreover, with the addition of same amount H_2O_2 in every interval of 100 s, the TEMPO- np-AuNPs nanocomposite on the GCE exhibited a good linear chronoamperometric response to H_2O_2 from 2.0 μM to 500 μM, which is shown in Figure 6. As what we have seen, when the concentration of H_2O_2 is below 10.0 μM, the slope of the

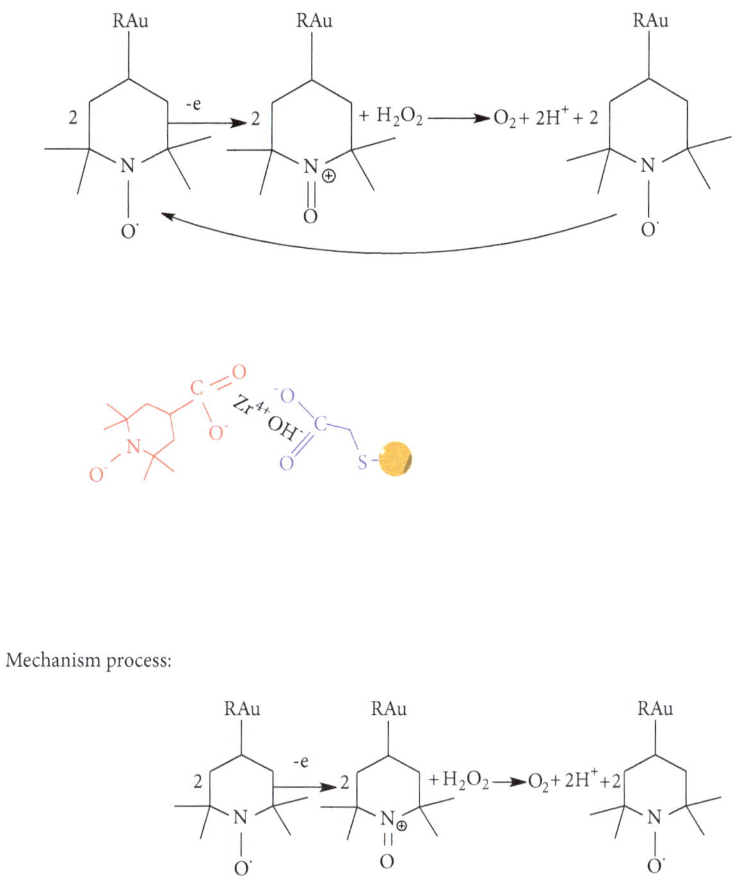

Mechanism process:

SCHEME 1: Equilibrium of single electron between TEMPO-np-AuNPs and H_2O_2.

linear (K_b= 0.66608, R^2=0.98185) is lower than that in a higher concentration range. In other words, the sensitivity of electrocatalytic oxidation activity to H_2O_2 is improved with the concentration of H_2O_2 increased in the solution and there are no substrate inhibition effects occurring at high concentration of H_2O_2. With successive addition of H_2O_2 (n = 5) as shown in Figures 6(b) and 6(c), we can obtain two linear regression equations: (1) below 10.0 μM, y = 0.66608x+4.51516 (R^2 = 0.98185); (2) 10.0 μM to 500 μM, y = 0.1078x+0.6047(R^2 = 0.9998). Error bars represent the standard deviations of five independent measurements. The repeatability of the system is assured by a relative standard deviation (RSD) of 2.8%. We use conventional three times the standard deviation of [LOD = 3(RSD/slope)] to estimate the limit of detection (LOD) of H_2O_2, which is 0.78 μM in our work and lower than some Au nanomaterials and nitroxide derivatives for H_2O_2 detection as listed in Table 1. The results of the stability measurements indicated that np-AuNPs still keep its stable original roughened surface and porous structure after 30 days of storage at 4°C. Moreover, compared with previous results, the TEMPO-np-AuNPs/GCE can retain 98.6% of its initial current response results in the same measurement conditions. Therefore, the TEMPO-np-AuNPs nanocomposite has a remarkable superiority for the electrochemical detection of H_2O_2 over the conventional electrochemical sensing materials and most of the reported Au nanomaterials and nitroxide derivatives probes (Table 1).

3.7. Interference Study. Some coexisting potential electroactive species may affect the sensor response, such as sucrose (SC), glucose (GC), dopamine (DA), and ascorbic acid (AA) [3, 48]. Good selectivity is crucial to ensure and facilitate the accurate assessment for biosensor in a particular application. For better detection *in vivo*, the interference study of TEMPO-np-AuNPs/GCE for the electrochemical detection of H_2O_2 was carried out to evaluate its practical feasibility. Results were shown in Figure 6(d); a discernible slight fluctuation is hard to see after the abundant successive addition of each interfering species (SC/GC/DA/AA), while no obvious interference signal was observed. Notably, the TEMPO-np-AuNPs/GCE had an 18μA response (E_{app} = + 0.95 V) as soon as another 0.1mM H_2O_2 was injected into the complex system of interference. The initial small responses caused by SC, GC, DA, and AA belong to normal current fluctuations, which is a deductible interference compared with that caused by H_2O_2. TEMPO-np-AuNPs/GCE exhibited an acceptable selectivity towards the practical *in vivo* electrochemical detection of H_2O_2.

Table 1: Comparison of recent Au nanomaterials and nitroxide derivatives for H_2O_2 detection.

Electrode design	L. R. (M)	D.L. (μM)	Stability	Reference
HRP/Cys/AuNP/ITO	$8.0\times10^{-6}\sim3.0\times10^{-3}$	2.00	83% (12 weeks)	[6]
HRP/$CaCO_3$-AuNPs/ATP/Au	$5.0\times10^{-7}\sim5.2\times10^{-3}$	0.10	96.4% (30 days)	[36]
HRP-nano-Au	$1.2\times10^{-5}\sim1.1\times10^{-3}$	6.10	75% (5 weeks)	[37]
HRP/AuNPs/poly(St-co-AA)	$8.0\times10^{-6}\sim7.0\times10^{-3}$	4.00	97.8 % (60 days)	[38]
Au/CeO_2 nanocomposite	$0\sim3.0\times10^{-4}$	5.00	---	[9]
AuNPs-N-GQDs/GC	$2.5\times10^{-5}\sim1.3\times10^{-2}$	0.12	89%, 3 weeks	[10]
Poly(BCB)/Au-NPs/GCE	$6.0\times10^{-5}\sim1.0\times10^{-2}$	0.23	95 %(2 week)	[39]
GC/MTMOS-$Au^{73}Ag^{27}$	$1.0\times10^{-5}\sim7.0\times10^{-5}$	1.00	---	[40]
Hb/Au nanoflowers/CNTs/GCE	$1.0\times10^{-6}\sim6.0\times10^{-4}$	7.30	---	[41]
TEMPO/GCE	$1.0\times10^{-7}\sim1.0\times10^{-8}$	0.05	100%,3 months	[22]
ChOx/TEMPO/GCE	$2\times10^{-5}\sim 2.5\times10^{-3}$	20.0	---	[23]
TEMPO-MA-np-AuNPs/GCE	$2.5\times10^{-6}\sim5.0\times10^{-4}$	0.78	98.6%(30 days)	**This work**

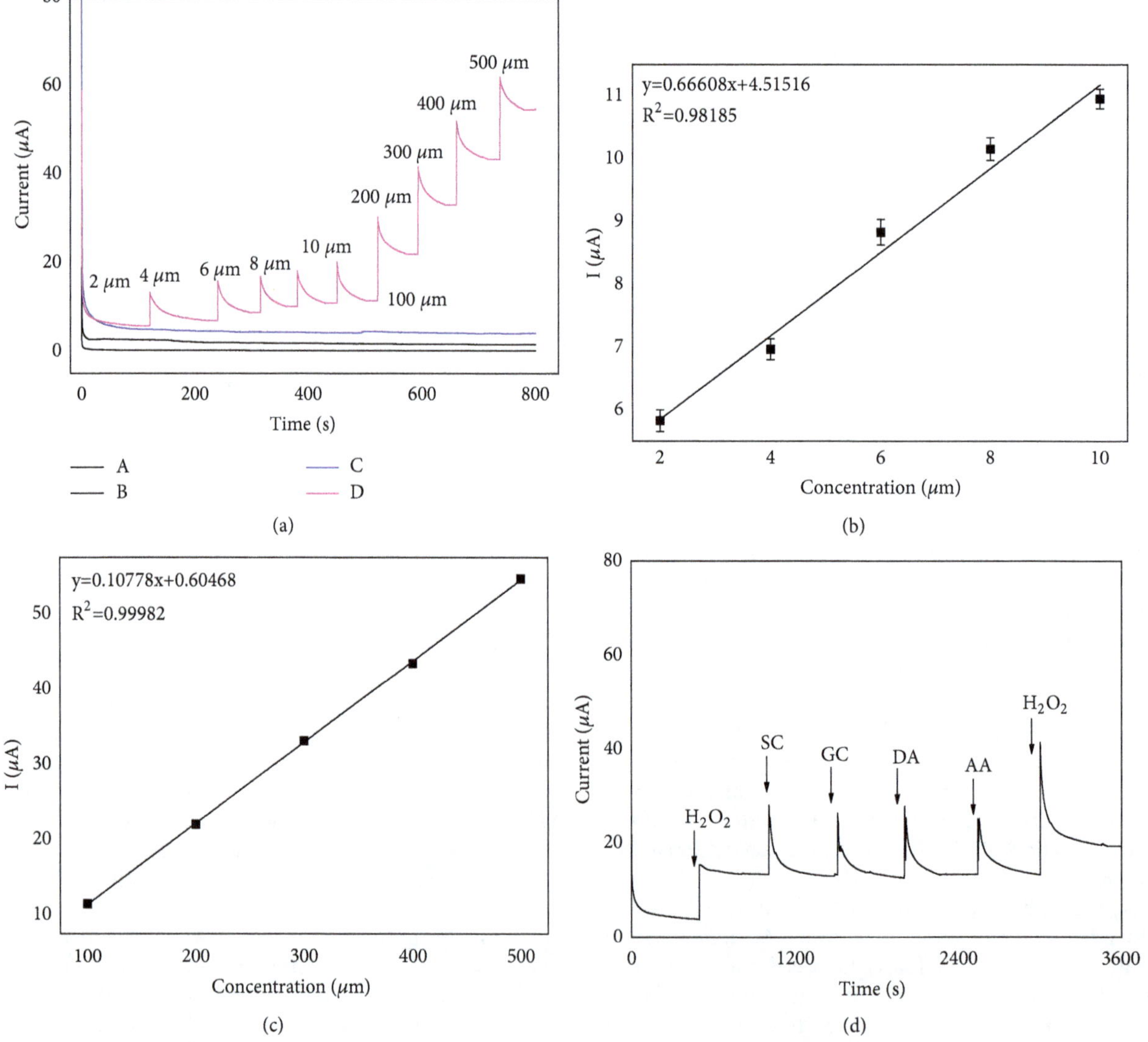

Figure 6: **(a)** Chronoamperometric responses observed at (A) bare GCE, (B) np-AuNPs/GCE, (C) TEMPO-np-AuNPs/GCE, and (D) TEMPO-np-AuNPs/GCE after successively injecting H_2O_2, in 0.1 M PBS (pH=7.0). Applied potential: 1.10 V and the calibration plot between the oxidation current and the H_2O_2 concentration **(b and c). (d)** Amperometric response of H_2O_2 and interferants at TEMPO-np-AuNPs/GCE at 1.10V in PBS (0.1 M, pH = 7.0). Injection sequence: 0.1mM H_2O_2, 100 mM SC, 100 mM GC, 100 mM DA, 100 mM AA, and 0.2 mM H_2O_2.

4. Conclusions

In this work, we demonstrated a strategy for producing large specific surface area nanoporous gold nanoparticles and manufactured the TEMPO-functionalized np-AuNPs nanocomposite; the electrooxidation to H_2O_2 with a high density of radicals on the TEMPO-np-AuNPs surface was also investigated. It is worth noting that this new nonenzymatic H_2O_2 probe is prepared under a gentle, secure, low-cost, and simple procedure. When the np-AuNPs are combined with 4-carboxy-TEMPO, the advantages of their unique properties for the electrochemical detection of H_2O_2 come to a double effective enhancement. Compared the peroxidase-like activity based on the direct electron transfer of the TEMPO-np-AuNPs/GCE to the electrochemical hydrogen peroxide biosensors of TEMPO-based ligand, we obtained a wide linear range for H_2O_2 detection. The enzyme-like activity with low detection limit, high sensitivity, low-cost, anti-interference, good reproducibility, and stability of this nanocomposite with TEMPO-based ligand make a contribution to improve the detection current signal of H_2O_2. Furthermore, this TEMPO-functionalized np-AuNPs nanocomposite study plays a significant role in facilitating the research of biosensors, gold nanomedicine, catalysis, or cancer therapy.

Acknowledgments

This work was supported by the National Natural Science Foundation of China (Grant no. 21575066).

References

[1] W. Chen, S. Cai, Q.-Q. Ren, W. Wen, and Y.-D. Zhao, "Recent advances in electrochemical sensing for hydrogen peroxide: A review," *Analyst*, vol. 137, no. 1, pp. 49–58, 2012.

[2] S. Chen, R. Yuan, Y. Chai, and F. Hu, "Electrochemical sensing of hydrogen peroxide using metal nanoparticles: A review," *Microchimica Acta*, vol. 180, no. 1-2, pp. 15–32, 2013.

[3] Z. Miao, D. Zhang, and Q. Chen, "Non-enzymatic hydrogen peroxide sensors based on multi-wall carbon nanotube/Pt nanoparticle nanohybrids," *Materials* , vol. 7, no. 4, pp. 2945–2955, 2014.

[4] M. Giorgio, M. Trinei, E. Migliaccio, and P. G. Pelicci, "Hydrogen peroxide: a metabolic by-product or a common mediator of ageing signals?" *Nature Reviews Molecular Cell Biology*, vol. 8, no. 9, pp. 722–728, 2007.

[5] K.-J. Huang, D.-J. Niu, X. Liu et al., "Direct electrochemistry of catalase at amine-functionalized graphene/gold nanoparticles composite film for hydrogen peroxide sensor," *Electrochimica Acta*, vol. 56, no. 7, pp. 2947–2953, 2011.

[6] J. Wang, L. Wang, J. Di, and Y. Tu, "Electrodeposition of gold nanoparticles on indium/tin oxide electrode for fabrication of a disposable hydrogen peroxide biosensor," *Talanta*, vol. 77, no. 4, pp. 1454–1459, 2009.

[7] Y.-D. Lee, C.-K. Lim, A. Singh et al., "Dye/peroxalate aggregated nanoparticles with enhanced and tunable chemiluminescence for biomedical imaging of hydrogen peroxide," *ACS Nano*, vol. 6, no. 8, pp. 6759–6766, 2012.

[8] L. Gu, N. Luo, and G. H. Miley, "Cathode electrocatalyst selection and deposition for a direct borohydride/hydrogen peroxide fuel cell," *Journal of Power Sources*, vol. 173, no. 1, pp. 77–85, 2007.

[9] C. Ampelli, S. G. Leonardi, A. Bonavita et al., "Electrochemical H2O2 sensors based on Au/CeO2 nanoparticles for industrial applications," *Chemical Engineering Transactions*, vol. 43, pp. 733–738, 2015.

[10] J. Jian and C. Wei, "In situ growth of surfactant-free gold nanoparticles on nitrogen-doped graphene quantum dots for electrochemical detection of hydrogen peroxide in biological environments," *Analytical Chemistry*, vol. 87, no. 3, pp. 1903–1910, 2015.

[11] A. A. Ensafi, N. Ahmadi, B. Rezaei, and M. M. Abarghoui, "A new electrochemical sensor for the simultaneous determination of acetaminophen and codeine based on porous silicon/palladium nanostructure," *Talanta*, vol. 134, pp. 745–753, 2015.

[12] S. Pedireddy, H. K. Lee, W. W. Tjiu et al., "One-step synthesis of zero-dimensional hollow nanoporous gold nanoparticles with enhanced methanol electrooxidation performance," *Nature Communications*, vol. 5, article no. 4947, 2014.

[13] C. Boyer, M. R. Whittaker, C. Nouvel, and T. P. Davis, "Synthesis of hollow polymer nanocapsules exploiting gold nanoparticles as sacrificial templates," *Macromolecules* , vol. 43, no. 4, pp. 1792–1799, 2010.

[14] C. Zhang, A. Zhu, R. Huang, Q. Zhang, and Q. Liu, "Hollow nanoporous Au/Pt core-shell catalysts with nanochannels and enhanced activities towards electro-oxidation of methanol and ethanol," *International Journal of Hydrogen Energy*, vol. 39, no. 16, pp. 8246–8256, 2014.

[15] Z. Liu, A. Nemec-Bakk, N. Khaper, and A. Chen, "Sensitive Electrochemical Detection of Nitric Oxide Release from Cardiac and Cancer Cells via a Hierarchical Nanoporous Gold Microelectrode," *Analytical Chemistry*, vol. 89, no. 15, pp. 8036–8043, 2017.

[16] R. A. Green, J. T. Hill-Cousins, R. C. D. Brown, D. Pletcher, and S. G. Leach, "A voltammetric study of the 2,2,6,6-tetramethylpiperidin-1-oxyl (TEMPO) mediated oxidation of benzyl alcohol in tert-butanol/water," *Electrochimica Acta*, vol. 113, pp. 550–556, 2013.

[17] M. Rafiee, K. C. Miles, and S. S. Stahl, "Electrocatalytic Alcohol Oxidation with TEMPO and Bicyclic Nitroxyl Derivatives: Driving Force Trumps Steric Effects," *Journal of the American Chemical Society*, vol. 137, no. 46, pp. 14751–14757, 2015.

[18] B. Karimi, M. Rafiee, S. Alizadeh, and H. Vali, "Eco-friendly electrocatalytic oxidation of alcohols on a novel electro generated TEMPO-functionalized MCM-41 modified electrode," *Green Chemistry*, vol. 17, no. 2, pp. 991–1000, 2015.

[19] P.-Y. Blanchard, O. Alévêque, T. Breton, and E. Levillain, "TEMPO mixed SAMs: Electrocatalytic efficiency versus surface coverage," *Langmuir*, vol. 28, no. 38, pp. 13741–13745, 2012.

[20] R. Ciriminna, G. Palmisano, and M. Pagliaro, "Electrodes functionalized with the 2,2,6,6-tetramethylpiperidinyloxy radical for the waste-free oxidation of alcohols," *ChemCatChem*, vol. 7, no. 4, pp. 552–558, 2015.

[21] S. Eken Korkut, D. Akyüz, K. Özdoğan, Y. Yerli, A. Koca, and M. K. Şener, "TEMPO-functionalized zinc phthalocyanine: Synthesis, magnetic properties, and its utility for electrochemical

sensing of ascorbic acid," *Dalton Transactions*, vol. 45, no. 7, pp. 3086–3092, 2016.

[22] B. Limoges and C. Degrand, "Electrocatalytic oxidation of hydrogen peroxide by nitroxyl radicals," *Journal of Electroanalytical Chemistry*, vol. 422, no. 1-2, pp. 7–12, 1997.

[23] S. Abdellaoui, K. L. Knoche, K. Lim, D. P. Hickey, and S. D. Minteer, "TEMPO as a Promising Electrocatalyst for the Electrochemical Oxidation of Hydrogen Peroxide in Bioelectronic Applications," *Journal of The Electrochemical Society*, vol. 163, no. 4, pp. H3001–H3005, 2016.

[24] V. Lloveras, E. Badetti, V. Chechik, and J. Vidal-Gancedo, "Magnetic interactions in Spin-labeled Au nanoparticles," *The Journal of Physical Chemistry C*, vol. 118, no. 37, pp. 21622–21629, 2014.

[25] L. Zhang, Y. B. Vogel, B. B. Noble et al., "TEMPO Monolayers on Si(100) Electrodes: Electrostatic Effects by the Electrolyte and Semiconductor Space-Charge on the Electroactivity of a Persistent Radical," *Journal of the American Chemical Society*, vol. 138, no. 30, pp. 9611–9619, 2016.

[26] Y. Sun and Y. Xia, "Increased sensitivity of surface plasmon resonance of gold nanoshells compared to that of gold solid colloids in response to environmental changes," *Analytical Chemistry*, vol. 74, no. 20, pp. 5297–5305, 2002.

[27] W. S. Chew, S. Pedireddy, Y. H. Lee et al., "Nanoporous Gold Nanoframes with Minimalistic Architectures: Lower Porosity Generates Stronger Surface-Enhanced Raman Scattering Capabilities," *Chemistry of Materials*, vol. 27, no. 22, pp. 7827–7834, 2015.

[28] A. Wittstock, J. Biener, and M. Bäumer, "Nanoporous gold: A new material for catalytic and sensor applications," *Physical Chemistry Chemical Physics*, vol. 12, no. 40, pp. 12919–12930, 2010.

[29] P. E. Laibinis, M. A. Fox, J. P. Folkers, and G. M. Whitesides, "Comparisons of Self-Assembled Monolayers on Silver and Gold: Mixed Monolayers Derived from HS(CH2)21X and HS(CH2)10Y (X, Y = CH3, CH2OH) Have Similar Properties," *Langmuir*, vol. 7, no. 12, pp. 3167–3173, 1991.

[30] A. Badia, S. Singh, L. Demers, L. Cuccia, G. R. Brown, and R. B. Lennox, "Self-assembled monolayers on gold nanoparticles," *Chemistry - A European Journal*, vol. 2, no. 3, pp. 359–363, 1996.

[31] J. Kong, A. R. Ferhan, X. Chen, L. Zhang, and N. Balasubramanian, "Polysaccharide templated silver nanowire for ultrasensitive electrical detection of nucleic acids," *Analytical Chemistry*, vol. 80, no. 19, pp. 7213–7217, 2008.

[32] M. Mazur, P. Krysiński, and G. J. Blanchard, "Use of zirconium-phosphate-carbonate chemistry to immobilize polycyclic aromatic hydrocarbons on boron-doped diamond," *Langmuir*, vol. 21, no. 19, pp. 8802–8808, 2005.

[33] Q. Hu, W. Hu, J. Kong, and X. Zhang, "PNA-based DNA assay with attomolar detection limit based on polygalacturonic acid mediated in-situ deposition of metallic silver on a gold electrode," *Microchimica Acta*, vol. 182, no. 1-2, pp. 427–434, 2015.

[34] D. Wan, H.-L. Chen, Y.-S. Lin, S.-Y. Chuang, J. Shieh, and S.-H. Chen, "Using spectroscopic ellipsometry to characterize and apply the optical constants of hollow gold nanoparticles," *ACS Nano*, vol. 3, no. 4, pp. 960–970, 2009.

[35] P. Silvert, R. Herrera-Urbina, and K. Tekaia-Elhsissen, "Preparation of colloidal silver dispersions by the polyol process," *Journal of Materials Chemistry*, vol. 7, no. 2, pp. 293–299.

[36] F. Li, Y. Feng, Z. Wang, L. Yang, L. Zhuo, and B. Tang, "Direct electrochemistry of horseradish peroxidase immobilized on the layered calcium carbonate-gold nanoparticles inorganic hybrid composite," *Biosensors and Bioelectronics*, vol. 25, no. 10, pp. 2244–2248, 2010.

[37] C. X. Leia, S. Q. Huc, N. Gao, G. L. Shen, and R. Q. Yu, "An amperometric hydrogen peroxide biosensor based on immobilizing horseradish peroxidase to a nano-Au monolayer supported by sol-gel derived carbon ceramic electrode," *Bioelectrochemistry*, vol. 65, no. 1, pp. 33–39, 2004.

[38] S. Xu, G. Tu, B. Peng, and X. Han, "Self-assembling gold nanoparticles on thiol-functionalized poly(styrene-co-acrylic acid) nanospheres for fabrication of a mediatorless biosensor," *Analytica Chimica Acta*, vol. 570, no. 2, pp. 151–157, 2006.

[39] S. A. Kumar, S.-F. Wang, and Y.-T. Chang, "Poly(BCB)/Au-nanoparticles hybrid film modified electrode: Preparation, characterization and its application as a non-enzymatic sensor," *Thin Solid Films*, vol. 518, no. 20, pp. 5832–5838, 2010.

[40] S. Manivannan and R. Ramaraj, "Core-shell Au/Ag nanoparticles embedded in silicate sol-gel network for sensor application towards hydrogen peroxide," *Journal of Chemical Sciences*, vol. 121, no. 5, pp. 735–743, 2009.

[41] Y.-C. Gao, K. Xi, W.-N. Wang, X.-D. Jia, and J.-J. Zhu, "A novel biosensor based on a gold nanoflowers/hemoglobin/carbon nanotubes modified electrode," *Analytical Methods*, vol. 3, no. 10, pp. 2387–2391, 2011.

[42] J.-J. Feng, G. Zhao, J.-J. Xu, and H.-Y. Chen, "Direct electrochemistry and electrocatalysis of heme proteins immobilized on gold nanoparticles stabilized by chitosan," *Analytical Biochemistry*, vol. 342, no. 2, pp. 280–286, 2005.

[43] F. Meng, X. Yan, J. Liu, J. Gu, and Z. Zou, "Nanoporous gold as non-enzymatic sensor for hydrogen peroxide," *Electrochimica Acta*, vol. 56, no. 12, pp. 4657–4662, 2011.

[44] F. Jia, C. Yu, K. Deng, and L. Zhang, "Nanoporous metal (Cu, Ag, Au) films with high surface area: general fabrication and preliminary electrochemical performance," *The Journal of Physical Chemistry C*, vol. 111, no. 24, pp. 8424–8431, 2007.

[45] T.-M. Cheng, T.-K. Huang, H.-K. Lin et al., "(110)-Exposed gold nanocoral electrode as low onset potential selective glucose sensor," *ACS Applied Materials & Interfaces*, vol. 2, no. 10, pp. 2773–2780, 2010.

[46] J. Yun, B. Li, and R. Cao, "Positively-charged gold nanoparticles as peroxidiase mimic and their application in hydrogen peroxide and glucose detection," *Chemical Communications*, vol. 46, no. 42, pp. 8017–8019, 2010.

[47] H. Cui, Z.-F. Zhang, M.-J. Shi, Y. Xu, and Y.-L. Wu, "Light emission of gold nanoparticles induced by the reaction of bis(2,4,6-trichlorophenyl) oxalate and hydrogen peroxide," *Analytical Chemistry*, vol. 77, no. 19, pp. 6402–6406, 2005.

[48] G. Yin, L. Xing, X.-J. Ma, and J. Wan, "Non-enzymatic hydrogen peroxide sensor based on a nanoporous gold electrode modified with platinum nanoparticles," *Chemical Papers*, vol. 68, no. 4, pp. 435–441, 2014.

15

Development of RP-HPLC Method for the Simultaneous Quantitation of Levamisole and Albendazole: Application to Assay Validation

S. Sowjanya[1] **and Ch. Devadasu**[2]

[1]*Department of Pharmaceutical Chemistry, Vishwa Bharathi College of Pharmaceutical Sciences, Perecherla, Guntur 522 009, Andhra Pradesh, India*

[2]*Department of Pharmaceutical Analysis & Quality Assurance, Vignan Pharmacy College, Vadlamudi, Guntur 522 213, Andhra Pradesh, India*

Correspondence should be addressed to S. Sowjanya; sowji.pharmacist@gmail.com

Academic Editor: Federica Pellati

A reverse phase high-performance liquid chromatographic (RP-HPLC) method was developed and validated for simultaneous estimation of levamisole and albendazole in drug substance and in its combinational dosage form. The analysis was carried out using *Inertsil ODS* C_{18} (4.6 x 150 mm, 5 μm) column, and the separation was carried out using a mobile phase containing a buffer of pH 3.5 and acetonitrile (70:30 v/v) pumped at a flow rate of 1.0 mL/min with variable wavelength UV-detection at 224 nm. Both the drugs were well resolved in the stationary phase and the retention times were 2.350 min and 4.055 for levamisole and albendazole, respectively. The method was validated and shown to be linear in the concentration range of 15-45μg/ml and 40-120μg/ml for levamisole and albendazole, respectively. The limit of detection (LOD) and limit of quantification (LOQ) were determined based on standard deviation of the y-intercept and the slope of the calibration curve. LOD and LOQ values were 2.08μg/ml and 6.03μg/ml for levamisole and 3.15μg/ml and 10.40μg/ml for albendazole, respectively. The accuracy of the method was assessed by adding known amount of standard solution (75 %, 100 %, and 125% of the sample concentration) to the preanalyzed sample solution of 100% concentration. All the samples were prepared and analyzed in triplicate. The percentage mean recovery by standard addition experiments of levamisole and albendazole is 99.66% and 98.73%, respectively.

1. Introduction

Levamisole hydrochloride salt is an anthelmintic and immunomodulator belonging to a class of synthetic imidazothiazole derivatives. It was discovered at Janssen Pharmaceutica in 1966. Levamisole has been used in humans to treat parasitic worm infections and has been studied in combination with other forms of chemotherapy for colon cancer, melanoma, and head and neck cancer. The drug was withdrawn from the US and Canadian markets in 2000 and 2003, respectively, due to the risk of serious side effects and the availability of more effective replacement medications. Levamisole is chemically (6S)-6-phenyl-2H,3H,5H,6H-imidazo[2,1-b][1,3]thiazole. It is indicated for adjuvant treatment in combination with fluorouracil after surgical resection in patients with Dukes' stage C colon cancer and also used to treat malignant melanoma and head/neck cancer. Levamisole was originally used as an antihelminthic to treat worm infestations in both humans and animals. Levamisole is official in Indian Pharmacopoeia[1], European Pharmacopoeia [2], British Pharmacopoeia[3], and United States Pharmacopoeia [4].

Albendazole is a member of the benzimidazole compounds used as a drug indicated for the treatment of a variety of worm infestations. Albendazole was first discovered at the SmithKline Animal Health Laboratories in 1972. It is a broad spectrum anthelmintic that is effective against roundworms, tapeworms, and flukes of domestic animals

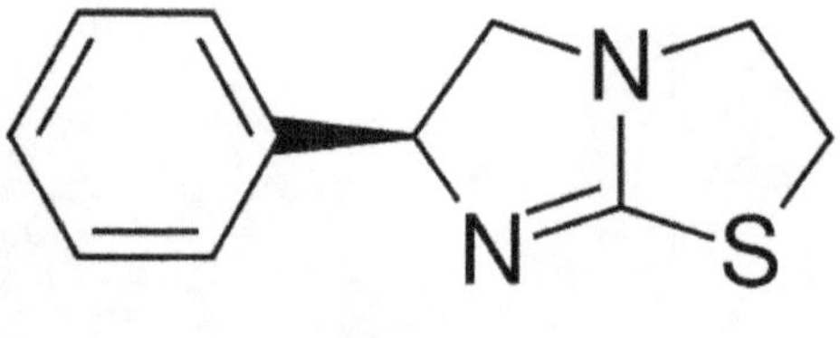

(a) 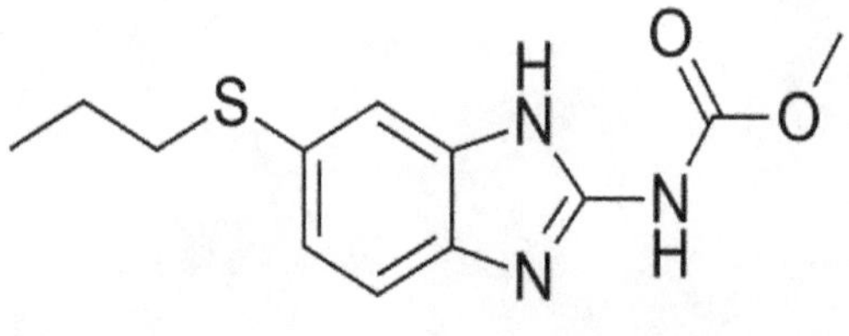 (b)

FIGURE 1: Chemical structures of (a) levamisole hydrochloride and (b) albendazole.

and humans. Albendazole, methyl N-[6-(propylsulfanyl)-1H-1,3-benzodiazol-2-yl]carbamate, is used for the treatment of parenchymal neurocysticercosis due to active lesions caused by larval forms of the pork tapeworm, *Taenia solium*, and for the treatment of cystic hydatid disease of the liver, lung, and peritoneum, caused by the larval form of the dog tapeworm, *Echinococcus granulosus*. Albendazole is official in the British Pharmacopoeia [5], the European Pharmacopoeia [6], and the United States Pharmacopoeia [7]. The chemical structures of the drug candidates selected for study were presented in Figure 1.

A thorough literature survey reveals that there were few analytical methods for estimation of albendazole, with other anthelmintic agents in various samples depending on the study employed. Some of the advanced instrumental methods have been developed earlier to determine the levamisole residues in cattle and swine livers by capillary gas chromatography-electron impact mass spectrometry by Stout SJ et al. [8], high-performance liquid chromatographic/gas chromatographic/tandem ion trap mass spectrometric determination of levamisole in milk by Chappell CG et al. [9], determination of levamisole in animal tissues using liquid chromatography-thermospray mass spectrometry by Cannavan A et al. [10], a liquid chromatographic-electrospray tandem mass spectrometric multiresidue method for anthelmintics in milk by Ruyck H et al. [11], quantitative analysis of levamisole in porcine tissues by high-performance liquid chromatography combined with atmospheric pressure chemical ionization mass spectrometry by Marc C et al. [12], liquid chromatographic determination of levamisole in animal plasma by Baere S et al. [13], determination of benzimidazoles and levamisole residues in milk by liquid chromatography-mass spectrometry by Piotr J et al. [14], a sensitive LC-MS/MS method for determination of levamisole in human plasma by Liping T et al. [15], development of a liquid chromatography-tandem mass spectrometry with pressurized liquid extraction method for the determination of benzimidazole residues in edible tissues by Chen D et al. [16], and determination of levamisole residue in animal livers by two liquid-liquid extraction steps, gas chromatography-mass spectrometry, by Xu J Xiao S et al. [17].

Some of the methods for determination of levamisole, along with some antihelminthics in various biological samples by HPLC, have been developed, including determination of the anthelmintic levamisole in plasma and gastrointestinal fluids by high-performance liquid chromatography by Marriner S et al. [18], determination of levamisole by HPLC in plasma samples in the presence of heparin and pentobarbital by Garcia JJ et al. [19], determination of levamisole and thiabendazole in meat by de Bukanski BW et al. [20], quantitation of levamisole in plasma using high-performance liquid chromatography by Vandamme TF et al. [21], solid-phase extraction and HPLC determination of levamisole hydrochloride in sheep plasma by Du Preez JL et al. [22], determination of levamisole in animal tissues using liquid chromatography by E. Dreassi et al. [23], quantitative chromatographic determination of several benzimidazole anthelmintic molecules in parasite material by Mottier L et al. [24], liquid chromatographic method with ultraviolet absorbance detection for measurement of levamisole in chicken tissues, eggs, and plasma done by El-Kholy H et al. [25], quantitative determination of albendazole metabolites in sheep spermatozoa and seminal plasma by Batzias GC et al. [26], a rapid HPLC method for analysis of ricobendazole and albendazole sulfone in sheep plasma by Wu Z et al. [27], HPLC assay of levamisole and abamectin in sheep plasma for application to pharmacokinetic studies by Sari P et al. [28], determination of levamisole in sheep muscle tissue by Tyrpenou AE et al. [29], and investigation of the persistence of levamisole and oxyclozanide in milk and fate in cheese by M. Whelan et al. [30]. Determination of levamisole with other anthelmintic agents in pharmaceuticals including veterinary preparations by HPLC includes high-performance liquid chromatography method for the analysis of several anthelmintics in veterinary formulations by E. C. Van-Tonder et al. [31], simultaneous determination of levamisole hydrochloride and mebendazole in tablets by R. Raman et al. [32], separation and determination of the process related impurities of mebendazole, fenbendazole, and albendazole in bulk drugs by Gomes AR et al. [33], simultaneous determination of levamisole and abamectin in liquid formulations by P. Sari et al. [34], simultaneous estimation of levamisole, mebendazole, and albendazole in pharmaceutical products by Kullai Reddy Ulavapalli et al. [35], method for determination of levamisole in bulk and dosage form by Ravisankar P et al. [36], enantioseparation of tetramisole by capillary electrophoresis and high-performance liquid chromatography and application of these techniques to enantiomeric purity determination of a veterinary drug formulation of L-levamisole done by B. Chankvetadze et al. [37], UPLC method for the determination of albendazole residues pharmaceutical manufacturing equipment surfaces by R. S. Chandan et al. [38], and HPTLC method for simultaneous estimation of levamisole hydrochloride and oxyclozanide in its bulk and pharmaceutical dosage form by Patel MB et al. [39]. Some of the GC methods for estimation of levamisole [40, 41] and determination of albendazole and its metabolites [42] in cheese [43] and in milk [44] have been developed. Few spectrophotometric methods have been developed, including the quantitative determination of albendazole with Chloramine-T and acid dyes by Basavaiah K et al. [45], ion-pair complex method by El-Didamony AM et al. [46], UV spectrophotometric method [47], derivative spectrophotometry method for determination of levamisole hydrochloride tablets by L. Liang et al. [48], extractive spectrophotometric method by Johnson Misquith et al. [49], mixture containing levamisole and triclabendazole in veterinary tablets by E. Dinc et al. [50], and estimation of mebendazole and levamisole hydrochloride in pharmaceutical formulations by Umang Shah et al. [51]. Some of the miscellaneous methods such as linear titrations for various samples have been developed [52–57].

2. Materials and Methods

Levamisole and albendazole gift samples were obtained from Chandra labs, Pvt. Ltd., Hyderabad. All chemicals and solvents were of analytical reagent grade. Acetonitrile and water (HPLC grade) were obtained from Merck Pvt. Ltd., Mumbai. Quantitative HPLC was performed on a high pressure gradient high-performance liquid chromatograph (Shimadzu HPLC, Class VP series) with two pumps, manual injector with loop volume of 20 μL (Rheodyne), programmable variable wavelength UV detector, and a reversed phase column (Inertsil column, C_{18} (150 x 4.6 ID) 5μm). The output signal was monitored and integrated using "Spincotech" software. A Systronics double beam UV-visible spectrophotometer 2203 with 1 cm matched quartz cells was used for all spectral and absorbance measurements and solutions were prepared in double distilled water.

2.1. Preparation of Mobile Phase. About 30 parts by volume of phosphate buffer were mixed with 70 parts by volume of acetonitrile and the resulting solution was filtered through a nylon membrane filter (pore size is 0.22 μm). The mobile phase was degassed through ultrasonication.

2.2. Preparation of Standard Stock Solution of Levamisole. 50 mg of levamisole was weighed, transferred into 500 ml volumetric flask, dissolved in methanol, and the resulting solution was diluted with methanol up to the mark. From the solution thus obtained, we carefully transferred 1 ml into a 10 ml volumetric flask and the volume was made up to the mark with methanol.

2.3. Preparation of Standard Stock Solution of Albendazole. 50mg of albendazole was weighed into 500ml volumetric flask, dissolved in methanol, and then diluted up to the mark with methanol; 10μg /ml of solution was prepared by diluting 1ml to 10ml with methanol.

2.4. Preparation of Mixed Standard Solution. Weigh accurately 150 mg of levamisole and 400 mg of albendazole in 100 ml of volumetric flask, dissolve them in 10ml of mobile phase, and make up the volume with mobile phase. From above stock solution 100μg/ml of levamisole and 4μg/ml of albendazole are prepared by diluting 1ml to 10ml with mobile phase. This solution is used for recording chromatogram.

2.5. Preparation of Sample Solution. Five tablets (each containing 150 mg of levamisole and 400 mg of albendazole) were weighed, taken into a mortar, crushed to fine powder, and uniformly mixed. Tablet stock solutions of levamisole (30μg/ml) and albendazole 80μg/ml) were prepared by dissolving weight equivalent to 150 mg of levamisole and 400 mg of albendazole dissolved in sufficient mobile phase. After that the solution was filtered using 0.45-micron syringe filter, sonicated for 5 min, and diluted to 100ml with mobile phase. Further dilutions were prepared in 5 replicates of 30μg/ml of levamisole and 400μg/ml of albendazole was made by adding 0.3ml and 0.8ml of stock solution to 10 ml of mobile phase.

2.6. Recommended Chromatographic Procedure for Assay. 20 μL of the standard solution was injected into the chromatographic system and chromatogram was recorded. 20 μL of the standard solution was injected five times into the chromatographic system, chromatograms were recorded, and peak areas were measured. 20 μL of the sample solution was injected two times into the chromatographic system, chromatograms were recorded, and peak areas were measured. The amount of the drug present in the formulation was computed by the following formula:

$$Amount = \frac{A_{T1}}{A_{S1}} \times \frac{D_{S1}}{D_{T1}} \times \frac{P1}{100} AW \quad (1)$$

where

A_{T1} is the average area counts of levamisole peak in chromatogram of sample solution,

A_{S1} is the average area counts of levamisole peak in chromatogram of standard solution,

D_{S1} is the dilution factor for the standard solution,

D_{T1} is the dilution factor for the sample solution,

P_1 is the percentage potency of levamisole working standard used (as is basis),

AW is the average weight of tablet.

3. Results and Discussions

3.1. Method Development and Optimisation. The present study was carried out to develop a simple, sensitive, precise, accurate, rapid, and economical HPLC method for the quantification of levamisole and albendazole present in its formulation. Preliminary trails were conducted with these solvents to find out a suitable mobile phase composition where all the analytes were resolved with good system suitability. Mixture of phosphate buffer pH 3.5 and acetonitrile (30:70 v/v) was selected as mobile phase to enhance compatibility of the sample with the mobile phase and to avoid baseline disturbances. About 20μg/ml of standard solution in methanol was scanned in the region between 200 and

Table 1: Optimized chromatographic conditions.

Mobile phase	Mixed Phosphate buffer (KH_2PO_4+K_2HPO_4): Acetonitrile 30:70
pH	3.5
Column	INERTSIL column, C18 (150x4.6 ID) 5μm
Flow rate	1.0 ml/min
Column temperature	Room temperature (20-25°C)
Sample temperature	Room temperature (20-25°C)
Wavelength	224nm
Injection volume	20 μl
Run time	10 min
Retention time	About 2.350min for Levamisole and 4.055min for Albendazole

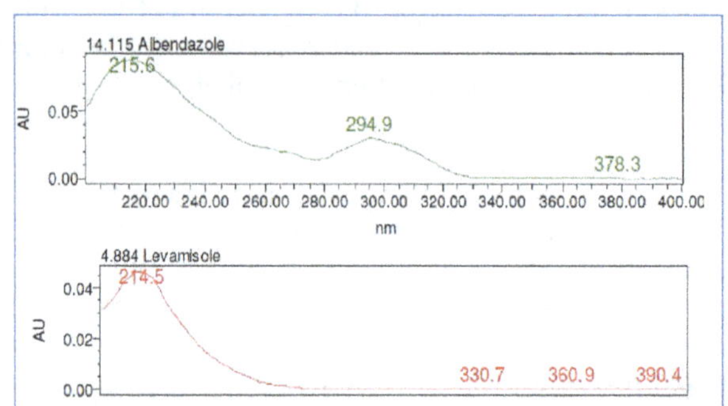

Figure 2: Individual UV spectrum of levamisole and albendazole.

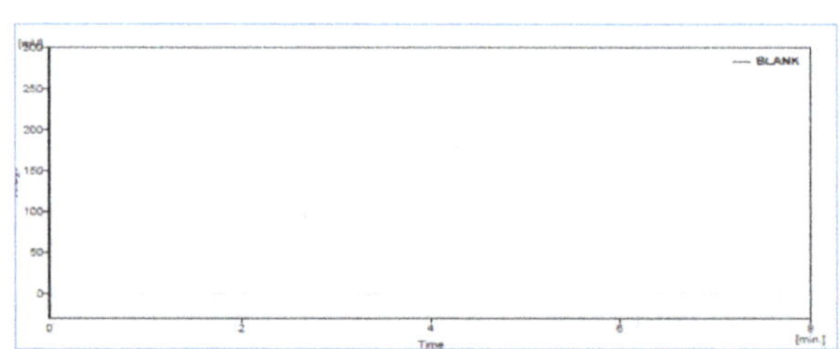

Figure 3: Blank chromatogram for specificity by using mobile phase.

400 nm. In simultaneous estimation of two drugs, isobestic wavelength is used. Isobestic point is the wavelength where the molar absorptivity is the same for two substances that are interconvertible. Therefore, this wavelength is used to estimate both drugs accurately. The isobestic point was found to be 224 nm for levamisole and albendazole; therefore, 224 nm was selected as detection wavelength. The UV absorption spectrum showing isobestic point of the two drug molecules was given in Figure 2, and the optimised chromatographic conditions were presented in Table 1.

3.2. Method Validation. The proposed method was validated as per the guidelines of ICH Q2R1. The assay method developed was subjected to validation by performing specificity, linearity, limit of detection, limit of quantification, precision, accuracy, and robustness. Specificity of the method was proved by demonstrating no excipients interference at the

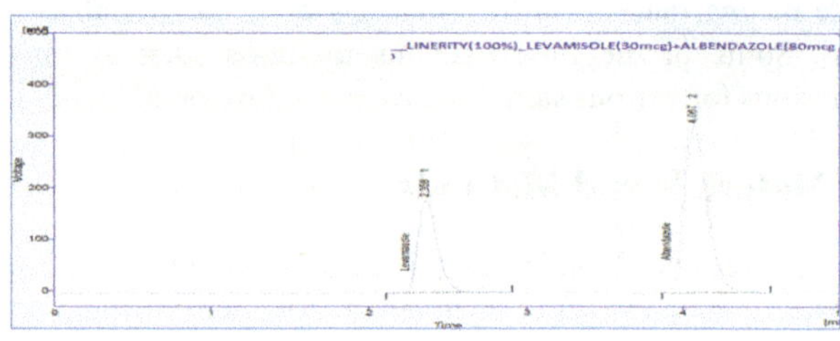

Figure 4: Chromatogram of levamisole and albendazole preparation-3.

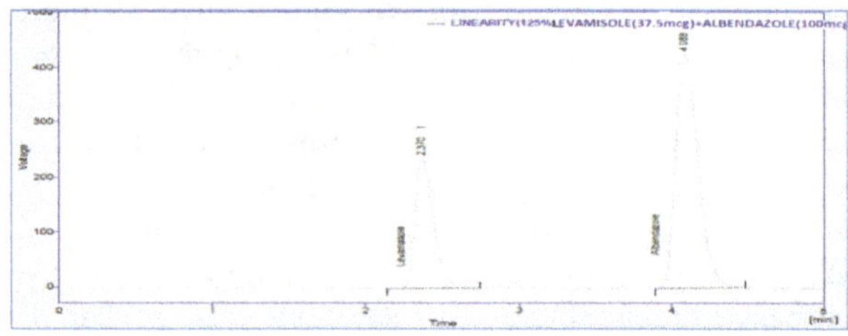

Figure 5: Chromatogram of levamisole and albendazole preparation-4.

retention time of both the drugs in chromatogram of assay sample. The results of specificity were given in Figure 3.

Linearity graph displayed good linearity over the concentration range examined. The polynomial regression for the linearity plot showed good linear relationship with coefficient of correlation. The relationship between the concentration of the drugs and their area was linear in the range examined (15-45μg/ml and 40-120μg/ml for levamisole and albendazole, respectively). Since all points lie in a straight line and the correlation coefficients were well within limits. The correlation coefficients were found to be 0.998 and 0.999 obtained between concentration and area of standard preparations of levamisole and albendazole, respectively. The chromatograms concerned with linearity testing were given (for the two given concentrations only) in Figures 4–6. The results of linearity were presented in Table 2 and the corresponding calibration curves were given in Figures 7 and 8 for levamisole and albendazole, respectively. The limit of detection (LOD) and limit of quantification (LOQ) were determined based on standard deviation of the y-intercept and the slope of the calibration curve. LOD and LOQ values were 2.08μg/ml and

TABLE 2: Linearity of levamisole and albendazole.

S. No.	Levamisole		Albendazole	
	Concentration (μg/ml)	Area	Concentration (μg/ml)	Area
1	15	768.60	40	1679.72
2	22.5	1163.20	60	2554.72
3	30	1522.81	80	3344.73
4	37.5	1996.49	100	4406.95
5	45	2368.14	120	5182.27

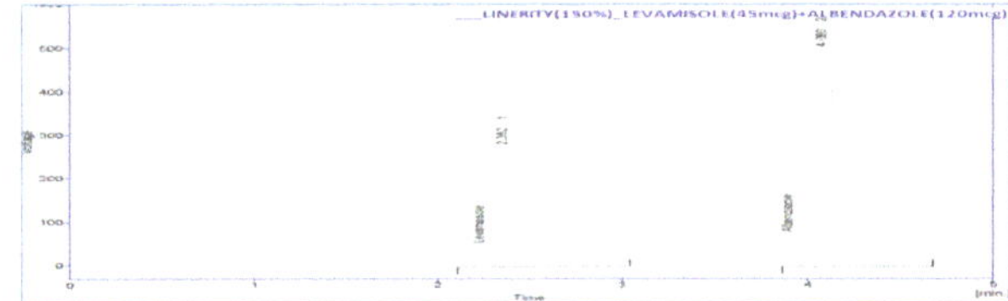

FIGURE 6: Chromatogram of levamisole and albendazole for preparation-5.

6.03μg/ml for levamisole and 3.15μg/ml and 10.40μg/ml for albendazole, respectively.

The accuracy of the method was assessed by adding known amount of standard solution (75 %, 100 %, and 125 % of the sample concentration) to the preanalyzed sample solution of 100% concentration. All the samples were prepared and analyzed in triplicate. The percentage mean recovery by standard addition experiments of levamisole and albendazole is 101.33 % and 98.73 %, respectively. High recovery values indicated proposed method for assay was highly accurate. The detailed results concerning the method accuracy were given in Figures 9–11 and Table 3. System precision, method precision, and intermediate precision were performed for the assay to evaluate precision of the method. % RSD of method precision was found to be **0.46** and **0.22** for levamisole and albendazole, respectively. Lower % RSD of precisions indicated that the method is precise. The results of precision are presented in Figure 12 and Table 4. The robustness of the method was evaluated to determine the capacity of the intended method to remain unaffected by changing organic phase composition of mobile phase, flow rate, and wavelength of detection. The developed method was found to be robust as these changes did not show significant effect on theoretical plates and tailing of the two drug candidates in assay and system suitability. Chromatogram showing the study of robustness is given in Figure 13, and the results of robustness and ruggedness are shown in Tables 5 and 6.

4. Conclusion

A simple and selective LC method is described for the determination of levamisole and albendazole dosage forms. Chromatographic separation was achieved on a C_{18} column using mobile phase consisting of a mixture of 20 Mm phosphate buffer (KH_2PO_4) pH and 3.5 acetonitrile (30:70 v/v), with

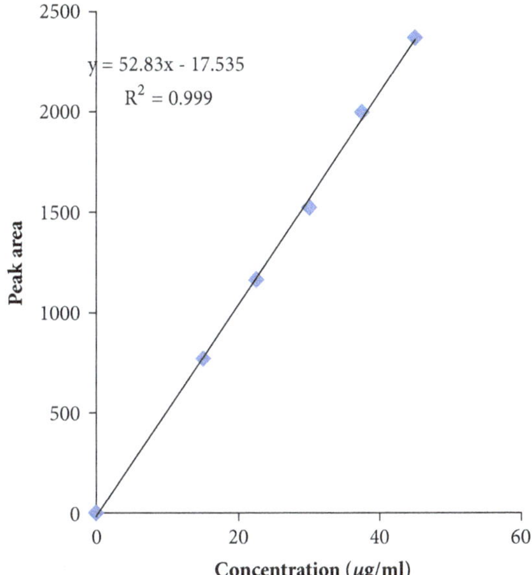

FIGURE 7: Calibration curve of levamisole.

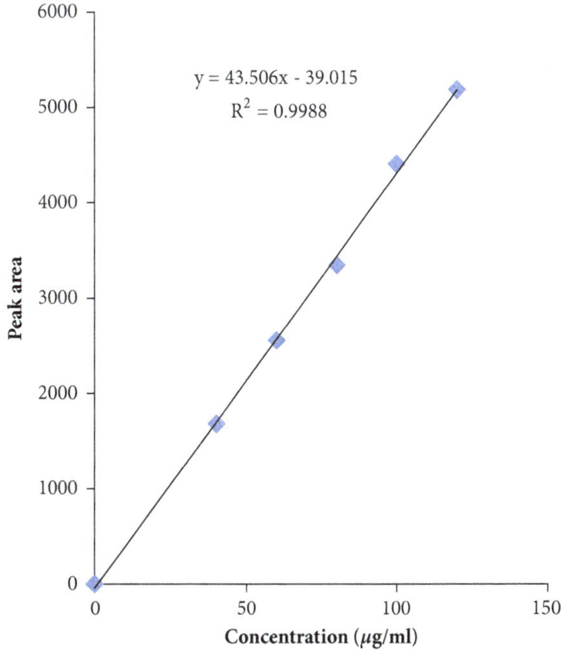

FIGURE 8: Calibration curve of albendazole.

TABLE 3: Results of accuracy by recovery studies.

Recovery level	Amount added	Area	Amount Recovered	% Recovery	Average % recovery
		Levamisole			
75 %	22.5	1163.20	22.8	101.33	100.31
	22.5	1156.99			
	22.5	1158.28			
100 %	30	1522.81	29.82	99.41	
	30	1509.63			
	30	1515.34			
125 %	37.5	1996.49	37.57	100.19	
	37.5	1986.18			
	37.5	1990.97			
		Albendazole			
75 %	60	2554.72	59.34	98.90	98.73
	60	2533.53			
	60	2548.31			
100 %	80	3344.73	78.70	98.37	
	80	3342.78			
	80	3342.22			
125 %	100	4406.95	98.92	98.92	
	100	4394.54			
	100	4427.63			

TABLE 4: Results of intra-assay precision studies of levamisole and albendazole.

Trail no.	Rt	**Levamisole**		**Albendazole**		
		Area	Amount (mg)	Rt	Area	Amount (mg)
1	2.357	1768.18	151.37	4.060	3628.24	402.02
2	2.337	1767.59	151.25	4.028	3600.32	401.19
3	2.337	1757.73	149.89	4.028	3595.94	399.45
4	2.357	1772.34	151.87	4.060	3625.39	401.52
5	2.342	1769.77	151.68	4.035	3605.63	401.05
6	2.3460	1767.12	151.15	4.042	3611.10	401.76
		% RSD	**0.46**		%RSD	**0.22**

TABLE 5: Result of robustness study.

Parameter	**Levamisole**		**Albendazole**	
	Retention time(min)	**Tailing factor**	**Retention time(min)**	**Tailing factor**
Flow				
0.8ml/min	2.817	1.585	4.860	1.585
1.0ml/min	2.355	1.571	4.062	1.571
1.2ml/min	2.022	1.574	3.487	1.574
Wavelength				
222nm	2.367	1.571	4.080	1.571
224nm	2.355	1.571	4.062	1.571
226nm	2.367	1.535	4.082	1.535

TABLE 6: Results for ruggedness.

Levamisole	%Assay	Albendazole	%Assay
Analyst 01	97.99	Analyst 01	99.96
Analyst 02	98.37	Analyst 02	97.59
%RSD	**0.27**	%RSD	**1.69**

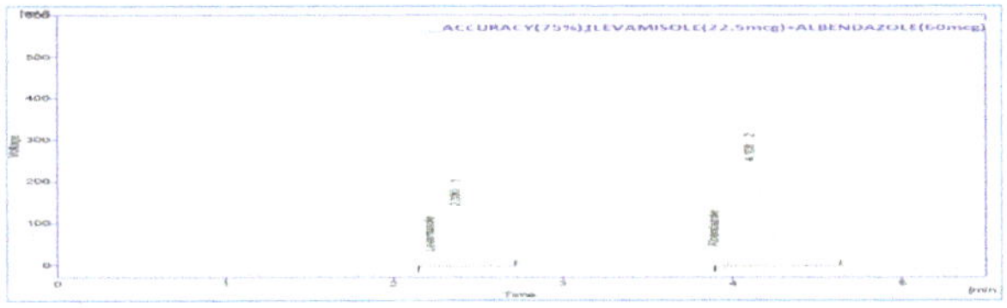

FIGURE 9: Chromatogram of 75% recovery.

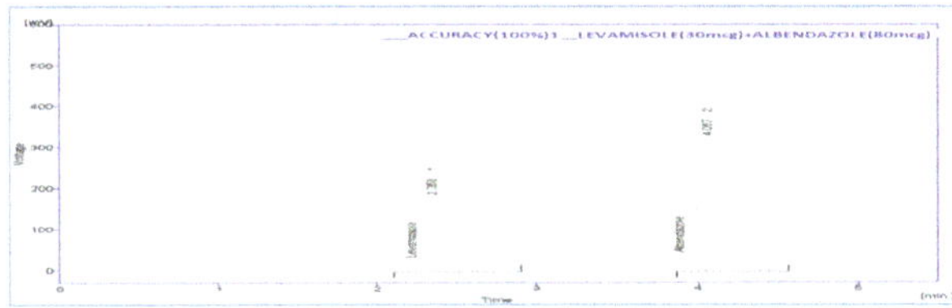

FIGURE 10: Chromatogram of 100% recovery.

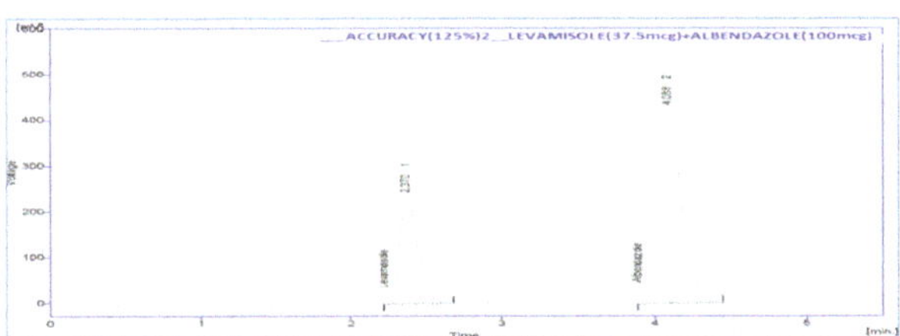

FIGURE 11: Chromatogram of 125% recovery.

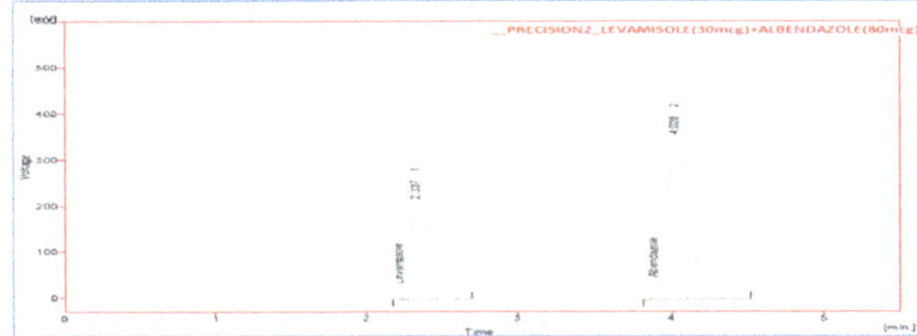

FIGURE 12: Chromatogram of precision injection 3.

detection of 224 nm. Linearity was observed in the range 15-45μg /ml for levamisole (r^2 =0.9975) and 40-120μg /ml for albendazole (r^2 =0.9994), as the amount of drugs estimated by the proposed methods was in good agreement with the label claim. The proposed methods were validated. The accuracy of the methods was assessed by recovery studies at three different levels. Recovery experiments indicated the absence of interference from commonly encountered pharmaceutical additives. The method was found to be precise as indicated by the repeatability analysis, showing % RSD less than 2. All statistical data proves validity of the methods and can be used

FIGURE 13: Chromatogram of levamisole and albendazole robustness (0.8 ml/min).

for routine analysis of pharmaceutical dosage form. From the above experimental results and parameters, it was concluded that this newly developed method for the simultaneous estimation of levamisole and albendazole was found to be simple, precise, and accurate; high resolution and shorter retention time make this method more acceptable, and it can be effectively applied for routine analysis in quality control department in industries.

References

[1] "Indian Pharmacopoeia, Govt. of India, Vol. 2, Ministry Of Health and Family Welfare, New Delhi, 2010, p. 1635-1636".

[2] European pharmacopoeia, "council of Europe, 6th Edition, 2008, volume 2, p. 2434".

[3] British Pharmacopoeia, *British Pharmacopoeial commission*, vol. 2, The Department Of Health, Social Services and Public Safety, 2011.

[4] United States pharmacopoeia, "United States Pharmacopoeial convention, Inc. Rockville, M.D., 2000, p. 1141".

[5] "The British Pharmacopoeia (Veterinary), Her Majesty's Stationary Office, London, UK, 2011, 91, 78".

[6] The European Pharmacopoeia, *The European Pharmacopoeia, Council of Europe*, Strasbourg, 7th edition, 2008.

[7] "The United States pharmacopeia 34, NF 29, Asian Ed. Rand Mc Nally, USA, 2011, 245, 3284".

[8] S. J. Stout, A. R. daCunha, R. E. Tondreau, and J. E. Boyd, "Confirmation of levamisole residues in cattle and swine livers by capillary gas chromatography-electron impact mass spectrometry," *Journal - Association of Official Analytical Chemists*, vol. 71, no. 6, pp. 1150–1153, 1988.

[9] C. G. Chappell, C. S. Creaser, J. W. Stygall, and M. J. Shepherd, "On-line high-performance liquid chromatographic/gas chromatographic/tandem ion trap mass spectrometric determination of levamisole in milk," *Biological Mass Spectrometry*, vol. 21, no. 12, pp. 688–692, 1992.

[10] A. Cannavan, W. J. Blanchflower, and D. G. Kennedy, "Determination of levamisole in animal tissues using liquid chromatography–thermospray mass spectrometry," *Analyst*, vol. 120, no. 2, pp. 331–333, 1995.

[11] H. De Ruyck, E. Daeseleire, H. De Ridder, and R. Van Renterghem, "Development and validation of a liquid chromatographic-electrospray tandem mass spectrometric multiresidue method for anthelmintics in milk," *Journal of Chromatography A*, vol. 976, no. 1-2, pp. 181–194, 2002.

[12] M. Cherlet, S. De Baere, S. Croubels, and P. De Backer, "Quantitative analysis of levamisole in porcine tissues by high-performance liquid chromatography combined with atmospheric

pressure chemical ionization mass spectrometry," *Journal of Chromatography B: Biomedical Sciences and Applications*, vol. 742, no. 2, pp. 283–293, 2000.

[13] S. De Baere, M. Cherlet, S. Croubels, K. Baert, and P. De Backer, "Liquid chromatographic determination of levamisole in animal plasma: Ultraviolet versus tandem mass spectrometric detection," *Analytica Chimica Acta*, vol. 483, no. 1-2, pp. 215–224, 2003.

[14] P. Jedziniak, T. Szprengier-Juszkiewicz, and M. Olejnik, "Determination of benzimidazoles and levamisole residues in milk by liquid chromatography-mass spectrometry: Screening method development and validation," *Journal of Chromatography A*, vol. 1216, no. 46, pp. 8165–8172, 2009.

[15] L. Tong, L. Ding, Y. Li et al., "A sensitive LC-MS/MS method for determination of levamisole in human plasma: Application to pharmacokinetic study," *Journal of Chromatography B*, vol. 879, no. 5-6, pp. 299–303, 2011.

[16] D. Chen, Y. Tao, H. Zhang et al., "Development of a liquid chromatography-tandem mass spectrometry with pressurized liquid extraction method for the determination of benzimidazole residues in edible tissues," *Journal of Chromatography B*, vol. 879, no. 19, pp. 1659–1667, 2011.

[17] J. Xu, S. Xiao, W. Dong et al., "Determination of levamisole residue in animal livers by two liquid-liquid extraction steps-gas chromatographymass spectrometry," *Chinese Journal of Chromatography*, vol. 30, no. 9, pp. 922–925, 2012.

[18] S. Marriner, E. A. Galbraith, and J. A. Bogan, "Determination of the anthelmintic levamisole in plasma and gastro-intestinal fluids by high-performance liquid chromatography," *Analyst*, vol. 105, no. 1255, pp. 993–996, 1980.

[19] J. J. Garcia, M. J. Diez, M. Sierra, and M. T. Teran, "Determination of levamisole by hplc in plasma samples in the presence of heparin and pentobarbital," *Journal of Liquid Chromatography*, vol. 13, no. 4, pp. 743–749, 1990.

[20] B. Wyhowski de Bukanski, J.-M. Degroodt, and H. Beernaert, "Determination of levamisole and thiabendazole in meat by HPLC and photodiode array detection," *Zeitschrift für Lebensmittel-Untersuchung und -Forschung*, vol. 193, no. 6, pp. 545–547, 1991.

[21] T. F. Vandamme, M. Demoustier, and B. Rollmann, "Quantitation of levamisole in plasma using high performance liquid chromatography," *European Journal of Drug Metabolism and Pharmacokinetics*, vol. 20, no. 2, pp. 145–149, 1995.

[22] J. L. Du Preez and A. P. Lötter, "Solid-phase extraction and HPLC determination of levamisole hydrochloride in sheep plasma," *Onderstepoort Journal of Veterinary Research*, vol. 63, no. 3, pp. 209–211, 1996.

[23] E. Dreassi, G. Corbini, C. La Rosa, N. Politi, and P. Corti, "Determination of levamisole in animal tissues using liquid chromatography with ultraviolet detection," *Journal of Agricultural and Food Chemistry*, vol. 49, no. 12, pp. 5702–5705, 2001.

[24] L. Mottier, L. Alvarez, and C. Lanusse, "Quantitative chromatographic determination of several benzimidazole anthelmintic molecules in parasite material," *Journal of Chromatography B*, vol. 798, no. 1, pp. 117–125, 2003.

[25] H. El-Kholy and B. W. Kemppainen, "Liquid chromatographic method with ultraviolet absorbance detection for measurement of levamisole in chicken tissues, eggs and plasma," *Journal of Chromatography B*, vol. 796, no. 2, pp. 371–377, 2003.

[26] G. C. Batzias, E. Theodosiadou, and G. A. Delis, "Quantitative determination of albendazole metabolites in sheep spermatozoa and seminal plasma by liquid chromatographic analysis with fluorescence detection," *Journal of Pharmaceutical and Biomedical Analysis*, vol. 35, no. 5, pp. 1191–1202, 2004.

[27] Z. Wu, N. J. Medlicott, M. Razzak, and I. G. Tucker, "Development and optimization of a rapid HPLC method for analysis of ricobendazole and albendazole sulfone in sheep plasma," *Journal of Pharmaceutical and Biomedical Analysis*, vol. 39, no. 1-2, pp. 225–232, 2005.

[28] P. Sari, J. Sun, M. Razzak, and I. Tucker, "HPLC assay of levamisole and abamectin in sheep plasma for application to pharmacokinetic studies," *Journal of Liquid Chromatography & Related Technologies*, vol. 29, no. 15, pp. 2277–2290, 2006.

[29] A. E. Tyrpenou and E. M. Xylouri-Frangiadaki, "Determination of levamisole in sheep muscle tissue by high-performance liquid chromatography and photo diode array detection," *Chromatographia*, vol. 63, no. 7-8, pp. 321–326, 2006.

[30] M. Whelan, C. Chirollo, A. Furey, M. L. Cortesi, A. Anastasio, and M. Danaher, "Investigation of the persistence of levamisole and oxyclozanide in milk and fate in cheese," *Journal of Agricultural and Food Chemistry*, vol. 58, no. 23, pp. 12204–12209, 2010.

[31] E. C. Van Tonder, M. M. De Villiers, J. S. Handford, C. E. P. Malan, and J. L. Du Preez, "Simple, robust and accurate high-performance liquid chromatography method for the analysis of several anthelmintics in veterinary formulations," *Journal of Chromatography A*, vol. 729, no. 1-2, pp. 267–272, 1996.

[32] R. Ramesh, S. K. Yadgire, and R. Bhanu, "Simultaneous determination of Levamisole HCL and Mebendazole in tablets by reverse phase HPLC," *Indian Drugs*, vol. 38, no. 2, pp. 67-68, 2001.

[33] A. R. Gomes and V. Nagaraju, "High-performance liquid chromatographic separation and determination of the process related impurities of mebendazole, fenbendazole and albendazole in bulk drugs," *Journal of Pharmaceutical and Biomedical Analysis*, vol. 26, no. 5-6, pp. 919–927, 2001.

[34] P. Sari, M. Razzak, and I. G. Tucker, "Rapid, Simultaneous Determination of Levamisole and Abamectin in Liquid Formulations Using HPLC," *Journal of Liquid Chromatography & Related Technologies*, vol. 27, no. 2, pp. 351–364, 2004.

[35] U. Kullai Reddy, J. Sriramulu, and U. R. Mallu, "RP-HPLC Method for Simultaneous Estimation of Levamisole, Mebendazole and Albendazole in Pharmaceutical Products," *Indian Journal of Novel Drug delivery*, vol. 3, no. 2, pp. 134–142, 2011.

[36] P. Ravisankar and G. Devala Rao, "Development and validation of RP-HPLC method for determination of levamisole in bulk and dosage form," *Asian Journal of Pharmaceutical and Clinical Research*, vol. 6, no. 3, pp. 169–173, 2013.

[37] B. Chankvetadze, N. Burjanadze, M. Santi, G. Massolini, and G. Blaschke, "Enantioseparation of tetramisole by capillary electrophoresis and high performance liquid chromatography and application of these techniques to enantiomeric purity determination of a veterinary drug formulation of L-levamisole," *Journal of Separation Science*, vol. 25, no. 12, pp. 733–740, 2002.

[38] R. S. Chandan, M. Vasudevan, and B. M. Deecaraman, "Development and Validation of A UPLC method the Determination of Albendazole Residues Pharmaceutical Manufacturing Equipment Surfaces," *International Scholarly and Scientific Research Innovation*, vol. 7, no. 12, pp. 910–914, 2013.

[39] M. B. Patel, R. K. Patel, S. G. Patel, and A. J. Patel, "Development and Validation of HPTLC Method for Simultaneous Estimation of Levamisole Hydrochloride and Oxyclozanide in its Bulk and

Pharmaceutical Dosage Form," *Austin Chromatogr*, vol. 4, no. 1, p. 1045, 2017.

[40] F. J. Schenck, L. V. Podhorniak, and R. Wagner, "A highly sensitive gas chromatographic determination of levamisole in milk," *Food Additives & Contaminants: Part A*, vol. 15, no. 4, pp. 411–414, 1998.

[41] S. Loussouarn, G. Blanc, and L. Pinault, "Quim., Determination of levamisole in European eel plasma by gas chromatography using liquid and solid-phase extraction," *Anal.(Barcelona)*, vol. 14, no. 4, pp. 223–227, 1995.

[42] J. J. Garcia, F. Bolás-Fernández, and J. J. Torrado, "Quantitative determination of albendazole and its main metabolites in plasma," *Journal of Chromatography B: Biomedical Sciences and Applications*, vol. 723, no. 1-2, pp. 265–271, 1999.

[43] D. J. Fletouris, N. A. Botsoglou, I. E. Psomas, and A. I. Mantis, "Determination of the Marker Residue of Albendazole in Cheese by Ion-Pair Liquid Chromatography and Fluorescence Detection," *Journal of Dairy Science*, vol. 80, no. 11, pp. 2695–2700, 1997.

[44] D. Fletouris, N. Botsoglou, I. Psomas, and A. Mantis, "Rapid Quantitative Screening Assay of Trace Benzimidazole Residues in Milk by Liquid Chromatography," *Journal of AOAC International*, vol. 79, no. 6, pp. 1281–1287, 1996.

[45] K. Basavaiah and H. C. Prameela, "Use of an oxidation reaction for the quantitative determination of albendazole with chloramine-T and acid dyes," *Analytical Sciences*, vol. 19, no. 5, pp. 779–784, 2003.

[46] A. M. El-Didamony, "Spectrophotometric determination of benzydamine HCl, levamisole HCl and Mebeverine HCl through ion-pair complex formation with methyl range," *Spectrochimica Acta—Part A: Molecular and Biomolecular Spectroscopy*, vol. 69, no. 3, pp. 770–775, 2008.

[47] J. Misquith and A. Dias, "Validated ultra violet spectroscopic methods for the determination of Levamisole Hydrochloride," *International Journal of PharmTech Research*, vol. 4, no. 4, pp. 1631–1637, 2012.

[48] J. Misquith, P. P. Prabhu, E. V. S. Subrahmanyam, and A. R. Shabaraya, "Extractive spectrophotometric methods for the determination of levamisole hydrochloride," *International Journal of PharmTech Research*, vol. 4, no. 3, pp. 1215–1220, 2012.

[49] L. Liang and Z. Tao, "First-derivative spectrophotometric determination of the content of levamisole hydrochloride tablets Yaowu," *Fenxi. Zazhi*, vol. 12, no. 4, pp. 238-239, 1992.

[50] E. Dinç, G. Pektş, and D. Baleanu, "Continuous wavelet transform and derivative spectrophotometry for the quantitative spectral resolution of a mixture containing levamizole and triclabendazole in veterinary tablets," *Reviews in Analytical Chemistry*, vol. 28, no. 2, pp. 79–92, 2009.

[51] U. Shah, T. Talaviya, and A. Gajjar, "Development and validation of derivative spectroscopic Method for the simultaneous estimation of mebendazole and levamisole hydrochloride in pharmaceutical formulations," *International Journal of Pharmaceutical Chemistry and Analysis*, vol. 2, pp. 108–112, 2015.

[52] S. Sadeghi, F. Fathi, and J. Abbasifar, "Potentiometric sensing of levamisole hydrochloride based on molecularly imprinted polymer," *Sensors and Actuators B: Chemical*, vol. 122, no. 1, pp. 158–164, 2007.

[53] G. Cao, H. Zhang, G. Yin, and Y. Pi, "Through Analytical Abstract C.D. Assay of ethambutol hydrochloride, mecillinam and levamisole hydrochloride tablets by linear titration," *Zhongguo.Yaoxue. Zazhi*, vol. 27, no. 2, pp. 96–98, 1992.

[54] J. J. Zhu and H. Gao, "Oscillographic potentiometric titration on a membrane by a parallel connection capacity method and a controlled current method," *Fenxi. Huaxue*, vol. 23, no. 5, pp. 506–511, 1995.

[55] J. P. Billon, M. Tissieres, and M. Courtier, "Determination of halides of [pharmaceutical] organic bases," *Analusis*, vol. 13, no. 9, pp. 430-431, 1985.

[56] R. C. Zhang, R. R. Zhan, and Y. F. Zhan, "Oscillographic titration with sodium tetraphenylborate for pharmaceutical and metallurgical analyses," *Fenxi. Ceshi. Xuebao*, vol. 212, pp. 39–41, 2002.

[57] W. Li and F. Yaowu, "Determination of levamisole hydrochloride in tablets by titration with ananionic surfactant," *Zazhi*, vol. 12, no. 2, pp. 117-118, 1992.

16

Biomarker Profiling for Pyridoxine Dependent Epilepsy in Dried Blood Spots by HILIC-ESI-MS

Elizabeth Mary Mathew,[1] Sudheer Moorkoth,[1] Leslie Lewis,[2] and Pragna Rao[3]

[1]*Department of Pharmaceutical Quality Assurance, Manipal College of Pharmaceutical Sciences, Manipal Academy of Higher Education, Manipal 576104, India*

[2]*Department of Paediatrics, Kasturba Medical College, Manipal Academy of Higher Education, Manipal 576104, India*

[3]*Department of Biochemistry, Kasturba Medical College, Manipal Academy of Higher Education, Manipal 576104, India*

Correspondence should be addressed to Sudheer Moorkoth; sudheermoorkoth@gmail.com

Academic Editor: Neil D. Danielson

Pyridoxine dependent epilepsy is a condition where the affected infant or child has prolonged seizures (status epilepticus), which are nonresponsive to anticonvulsant therapy but can be treated with pharmacological doses of pyridoxine. If identified earlier and treated prophylactically with pyridoxine, severe brain damage due to seizures can be prevented. Alpha-amino adipic semialdehyde (AASA), piperidine-6-carboxylic acid (P6C), and pipecolic acid (PA) are known biomarkers of pyridoxine dependent epilepsy. We report the development and validation of a hydrophilic interaction liquid chromatography (HILIC) hyphenated with mass spectroscopy for the quantification of the above analytes from dried blood spot samples. The samples were extracted using methanol and analysed on a iHILIC fusion plus column with formic acid buffer (pH 2.5): acetonitrile (20:80) at a flow rate of 0.5 mL/min within 3 minutes. The method demonstrated a LOD of 10 ng/mL, LOQ of 50 ng/mL, linearity of $r^2 \geq 0.990$, and recovery of 92-101.98% for all analytes. The intra- and interday precision CVs were < 8% and 6%, respectively. Extensive stability studies demonstrated that the analytes were stable in stock solution and in matrix when stored at -80°C. We performed method comparison studies of the developed method with the literature reported method using normal samples and matrix matched spiked samples at pathological concentrations to mimic clinical validity. The Bland-Altman analysis for comparison of the analytical suitability of the method for the biomarkers in healthy and spiked samples with the literature reported method revealed a bias which suggested that the method was comparable. The newly developed method involves no derivatisation and has a simple sample preparation and a low run time enabling it to be easily automated with a high sample throughput in a cost-effective manner.

1. Introduction

Pyridoxine dependent epilepsy (PDE) is an autosomal recessive metabolic encephalopathy that presents in affected newborns within the first month of life with myoclonic, tonic clonic seizures in the neonatal period and partial seizures in the early infantile period. These seizures are known to respond to pyridoxine and reoccur on pyridoxine withdrawal [1, 2]. The exact incidence of PDE is still unknown. Studies conducted in Netherlands, United Kingdom, and Ireland report the incidence as 1:100000 to 700000 individuals and more than 100 cases have been reported worldwide [3, 4]. This disorder arises due to the mutation in the antiquitin gene which codes for the alpha-amino adipic semialdehyde dehydrogenase enzyme that is responsible for the conversion of alpha-amino adipic semialdehyde to alpha-amino adipic acid. Defect in the enzyme leads to the accumulation of alpha-amino adipic semialdehyde (α-AASA) that exists in equilibrium with piperidine-6-carboxylic acid (P6C). The formed P6C undergoes Knoevenagel condensation with pyridoxal phosphate in the body leading to its deficiency. P6C is also known to be an intermediate in the pipecolic acid pathway and its accumulation leads to enzymatic block which in turn results in toxic levels of pipecolic acid [5].

Analytical platforms like LC-MS have been explored for the quantification of AASA, P6C, and PA in various body fluids [6–11]. These methods employ derivatisation using fluorenylmethyloxycarbonyl chloride or butanolic hydrogen chloride [12] in order to alter the retention on C18 columns. Recently the hyphenation of hydrophilic interaction liquid chromatography with mass spectrometry have demonstrated suitability in the quantification of pipecolic acid in plasma [11], but reports on the suitability of this chromatographic technique for AASA or P6C are unknown. At present the diagnosis of this disorder involves the quantification of these biomarkers from plasma, serum, and urine samples. A recent investigation has demonstrated the suitability of dried blood spots as a suitable matrix for these analytes especially for infants and newborn [9]. The objective of the current study was to develop and validate a novel derivatisation free and sensitive bioanalytical LC-MS method for simultaneous quantification of AASA/P6C and PA using HILIC-ESI-MS from dried blood spot samples.

2. Materials and Methods

2.1. Materials and Reagents. L-allysine ethylene acetal (AEA) (>98%), Amberlyst 15 hydrogen form, and pipecolic acid (PA) (99%) were purchased from Sigma-Aldrich (St. Louis, US). DL-pipecolic acid-d9 (internal standard) was purchased from Toronto Research Chemicals (Ontario, Canada). LC-MS grade acetonitrile was purchased from Biosolve Chimie SARL (Dieuze, France). Formic acid, 85% (AR grade), was purchased from Merck (Kenilworth, US). In-house Milli Q water (Siemens Ultra Clear) was used. iHILIC fusion (+) column (100 × 2.1 mm, 3.5 μm) was purchased from HILLICON AB (Umea, Sweden).

2.2. Instrumentation. A Thermo Scientific (Massachusetts, US) LC-mass spectrometer with Dionex Ultimate 3000 liquid chromatograph interfaced with a linear ion trap analyser by an electron spray ionisation source was used. MS/MS and chromatographic method development was performed using LTQ XL (Massachusetts, US) and Chromeleon (Massachusetts, US) software, respectively. Batch analysis was done using the XCalibur software (Massachusetts, US) and quantification was done using LC Quan (Massachusetts, US).

2.3. Sample Collection. Ethical clearance (MUEC/010/2017 dated 08.05.2017) was obtained from the Ethics Committee of Manipal Academy of Higher Education, Manipal. Dried blood spot samples were collected from neonates in Kasturba Hospital, Manipal, Karnataka, India, for a period of six months from July 2017 by a certified nurse. The samples were dried at room temperature for three hours followed by which the samples were transferred to envelopes with desiccants. The envelopes were then transferred to zip lock plastic bags and were stored at -80°C until further analysis.

2.4. AASA and P6C Synthesis. Due to the lack of a true reference standard for AASA, we proceeded to the synthesis of AASA in the laboratory based on reported methods [8] with some modifications. 5 mg AEA and 15 mg Amberlyst beads were mixed in 1 mL of water for 10 min using a shaker at 1320 rpm. The resultant solution was filtered and the remaining Amberlyst beads were washed with 0.5 mL water and transferred to the same filter. The washing and filtering steps were repeated with 1 mL of water twice. Deblocking efficiency was checked by comparing the mass spectra obtained before and after reaction. The obtained solution was further diluted with 3.1 mL water. The MS intensities observed for AASA and P6C were 1:3. The reproducibility of the procedure was checked by multiple injections using different stock preparations. The resultant stock solution had a concentration of 1 mmol/L of AASA and 3 mmol/L of P6C. This solution was used for the preparation of calibrators and controls. From this stock solution, a working stock containing 0.3μg/mL of AASA and 1 μg/mL of P6C (henceforth referred to as equilibrium mixture in the manuscript) was prepared for optimizing MS parameters by direct infusion.

2.5. Calibrators and Quality Controls. The calibrators and quality controls samples were prepared in leftover blood from healthy controls after adjusting the haematocrit to 50% similar to that of a newborn. Stock solutions of PA, d9-PA (1000 μg/mL), and AASA/P6C (321 μg/mL) were prepared in water. Working stocks were prepared in 50:50%v/v acetonitrile: water and spiked into aliquots of haematocrit adjusted blood. 40μl of the spiked blood was pipetted onto Whatman 903 filter paper to prepare calibrators and quality controls (LLQC, LQC, MQC, and HQC) as per Table 1.

2.6. LC-MS Method and Sample Analysis. Optimization of MS conditions was performed by infusing solutions of PA (1 μg/mL) and the equilibrium mixture (0.3 μg/mL AASA and 1 μg/mL P6C) independently at a rate of 10 μL/min through the direct infusion pump in ESI (+) polarity mode. During optimization, the mass scan filters were set at a centre mass of m/z 128, 146, 130, and 139 for P6C, AASA, PA, and d9-pipecolic acid, respectively with a width of m/z 10. The iHILIC fusion (+) column facilitated the retention of PA and AASA/P6C without derivatisation. The mobile phase consisted of 80 volumes of acetonitrile and 20 volumes of formic acid buffer (pH 2.5) delivered isocratically at a flow rate of 0.5 mL/min. The sample injection volume was optimized at 10μL. The retention times of PA, AASA/ P6C, and d9-PA were 1.79, 2.59, and 1.80 minutes, respectively. The total run time of the method was 3 minutes. The optimized mass spectrometer parameters were as follows: spray voltage: 5V, vaporizer temperature: 300°C, nitrogen sheath gas flow: 55 arbitrary units, auxiliary gas flow: 12 arbitrary units, sweep gas flow: 2 arbitrary units, ion transfer capillary temperature: 275°C with a voltage of 15.00 V and tube lens: 90 V, multipole 00 offset: -5.75 V, lens 0: -9.50V, multipole 0 offset: -9.75V, lens voltage: -15V, gate lens: -70,00V, multipole 1 offset: -15.50 V, multipole RF amplitude: 400 V, and front lent lens: -14.75 V. The optimized SRM transition for each analyte was as follows: AASA (*m/z* 146 ⟶ 128.13, CE: 35), P6C (*m/z* 127.9 ⟶ 81.8, CE: 35), PA (*m/z* 130 ⟶ 83.94, CE: 40), and d9-PA (*m/z* 139 ⟶ 93.1 CE: 40) in SRM mode. The optimized chromatogram is presented in Figure 1.

2.7. Extraction Optimization from Dry Blood Spot (DBS). 3.2 mm DBS were made using standard leather punch and

Table 1: Method performance specifications for AASA/P6C and pipecolic acid.

Analyte	Performance specifications								
	Calibration range (μg/mL)Linearity	LOQ (μg/mL)	LOD (μg/mL)	QC Levels (μg/mL)	Accuracy (%) Inter	Accuracy (%) Intra	Precision (%) Inter	Precision (%) Intra	Percent Recovery (%)
AASA/P6C	0.05-10 (r^2=0.999)	0.05	0.01	LLOQ (0.05)	92.06	99.71	3.43	3.12	-
				LQC (0.5)	101.98	100.15	5.16	5.31	80.22
				MQC (1)	100.78	96.78	3.60	4.16	80.88
				HQC (8)	101.32	94.12	7.87	5.12	85.12
Pipecolic acid	0.05-9 (r^2=0.997)	0.05	0.01	LLOQ (0.05)	100.10	99.22	2.25	3.78	-
				LQC (0.5)	98.10	100.87	1.66	5.00	90.33
				MQC (1)	100.32	100.21	3.38	2.84	93.23
				HQC (8)	98.17	98.74	2.34	2.31	92.16

LOQ: limit of quantification, LOD: limit of detection, LLOQ: lower limit of quantification, LQC: low quality control, MQC: medium quality control, HQC: high quality control, AASA: alpha-amino adipic semialdehyde, and P6C: piperidine-6-carboxylic acid.

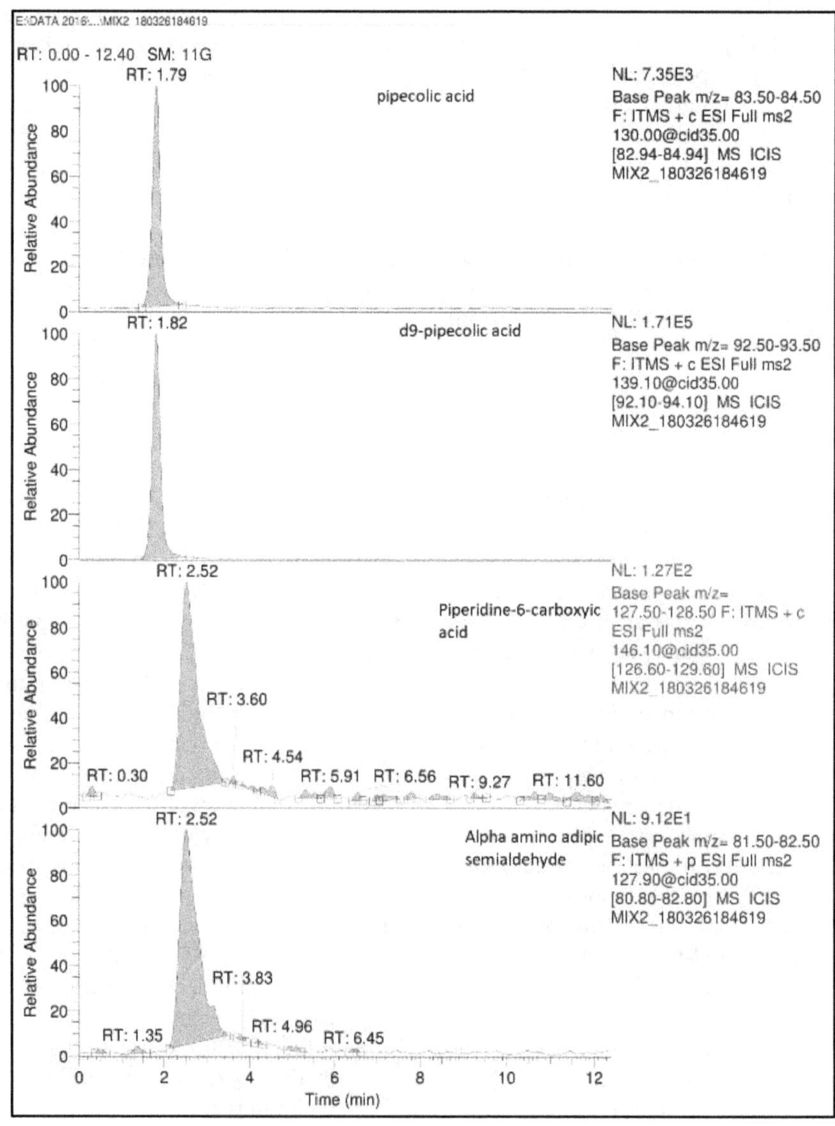

Figure 1: Optimized LC-MS chromatogram of α-amino adipic semialdehyde, piperidine-6-carboxylic acid, and pipecolic acid in SRM mode.

transferred to a 0.5 mL Eppendorf tube followed by the addition of 100μL of methanol containing the internal standard. The tubes were vortex mixed for 30 min and the resultant solutions were evaporated to dryness at 30°C using a nitrogen evaporator for 5 min. The dried residue was reconstituted in the mobile phase and the subsequent solution was injected to LC-MS.

For the quantification of AASA/P6C and pipecolic acid in the dried blood spot, the method reported by Yuzyuk [12] was adopted. A ten-point calibration curve was constructed based on the peak area ratio of the analyte to internal standard (d9-pipecolic acid). Since AASA and P6C exist in equilibrium in the body, the calibration curve of AASA/P6C was constructed based on the sum of the area ratios of AASA and P6C on the Y axis and concentration on the X axis. Each batch consisted of blank (mobile phase), zero standard (with internal standard), a set of ten calibrators, and QC samples (bracketed and placed at regular intervals among the unknown samples).

2.8. Method Validation. Validation studies were performed as per Food and Drug Administration (FDA) guidelines [13]. A batch of six DBS samples from healthy neonates was processed with internal standard to assess the specificity of the method. To determine the LOD and LOQ we followed the procedure of Gachet et al. [14]. The LOD was determined using standard solutions serially diluted until a signal to noise ratio (S/N) of 3 was obtained. LLOQ was determined using six spiked DBS samples processed with the internal standard at which the S/N was 10 and showed positive values after blank subtraction. Linearity, accuracy, and recovery were determined by the blank subtraction technique [14, 15]. To perform linearity, endogenous levels of the analyte in the matrix were first determined by the standard addition method and this endogenous analyte to IS ratio was subtracted from the spiked DBS analyte to IS ratio. The subtracted peak area ratio was plotted versus analyte concentration and regression analysis was performed. Five replicates at each concentration for PA and AASA/P6C were performed for linearity. The accuracy of the method was determined by comparing the mean calculated responses of six replicate QC samples (after subtraction of endogenous level) to the nominal concentration values at each level. The absolute extraction recoveries of analytes were determined by comparing the responses (after subtraction of endogenous level) of six replicates at LQC, MQC, and HQC levels with neat standard solutions of same concentrations prepared in the mobile phase. The matrix effect was determined by comparing the response (after subtraction of endogenous level) obtained by the postextraction spike method at LQC and HQC levels (six replicates) with corresponding standard solutions at each QC level prepared in the mobile phase. The carry-over was measured using blank samples injected after analysis of the upper limit of quantification (ULOQ) samples with internal standard. Intraday and interday precision were performed in two batches by analysing six replicates at four QC levels on a single day and over three separate days, respectively.

2.9. Stability Studies. Extensive stability studies for the biomarkers at different conditions recommended by the guidelines were performed [13]. Stock solution stability was performed by comparing six replicates of a freshly prepared neat MQC with that of the stability samples at 2, 4, 6, and 8 h at room temperature. Stability of analytes in matrix was evaluated using six replicates of LQC and HQC at bench top (0, 0.5, 1, 1.5, 2, 4, and 7 h, room temperature) and freeze thaw (3 cycles, -80°C) and in autosampler (6, 12, 24, and 48 h at 4°C). Long-term stability on storage at -80°C was assessed on 7, 15, 30, 60, 90, 180, and 360 days using six replicates of LQC and HQC. On the day of the analysis samples were thawed unassisted at room temperature and compared with initial results at each QC level. The results obtained were computed for mean, standard deviation, and percent change. The results obtained were analysed by "repeated measures ANOVA" and "paired t-test". A p value < 0.05 was considered statistically significant.

2.10. Method Comparison Studies. Method comparison study was performed to assess how the newly developed method correlates with the method reported in literature. 50 samples were coded, and a blind analysis was carried out on the developed HILIC method. These DBS samples were then subjected to analysis by the method reported by Yuzuk et al. [12]. The results were analysed by Bland-Altman analysis using SPSS 16.0 for method comparison.

Since there was no true clinical sample available to demonstrate the clinical validity of the assay, we followed the approach recommended by Taverniers et al. [16] and Jung et al. [9]. To mimic positive samples, we first determined the endogenous levels of the analytes in blood by the standard addition approach and eleven spiked DBS samples at pathological levels (matrix matched) were prepared by a separate analyst at levels of 1-6μg/mL and 0.5-3 μg/mL for AASA/P6C and pipecolic acid, respectively, based on reports of Sadilkova et al. [8]. All the samples were coded with numbers and a blind random analysis was conducted by a separate analyst on five different days on the developed HILIC method and by the method reported by Yuzuk et al. [12]. The result obtained was analysed by Bland-Altman analysis using SPSS 16.0.

3. Results and Discussion

PA, AASA, and P6C are intermediates in the catabolism of lysine. These analytes are polar in nature and their retention on reverse phase columns is challenging. We have explored the ability of hydrophilic interaction liquid chromatography and LC-MS for the determination of these analytes in dried blood spot. Three HIILIC columns, namely, Venusil amide (100 X 4.6 mm, 3μ), iHILIC fusion (100 X 2.1 mm, 3μ), and iHILIC fusion plus (100 X 2.1 mm, 3μ), were investigated for their suitability at pH 2.5 and 4.5 using 90 parts of acetonitrile and 10 parts of buffer. At pH 4.5 the iHILIC column and the iHILIC fusion plus column demonstrated a capacity factor greater than 15 for all the analytes. At pH 2.5 the iHILIC fusion plus demonstrated a capacity factor of less than 8. Hence the iHILIC fusion plus column at pH 2.5 was seen to be suitable for further experimental studies. The retention times of PA, AASA/P6C, and d9-PA were 1.79, 2.52, and 1.82 min,

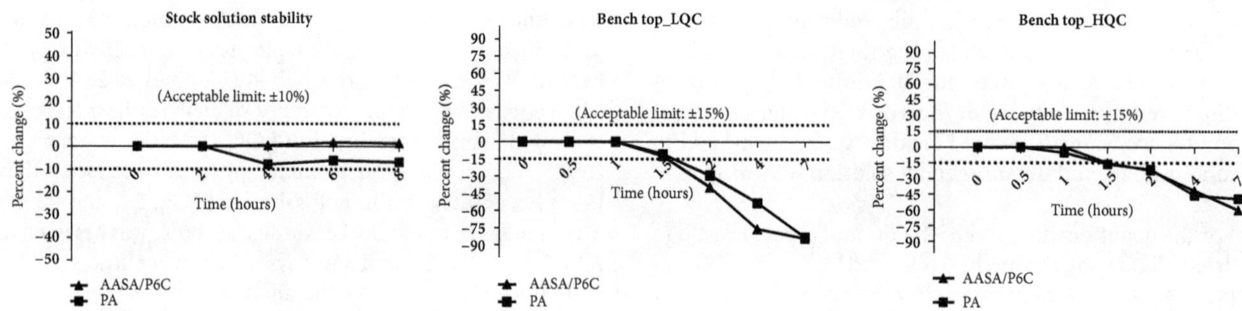

FIGURE 2: Stock solution stability studies and bench top stability studies of low quality controls (LQC) and high-quality controls (HQC) of α-amino adipic semialdehyde, piperidine-6-carboxylic acid, and pipecolic acid.

TABLE 2: Results of stock solution stability for alpha-amino adipic semialdehyde/piperidine-6-carboxylic acid and pipecolic acid.

	Stock solution stability			
	AASA/P6C		PA	
Time points (hours)	Mean back calculated concentration (SD) (mcg/mL)[1]	% change	Mean back calculated concentration (SD) (mcg/mL)[1]	% change
0	0.89 (0.04)	0.00	0.85 (0.01)	0.00
2	0.89 (0.06)	0.05	0.85 (0.01)	0.00
4	0.90 (0.01)	0.45	0.78 (0.03)	-7.84
6	0.91 (0.1)	1.79	0.80 (0.03)	-6.10
8	0.91 (0.007)	1.43	0.79 (0.03)	-6.87
P value[2]	0.138		0.007	

[1]Values expressed as mean (SD); [2]statistical treatment performed using repeated measures ANOVA. p value < 0.05 was considered significant. AASA: alpha-amino adipic semialdehyde, P6C: piperidine-6-carboxylic acid, and PC: pipecolic acid.

respectively (Figure 1). The ability of mass spectrometry for selective reaction monitoring (SRM) enabled us to achieve unique advantage in terms of specificity. The MS^2 function in linear ion traps enabled different SRM transitions for different analytes (PA: m/z 130 ⟶ 83.94, CE: 40; d9-PA: m/z 139 ⟶ 93.1, CE: 40). These differences in SRM transitions enabled the quantification of the analytes even though there was coelution in chromatography.

The DBS extraction procedure was optimized for extraction solvent, rotation speed (rpm), and rotation time. 100% methanol was seen to produce maximum recovery. An increasing trend in recovery was observed with increasing the rotation time with no further increase after 30 min at 400 rpm. The method was seen to be cost-effective and demonstrated recoveries of 90.33-92.16 % and 80.22-85.12% for PA and AASA/P6C, respectively (Table 1).

The lower limit of detection and limit of quantification was 10 ng/mL (S/N of 3) and 50 ng/mL (S/N of 10), respectively for all the analytes. No significant endogenous interferences were observed at the retention time of the analytes and ISTD proving the method to be specific. The method demonstrated good linearity (0.05-10 μg/mL) with regression coefficient (r^2) value at or above 0.997. Our LC-MS method achieved good precision with a total CV of 3.12-7.8% (AASA/P6C) and 1.6- 5.0% (PA) at the various quality control levels. The accuracy of the method was evaluated through spike recovery and ranged from 92.06 to 101.98 % and 98.10 to 100.87 % for AASA/P6C and PA, respectively, at various quality control levels. The validation summary is provided in Table 1.

Results of the stability studies represented by mean back calculated concentrations and the respective % changes are as presented in Tables 2 and 3. PA and AASA/P6C were found to be stable in stock solution prepared in methanol with a percent change of 1.43% and -6.87 %, respectively, at eight hours from the baseline value (Table 2 and Figure 2). These levels are well within the recommended limit of ±10%. The bench top stability studies undertaken at LQC and HQC levels for each analyte to study the influence of laboratory conditions demonstrated a decreasing trend in the levels of the analytes with time. We observed a % change of -12.44% and 10.73% from initial value at 1.5 h for AASA-P6C and pipecolic acid, respectively, at LQC levels. Additionally, at HQC levels we observed a % change of -15% from baseline for AASA-P6C and -16.13% for pipecolic acid at 1.5 h. More than 80% change from baseline value was observed within 7 h. Our results align parallel to the earlier reports by other workers in urine and serum [6, 8, 9]. To study the influence of freeze thaw cycles on analyte integrity, we measured the basal levels of AASA/P6C and PA at LQC and HQC levels. The second measurement was made after subjecting the QC's to three freeze thaw cycles. AASA/P6C demonstrated stability for three cycles with a percent change of -6.3% and -2.45% at LQC and HQC levels, respectively. At LQC levels PA demonstrated a percent change of -7.15% and at HQC it was -0.53%. These observations are well within the acceptance

TABLE 3: Results of bench top, freeze thaw, autosampler, and long term stability for alpha-amino adipic semialdehyde/piperidine-6-carboxylic acid and pipecolic acid.

	Stability expressed as "mean back calculated concentration" (μg/mL)[1] ± SD, (% change)			
	Bench top stability (room temperature, 7 hours)			
Time point	LQC		HQC	
	AASA/P6C	Pipecolic acid	AASA/P6C	Pipecolic acid
0 hr	0.44±0.01 (0.00)	0.44±0.01 (0.00)	7.45±0.10 (0.00)	7.54±0.05 (0.00)
0.5 hr	0.44±0.01 (0.00)	0.43±0.00 (0.00)	7.21±0.31 (0.00)	7.49±0.14 (0.00)
1 hr	0.44±0.01 (0.00)	0.43±0.00 (0.00)	7.16±0.21 (0.00)	7.15±0.21 (-5.21)
1.5 hr	0.38±0.02 (-12.44)	0.38±0.01 (-10.73)	6.40±0.02 (-15.00)	6.39±0.21 (-16.13)
2 hr	0.26±0.02 (-39.40)	0.31±0.02 (-29.11)	5.80±0.20 (-22.20)	5.91±0.16 (-21.19)
4 hr	0.10±0.01 (-75.91)	020±0.02 (-53.33)	4.36±0.55 (-41.43)	4.04±1.29 (-46.17)
7 hr	0.06±0.01 (-84.20)	0.07±0.008 (-83.30)	2.91±0.94 (-60.00)	3.79±0.15 (-49.37)
p value[2]	<0.05	<0.05	<0.05	<0.05
	Freeze thaw stability (three freeze thaw cycles)			
0 cycle	0.44±0.01 (0.00)	0.44±0.01 (0.00)	7.45±0.11 (0.00)	7.50±0.05 (0.00)
3 cycles	0.41±0.02 (-6.31)	0.41±0.02 (-7.15)	7.27±0.38 (-2.45)	7.46±0.06 (-0.53)
p value[3]	0.11	0.08	0.36	0.36
	Autosampler stability (4°C, 48 hours)			
0 hr	0.45±0.01 (0.00)	0.44±0.01 (0.00)	7.75±0.10 (0.00)	7.50±0.05 (0.00)
6 hr	0.46±0.01 (0.59)	0.45±0.01 (0.52)	7.74±0.11 (-0.14)	7.41±0.14 (-1.15)
12 hr	0.44±0.01 (-3.39)	0.43±0.01 (-3.61)	7.74±0.14 (-0.16)	7.33±0.21 (-2.31)
24 hr	0.42±0.01 (-6.65)	0.42±0.01 (-5.24)	7.71±0.13 (-0.50)	7.28±0.17 (-3.01)
48 hr	0.42±0.03 (-8.35)	0.48±0.03 (-9.23)	7.65±0.14 (-1.23)	7.25±0.16 (-3.30)
p value[2]	0.05	0.04	0.68	0.06
	Long term stability (-80°C, 360 days)			
0 day	0.45±0.01 (0.00)	0.44±0.01 (0.00)	8.47±0.20 (0.00)	7.18±0.23 (0.00)
7 days	0.46±0.02 (1.13)	0.44±0.02 (0.28)	7.95±0.49 (-6.16)	6.80±0.42 (-5.30)
15 days	0.46±0.03 (2.10)	0.45±0.02 (1.57)	7.94±0.10 (-6.22)	6.73±0.12 (-6.21)
30 days	0.43±0.02 (-6.24)	0.42±0.02 (-6.14)	7.63±0.06 (-9.96)	6.40±0.08 (-10.81)
60 days	0.41±0.01 (-9.89)	0.39±0.02 (-10.95)	7.82±0.34 (-7.69)	6.54±0.30 (-8.91)
90 days	0.41±0.02 (-9.70)	0.42±0.02 (-5.41)	7.86±0.15 (-7.22)	6.72±0.18 (-6.34)
180 days	0.42±0.02 (-7.76)	0.42±0.02 (-5.75)	7.91±0.14 (-6.61)	6.63±0.06 (-7.50)
360 days	0.42±0.00 (-6.96)	0.42±0.00 (-4.05)	7.90±0.12 (-6.77)	6.5±0.11 (-8.43)
p value	<0.01	0.08	0.02	<0.05

[1]Values expressed as mean (standard deviation); [2]statistical treatment performed using repeated measures ANOVA (p value less than 0.05 was considered significant); [3]statistical treatment performed using paired t-test and a p value less than 0.05 which was considered significant. AASA: alpha-amino adipic semialdehyde; P6C: piperidine-6-carboxylic acid.

limit of ±15% denoting stability. The stability study on the influence of resident time in the autosampler maintained at 4°C at LQC levels demonstrated a percent change of -8.35% and -9.23 % within 48 hours for AASA/P6C and PA, respectively. At HQC levels these changes were -1.23% and -3.3%. Long-term stability studies at -80°C for a period of 360 days demonstrated a change of only -6.96 % and -4.05 % at LQC levels and a change of -6.77 % and -8.43 % at HQC levels for AASA/P6C and PA, respectively. These results are within the recommended limit of ± 15, indicating long-term stability for these analytes similar to earlier reports in plasma [8]. Graphical representations of the stability data are presented in Figures 2 and 3.

Results of the comparison study of the developed LC-MS method to the literature reported method showed that the mean ± SD for normal sample was 279.64 ±120.90 and 306.12 ±128.88 ng/mL by the developed HILIC-MS method compared to 280.92 ±121.77 and 307.85 ±128.22 ng/mL for AASA/P6C and PA, respectively, by the literature reported method. The Bland-Altman analysis (Figure 4) revealed a bias of -0.99 and -1.91 for AASA/P6C and PA, respectively, proving that the methods are comparable.

The study conducted to mimic the clinical validity using spiked positive control samples demonstrated a recovery of 82-106% and 92.8-119% for AASA/P6C and pipecolic acid, respectively. Additionally the Bland-Altman analysis of the

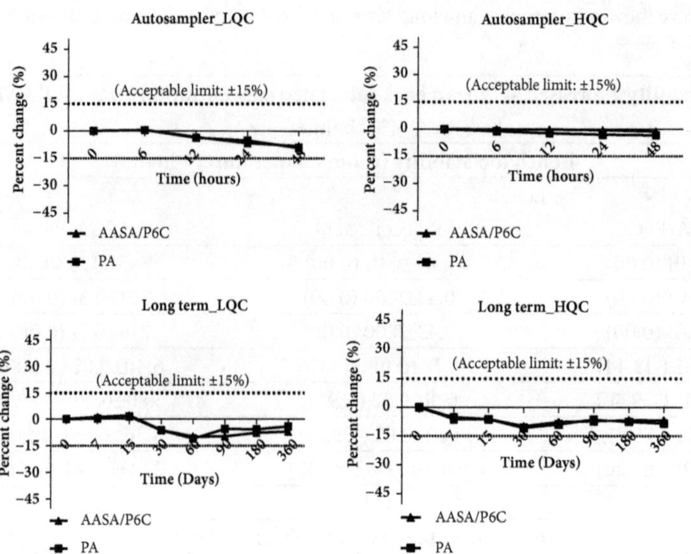

FIGURE 3: Autosampler and long-term stability studies of low quality controls (LQC) and high-quality controls (HQC) of α-amino adipic semialdehyde, piperidine-6-carboxylic acid, and pipecolic acid.

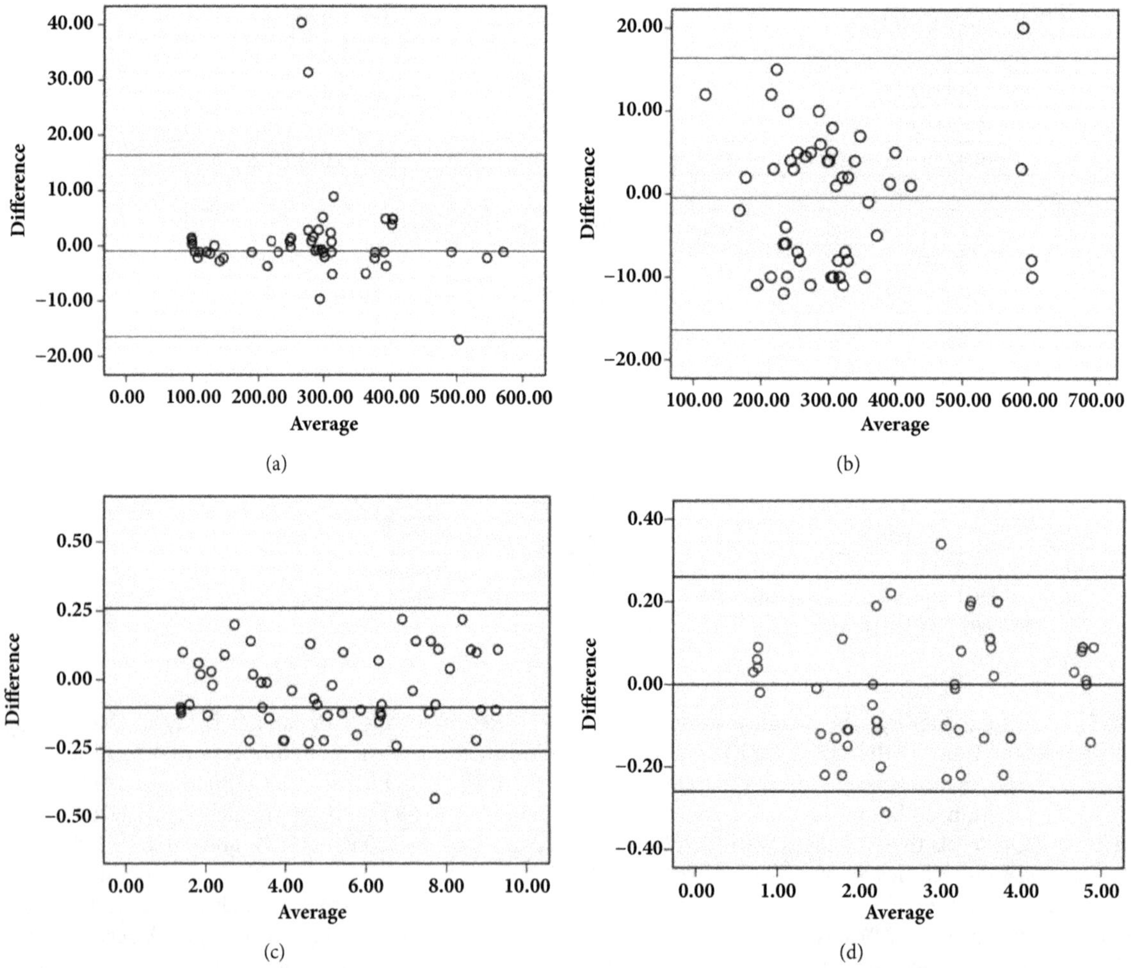

(a)

(b)

(c)

(d)

FIGURE 4: Bland-Altman analysis of the developed HILIC-MS method with literature reported LC-MS method for healthy and matrix matched spiked samples. (a) Healthy: α-amino adipic semialdehyde/piperidine-6-carboxylic acid. (b) Healthy: pipecolic acid. (c) Matrix matched spiked: α-amino adipic semialdehyde/piperidine-6-carboxylic acid. (d) Matrix matched spiked: pipecolic acid.

results of % recovery using SPSS 16.0 showed a mean±SD of 3.40±1.62 and 1.94±0.84 μg/mL by the developed HILIC-MS method compared to 3.44±1.62 and 1.95±0.81 μg/mL for AASA/P6C and pipecolic acid, respectively, by the literature reported method with a bias of -0.04 and -0.10 for AASA-P6C and pipecolic acid, respectively (Figure 4).

4. Conclusions

A simple, rapid, and sensitive method for the simultaneous determination of AASA/P6C and pipecolic acid has been developed and validated on HILIC-ESI-MS/MS. The method has a low run time and lacks derivatisation. These advantages permit us to handle a large sample throughput in a cost-effective manner. Further work in this direction can be undertaken to confirm the clinical validity of the method with true clinical samples of PDE.

Acknowledgments

The authors acknowledge Manipal University for providing infrastructure facilities to carry out this work. The work was supported by Vision Group of Science and Technology, Karnataka [VGST/CESEM/GRD-308 (2014-15)/53/17-18/218], and Grand Challenges Canada [R-ST-POC-1707-07275].

References

[1] R. Surtees and N. Wolf, "Treatable neonatal epilepsy," *Archives of Disease in Childhood*, vol. 92, no. 8, pp. 659–661, 2007.

[2] P. Baxter, "Pyridoxine-dependent and pyridoxine-responsive seizures," *Developmental Medicine & Child Neurology*, vol. 43, no. 6, pp. 416–420, 2001.

[3] P. Baxter, "Epidemiology of pyridoxine dependent and pyridoxine responsive seizures in the UK," *Archives of Disease in Childhood*, vol. 81, no. 5, pp. 431–433, 1999.

[4] J. V. Been, L. A. Bok, P. Andriessen, and W. O. Renier, "Epidemiology of pyridoxine dependent seizures in the Netherlands," *Archives of Disease in Childhood*, vol. 90, no. 12, pp. 1293–1296, 2005.

[5] S. Stockler, B. Plecko, S. M. Gospe et al., "Pyridoxine dependent epilepsy and antiquitin deficiency. Clinical and molecular characteristics and recommendations for diagnosis, treatment and follow-up," *Molecular Genetics and Metabolism*, vol. 104, no. 1-2, pp. 48–60, 2011.

[6] E. A. Struys, L. A. Bok, D. Emal, and M. A. Wielemsen, "Applicability of urinary measurement of Δ 1 -piperideine-6-carboxylate, the alter ego of α-aminoadipic semialdehyde, in Antiquitin deficiency," *Journal of Inherited Metabolic Disease*, vol. 34, supplement 3, pp. S49–S286, 2011.

[7] I. A. Pena, L. A. Marques, A. B. A. Laranjeira, J. A. Yunes, M. N. Eberlin, and P. Arruda, "Simultaneous detection of lysine metabolites by a single LC–MS/MS method: monitoring lysine degradation in mouse plasma," *SpringerPlus*, vol. 5, no. 1, pp. 1–9, 2016.

[8] K. Sadilkova, S. M. Gospe Jr., and S. H. Hahn, "Simultaneous determination of alpha-aminoadipic semialdehyde, piperideine-6-carboxylate and pipecolic acid by LC-MS/MS for pyridoxine-dependent seizures and folinic acid-responsive seizures," *Journal of Neuroscience Methods*, vol. 184, no. 1, pp. 136–141, 2009.

[9] S. Jung, N.-T. B. Tran, S. M. Gospe, and S. H. Hahn, "Preliminary investigation of the use of newborn dried blood spots for screening pyridoxine-dependent epilepsy by LC-MS/MS," *Molecular Genetics and Metabolism*, vol. 110, no. 3, pp. 237–240, 2013.

[10] I. Ferrer-López, P. Ruiz-Sala, B. Merinero, C. Pérez-Cerdá, and M. Ugarte, "Determination of urinary alpha-aminoadipic semialdehyde by LC-MS/MS in patients with congenital metabolic diseases," *Journal of Chromatography B*, vol. 944, pp. 141–143, 2014.

[11] M. Semeraro, M. Muraca, G. Catesini et al., "Determination of plasma pipecolic acid by an easy and rapid liquid chromatography-tandem mass spectrometry method," *Clinica Chimica Acta*, vol. 440, pp. 108–112, 2015.

[12] T. Yuzyuk, A. Liu, A. Thomas et al., "A novel method for simultaneous quantification of alpha-aminoadipic semialdehyde/piperideine-6-carboxylate and pipecolic acid in plasma and urine," *Journal of Chromatography B*, vol. 1017-1018, pp. 145–152, 2016.

[13] US Department of Health and Human Services, "Bioanalytical Method Validation, Guidance for Industry," 2001, http://www.moh.gov.bw/Publications/drug_regulation/Bioanalytical%20Method%20Validation%20FDA%202001.pdf.

[14] M. S. Gachet, P. Rhyn, O. G. Bosch, B. B. Quednow, and J. Gertsch, "A quantitiative LC-MS/MS method for the measurement of arachidonic acid, prostanoids, endocannabinoids, N-acylethanolamines and steroids in human plasma," *Journal of Chromatography B*, vol. 976-977, pp. 6–18, 2015.

[15] R. Thakare, Y. S. Chhonker, N. Gautam, J. A. Alamoudi, and Y. Alnouti, "Quantitative analysis of endogenous compounds," *Journal of Pharmaceutical and Biomedical Analysis*, vol. 128, pp. 426–437, 2016.

[16] I. Taverniers, M. De Loose, and E. Van Bockstaele, "Trends in quality in the analytical laboratory. II. Analytical method validation and quality assurance," *TrAC Trends in Analytical Chemistry*, vol. 23, no. 8, pp. 535–552, 2004.

17

Quantum Cascade Laser Infrared Spectroscopy for Online Monitoring of Hydroxylamine Nitrate

Marissa E. Morales-Rodriguez,[1,2] Joanna McFarlane,[1] and Michelle K. Kidder[1]

[1] *Oak Ridge National Laboratory, USA*

[2] *The Bredesen Center at the University of Tennessee Knoxville, USA*

Correspondence should be addressed to Joanna McFarlane; mcfarlanej@ornl.gov

Academic Editor: Charles L. Wilkins

We describe a new approach for high sensitivity and real-time online measurements to monitor the kinetics in the processing of nuclear materials and other chemical reactions. Mid infrared (Mid-IR) quantum cascade laser (QCL) high-resolution spectroscopy was used for rapid and continuous sampling of nitrates in aqueous and organic reactive systems, using pattern recognition analysis and high sensitivity to detect and identify chemical species. In this standoff or off-set method, the collection of a sample for analysis is not required. To perform the analysis, a flow cell was used for in situ sampling of a liquid slipstream. A prototype was designed based on attenuated total reflection (ATR) coupled with the QCL beam to detect and identify chemical changes and be deployed in hostile environments, either radiological or chemical. The limit of detection (LOD) and the limit of quantification (LOQ) at 3σ for hydroxylamine nitrate ranged from 0.3 to 3 and from 3.5 to 10 $g{\cdot}L^{-1}$, respectively, for the nitrate system at three peaks with wavelengths between 3.8 and 9.8 μm.

1. Introduction

The monitoring of chemical processing in hazardous or extreme conditions challenges methods that rely on sampling followed by offline analysis. Continuous processes with reactive species are particularly difficult to control and would benefit from active online monitoring of reagents or products or both. Nuclear isotope separations depend on careful control of redox chemistry, using reactive species such as hydroxylamine nitrate, HAN, to change the oxidation state of actinides dissolved in aqueous solution. Hence, we describe a spectroscopic method that could be used to monitor HAN reactions in real time. Because its flexibility, the method could be applied to any aqueous species with absorption in the mid-infrared.

Vibrational IR spectroscopy is a tool that offers the selectivity required for identifying molecular species as IR absorptions are characteristic and specific to molecular groups. Vibrational spectra can be interpreted to give thermal energies of IR-active compounds, allowing these to be included in chemical kinetic and dynamical models. Traditionally, IR transmittance is not utilized to characterize aqueous solutions because of the absorption of H_2O, but advances in Mid-IR FTIR and the incorporation of the attenuated total reflection accessory (ATR) make it usable for aqueous solution chemistry [1–3]. The same principle of minimizing matrix absorption by using ATR was employed here. As an IR source, we used a set of four quantum cascade lasers (QCL). Potential advantages of the QCL system over a broadband source such as that used in an FTIR include portability because an evacuated light path is not required, spectral resolution based on the laser linewidth, and enhanced sensitivity through high peak power of the excitation laser [4–7]. A recent study by Pengel and colleagues [8] demonstrated the feasibility of using QCLs to monitor chemicals in solution in a static system and Alcaráz and colleagues used an external cavity QCL for measurements of proteins in the mid-IR [9]. In the work described here, the goal was to demonstrate the capability of a QCL-ATR compact and off-set system to continuously monitor (with samples taken every minute) an aqueous phase reaction in a nuclear application.

The utility of QCL standoff detection of molecules has been demonstrated in the solid and gas phase at ORNL, e.g., methane in field experiments and in the detection of explosive dust collected on solid surfaces [10–16]. However, many chemical processes occur in solution phase, and involve different molecules with distinguishing functional groups. In aqueous solution, there are two issues that need to be addressed, the high background and spectral selectivity. Hence, Raman is usually the method of choice for vibrational spectroscopy as it avoids background absorption from H_2O. Because of selection rules, Raman is generally much less sensitive than IR absorption, unless methods such as surface enhanced Raman are used [17, 18]. For instance, Raman has been used to monitor the degradation of anion-exchange resins used for the separation of plutonium isotopes in highly acidic conditions, e.g., Buscher et al. [19]. Van Staden and colleagues cite a detection limit for both nitrate and nitrate as 500 mg/L [20]. Resonance Raman has been used to study nitrate and nitrite in wastewater treatment processes, with detection limits of 7 μg [21]. This method depends on far UV excitation; however, this method becomes unfeasible for use in applications involving high concentrations of nitrate because self-absorption becomes problematic at concentrations above 3.5 mM.

The QCL-ATR system was used to monitor and assay nitrate-nitrite chemistry representative of the process for plutonium-238 production for NASA deep space missions. The chemical processing of neptunium-237 targets after irradiation involves several steps to (a) separate fission products, (b) separate the neptunium and plutonium, and (c) make purification and polishing. This process achieves separation of neptunium and plutonium through redox chemistry and selective liquid-liquid extraction from nitric acid solution (where the target is dissolved) in a tributyl phosphate (TBP)-organic mixture. Recovery of the plutonium from the organic phase needs introduction of hydroxylamine nitrate ($NH_3OH^+{\cdot}NO_3^-$) or HAN that is used to reduce Pu(IV) to Pu(III). Hydroxylamine, or HA, is classified as a self-reactive substance [22]. The autocatalytic reaction scheme that takes place in nitric acid solution is given in Reaction (1), showing the conversion of nitric to nitrous acid [23, 24], and the decomposition of NH_3OH^+, Reaction (2).

$$2HNO_3 + NH_2OH \longrightarrow 3HNO_2 + H_2O \quad (1)$$

$$HNO_2 + NH_3OH^+ \longrightarrow N_2O + 2H_2O + H^+ \quad (2)$$

As Reactions (1) and (2) progress consuming NH_3OH^+, in strong nitric acid the amount of HNO_2 can increase causing an uncontrolled reaction that can affect the recovery of the plutonium. Hence, it is important to be able to monitor the processes continuously, which is not possible with offline methods of analysis. Commonly employed methods to analyze hydroxylamine nitrate include reacting it with compounds that are spectroscopically active, such as a titration with Griess reagent, [25] or by high performance liquid chromatography [26].

Spectroscopy promises to fulfill the requirements of online analysis and selectivity for nitrate and nitrite species [27, 28]. However, there is no direct spectroscopic measurement in the UV-visible region or mid-IR for HAN in water [29]. We have developed a new method of in situ or online measurement in aqueous solution, which can be used to effectively monitor key species such as NO_3^- and HNO_2 involved in the redox process chemistry of actinides. This prototype can be incorporated into the reaction system for continuous monitoring of the reaction progress to provide rapid quantitative and qualitative analysis. The mechanisms of the prototype, data collection, and data analysis are discussed in the next section.

IR spectra of HA and derivatives have been observed in inert gas matrices and calculated by self-consistent field methods [30], observed in the gas phase as n-ethyl hydroxylamine and hydroxylamine chloride [27], and observed in molecular beams [31]. The reported spectrum of hydroxylamine chloride shows that hydroxylamine exhibits active absorption behavior with multiple peaks in the regions of ~4 and 7-12 μm, coincident with the spectral region covered by the QCL system utilized in this project. Hence, the QCL spectroscopic method was employed online to observe simultaneously the spectral signatures of NH_3OH^+ and HNO_2.

2. Equipment and Materials

To monitor a chemical process a liquid flow system was established between the reactor vessel and the ATR cell. As the flow passes across the crystal of the ATR cell, the QCL-IR beam irradiates the ATR cell crystal generating an evanescent wave that is in contact with the solution. For this prototype, we used a commercially available Varian HATR base assembly cell with heated flow cell using a Ge 45° coated window from Pike Technologies. Germanium was chosen because it is little affected by HNO_3. This ATR cell model permits introduction of light from a nonstandard source (i.e., the QCL). The temperature in the ATR work plate was controlled to maintain a constant temperature (±1°C) from the reaction vessel to the ATR. The QCL beam was internally reflected 10 or 20 times, depending on the thickness of the crystal being 4 or 2 mm, respectively.

The QCL system (Daylight Solutions, CA) has four laser modules with wavelength tuning ranges of 3.77-4.49 μm and 6.87-12.50 μm. The QCL sources are broadly tunable and pulsed and provide peak power of up to 400 mW. The QCL module was operated in a pulsed mode of 500 ns, 100 kHz repetition rate, 5% duty cycle, and a scan speed of 0.5 μm/sec. The broadly tunable wavelength range allows the system to cover multiple spectral absorption features or peaks of the chemicals of interest, increasing the sensitivity and selectivity of the method. The outgoing beam from the ATR cell is directed onto an MCT detector from VIGO systems model PCI-3TE. The MCT is a compact three stage thermoelectrically cooled detector optimized to provide high performance in the spectral range from 2 to 13 μm, D∗ being $\geq 4.5\times10^8$ cmHz$^{0.5}$/W at λopt. The detector converts the laser light to an electrical signal with corresponding amplitude per wavelength; this is the QCL power profile used as background. In the presence of chemical species, changes in signal amplitude per wavelength of the QCL give a characteristic

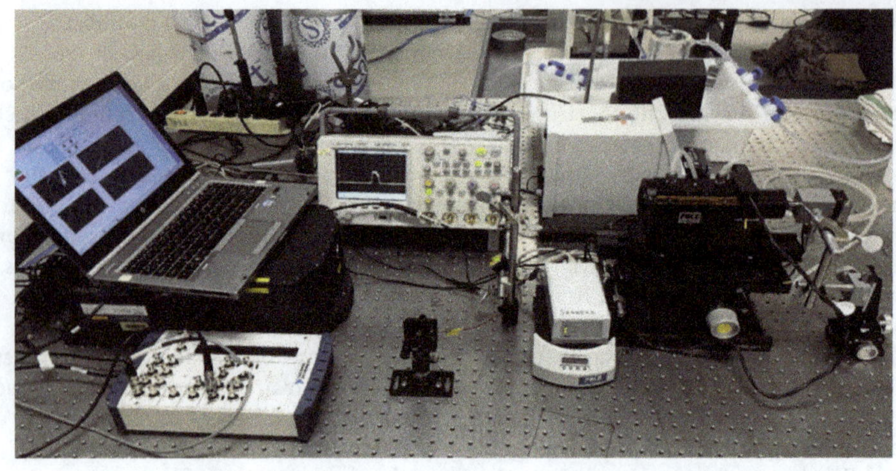

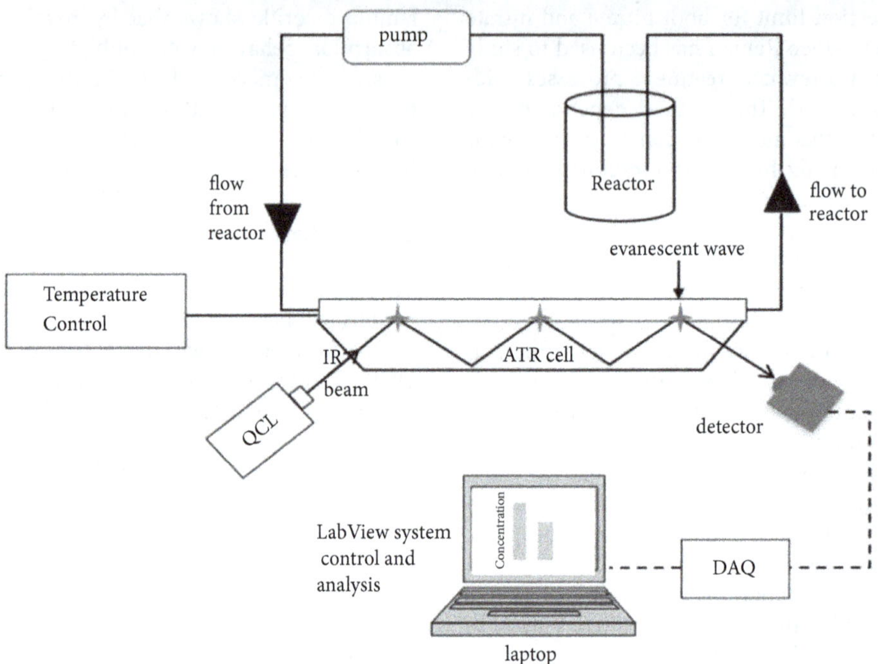

FIGURE 1: (a) Picture of the ATR-QCL laboratory set up and (b) Schematic of the ATR-QCL apparatus.

absorption spectrum of each analyte and therefore a definitive fingerprint for molecular identification. Using this prototype, chemical identification exhibits as attenuation at specific wavelengths of the QCL-IR beam exiting the ATR cell.

A LabVIEW™ [32] interface communicates with the QCL module to control the start/stop of the wavelength scan, data collection, display, and storage. The data is collected using the multifunctional data acquisition NI USB-6251 BNC from National Instruments. The laser module sends a trigger pulse to the DAQ board to begin data capture of each wavelength scan. The start and stop of each scan is determined by the time required for each laser module to tune through its wavelength range. As the QCL emits tuning wavelengths, the LabVIEW interface collects the output voltage from the detector as a function of time. The output voltage of the detector is directly related to the intensity or attenuation of the QCL source per wavelength. The LabVIEW interface uses a Fourier transform to extract the spectral information from the light intensity collected as a function of time at the frequency of the laser pulse rate. The total time from start to end to acquire one spectrum is about 45 seconds with a resolution of 0.6 to 2 nm, about 5 cm^{-1}. Thus the spectral resolution of this system is comparable to or better than reported systems for online FTIR spectroscopy [33]. The laboratory setup and schematic diagram are shown in Figures 1(a) and 1(b), respectively.

Solutions of sodium nitrate (10 mM to 2 M), hydroxylamine nitrate (10 μM to 2 M), nitric acid (1-4 M), and sodium nitrite (50 μM to 1 M) were prepared from standard lab reagents (JT Baker batch 45228, Aldrich Lot # 16412DOV, Fisher Lot # 1212072, respectively) without additional purification. Solutions were made up to the mark with deionized water (DI, 18.0 MΩ·cm) in 100 mL volumetric flasks. Solutions were either pipetted directly into the ATR trough or were circulated from a reaction vessel using a peristaltic pump running at $16.5\,cm^3{\cdot}min^{-1}$, a flow rate that did not affect the spectra. The reaction vessel, 500 mL, was a custom double-walled glass vessel that could be independently heated using a circulating water bath and was stirred at 200 rpm. Experiments started by preheating reagents to the set temperature before mixing in the reaction vessel. In most cases, the ATR window was also preheated. Reaction times were recorded from the time at which the reagents were mixed in the magnetically stirred reaction vessel. Transit time between the reaction vessel and the ATR was 40s, and so gives the lower bound on time resolution with this configuration. Temperatures ranged from 35 to 80°C, as determined by a NIST standardized SPER Type K thermocouple.

3. Results and Discussion

As this is a spectroscopic technique using an ATR liquid cell, a modified Beer-Lambert law can explain the absorption behavior [34]:

$$A = \varepsilon c b' \tag{3}$$

where A is the absorbance, ε is the molar absorptivity, and b' is the effective path length. For ATR-FTIR spectroscopy, the effective path length is equal to the number of reflections of the QCL beam in the ATR crystal times the penetration depth.

$$b' = N d_p \tag{4}$$

In (4), N is defined as the number of reflections and d_p is the penetration depth at each reflection per wavelength. The penetration depth of the QCL beam is defined as

$$d_p = \frac{\lambda_1}{2\pi n_1 \left[\sin^2\theta - n_{21}{}^2\right]^{1/2}} \tag{5}$$

where n_1 is the index of refraction of the crystal [35, 36], n_2 is the index of refraction of the sample medium in contact with the crystal [37], n_{21} equals n_2/n_1, and $\lambda_1 = \lambda_{\text{vaccum}}/n_1$. Both two and four mm thick germanium windows were used in this experiment. Figure 2 shows representative penetration depths for a Ge window at room temperature, for our wavelength range of 3.77-12.50 μm.

To analyze the spectral data, five individual wavelength scans of each sample were taken for averaging. Using the average value, the transmission of the sample was corrected for matrix absorption, I/I_o, where I_o is the transmittance of water (blank) and I is the uncorrected transmittance of the sample. From the corrected transmittance value, the absorption is calculated by

$$A = -\log\frac{I}{I_0} \tag{6}$$

Absorption spectra of hydroxylamine nitrate (1 mM to 1 M) across the four QCLs are shown in Figure 3. Calibration curves for analytes of interest were prepared using characteristic peak heights or peak areas for various concentrations, as shown in Figure 4. The correlation with concentration

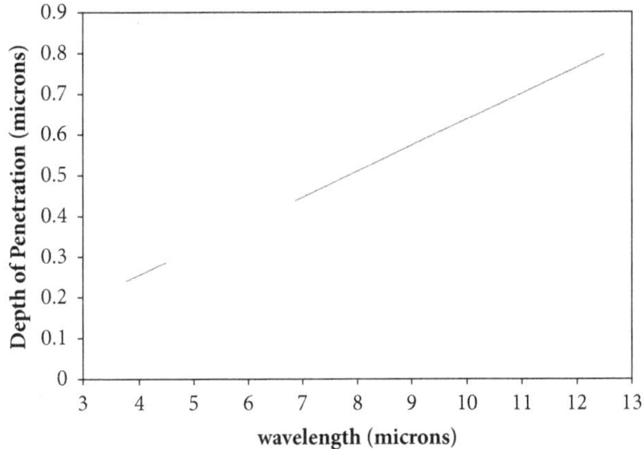

FIGURE 2: Calculated penetration depth of the QCL wavelength regions, 3.77-12.50 μm for a Ge window.

was much better for the peak at 7.47 μm than the other absorptions, suggesting that this spectral feature had the least overlap with peaks of other species in solution, although the slopes of the peaks at 7.47 and 3.77 μm are very similar. In this case, with the baseline being pure water, the difference in correlation reflects the signal-to-noise ratio for the peaks at 7.47 and 3.77 μm. The peak at 3.77 μm is at the far blue end of the QCL spectrum where the laser power density is much lower than in the center of the spectrum. The signal-to-noise ratio may also be influenced by the subtraction of the nearby strong absorption for CO_2. The peak at 9.81 μm appears to have a lower sensitivity to HAN than the other peaks and may show increased noise being located at the limits of the power spectra of two QCLs. It is possible that the choice of a different reference solution, such as HNO_3 solution, would provide a background spectrum that would give an improved correlation for the peaks at 3.77 and 9.81 μm, but it would not be appropriate for the peak at 7.47 μm.

To calculate the standard deviation on absorption, five intensity ratios from separate scans were used to calculate absorption values using (6). The standard deviation of the noise of the blank at the peak absorption wavelength was then used to calculate the limit of detection (LOD) and the limit of quantification (LOQ) according to IUPAC [38]. The LOD and LOQ are defined as 3.3 σ/m and 10 σ/m, respectively, where σ is the standard deviation and m is the slope of the calibration curve [39]. It has been reported in the literature [40] that the hydroxylamine vibrational spectrum absorbs in the wavelength region covered by our ATR-QCL system, including: the N-OH stretch at 9.7 μm, the NO_3 combination stretch at 7.4 μm, and the N-H stretch at 3.8-4.1 μm.

Table 1 shows the corresponding calculated values of LOD and LOQ, in grams per liter, for the hydroxylamine nitrate system using the ATR-QCL measurement system.

To demonstrate molecular species detection in high concentration nitrate, we studied the infrared absorption of the aqueous chemical system of $NaNO_2$ in the presence of $NaNO_3$. Spectra of different concentrations of $NaNO_2$ were collected using the QCL-ATR prototype and are shown in Figure 5. The concentration of $NaNO_2$ in $NaNO_3$ was

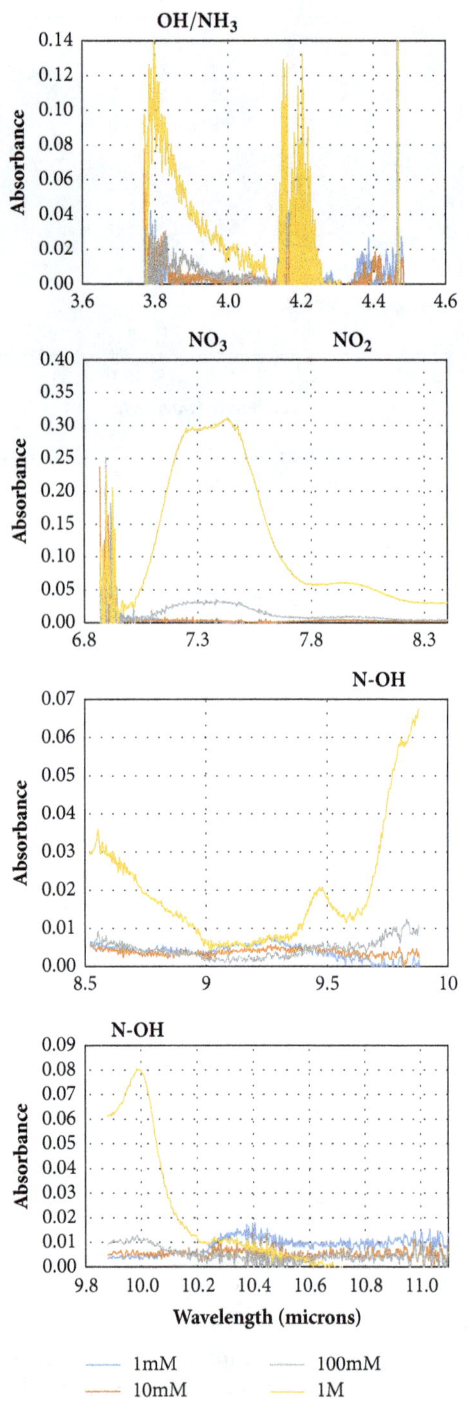

FIGURE 3: ATR-QCL absorption spectra of hydroxylamine nitrate (1 mM to 1 M) across the spectral range of the four QCL lasers used in this study.

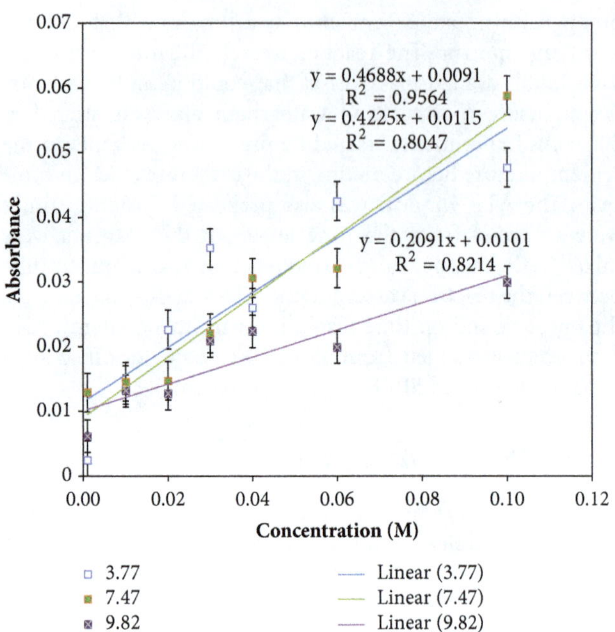

FIGURE 4: Calibration curves for hydroxylamine nitrate using the QCL-ATR system. The response was very similar for the peaks at 3.77 and 7.46 μm, with the latter giving better linearity. The sensitivity at 9.81 μm was half that seen at the shorter wavelengths.

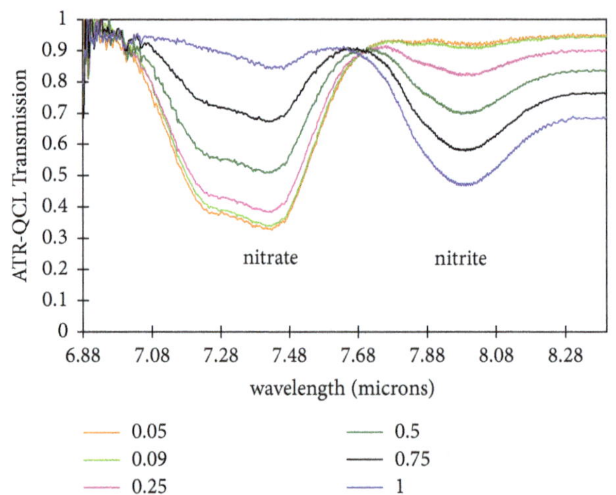

FIGURE 5: ATR-QCL transmission spectra of nitrite molar fraction range of 0.05-1 in presence of nitrate 1 M solution. The isobestic point at ~7.7 μm is evident from the data.

varied from 0.05 to 1 molar fraction. The absorption peaks correspond well to reference spectra downloaded from the database of the National Institute of Standards and Technology (NIST) [41, 42], although the isobestic point is shifted from 7.60 to 7.65 μm with the QCL system. Regression analysis gave a linear correlation between peak height and concentration, as well as peak area and concentration. Thus, changes in transmission profile can be related to the changes in concentrations in the flow cell. This confirms the capability of the ATR-QCL system to identify multiple species in a chemical system at different concentrations.

To validate the performance QCL-ATR flow system in terms of spectral absorption, an ATR-FTIR spectrum was taken using an FTIR Digilab FTS 7000, equipped with a diamond ATR from Harrick, KBr beam-splitter (Mid-IR), and DTGS detector. An aqueous solution of HAN with concentration of 100 mM was used for this test and the results of both techniques compared. Differences in signal

TABLE 1: Calculated LOD and LOQ (g·L^{-1}).

Wavelength (μm)	peak	LOD (g·dm^{-3})	LOQ (g·dm^{-3})
3.778	OH/NH_3	0.5-1.14	3.45
7.467	NO_3	0.31-0.634	2.11
9.754	N-OH	1.5-2.94	9.81

LOD=3.3 σ/slope, where σ = STDEV of the blank.
LOQ=10 σ/slope.

strength were observed for the absorption peaks for the two measurement methodologies. For example, the NO_3 peak in the FTIR spectra showed a maximum absorbance of 0.007 at 7.0 μm meanwhile the ATR-QCL system showed an absorbance of 0.035 for the same concentration, a factor of five increases in sensitivity. In the case of the NH_3 peak at 3.7 μm, the FTIR shows an absorbance of 0.002 whereas the ATR- QCL spectroscopy system has an absorbance of about 0.04, a factor of 20 differences in sensitivity.

Sensitivity differences can be attributed to the power of the excitation light source, the optical configuration of the sampling cell, and the choice of detector, i.e., MCT versus DTGS. The sensitivity in the QCL system was also affected by the power spectrum of each laser, giving rise to much lower signal-to-noise at low and high wavelengths. As it was not possible in either case to measure the power entering the cell and that absorbed by the analyte, it is not possible to assign which factor played the most important role. However, the comparison with FTIR-ATR does provide an indication of which factors are important in determining the sensitivity of the QCL-ATR method.

To demonstrate the capability of the system to monitor reactions in real time, the reaction between hydroxylamine nitrate and nitric acid was studied. These species will react quickly at high acid concentrations and at high temperatures [29]. In one configuration, 2 mL of HAN (1M) was placed directly on the ATR work plate preheated to 80°C. Nitric acid was added to the plate via pipette (0.5 mL of 4 M HNO_3) and the spectrum was monitored over time as the system reacted. However, the use of small volumes and the open plate led to spectral changes that could be mainly attributed to evaporation from the ATR trough.

The tests were repeated in a flowing system as described earlier, involving 200 mL total volume of reagents, with the nitric acid being added to the hydroxylamine nitrate at the start of the experiment. In this case, a cap on the ATR trough minimized evaporation and the reaction was followed for an hour. The obscuring effects seen with the heated trough plate configuration were not observed. A spectrum of HAN 1 M solution was taken before combining reagents, to enable comparison of absorption behavior before and after addition of HNO_3. After the HAN spectrum was taken, the reaction was started by the addition of HNO_3 and the reaction mixture was monitored frequently by taking spectra (up to every 45 s) for an hour. Changes in the spectra observed at specific wavelengths, associated with the N-H and N-OH bonds of the HAN molecule, indicate changes in HAN concentration. Figure 6 gives raw data from the flowing reaction, in the spectral region from 8.5 to 9.9 μm, or the stretching region in which N-OH bonds absorb.

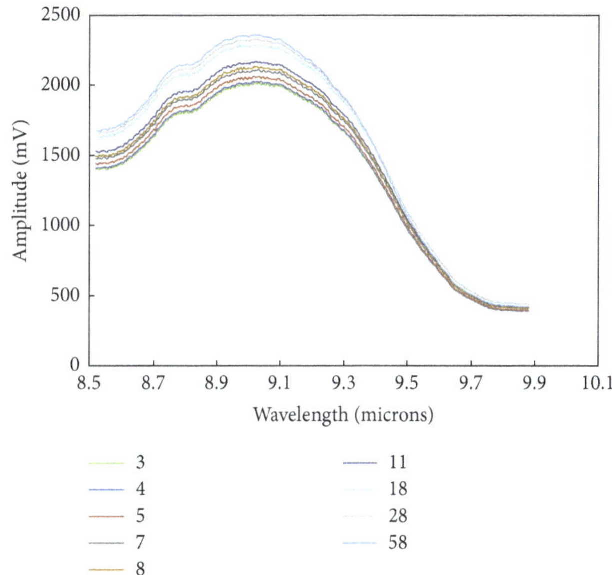

FIGURE 6: ATR-QCL raw data, signal amplitude (mV) in the absorption region of hydroxylamine nitrate centered at 8.95 μm, taken during the reaction of HAN 0.5M and HNO_3 2 M at 80°C. The legend refers to the reaction time in minutes.

Figure 7 shows changes in hydroxylamine nitrate concentrations derived from peak heights for the feature at 3.77 μm attributed to the N-H stretch of the amines, directly related to the HAN molecular structure. Final conditions have 0.5 M hydroxylamine nitrate reacting with 2 M HNO_3 at 80°C. A steady decrease in the concentration of hydroxylamine nitrate was observed from 3 to 60 min, showing consumption of HAN as it reacted with HNO_3. A lag period and even a slight increase in hydroxylamine concentration were observed, before the expected decrease. The increases could have arisen from effects due to mixing coupled with short-term autocatalysis [23]. These effects may also explain the slightly larger than expected overall decrease in hydroxylamine concentration, but this most likely arose as an artifact from the lag in sampling from the reaction vessel to the ATR cell. The peak at 9.95 μm also showed an abrupt increase in absorption in the first six minutes under the influence of the shoulder of the strong nitrate absorbance from the increase in nitric acid. The peak height then slowly decreased under the influence of hydroxylamine reaction, but the overall effect was much

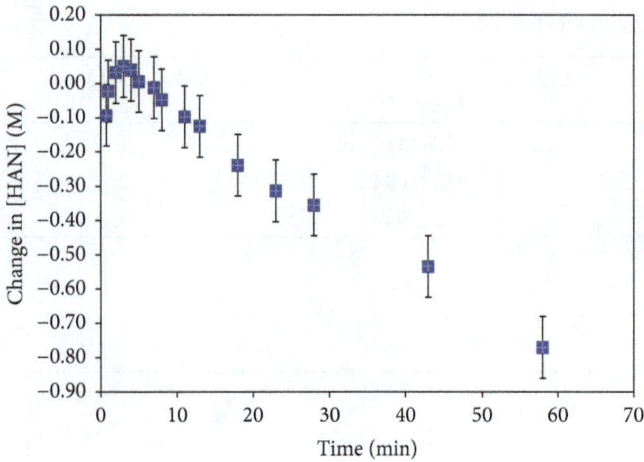

FIGURE 7: Depletion of HAN as a function of time, from ATR-QCL spectra at 3.77 μm. Final conditions are HAN 0.5M and HNO_3 2 M aqueous solution at 80°C.

smaller than that of the 3.77 μm feature. Thus, it is useful to be able to compare multiple absorptions, as some may be more appropriate for following a reaction in a particular chemical environment than others. In addition, choice of which background spectrum should be used as I_0 in the equation (6) is critical in determining which spectral effects arise from reagent mixing and sampling and which are representative of the kinetics. Details on the chemical kinetics of hydroxylamine nitrate decomposition will be discussed in a future publication.

4. Conclusion

We have developed an approach for high sensitivity and real-time online measurements to monitor chemical processes in aqueous systems. An aqueous reactive system was continuously sampled using mid infrared (Mid-IR) external cavity quantum cascade laser (QCL) high-resolution spectroscopy to detect and identify chemical species in strong nitric acid. The analytical method provides high selectivity, wide dynamic range, and flexibility and can be deployable at a location close to a shielded chemical reactor or "hot cell".

The sensitivity of the technique to detect small changes in the spectral region due to changes in the chemical molecular structure has been demonstrated. Single or multiple wavelengths can be used to monitor specific absorption behavior of chemical species as a function of time increasing the selectivity of the method. Because infrared spectroscopy provides specific information about the molecular structure, changes observed are specific to each reagent and product and its concentration. This technique can potentially be highly selective to the chemical composition of the aqueous system under investigation, particularly if QCL power spectra are centered on the absorptions of interest. The QCL-ATR configuration was found to provide resolution comparable to an FTIR and provides sufficient sensitivity to detect reactive chemicals, such as hydroxylamine nitrate, in strong (4M) nitric acid.

To perform an analysis of a target nitrate/nitrite system related to the separation of Np and Pu for Pu-238 production, a flow cell was used for continuous sampling of a liquid slipstream. IR spectroscopy was used to detect and identify chemical changes related to the decomposition of HAN in strong nitric acid, an important redox reagent used in the process. Unlike current analytical methods employed in Pu-238 production, this standoff or off-set method does not require the collection of a sample for analysis. Hence, this method can provide online monitoring of the concentration and chemical reactions of HAN, providing information needed for process control.

Disclosure

Acknowledgments

The work performed was supported by the Laboratory Director's Research and Development Program of Oak Ridge National Laboratory. Oak Ridge National Laboratory is operated for the U.S. Department of Energy by UT-Battelle under contract no. DE-AC05-00OR22725.

References

[1] A. Edelmann, J. Diewok, J. R. Baena, and B. Lendl, "High-performance liquid chromatography with diamond ATR-FTIR detection for the determination of carbohydrates, alcohols and organic acids in red wine," *Analytical and Bioanalytical Chemistry*, vol. 376, no. 1, pp. 92–97, 2003.

[2] V. Acha, M. Meurens, H. Naveau, and S. N. Agathos, "ATR-FTIR sensor development for continuous on-line monitoring of chlorinated aliphatic hydrocarbons in a fixed-bed bioreactor," *Biotechnology and Bioengineering*, vol. 68, no. 5, pp. 473–487, 2000.

[3] Y. Chen, Y.-H. Zhang, and L.-J. Zhao, "ATR-FTIR spectroscopic studies on aqueous $LiClO_4$, $NaClO_4$, and $Mg(ClO_4)_2$ solutions," *Physical Chemistry Chemical Physics*, vol. 6, no. 3, pp. 537–542, 2004.

[4] P. Jouy, M. Mangold, B. Tuzson et al., "Mid-infrared spectroscopy for gases and liquids based on quantum cascade technologies," *Analyst*, vol. 139, no. 9, pp. 2039–2046, 2014.

[5] B. Mizaikoff, "Waveguide-enhanced mid-infrared chem/bio sensors," *Chemical Society Reviews*, vol. 42, no. 22, pp. 8683–8699, 2013.

[6] M. Brandstetter, A. Genner, K. Anic, and B. Lendl, "Tunable external cavity quantum cascade laser for the simultaneous determination of glucose and lactate in aqueous phase," *Analyst*, vol. 135, no. 12, pp. 3260–3265, 2010.

[7] S. M. Eid, M. K. Abd El-Rahman, M. R. Elghobashy, and K. M. Kelani, "Attenuated Total Reflectance Fourier Transformation Infrared spectroscopy fingerprinted online monitoring of the kinetics of circulating Butyrylcholinesterase enzyme during metabolism of bambuterol," *Analytica Chimica Acta*, 2018.

[8] S. Pengel, B. Schönberger, S. Nayak, and A. Erbe, "Attenuated Total Reflection Mid-IR-spectroscopy for Electrochemical Applications using a QCL," in *Proceedings of the Laser Applications to Chemical, Security and Environmental Analysis*, p. LT6B.1, San Diego, CA, 2012.

[9] M. R. Alcaráz, A. Schwaighofer, C. Kristament et al., "External-Cavity Quantum Cascade Laser Spectroscopy for Mid-IR Transmission Measurements of Proteins in Aqueous Solution," *Analytical Chemistry*, vol. 87, no. 13, pp. 6980–6987, 2015.

[10] C. W. Van Neste, M. E. Morales-Rodríguez, L. R. Senesac, S. M. Mahajan, and T. Thundat, "Quartz crystal tuning fork photoacoustic point sensing," *Sensors and Actuators B: Chemical*, vol. 150, no. 1, pp. 402–405, 2010.

[11] C. W. Van Neste, L. R. Senesac, and T. Thundat, "Standoff photoacoustic spectroscopy," *Applied Physics Letters*, vol. 92, 234102 pages, 2008.

[12] C. W. Van Neste, L. R. Senesac, and T. Thundat, "Standoff spectroscopy of surface adsorbed chemicals," *Analytical Chemistry*, vol. 81, no. 5, pp. 1952–1956, 2009.

[13] P. Bingham, M. E. Morales-Rodriguez, P. Datskos, and D. Graham, "Multi-spectral infrared computed tomography," in *Proceedings of the Computational Imaging XIV 2016*, USA, February 2016.

[14] R. Maamary, X. Cui, E. Fertein et al., "A quantum cascade laser-based optical sensor for continuous monitoring of environmental methane in Dunkirk (France)," *Sensors*, vol. 16, no. 224, 2016.

[15] L. R. Senesac, M. E. Morales Rodriguez, T. Thundat, M. K. Rafailov, and P. Datskos, "Standoff imaging of chemicals using," in *IR spectroscopy, SPIE Micro- and Nanotechnology Sensors, Systems, and Applications III*, pp. 2D1–8, Orlando, FL, 2011.

[16] M. E. Morales-Rodríguez, L. R. Senesac, S. Rajic, N. V. Lavrik, D. B. Smith, and P. G. Datskos, "Infrared microcalorimetric spectroscopy using quantum cascade lasers," *Optics Expresss*, vol. 38, no. 4, pp. 507–509, 2013.

[17] D. Cialla, A. März, R. Böhme et al., "Surface-enhanced Raman spectroscopy (SERS): progress and trends," *Analytical and Bioanalytical Chemistry*, vol. 403, no. 1, pp. 27–54, 2012.

[18] D.-W. Li, W.-L. Zhai, Y.-T. Li, and Y.-T. Long, "Recent progress in surface enhanced Raman spectroscopy for the detection of environmental pollutants," *Microchimica Acta*, vol. 181, no. 1-2, pp. 23–43, 2014.

[19] C. T. Buscher, R. J. Donohoe, S. L. Mecklenburg et al., "Raman spectroscopic study of the aging and nitration of actinide processing anion-exchange resins in concentrated nitric acid," *Applied Spectroscopy*, vol. 53, no. 8, pp. 943–953, 1999.

[20] J. F. van Staden, M. A. Makhafola, and D. de Waal, "Kinetic Study of the Decomposition of Nitrite to Nitrate in Acid Samples Using Raman Spectroscopy," *Applied Spectroscopy*, vol. 50, no. 8, pp. 991–994, 2016.

[21] A. Ianoul, T. Coleman, and S. A. Asher, "UV resonance raman spectroscopic detection of nitrate and nitrite in wastewater treatment processes," *Analytical Chemistry*, vol. 74, no. 6, pp. 1458–1461, 2002.

[22] United Nations, Recommendations on the Transport of Dangerous Goods, Manual of Tests and Criteria, United Nations, New York and Geneva, 2003.

[23] R. J. Gowland and G. Stedman, "Kinetic and product studies on the decomposition of hydroxylamine in nitric acid," *Journal of Inorganic and Nuclear Chemistry*, vol. 43, no. 11, pp. 2859–2862, 1981.

[24] C. Koch, M. Brandstetter, P. Wechselberger et al., "Ultrasound-enhanced attenuated total reflection mid-infrared spectroscopy in-line probe: Acquisition of cell spectra in a bioreactor," *Analytical Chemistry*, vol. 87, no. 4, pp. 2314–2320, 2015.

[25] G. Imorwitz and P. I. M. Keliher, "Spectrophotometric determination of nitrite with composite reagents containing sulphanilamide, sulphanilic acid or 4-nitroaniline as the diazotisable aromatic amine and N-(1-Naphthyl)ethylenediamine as the coupling agent," *Analyst*, vol. 109, pp. 1281–1286, 1984.

[26] L. Liu, M. Papadaki, E. Pontiki, P. Stathi, W. J. Rogers, and M. S. Mannan, "Isothermal decomposition of hydroxylamine and hydroxylamine nitrate in aqueous solutions in the temperature range 80-160 ∘C," *Journal of Hazardous Materials*, vol. 165, no. 1-3, pp. 573–578, 2009.

[27] P. A. Giguère and I. D. Liu, "Infrared spectrum, molecular structure, and thermodynamic functions of hydroxylamine," *Canadian Journal of Chemistry*, vol. 30, no. 12, pp. 948–962, 1952.

[28] M. J. Moorcroft, J. Davis, and R. G. Compton, "Detection and determination of nitrate and nitrite: a review," *Talanta*, vol. 54, no. 5, pp. 785–803, 2001.

[29] J. McFarlane, L. H. Delmau, D. W. DePaoli, C. H. Mattus, C. E. Phelps, and B. D. Roach, "Hydroxylamine Nitrate Decomposition under Non-radiological Conditions," Tech. Rep. ORNL/TM–2015/156, 2015.

[30] A. T. Kowal, "Anharmonic vibrational spectra of hydroxylamine and its 15N, 18O, and deuterium substituted analogs," *Spectrochimica Acta - Part A Molecular and Biomolecular Spectroscopy*, vol. 58, no. 5, pp. 1055–1067, 2002.

[31] J. M. Price, M. W. Crofton, and Y. T. Lee, "Vibrational spectroscopy of the ammoniated ammonium ions $NH_4^+(NH_3)_n$ (n = 1-10)," *The Journal of Physical Chemistry C*, vol. 95, no. 6, pp. 2182–2195, 1991.

[32] National Instruments, LabVIEW, http://www.ni.com/en-us/shop/labview.html.

[33] I. Marziano, D. C. A. Sharp, P. J. Dunn, and P. A. Hailey, "On-line mid-IR spectroscopy as a real-time approach in monitoring hydrogenation reactions," *Org Proc Res Devel*, vol. 4, pp. 357-61, 2000.

[34] J. D. Schuttlefield and V. H. Grassian, "ATR-FTIR spectroscopy in the undergraduate chemistry laboratory part I: Fundamentals and examples," *Journal of Chemical Education*, vol. 85, no. 2, pp. 279–281, 2008.

[35] B. J. Frey, D. B. Leviton, and T. J. Madison, *Temperature-Dependent Refractive Index of Silicon and Germanium*, CERN, 2006.

[36] A. J. Moses, *Refractive Index of Optical Materials in The Infrared Region*, Hughes Aircraft Company, Culver City, CA, 1970.

[37] S. T. Massie and M. Hervig, "HITRAN 2012 refractive indices," *Journal of Quantitative Spectroscopy & Radiative Transfer*, vol. 130, pp. 373–380, 2013.

[38] G. L. Long and J. D. Winefordner, "Limit of detection. A closer look at the IUPAC definition," *Analytical Chemistry*, vol. 55, no. 7, pp. 712A–724A, 2008.

[39] D. A. Armbruster and T. Pry, "Limit of blank, limit of detection and limit of quantitation," *The Clinical Biochemist Reviews*, vol. 29, pp. S49–S52, 2008.

[40] Z. Li, Q. Yang, X. Qi et al., "A novel hydroxylamine ionic liquid salt resulting from the stabilization of NH_2OH by a SO_3H-functionalized ionic liquid," *Chemical Communications*, vol. 51, no. 10, pp. 1930–1932, 2015.

[41] M. W. Chase Jr., "NIST-JANAF Thermochemical Tables, 4th Edition," *Journal of Physical and Chemical Reference Data*, 1996.

[42] National Institute of Standards and Technology, Chemistry Webbook, NIST, Gaithersburg, MA, 2017.

18

Comparative Glycopeptide Analysis for Protein Glycosylation by Liquid Chromatography and Tandem Mass Spectrometry: Variation in Glycosylation Patterns of Site-Directed Mutagenized Glycoprotein

Young Hye Hahm,[1] Sung Ho Hahm,[1] Hyoun Young Jo,[1] and Yeong Hee Ahn[2]

[1] *Rophibio Inc., Cheongju 28160, Republic of Korea*
[2] *Department of Biomedical Science, Cheongju University, Cheongju 28160, Republic of Korea*

Correspondence should be addressed to Yeong Hee Ahn; ahnyh@cju.ac.kr

Academic Editor: Günther K. Bonn

Glycosylation is one of the most important posttranslational modifications for proteins, including therapeutic antibodies, and greatly influences protein physiochemical properties. In this study, glycopeptide mapping of a reference and biosimilar recombinant antibodies (rAbs) was performed using liquid chromatography-electrospray ionization tandem mass spectrometry (LC-ESI-MS/MS) and an automated Glycoproteome Analyzer (GPA) algorithm. The tandem mass analyses for the reference and biosimilar samples indicate that this approach proves to be highly efficient in reproducing consistent analytical results and discovering the implications of different rAb production methods on glycosylation patterns. Furthermore, the comparative analysis of a mutagenized rAb glycoprotein proved that a single amino acid mutation in the Fc portion of the antibody molecule caused increased variations in glycosylation patterns. These variations were also detected by the mass spectrometry method efficiently. This mapping method, focusing on precise glycopeptide identification and comparison for the identified glycoforms, can be useful in differentiating aberrant glycosylation in biosimilar rAb products.

1. Introduction

Glycosylation involves the covalent attachment of glycans to specific amino acid residues on protein and is one of the important posttranslational protein modification processes [1]. Glycosylation alters the properties of proteins including pharmacokinetics, effector functions, solubility, and stability [2]. Aberrant glycosylation in glycoproteins has been related to the occurrence and progression of certain diseases [3]. Careful observation of protein glycosylation is crucial for the development of stable and effective drugs. Furthermore, they are crucial in comparing biosimilar products to reference drugs, as mandated by regulatory agencies [4]. Changes in manufacturing process conditions for biologics, such as process optimization, scale-up production, and site changes, may impact glycosylation patterns of the resulting recombinant antibody (rAb) [5, 6]. As such, glycosylation is considered a critical quality attribute (CQA) for rAb therapeutics as it has the potential to determine whether the biosimilar candidate is highly similar to the reference drug. The glycosylation pattern of a rAb affects a wide spectrum of biological processes. Therefore, consistent glycosylation is necessary to prove the drug's ability to maintain safety and efficacy [7].

Mass spectrometry is a core technology in the field of proteomics with high-throughput performance and accurate digitalized informatics [8]. One particular mass analysis technique is matrix assisted laser desorption/ionization mass spectrometry (MALDI-MS) equipped with time-of-flight (TOF) analyzer [9–12]. Another widely used method in mass spectrometry is liquid chromatography-electrospray ionization tandem mass spectrometry (LC-ESI-MS/MS) [13–17]. LC-ESI-MS/MS is a robust, high-throughput method in identifying and quantifying glycopeptides in enzymatic digests from various proteomic samples. High-throughput ESI-MS/MS for glycopeptides, initially separated by LC, allows for the identification and quantification of all detectable features, which, in turn, provides a more detailed account of protein glycosylation patterns. Data analysis for large amounts of raw mass data resulting from LC-ESI-MS/MS for the glycoproteome is a challenging task; therefore, various search engines have been developed such as Glycomaster DB [18], Byonic [19], MAGIC [20], and Glycoproteome Analyzer (GPA) [21]. GPA is capable of identifying site-specific N-glycopeptides efficiently and features the use of 3-top monoisotopic mass peak intensity of glycopeptides [21, 22]. The high-speed mapping of glycopeptides using GPA has proven to display analytical efficiency with a false display rate (FDR) ≤1%.

Here, we have introduced a practical method for mapping and comparing the glycosylation patterns in rAb glycoproteins, in which glycopeptide samples prepared by in-solution or in-gel protein digestion are analyzed by LC-ESI-MS-MS. The developed method is applied for comparative analysis of glycosylation patterns of biosimilar rAbs as well as a mutagenized rAb glycoprotein.

2. Experimental

2.1. Reagents and Chemicals. The reference rAb glycoprotein (Adalimumab, commercially known as Humira) was obtained from G Sam Hospital (Gyeonggi-do, Korea). HiTrap Mabselect SuRe columns were purchased from GE Healthcare and C_{18} trap column was purchased from Harvard Apparatus (Holliston, MA USA). Trypsin for protein digestion was obtained from Promega (Madison, WI, USA). 1,4-dithiothreitol (DTT), iodoacetamide (IAA), trifluoroacetic acid (TFA), and formic acid (FA) were purchased from Sigma Aldrich (St. Louis, MO, USA). HPLC-grade water and acetonitrile (ACN) were purchased from J.T. Baker (Phillipsburg, NJ, USA). CHO-k1 cells were purchased from ATCC (Manassas, VA, USA) and ExpiCHO-s cells were purchased from Thermo Fisher Scientific (Waltham, MA, USA).

2.2. Expression Vector for the Biosimilar. The light chain and heavy chain genes for biosimilar rAb were synthesized (Genscript, Piscataway, NJ, USA) and cloned into a custom expression vector, pCPp2-CMV. The expression vector contains both the heavy chain and the light chain genes for the biosimilar rAb controlled by the human CMV promoter, respectively. The vector also contains puromycin-resistance gene as a selectable marker under SV40 promoter.

2.3. Transient Expression of Biosimilar rAbs. For the transient expression of biosimilar rAbs, ExpiCHO-s cells were transfected with the expression vector and maintained in fed-batch culture following manufacturer's protocol for max titer. Briefly, 50 ml of Expi-CHO-s cells cultured in ExpiCHO culture medium in a 250 shaker flask was transfected with 50 μg of the expression vector DNA using Expifectamine CHO reagent. Transfected cells were maintained on shaker at 32°C in a CO_2 incubator and ExpiCHO enhancer and feed were added as recommended. Cells were monitored every 24 hours for the cell growth and viability. Cells were harvested 7-10 days after transfection before cell viability drops below 75%. Cell supernatant was collected by centrifugation at 3,000 g for 30 min and filtered through a vacuum filtration unit (0.22 um). Filtered cell supernatant was subjected to purification for biosimilar rAbs by protein A affinity chromatography using HiTrap Mabselect SuRe columns.

2.4. Mutagenesis of Biosimilar rAb Generation. A mutagenized heavy chain gene containing tryptophan to valine substitution within the CH2 domain of the heavy chain (at position C41 by IMGT codon numbering or 290 by Kabat numbering) was synthesized and cloned into the expression vector containing the light chain gene. The newly cloned DNA was transfected into CHO-k1 cells and mutagenized rAb was purified from cell supernatant using HiTrap Mabselect SuRe columns.

2.5. Stable Expression of the Biosimilar rAbs after Cell Line Generation. Cell lines for the biosimilar rAbs were generated using suspension-adapted CHO-k1 cells. CHO-k1 cells were preadapted in serum-free suspension culture and maintained in CDM4CHO (Hyclone, Little Chalfont, UK) chemically defined medium. 20 μg of the wild-type expression vector was used to transfect 2×10^6 cells by electroporation following a proprietary recombination-based transfection protocol. Transfected cells were monitored daily and subject to media change and antibiotic selection with puromycin (8 ug/mL) every 48 hours for 12-14 days. Stably transfected cell pool was then used for the selection of high titer cell lines by FACS cell sorting using MoFlo-XDP (Beckman Coulter, Brea, CA, USA). FACS cell sorting was performed after staining the cells using R-PE conjugated $F(ab')_2$ fragment goat anti-human IgG Fcγ fragment specific antibody (Jackson ImmunoResearch Laboratories, Inc., West Grove, PA, USA) following the protocol in Brezinsky et al. (2003) [23]. High titer cell lines were sorted directly into 96-well plates containing 200 μl of proprietary single cell growth medium per well, based on the R-PE fluorescence expression levels, and after excluding dead cells and doublets. Cell sorting was performed using the single cell sort mode to increase the efficiency sorting one cell per well. Cells sorted into 96 well plates were monitored for cell growth and clonality by scanning the plate using Clone Select Imager (Molecular Devices, San Jose, CA, USA). Approximately 15 days after the sorting, expression levels of the biosimilar rAb were determined with an ELISA assay and 20 clones with the highest expression levels were expanded for further analysis. A final clone with the highest cell titer was selected from the 20 clones, preserved in liquid nitrogen stocks, and then used for the expression of the biosimilar in a fed-batch condition. For the expression of the biosimilar from the cell line, cells were seeded in 30 ml of CDM4CHO with 8 μg/ml puromycin and 800 μl of Sheff Pulse II (Kerry Bioscience, Ithaca, NY, USA) feeding medium was added to the culture every 48 hours. Cells were maintained at 32°C for 10-12 days and harvested when cell viability dropped below 75%. Biosimilar rAb was purified from the cell culture supernatant using HiTrap Mabselect SuRe protein A columns.

2.6. Protein Quantitation with Indirect ELISA Assay. 96-well plates for ELISA assay were prepared by coating the wells with goat anti-human IgG whole molecule primary antibody (5 μg/mL, Sigma Aldrich) and blocking with tris-buffered saline with BSA pH 8. Purified IgG from human serum was used as a standard in serial dilutions in two folds starting from 50 ng/ ml. Biosimilar rAb samples were diluted so that the titer estimation can be made within the standard curve range. Standards and samples were added to the wells in duplicates. After extensive washing, the wells were probed with goat anti-human IgG- (gamma-chain specific-) peroxidase antibody (1:20,000, Sigma Aldrich) and colorized with 3,3′,5,5′-Tetramethylbenzidine (TMB) liquid substrate. Color change was induced with 1% HCl and the absorbance values were measured at 450 nm with VMax microplate reader (Molecular Devices).

2.7. Protein Characterization with SDS-PAGE and Western Blotting. The protein was characterized with sodium dodecyl sulfate polyacrylamide gel electrophoresis (SDS-PAGE) and Western blotting. Samples were reduced in a 1x sample reducing agent at 90-10°C for 5 min. The samples, along with a protein standard, were loaded into a 4-20% Tris-PAG precast gel (Komabiotech, Seoul, Korea) with 1x Tris-Glycine SDS running buffer for 90 min at a constant voltage of 120V. The resolved protein was either stained using Coomassie blue reagent overnight and destained with 50% ddH_2O, 40% methanol and 10% acetic acid before imaging or transferred to a nitrocellulose membrane using the TransBlot-Semidry Transfer System (Bio-rad) for the Western blotting. The membranes were blocked in 5% NFDM and incubated in a primary antibody solution containing goat anti-human IgG whole molecule antibody (1:100,000, Sigma Aldrich). The membrane was probed with rabbit anti-goat IgG HRP-conjugated antibody (1:20,000, Sigma Aldrich) and detected using a LumiGLO chemiluminescent substrate kit (KPL). The relative intensity of the protein bands was scanned using FusionFx (Vilber, Collegien, France).

2.8. In-Solution Enzymatic Digestion. 100 ug of protein sample obtained from protein A affinity chromatography was denatured with 50 mM ammonium bicarbonate (ABC) and 2 M urea. The sample was reduced by 100 mM DTT for 1 h at 35°C. After incubation, the sample was alkylated with 100 mM IAA for 1 h in a darkroom and digested with trypsin (0.125 ug/uL) overnight at 37°C. Digested sample was dried in a SpeedVac and stored in −20°C or reconstituted in 1% FA for desalting. SPE micro-spin column was equilibrated by centrifugation with wash buffer (99% H_2O, 1% FA) and elution buffer (80% ACN, 1% FA). Aliquot of the tryptic digests was applied to the column, washed twice, and eluted twice with 30 uL elution buffer. The eluted peptide sample was dried in a SpeedVac.

2.9. In-Gel Enzymatic Digestion. In-gel protein sample excised from SDS-PAGE was repeatedly dehydrated with 500 uL ACN for 10 min at room temperature. The gel slices were reduced with 75 uL DTT for 30 min at 56°C, alkylated with IAA for 20 min in a darkroom, and destained with 500 uL 100 mM ammonium bicarbonate/acetonitrile (ABC/ACN) for 30 min with periodic vortexing. After dehydration, the gel slices were rehydrated with 10 mM ABC and 10% ACN buffer containing trypsin and digested at 37°C overnight. 5% FA/ACN (1:2, vol/vol) was added to the digested mixture and incubated at 37°C for 15 min with slight vortex. The digested sample was dried in a SpeedVac and either stored in −20°C or reconstituted in 0.1% FA for MS/MS analysis. Aliquot of the tryptic digests was applied to SPE micro-spin column, washed twice, and eluted twice with 30 uL elution buffer. The eluted peptide sample was dried in a SpeedVac.

2.10. LC-ESI-MS/MS Analysis. Tryptic peptide sample reconstituted with water containing 0.1% formic acid was separated by a Nano Acquity UPLC system (Waters, USA). 5 uL aliquot of the peptide solution was applied to a C_{18} trap column (i.d. 180 um, length 20 mm, and particle size 5 um; Waters, USA) with an autosampler. The peptides were desalted in the trap column for 10 min at a 5 uL/min flow rate. The trapped peptides were back-flushed into a homemade C_{18} trap column (i.d. 100 um, length 200 mm, and particle size 3 um) for separation. Mobile phases A and B were composed with 100% H_2O containing 0.1% FA and 100% ACN containing 0.1% FA, respectively. The LC gradient began at 5% mobile phase B and was maintained for 15 min. The gradient was ramped up to 15% B for 5 min, 50% B for 75 min, and 95% B for 1 min. 95% B was maintained for 13 min and decreased to 5% B for 1 min. The column was finally re-equilibrated with 5% B for 10 min. LTQ Orbitrap Elite mass spectrometer (Thermo Fisher Scientific) equipped with a nanoelectrospray source was used to analyze the separated peptides. The electrospray voltage was set to 2.2 kV and the LTQ Orbitrap Elite mass spectrometer (Thermo Fisher Scientific) was operated in data-dependent mode during the chromatographic separation. The MS acquisition parameter for full-scan resolution was 60,000 in the Orbitrap for each sample. Data-dependent MS/MS scans were acquired by collision-induced dissociation (CID) and higher-energy collision dissociation (HCD). CID and HCD fragmentation scans were acquired in linear trap quadrupole (LTQ) mode with a 30-ms activation time and in Orbitrap at resolution 15,000 with a 20-ms activation time. A 35% normalized collision energy (NCE) and 5.0 Da isolation window were used for CID and HCD analyses. Previously fragmented ions were excluded for 300 s in all MS/MS scans. Raw mass data acquired were analyzed with the automated Glycoproteome Analyzer (GPA) algorithm to identify glycopeptides and then the resulting identified glycopeptides were further confirmed manually with careful investigation in retention time, chromatographic peak shape, isotopic mass distribution pattern, and fragmented ion matching.

Table 1: List of tryptic glycopeptides from triplicate analysis of reference rAb[a].

No.	Glycopeptides	Ion Charge (+)	Detected Mass (m/z)	Run 1		Run 2		Run 3	
				RT (min)	MS_Intensity	RT (min)	MS_Intensity	RT (min)	MS_Intensity
1	EEQYNSTYR_3_3_0_0	2	1142.959	27.70	92380184	28.39	3270910	28.40	10319768
2	EEQYNSTYR_3_3_1_0	2 3	1215.989 810.996	27.41	372193798	28.31	72807858	28.30	188794868
3	EEQYNSTYR_3_4_0_0	2 3	1244.499 830.000	27.70	91063963	28.39	15496872	28.42	15407082
4	EEQYNSTYR_3_4_1_0	2 3	1318.029 878.688	27.40	1541540913	28.28	398363214	28.30	815484054
5	EEQYNSTYR_3_5_1_0	3	946.380	27.60	7447300	28.32	5456578	28.30	7199469
6	EEQYNSTYR_4_2_0_0	2	1122.443	27.60	126933115	28.39	1113753	28.35	6721632
7	EEQYNSTYR_4_3_0_1	3	913.355	28.28	188888	28.88	2653427	28.90	1300713
8	EEQYNSTYR_4_3_1_0	2	1297.018	27.40	50607743	28.25	6312278	28.21	20746171
9	EEQYNSTYR_4_3_1_1	2 3	1442.557 962.043	28.10	29760350	28.81	55607576	28.80	28356277
10	EEQYNSTYR_4_4_1_0	2 3	1398.554 932.706	27.38	285767417	28.25	58136525	28.20	205614717
11	EEQYNSTYR_4_4_1_1	3	1029.738	28.10	5127014	28.80	36633653	28.85	12664730
12	EEQYNSTYR_4_5_1_0	3	1000.398	27.54	2426554	28.32	4640770	28.30	6085787
13	EEQYNSTYR_5_2_0_0	2 3	1203.473 802.652	27.50	542316083	28.28	16939021	28.29	69462798
14	EEQYNSTYR_5_4_1_0	3	986.723	27.36	3496656	28.10	2860860	28.10	10937706
15	EEQYNSTYR_5_4_1_1	3	1083.755	28.10	8174096	28.81	18861846	28.80	25603836
16	EEQYNSTYR_5_4_1_2	3	1180.784	28.52	1129042	29.20	3397473	29.29	1181060
17	EEQYNSTYR_6_2_0_0	2	1284.497	27.40	102132925	28.18	733668	28.21	14024518
18	EEQYNSTYR_6_3_1_0	3	973.049	27.36	4256849	28.18	638835	28.20	4721699
19	EEQYNSTYR_6_3_1_1	3	1070.081	28.04	5703122	28.74	18933930	28.70	8336957
20	EEQYNSTYR_7_2_0_0	2	1365.524	27.40	55356342	28.18	559700	28.21	5262829

[a]MS_Intensity is based on the summation of 3-top monoisotopic mass peak intensity [21]. If simultaneously detected, MS_Intensity values of 2^+ and 3^+ mass ions are summed.

3. Results and Discussion

3.1. LC-ESI-MS/MS Analysis of Reference rAb Glycoprotein. The reference rAb glycoprotein desalted with a HiTrap Mabselect SuRe protein A column (GE Healthcare) according to manufacturer's protocols was quantified by UV spectrometry, ELISA assay, and BCA assay and characterized by SDS-PAGE and Western blotting. The reference rAb glycoprotein sample desalted was reduced, alkylated in a darkroom, and digested with trypsin overnight. The digested sample was desalted with a SPE micro-spin column and analyzed by LC-ESI-MS/MS coupled with CID and HCD fragmentation techniques.

Figure 1 shows a base peak chromatogram (BPC) obtained from the analysis of the tryptic digests of the reference rAb glycoprotein. The obtained raw mass data was analyzed with GPA algorithm and the automatically assigned glycopeptides were further validated manually. Among the identified glycopeptides, EEQYNSTYR_3(hexose)_3(*N*-acetylhexosamine)_1(fucose)_0(sialic acid) was eluted at the retention time (RT), 28.25 min. The extracted ion chromatogram of the monoisotopic peak pattern of the precursor ion (2+) for glycopeptide EEQYNSTYR_3_3_1_0 was illustrated in Figure 1. Tandem mass spectra of the glycopeptide EEQYNSTYR_3_3_1_0, obtained by CID and HCD fragmentation techniques, were also shown in Figure 2. We confirmed that fragment ions were well-matched with their structures.

The analysis results of the reference rAb obtained by triplicate sample preparation and mass analyses were summarized in Table 1. LC-ESI-MS/MS with tryptic digestion detected 20 consistent glycosylation patterns with similar distributions. The glycosylation patterns normalized for each run were graphed together in Figure 3. We observed three major fucosylated glycopeptides, EEQYNSTYR_3_3_1_0, EEQYNSTYR_3_4_1_0, and EEQYNSTYR_4_4_1_0, which are reportedly predominant glycoforms found in recombinant antibodies produced in CHO cells [7, 24, 25]. In addition, high mannose forms with the relative ratios <13% were also observed—EEQYNSTYR_4_2_0_0, EEQYNSTYR_5_2_0_0, EEQYNSTYR_6_2_0_0, and EEQYNSTYR_7_2_0_0. These glycosylation patterns have been frequently observed in rAb

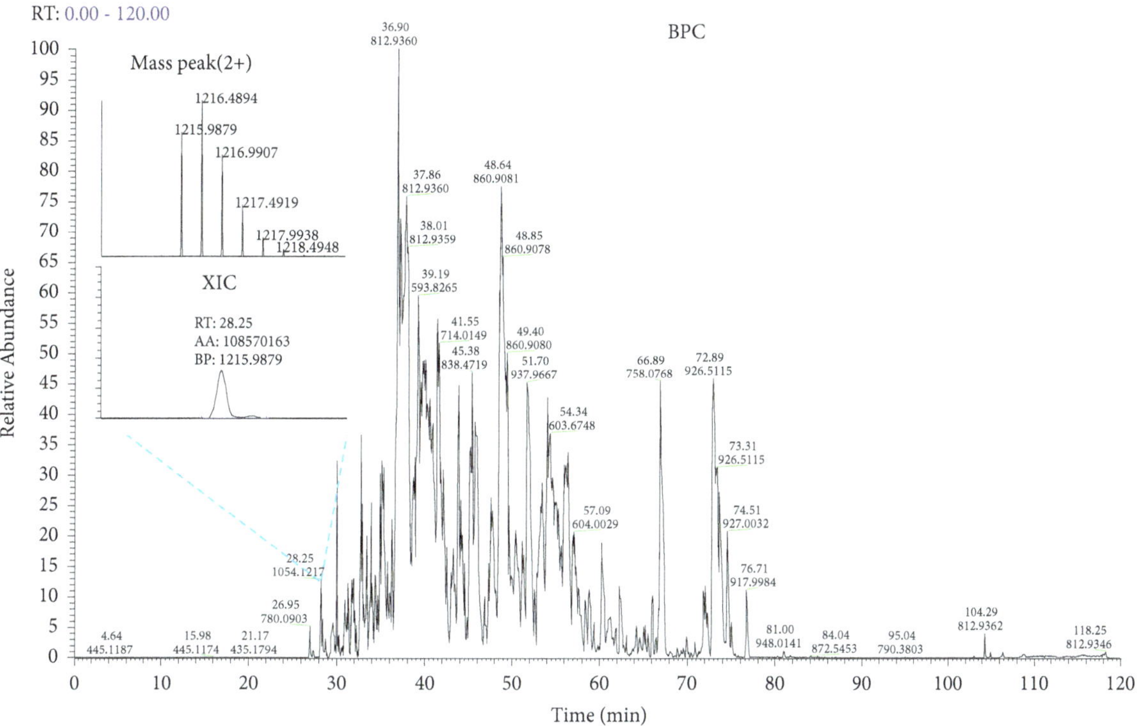

Figure 1: Base peak chromatogram (BPC) by LC-tandem MS analysis of the tryptic digests of reference rAb. The extracted ion chromatogram (XIC) of the glycopeptide EEQYNSTYR_3_3_1_0, identified as one of major components at RT 28.25 min, was also inserted together with the monoisotopic peak pattern of the glycopeptide.

therapeutics due to the fact that recombinant rAbs are not fully modified and substantially affected by posttranslational modifications [26, 27]. After confirming the reliability and repeatability in the analysis of the reference rAb glycoprotein, a comparative analysis was performed using biosimilar rAbs A and B, which were prepared by two different methods of production.

3.2. Comparative Analysis of Glycosylation Patterns in Biosimilar rAbs. Samples of biosimilar rAbs were obtained from either transiently transfected cells or a stably transfected cell line and referred to as biosimilar rAb A and B, respectively, for clarification purposes. The purified biosimilar rAbs A and B were digested with trypsin and analyzed with LC-ESI-MS/MS. The obtained raw mass data was analyzed with GPA algorithm and the identified glycopeptides were further validated manually. The glycosylation patterns acquired for the biosimilar rAb A and B were individually normalized and the mapping results were compared with those obtained from the analysis for the reference rAb sample (see Figure 4).

The major glycoforms for the biosimilar rAb A and rAb B as well as the reference rAb were consistent—EEQYNSTYR_3_3_1_0, EEQYNSTYR_3_4_1_0, and EEQYNSTYR_4_4_1_0. However, we observed variations in the minor glycosylation species not only between the reference rAb and biosimilar rAbs but also between the two biosimilar rAbs samples produced using different methods. The biosimilar rAb A and B displayed a decreased abundance for EEQYNSTYR_3_4_1_0 and an increased abundance for EEQYNSTYR_4_4_1_0 with the addition of galactose, in comparison to the reference rAb. The biosimilar rAb B displayed higher abundance for glycans with sialic acids—EEQYNSTYR_4_4_1_1, EEQYNSTYR_5_4_1_1, and EEQYNSTYR_5_4_1_2—in comparison to the biosimilar rAb A. These results suggest that LC-tandem mass spectrometry can be a highly sensitive tool for distinguishing subtle differences in minor glycosylation species.

3.3. Glycosylation Patterns in Mutagenized rAb Glycoprotein. We tested the potential validity of using LC-tandem mass spectrometry in the determination process for the similarity of biosimilar products to the reference materials in biosimilar samples with a single mutation in amino acid sequences. Unintentional mutations that arise during protein production can cause changes in glycosylation patterns, which in turn can impact the safety, quality, and potency of the recombinant rAbs [28]. A tryptophan to valine amino acid mutation was created in the vicinity of the glycosylation site of Adalimumab to test its effect on the pattern of glycosylation to be detected using LC-tandem mass spectrometry. Interestingly, the single amino acid mutation caused the inability for biosimilar rAb to bind to the protein A affinity column (data not shown). So, instead of purifying the mutagenized rAb using protein A columns, we performed the analysis using in-gel samples. Unpurified samples were analyzed by SDS-PAGE and the expected heavy chain and light chain bands were observed at 51 kDa and 26 kDa, respectively (see Figure 5). The bands excised from the gel were subjected to in-gel

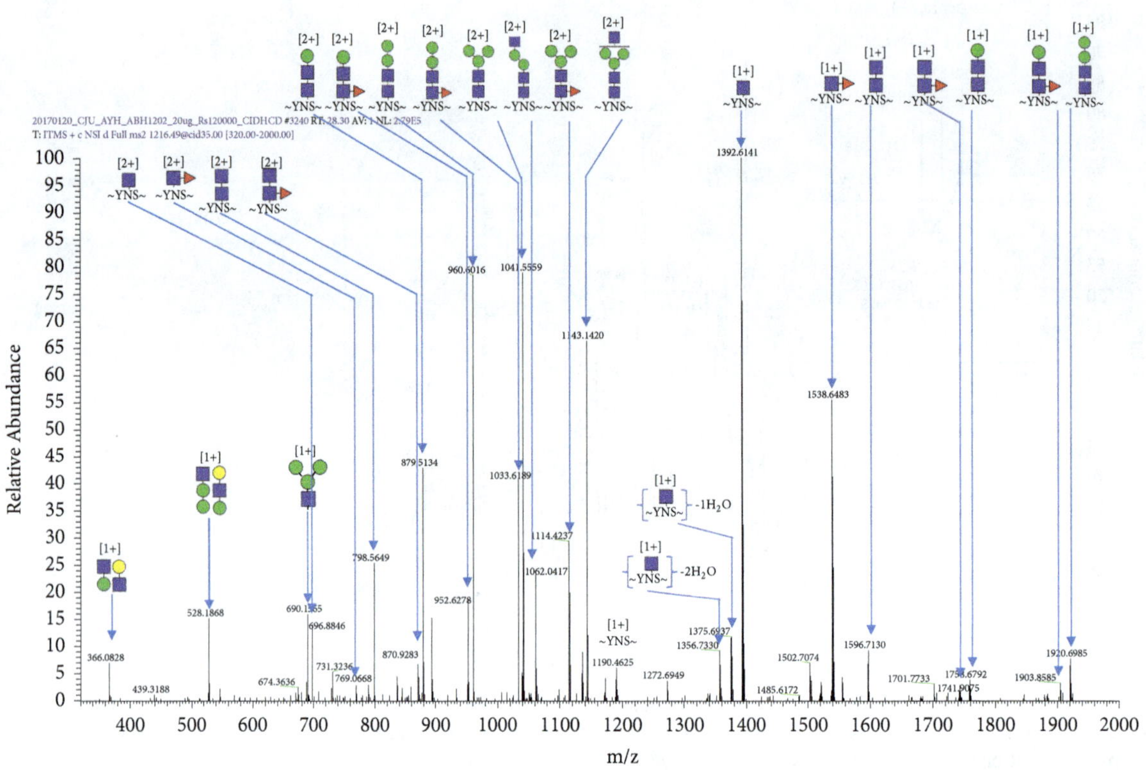

(a) MS/MS spectrum, EEQYNSTYR_3_3_1_0 : CID

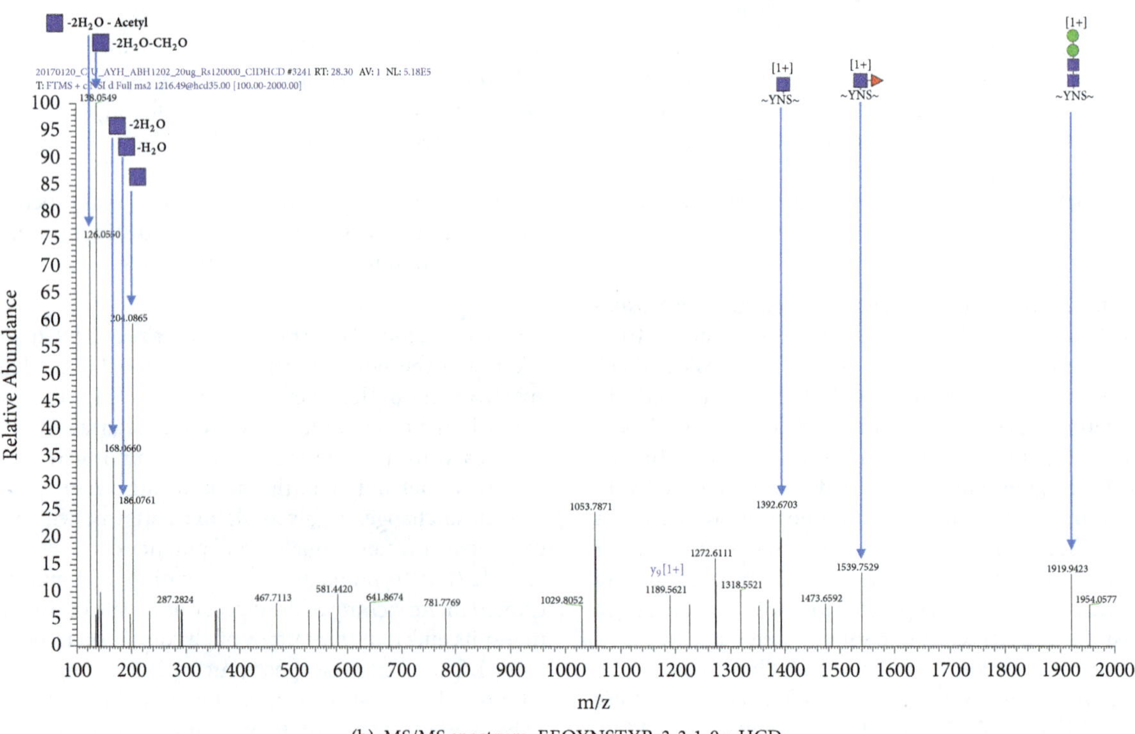

(b) MS/MS spectrum, EEQYNSTYR_3_3_1_0 : HCD

FIGURE 2: Tandem mass spectra of the glycopeptide EEQYNSTYR_3_3_1_0, obtained by (a) collision-induced dissociation and (b) higher-energy collision dissociation.

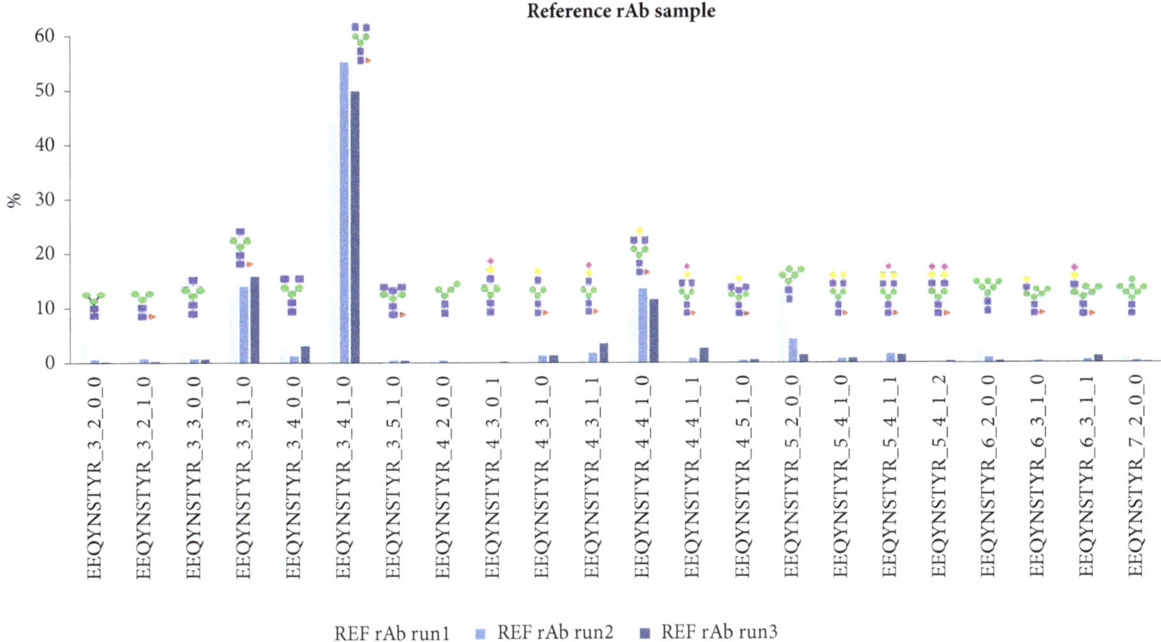

FIGURE 3: Triplicate analysis for glycosylation patterns of the reference rAb sample using LC-ESI-MS/MS. Each acquired mass dataset was normalized before comparative analysis. The sugar symbols: diamond (magenta) = sialic acid, circle (yellow) = galactose, square (blue) = GlcNAc, circle (green) = mannose, triangle (magenta) = fucose.

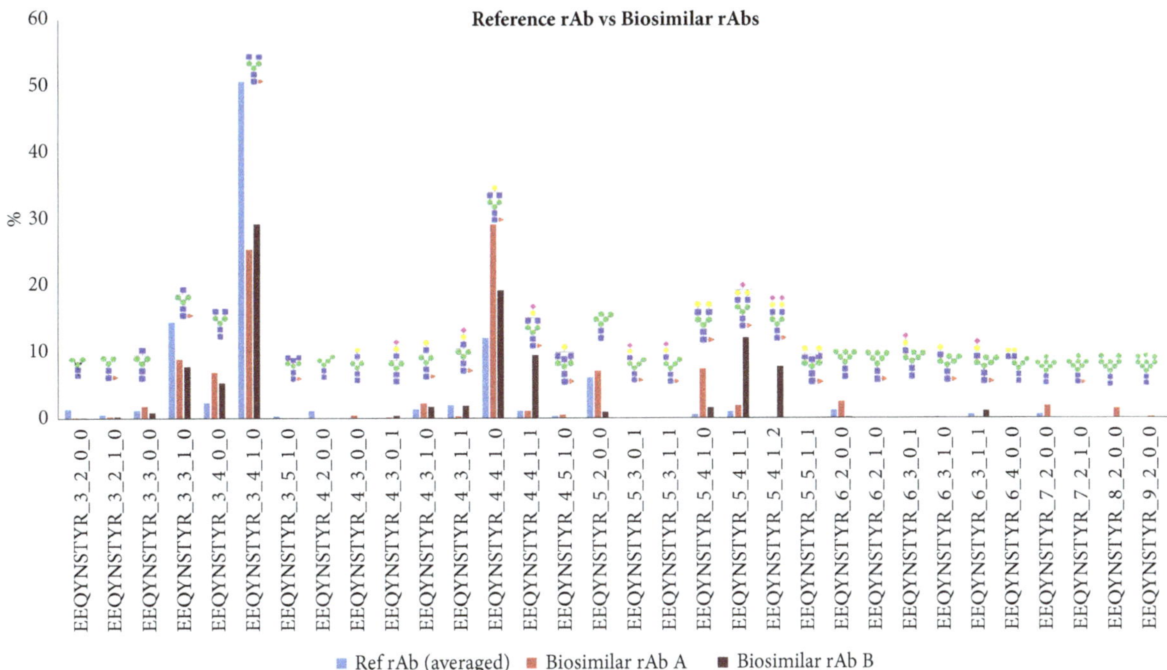

FIGURE 4: Comparison of glycosylation patterns of the reference rAb and the biosimilar rAb samples, obtained by LC-ESI-MS/MS. The glycosylation patterns for the reference rAb sample were obtained by triplicate analysis and averaged. All datasets acquired from each glycoprotein sample were normalized before comparative analysis.

tryptic digestion. The digested sample was desalted by ZipTip for mass spectrometry.

Prior to analysis of the tryptic digests of the mutagenized rAb glycoprotein, a comparative analysis was performed between the in-solution reference rAb and the in-gel reference rAb glycoprotein to ensure the reliability and repeatability of the developed glycopeptide mapping methods (see Figure 6). The results in Figure 7 show that the mass

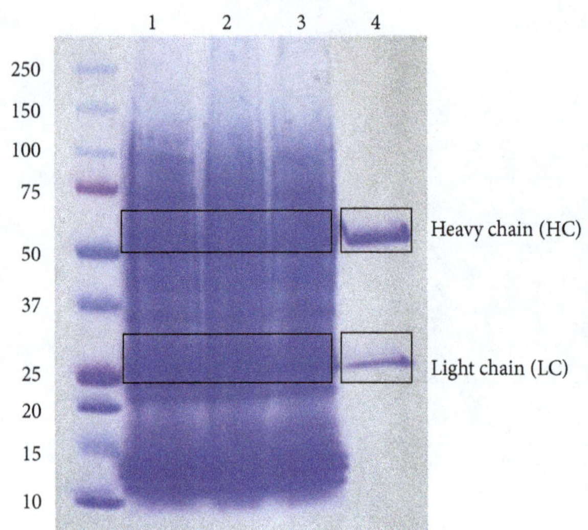

FIGURE 5: SDS-PAGE result for mutagenized rAb samples (lanes 1-3) and reference rAb glycoprotein (lane 4). Heavy chain and light chain bands were observed at 51 kDa and 26 kDa, respectively.

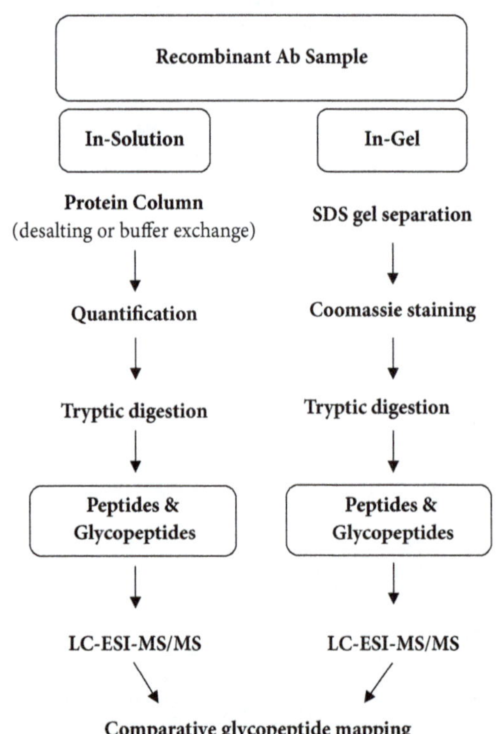

FIGURE 6: Comparative analysis for in-solution and in-gel rAb samples. In-solution sample is applied to a protein column and subject to tryptic digestion. In-gel sample is separated from SDS-PAGE gel and destained before tryptic digestion.

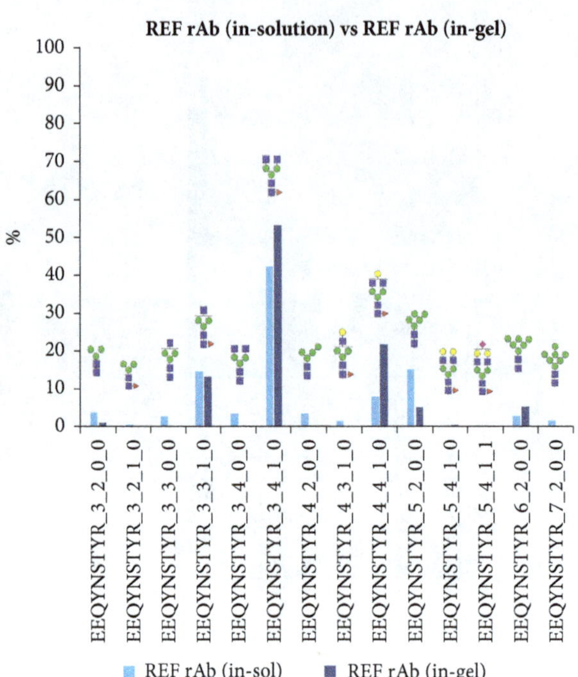

FIGURE 7: Comparison of the glycosylation patterns of the reference rAb, obtained, respectively, from the in-solution and the in-gel method. The two methods for sample preparation provided comparable results, implying that the in-gel method can be complimentary to the in-solution method for mapping protein glycosylation for our purposes.

spectrometry analysis of both the in-solution and the in-gel reference rAb samples yielded similar results with consistent glycosylation patterns, proving the reliability of the method.

We then conducted the analysis of the in-gel tryptic digests of the mutagenized rAb glycoprotein and compared its pattern against the in-gel tryptic digests of the reference rAb glycoprotein. As shown in Figure 8, we observed more substantiated variations in overall glycosylation patterns. Again, the major glycoforms (including EEQYNSTYR_3_3_1_0, EEQYNSTYR_3_4_1_0, and EEQYNSTYR_4_4_1_0) maintain similarity between samples to a certain extent, consistent with the previous analysis results of the reference rAb and the wild-type biosimilar samples prepared by the in-solution method. On the other hand, the differences in glycosylation patterns in minor species of the mutagenized rAb were more substantiated in comparison to those of the reference rAb. The mapping of the mutagenized rAb glycoprotein displayed highly sialylated and galactosylated glycoforms, such as EEQYNSTYR_5_4_1_1, EEQYNSTYR_5_4_1_2, EEQYNSTYR_6_5_1_1, and EEQYNSTYR_6_5_1_2. We also observed the increase in high-branched glycoforms, such as EEQYNSTYR_6_5_1_0, EEQYNSTYR_6_5_1_1, EEQYNSTYR_6_5_1_2, and EEQYNSTYR_7_6_1_2. Overall, we observed a significant increase of sialylated, galactosylated, and branched glycoforms in the mutagenized rAb glycoprotein. These substantial differences in glycosylation patterns can be attributable to the single amino acid mutagenesis and are as expected from the observation that the single amino acid change also caused the rAb's inability to bind to protein A column.

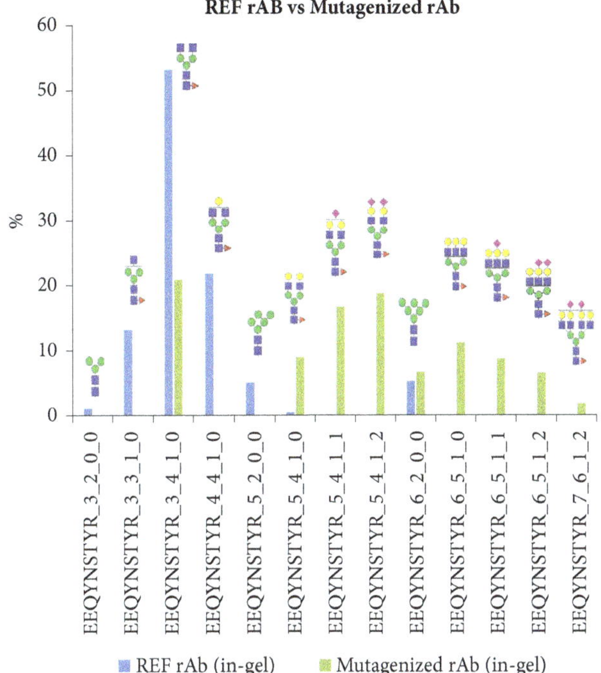

FIGURE 8: Comparative glycopeptide mapping of the reference rAb and the mutagenized rAb glycoprotein by in-gel digestion and LC-ESI-MS/MS. Each dataset was normalized before comparative analysis. Glycosylation patterns were notably different between the reference rAb and the mutagenized rAb samples. Increased abundance of sialylation, galactosylation, and branching was observed in the mutagenized rAb in comparison to the reference rAb glycoprotein.

4. Conclusions

The replicate analyses of the reference rAb glycoprotein allowed us to conclude that the comparative glycopeptide mapping method using LC-tandem mass spectrometry and automated Glycoproteome Analyzer (GPA) software is a highly robust method for monitoring glycosylation patterns. When applying this comparative mapping method to the analyses of the reference and biosimilar rAbs, variations in the distribution of galactosylated and sialylated glycoforms in minor glycoform species were detected without any significant changes in the major glycoform species. These variations on the minor glycoform species of the biosimilar rAb compared to the reference rAb likely reflect the inherent nature of variations in biosimilar development, such as variations in cell lines used and various conditions for culturing the cells in bioreactors for the production of biosimilars. The ability of LC-tandem mass spectrometry-based system to readily detect such delicate variations in minor glycoform species suggests that the method of analysis can be an effective tool to be used in determining similarities of biosimilars to the reference materials in terms of glycosylation patterns with increased accuracy and possibly supplementing results obtained from the more widely used 2AB-based methods. It is not clear at this point, however, if these subtle variations in the minor glycoform species without any significant variations on the major glycoform species can become the basis for the failure of the biosimilar rAb in proving similarity with the reference rAb within the scope of the FDA guideline. The possible variations in functionality and/or immunogenicity of the biosimilar rAb over the reference rAb can only be determined with additional tests following this report.

We also tested the effect of a single amino acid mutation on the overall pattern of glycosylation of the biosimilar rAb. Coincidently, this single amino acid substitution introduced within the CH2 domain of the heavy chain (at position C41 by IMGT codon numbering or 290 by Kabat numbering) of IgG molecule caused the inability of the mutagenized biosimilar rAb to bind to protein A column. Without a readily available method to purify the mutagenized biosimilar rAb, we used in-gel based analysis of the glycoproteins. The in-gel based method was first proven to be effective when compared with in-solution based method in the analyses performed using the reference rAb sample. The in-gel based analysis of the mutagenized rAb glycoprotein showed that a single amino acid mutation in the vicinity of the IgG Fc glycosylation site can cause an increased level of variations on the overall pattern of glycoforms and with more pronounced variations on the minor species. The fact that a single mutation in the amino acid sequence of biosimilar can cause a significant change in the resulting glycosylation patterns and/or physiochemical properties emphasizes the importance of detailed glycopeptide mapping methods, such as LC-tandem mass spectrometry-based method as described our experiment, which can distinguish even slight variations on the minor glycoform species as an added accuracy in biosimilar development.

Acknowledgments

This research was funded by the National Research Foundation [Grant no. NRF-2017R1D1A3B05029756] and the Korean Institute for Regional Program Evaluation of the Ministry of Trade, Industry, and Energy (Grant no. R0003808) through the Regional Promotion of R&D Technology Development Project.

References

[1] P. M. Rudd, T. Elliott, P. Cresswell, I. A. Wilson, and R. A. Dwek, "Glycosylation and the immune system," *Science*, vol. 291, no. 5512, pp. 2370–2376, 2001.

[2] J. Batra and A. S. Rathore, "Glycosylation of monoclonal antibody products: Current status and future prospects," *Biotechnology Progress*, vol. 32, no. 5, pp. 1091–1102, 2016.

[3] M. M. Fuster and J. D. Esko, "The sweet and sour of cancer: glycans as novel therapeutic targets," *Nature Reviews Cancer*, vol. 5, no. 7, pp. 526–542, 2005.

[4] A. Beck and J. M. Reichert, "Approval of the first biosimilar antibodies in Europe: a major landmark for the biopharmaceutical industry," *mAbs*, vol. 5, no. 5, pp. 621–623, 2013.

[5] A. Beck, S. Sanglier-Cianférani, and A. Van Dorsselaer, "Biosimilar, biobetter, and next generation antibody characteri-

zation by mass spectrometry," *Analytical Chemistry*, vol. 84, no. 11, pp. 4637–4646, 2012.

[6] R. Jefferis, "Recombinant antibody therapeutics: the impact of glycosylation on mechanisms of action," *Trends in Pharmacological Sciences*, vol. 30, no. 7, pp. 356–362, 2009.

[7] L. Liu, "Antibody glycosylation and its impact on the pharmacokinetics and pharmacodynamics of monoclonal antibodies and Fc-fusion proteins," *Journal of Pharmaceutical Sciences*, vol. 104, no. 6, pp. 1866–1884, 2015.

[8] J. Pan, S. Zhang, and C. H. Borchers, "Comparative higher-order structure analysis of antibody biosimilars using combined bottom-up and top-down hydrogen-deuterium exchange mass spectrometry," *Biochimica et Biophysica Acta (BBA) - Proteins and Proteomics*, vol. 1864, no. 12, pp. 1801–1808, 2016.

[9] P. Juhasz, C. E. Costello, and K. Biemann, "Matrix-assisted laser desorption ionization mass spectrometry with 2-(4-hydroxyphenylazo)benzoic acid matrix," *Journal of The American Society for Mass Spectrometry*, vol. 4, no. 5, pp. 399–409, 1993.

[10] B. Domon and R. Aebersold, "Mass spectrometry and protein analysis," *Science*, vol. 312, no. 5771, pp. 212–217, 2006.

[11] Y. Wada, M. Tajiri, and S. Yoshida, "Hydrophilic affinity isolation and MALDI multiple-stage tandem mass spectrometry of glycopeptides for glycoproteomics," *Analytical Chemistry*, vol. 76, no. 22, pp. 6560–6565, 2004.

[12] X. Yang, S. M. Kim, R. Ruzanski et al., "Ultrafast and high-throughput N-glycan analysis for monoclonal antibodies," *mAbs*, vol. 8, no. 4, pp. 706–717, 2016.

[13] J. B. Fenn, M. Mann, C. K. Meng, S. F. Wong, and C. M. Whitehouse, "Electrospray ionization for mass spectrometry of large biomolecules," *Science*, vol. 246, no. 4926, pp. 64–71, 1989.

[14] Y. H. Ahn, J. Y. Kim, and J. S. Yoo, "Quantitative mass spectrometric analysis of glycoproteins combined with enrichment methods," *Mass Spectrometry Reviews*, vol. 34, no. 2, pp. 148–165, 2015.

[15] J. Stadlmann, M. Pabst, D. Kolarich, R. Kunert, and F. Altmann, "Analysis of immunoglobulin glycosylation by LC-ESI-MS of glycopeptides and oligosaccharides," *Proteomics*, vol. 8, no. 14, pp. 2858–2871, 2008.

[16] M. Wuhrer, J. C. Stam, F. E. Van De Geijn et al., "Glycosylation profiling of immunoglobulin G (IgG) subclasses from human serum," *Proteomics*, vol. 7, no. 22, pp. 4070–4081, 2007.

[17] K. Hirayama, R. Yuji, N. Yamada, K. Kato, Y. Arata, and I. Shimada, "Complete and Rapid Peptide and Glycopeptide Mapping of Mouse Monoclonal Antibody by LC/MS/MS Using Ion Trap Mass Spectrometry," *Analytical Chemistry*, vol. 70, no. 13, pp. 2718–2725, 1998.

[18] L. He, L. Xin, B. Shan, G. A. Lajoie, and B. Ma, "GlycoMaster DB: Software to assist the automated identification of N-linked glycopeptides by tandem mass spectrometry," *Journal of Proteome Research*, vol. 13, no. 9, pp. 3881–3895, 2014.

[19] M. Bern, Y. J. Kil, and C. Becker, *Current Protocols in Bioinformatics*, A. D. Baxevanis, G. A. Petsko, L. D. Stein, and G. D. Stormo, Eds., John Wiley & Sons, Inc., Hoboken, NJ, USA, 2012.

[20] K. Lynn, C. Chen, T. M. Lih et al., "MAGIC: An Automated N-Linked Glycoprotein Identification Tool Using a Y1-Ion Pattern Matching Algorithm and," *Analytical Chemistry*, vol. 87, no. 4, pp. 2466–2473, 2015.

[21] G. W. Park, J. Y. Kim, H. Hwang et al., "Integrated GlycoProteome Analyzer (I-GPA) for Automated Identification and Quantitation of Site-Specific N-Glycosylation," *Scientific Reports*, vol. 6, no. 1, 2016.

[22] N. Y. Choi, H. Hwang, E. S. Ji et al., "Direct analysis of site-specific N-glycopeptides of serological proteins in dried blood spot samples," *Analytical and Bioanalytical Chemistry*, vol. 409, no. 21, pp. 4971–4981, 2017.

[23] S. Brezinsky, G. Chiang, A. Szilvasi et al., "A simple method for enriching populations of transfected CHO cells for cells of higher specific productivity," *Journal of Immunological Methods*, vol. 277, no. 1-2, pp. 141–155, 2003.

[24] M. Nakano, D. Higo, E. Arai et al., "Capillary electrophoresis-electrospray ionization mass spectrometry for rapid and sensitive N-glycan analysis of glycoproteins as 9-fluorenylmethyl derivatives," *Glycobiology*, vol. 19, no. 2, pp. 135–143, 2009.

[25] S. Sha, C. Agarabi, K. Brorson, D.-Y. Lee, and S. Yoon, "N-Glycosylation Design and Control of Therapeutic Monoclonal Antibodies," *Trends in Biotechnology*, vol. 34, no. 10, pp. 835–846, 2016.

[26] G. C. Flynn, X. Chen, Y. D. Liu, B. Shah, and Z. Zhang, "Naturally occurring glycan forms of human immunoglobulins G1 and G2," *Molecular Immunology*, vol. 47, no. 11-12, pp. 2074–2082, 2010.

[27] X. Chen and G. C. Flynn, "Analysis of N-glycans from recombinant immunoglobulin G by on-line reversed-phase high-performance liquid chromatography/mass spectrometry," *Analytical Biochemistry*, vol. 370, no. 2, pp. 147–161, 2007.

[28] Y. Li, T. Fu, T. Liu et al., "Characterization of alanine to valine sequence variants in the Fc region of nivolumab biosimilar produced in Chinese hamster ovary cells," *mAbs*, vol. 8, no. 5, pp. 951–960, 2016.

19

Determination of Trace Antimony (III) in Water Samples with Single Drop Microextraction using BPHA-[C_4mim] [PF_6]System followed by Graphite Furnace Atomic Absorption Spectrometry

Xiaoshan Huang, Mingxin Guan, Zhuliangzi Lu, and Yiping Hang

School of Chemistry and Chemical Engineering, South China University of Technology, Guangzhou 510640, China

Correspondence should be addressed to Yiping Hang; yphang@scut.edu.cn

Academic Editor: Marcela Z. Corazza

A new sensitive method for antimony (III) determination by graphite furnace atomic absorption spectrometry (GFAAS) has been developed by using N-benzoyl-N-phenylhydroxylamine (BPHA) and 1-butyl-3-methylimidazolium hexafluorophosphate ([C_4mim][PF_6]) single drop microextraction. The single drop microextraction (SDMM) system is more competitive compared with other traditional extraction methods. Under the optimized conditions, the limit of detection (signal-to-noise ratio is 3) and the enrichment factor of antimony (III) are 0.01 $\mu g{\cdot}L^{-1}$ and 112, respectively. The relative standard deviation of the 0.5 $\mu g{\cdot}L^{-1}$ antimony (III) is 4.2% (n=6). The proposed method is rather sensitive to determinate trace antimony (III) in water.

1. Introduction

Antimony (Sb) compounds have been extensively applied for various industrial materials such as glass, semiconductors, and pharmaceutical products [1–3]. As Sb can be easily exposed to the surface water, it has been considered as a priority pollutant in drinking water in the United States, Canada, Europe, and Japan while their action levels are ranged from 2 to 6 $\mu g{\cdot}L^{-1}$ [4]. Sb has two kinds of valence states including Sb (III) and Sb (V), and different chemical forms of them have different levers of toxicity. In inorganic chemistry, the compounds of Sb (III) are almost ten times more poisonous than the compounds of Sb (V) [5–7]. Many diseases including respiratory tract irritation, dermatitis, conjunctivitis, suppuration of the nasal septum, gastritis, or cellular damage in the lungs, heart, and kidneys will be triggered by excessive exposure to Sb (III) [8, 9]. Thus, a reliable and accurate method for the determination of Sb (III) is very necessary.

Several techniques including UV-visible spectrophotometry [10, 11], inductively coupled plasma mass spectrometry [12, 13], hydride generation atomic fluorescence spectrometry [14, 15], flame atomic absorption spectrometry [16–18], and electrothermal atomic absorption spectrometry [19–21] have been used for the determination of antimony species in various samples. Considering the poor sensitivities of flame atomic absorption spectrometry and UV-visible spectrophotometry, the more limited condition of hydride generation atomic fluorescence spectrometry, and the expensive price and analysis cost of inductively coupled plasma mass spectrometry, graphite furnace atomic absorption spectrometry (GFAAS) is an efficient alternative to determinate trace and ultratrace amounts of antimony. Besides the requirement of a relatively small injection volume, a partially eliminated matrix during the pyrolysis is another advantage of GFAAS. Despite the high sensitivity of GFAAS, it is still necessary to use separation techniques that allow preconcentration of antimony species, due to the complex matrix interferences and the low concentration of antimony species in water sample.

Many miniaturized techniques such as homogenous liquid-liquid microextraction [22, 23], solid phase extraction technology [24–26], cloud point extraction [2, 27], single drop microextraction [28–30], hollow-fiber liquid phase microextraction [17, 31], and dispersive liquid phase microextraction [32, 33] have been used as the processing methods of preconcentration. Also, these methods have been applied to preconcentration of antimony species. Compared with

Table 1: Graphite furnace atomizer temperature-rising program.

Steps	Temperature (°C)	Ramp time (s)	Hold time (s)	Argon flow rate ($mL\cdot min^{-1}$)
Drying	120	10	15	250
Pyrolysis	600	5	20	250
Atomization	2000	0	5	0
Cleaning	2400	1	3	250

other methods, SDMM is a new and environmentally friendly sample pretreatment technology. It has the advantages of low cost, simple device, easy operation, very low amounts of organic solvent, and high enrichment efficiency. For its striking advantages, SDMM was selected as the preconcentration methods in our study. Moreover, considering the toxicity and flammability of organic solvents, the ionic liquid of 1-butyl-3-methylimidazolium hexafluorophosphate ($[C_4mim][PF_6]$) has been employed in SDME because of its environmental friendliness. As reported, ionic liquids have been used as novel solvents for the extraction of metal ions at room temperature [34–37]. Ionic liquid does not have detectable vapor pressure and it can avoid environmental and safety problems. Up till now, few analytical applications of SDME method based on $[C_4mim][PF_6]$ for extraction and preconcentration of Sb (III) have been reported. Thus, further studies into the use of $[C_4mim][PF_6]$ in SDME are important in order to improve existing methods.

In this study, a method for Sb (III) determination in water samples by SDME combined with GFAAS was proposed. BPHA and $[C_4mim][PF_6]$ were employed as complexing agent and extraction solvent, respectively. The SDME system was fully characterized through optimizations of the relevant variables influencing the extraction of Sb (III).

2. Materials and Methods

2.1. Reagents. All the reagents were of analytical grade. The experiment water was double distilled deionized water purified by Millipore (Millipore, Bedford, MA, USA). Potassium antimony tartrate was from Tianjin Kemiou Chemical Reagent Co. Ltd., Tianjin, China. N-Benzoyl-N-phenylhydroxylamine (BPHA), sodium thiosulphate, hydrochloric acid, ammonia solutions oxine, dpy, dichloromethane (CH_2Cl_2), and trichloromethane ($CHCl_3$) were bought from Aladdin (Aladdin Industrial Co., Shanghai, China). 1-Butyl-3-methylimidazolium hexafluorophosphate ($[C_4mim][PF_6]$) was bought from J&K (J&K Scientific Ltd., Beijing, China). The 1000 $mg\cdot L^{-1}$ Sb (III) stock solution was prepared by amounts of potassium antimony tartrate dissolved in double distilled deionized water. The pH of the Sb (III) stock solution was adjusted to 2.0 with 0.1 M hydrochloric acid and 0.1 M ammonia solution. 1×10^{-3} M BPHA was dissolved in methanol.

2.2. Instruments. Atomic absorption spectra were performed by a Perkin-Elmer 900T atomic absorption spectrometer (Perkin-Elmer, USA) equipped with transverse heated graphite atomizer, pyrolytic graphite coated tubes (Beijing, China), and an antimony hollow cathode lamp (Perkin-Elmer, USA) recommended by the manufacturer. In the whole operation, except for atomization mode, argon 99.99% was used as protective and purge gas, and the flow rate was 250 $mL\cdot min^{-1}$. The pH of all solutions was performed by pHB-3C (Shanghai Weiye Instrument Factory). And all the stir in this study was performed by 85-2A magnetic stirrer (Changzhou Aohua Instrument Co. Ltd.). Some instrumental parameters of GFAAS were as follows: the lamp current was 10 mA, the wavelength was 217.6 nm, the spectral bandpass was 0.7 nm, and the background correction was Zeeman. Table 1 showed graphite furnace atomizer temperature-rising program.

2.3. Preparation of Samples. All the water samples were filtered through a 0.45 μm pore size membrane filter to remove the suspended particulate matters and the pH of all the water samples was adjusted to 2.0 by using 0.1 M hydrochloric acid and 0.1 M ammonia solution. Water samples including bottled mineral water, river water (pH=6.2, Beijiang River, Shaoguan, China), and tap water were collected locally. Each of the treated water samples was preserved at 1.9 mL for the later determination.

2.4. Process of Single Drop Microextraction. The apparatus of SDME was the same as what we studied before [39]. 1.9 mL treated water sample or 1 $\mu g\cdot L^{-1}$ Sb (III) standard solution and 100 μL 1×10^{-4} M BPHA solution were added to a 5 mL vial. Microsyringe with 5 μL of $[C_4mim][PF_6]$ was positioned above the vial, and the needle was inserted through the septum. The tip of syringe needle was attached to a flared polytetrafluoroethylene tube. Then, the needle tip was immersed into the sample solution, and the microsyringe was pushed slowly in order to make the microdrop hang under the needle tip steadily. The time of the extraction was 6 min under the stirring rate of 600 rpm. After extraction, microdrop was inhaled into the microsyringe and injected into the graphite furnace atomic absorption spectrometer for analysis manually.

3. Results and Discussion

3.1. Chelating Agent and Extraction Solvent. A chelating agent has a great influence on the extraction efficiency of Sb (III). Thus, a suitable chelating agent is very important. Figure 1(a) showed the pattern of the atomic absorbance of Sb (III) with different chelating agents (BPHA, Oxine, and Dpy) and their

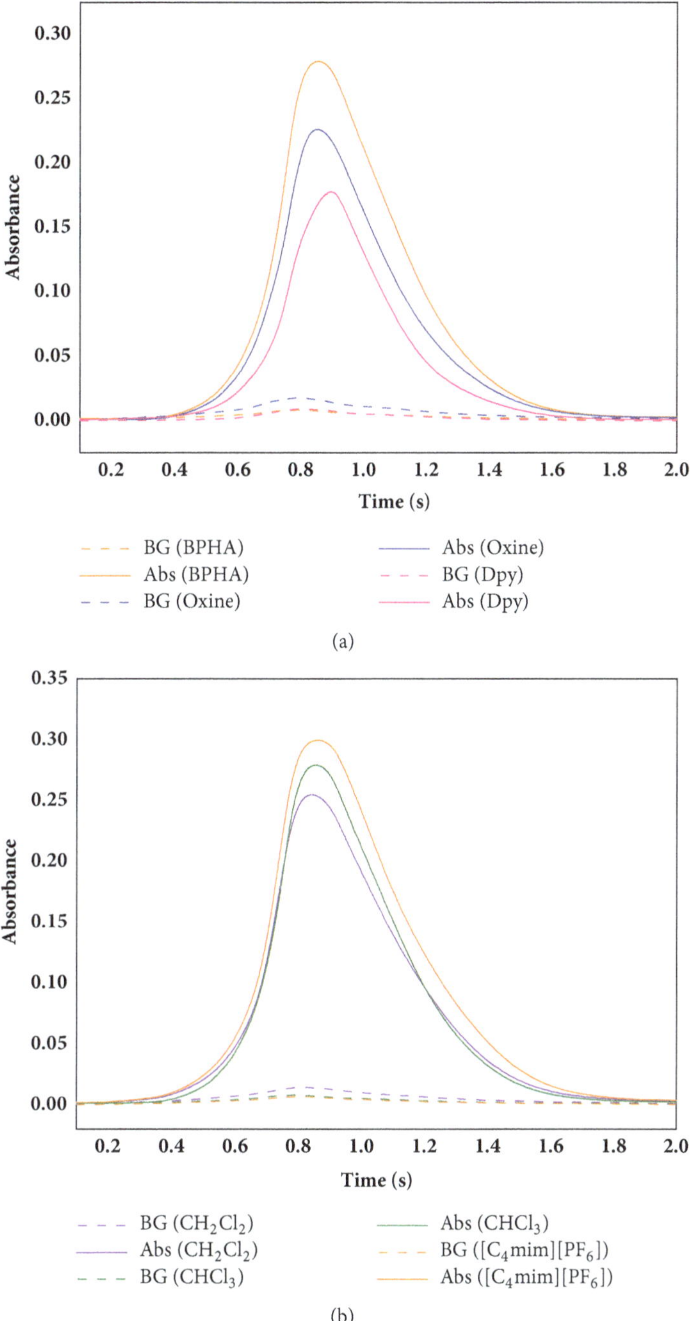

FIGURE 1: (a) The atomic absorbance of Sb (III) with different chelating agents such as BPHA, Oxine, and Dpy and their background absorbance. (b) The atomic absorbance of Sb (III) in different extraction solvents such as CH_2Cl_2, $CHCl_3$, and [C_4mim][PF_6] and their background absorbance; BG: background absorbance without Sb (III); Abs: absorbance.

background absorbance. It was found that the absorbance signal of Sb (III) with the chelating agent of BPHA was stronger than others. Although the absorbance signal of Sb (III) with Oxine was good, it had the stronger background interference. The absorbance signal of Sb (III) with Dpy had weaker signal compared with BPHA and Oxine. As a result, BPHA was selected as the chelating agent for SDME.

A suitable extraction solvent is also important for SDME. The density of the extraction solvent can be supposed to be higher than water so that it could keep the drop stable. CH_2Cl_2, $CHCl_3$, and [C_4mim][PF_6] that were used in liquid-liquid extraction were evaluated as the extraction solvents. Each extraction solvent was dealt with via three different chelating agents, and then the method of SDME-GFAAS was applied to determine the amounts of Sb (III), and the results were shown in Table 2. Whatever the chelating agent was, [C_4mim][PF_6] always had the strongest signal. Figure 1(b) described the atomic absorbance of Sb (III) in different extraction agents (CH_2Cl_2, $CHCl_3$, and [C_4mim][PF_6]) with BPHA as the chelating agent and their background

TABLE 2: The atomic absorbance of three chelating agents in three different extraction solvents.

Chelating agent	Extraction solvent		
	CH_2Cl_2	$CHCl_3$	$[C_4mim][PF_6]$
Oxine	0.287	0.274	0.294
Dpy	0.158	0.143	0.176
BPHA	0.322	0.355	0.364

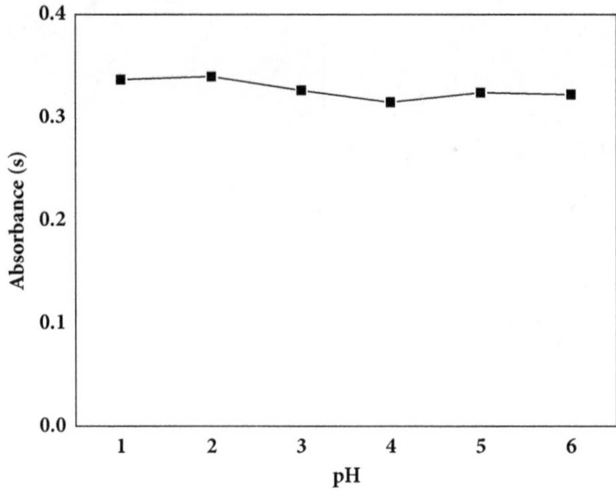

FIGURE 2: The atomic absorbance of Sb (III) in solutions with the pH value from 2.0 to 6.0.

absorbance. Therefore, $[C_4mim][PF_6]$ was selected as the extraction solvent for SDME.

3.2. Optimization of Single Drop Microextraction Conditions

3.2.1. pH. The pH of the solution played an important role in the formation of metal chelate and the influence of the stability of chemicals. Also, it could affect the extraction of Sb (III) in the BPHA-$[C_4mim][PF_6]$ system. The study of the pH was ranged from 1.0 to 6.0 and the results were shown in Figure 2. The absorbance signals of Sb (III) were high and less volatile in the range of 1.0-6.0. Considering that there may be interferences due to the competitive complexation reaction of other metal ions when the pH value was at a high level, pH=2.0 was chosen for the further study.

3.3. BPHA Concentration. It was also necessary to find the minimal concentration of BPHA. The effect of BPHA concentration on the extraction efficiency of Sb (III) was investigated. The results were illustrated in Figure 3(a) and, as can be seen from it, the absorbance signal of Sb (III) increased with the BPHA concentration from 6×10^{-5} to 8×10^{-5} M and remained constant when the concentration of BPHA was above 8×10^{-5} M. To make the treatment easier, the value of 1×10^{-4} M was chosen for the further study.

3.4. Solvent Drop Size. The effect of drop size was shown in Figure 3(b), and it was found that the absorbance of Sb (III) increased with the increase of the drop size from 2.0 to 6.0 μL. However, the drop size increasing usually resulted in the fall of the microdrop. In general, the stability of the microdrop depends on upward floating force, downward gravity, and adhesion forces [18]. In order to enhance the adhesion force of the microdrop, a flared polytetrafluoroethylene tube was attached to the tip of syringe needle. All these things were taken into account, and then 5.0 μL was chosen as the drop size for extraction.

3.5. Stirring Rate. It was well known that the stirring rate could affect the extracting speed by changing the mass transfer in the sample solution. The effect of stirring rates on extraction efficiency was studied in the range of 200 to 800 rpm. The results in Figure 3(c) showed that the increasing stirring rate of the sample greatly improved the absorbance of Sb (III). However, the microdrop easily fell off the needle of the microsyringe when the stirring rate was above 600 rpm. Increasing stirring rate could also cause a reduction of $[C_4min][PF_6]$ microdrop volume, because the dissolution of ionic liquid was enhancing. Thus, 600 rpm was selected as the best stirring rate in this study.

3.6. Extraction Time. The extraction efficiency depended on the length of the extraction time until the equilibrium was reached. Although the maximum sensitivity was achieved in equilibrium, complete equivalent was not necessary to obtain accurate analysis. Thus, the effect of extraction time on extraction efficiency had been studied from 2 to 10 min. The results were illustrated in Figure 3(d). There were a sharp increase from 2 to 6 min and a slow increase from 6 to 10 min. As the time went on, microdrop would fall into solutions. In order to avoid it, 6 min was selected as the extraction time which was enough for extracting Sb (III) for determination.

3.7. Optimization of Graphite Furnace Atomic Absorption Spectrometry. In order to reduce the chemical interference and the background signal, the work investigated the influence of pyrolysis temperature from 400°C to 800°C and atomization temperature from 1800°C to 2200°C. The 0.5 $\mu g{\cdot}L^{-1}$ Sb (III) standard solutions were dealt with via the pretreatment of SDME and determined by GFAAS. As shown in Figure 4(b), background signals were stronger when the pyrolysis temperature was lower because of the excessive vaporization of BPHA and ionic liquid at atomization stage, and the strongest signal appeared at 2000°C. It was found in Figure 4(a) that the matrix was sufficiently eliminated and maximum absorbance was achieved at the pyrolysis

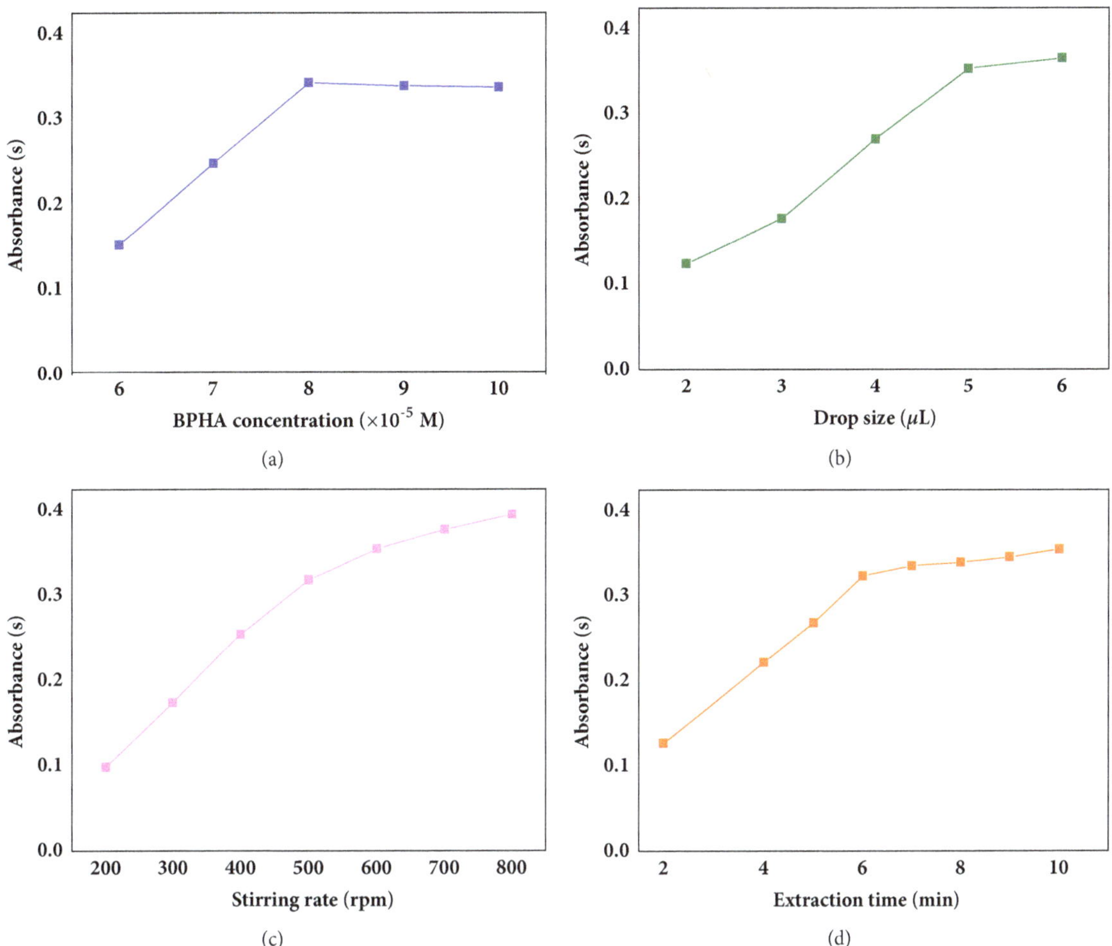

FIGURE 3: (a) The atomic absorbance of Sb (III) in solutions with the concentration of BPHA from 6×10^{-5} M to 10×10^{-5} M. (b) The atomic absorbance of Sb (III) in solutions with the drop size from 2.0 to 6.0 μL. (c) The atomic absorbance of Sb (III) in solutions with the stirring rate from 200 to 800 rpm. (d) The atomic absorbance of Sb (III) in solutions with the extraction time from 2 to 10 min.

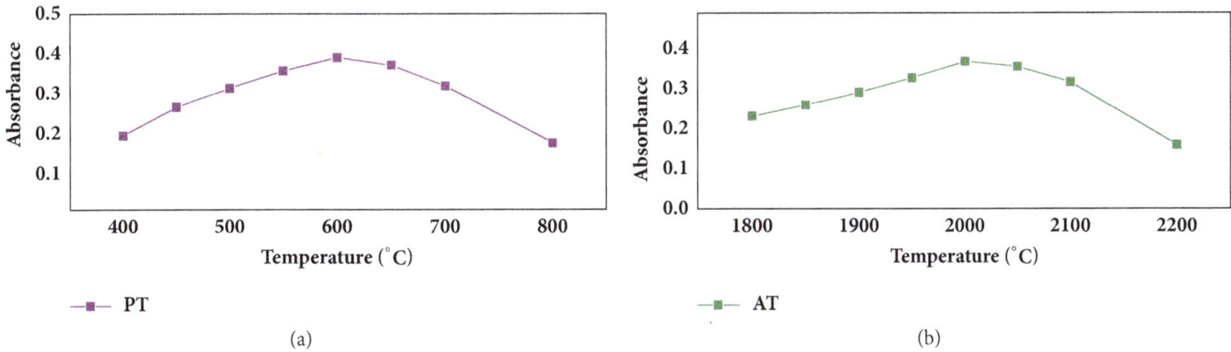

FIGURE 4: (a) The absorbance of Sb (III) with the pyrolysis temperature (PT) from 400°C to 800°C. (b) The absorbance of Sb (III) with the atomization temperature (AT) from 1800°C to 2200°C.

temperature of 600°C; however the absorbance decreased with increasing of the pyrolysis temperature due to the loss of Sb at higher temperature. In addition, the time of atomization was 5 s. As the results showed in Figure 4, 2000°C was chosen as the atomization temperature and 600°C as the pyrolysis temperature.

3.8. Effect of Interferences. One of the interferences was other metal ions reacting with chelating agents and the other was coextraction. In order to validate the selectivity of Sb (III) in microextraction system, different amounts of ions were added to the 1.0 $\mu g \cdot L^{-1}$ Sb (III) solutions, respectively. After determination, coexisting ions were considered to have

TABLE 3: Comparison of the proposed method with other methods for determination of antimony (III).

Method	Linear ranges ($ng \cdot mL^{-1}$)	Limits of detection ($ng \cdot mL^{-1}$)	Enrichment factor	Relative standard deviation	References
DLLME-ETAAS[a]	0.05-5	0.05	115	4.5%	[32]
CPE-ETAAS[b]	-	1.82	45	2.6%	[38]
VASEME-ETAAS[c]	0.4-8	0.09	53	5.4%	[20]
HFSLME-TAFFAAS[d]	5-200	0.8	160	7.8%	[17]
SDME-GFAAS	0.02-50	0.01	112	4.2%	This work

[a]Dispersive liquid-liquid microextraction-electrothermal atomic absorption spectrometry.
[b]Cloud point extraction- electrothermal atomic absorption spectrometry.
[c]Vortex-assisted surfactant-enhanced emulsification microextraction-electrothermal atomic absorption spectrometry.
[d]Hollow fiber supported liquid membrane extraction-thermospray flame furnace atomic absorption spectrometry.

TABLE 4: Determination of Sb (III) in water samples.

Samples	Added ($\mu g \cdot L^{-1}$)	Found ($\mu g \cdot L^{-1}$)	Recovery (%)
Bottle mineral water	0	< Limits of detection	-
	0.1	0.102 ± 0.01	102±1
	0.4	0.401 ± 0.008	100±2
River water	0	< Limits of detection	-
	0.1	0.104 ± 0.015	104±2
	0.4	0.407 ± 0.012	102±3
Tap water	0	< Limits of detection	-
	0.1	0.098 ± 0.011	98±1
	0.4	0.398 ± 0.009	99±2

interferences when the change of Sb (III) absorption value was more than 5%. As shown in the Results, the tolerance limit of coexisting ions including Na^+, K^+, Ca^{2+}, Mg^{2+}, Cl^-, SO_4^{2-}, and PO_4^{3-} was 2000 $mg \cdot L^{-1}$; of coexisting ions including Co^{2+}, Cd^{2+}, Cu^{2+}, Mn^{2+}, Zn^{2+}, and Ag^+ was 200 $mg \cdot L^{-1}$; of coexisting ions including Al^{3+}, Cr^{3+}, Fe^{3+}, Ni^{2+}, and Pb^{2+} was 50 $mg \cdot L^{-1}$.

3.9. Standard Analysis. As shown in the Results, the low limit of detection (LOD, 3σ) was 0.01 $ng \cdot mL^{-1}$, the relative standard deviation of 0.5 $ng \cdot mL^{-1}$ (n=6) was 4.2%, and the linear value ranged from 0.02 to 5 $ng \cdot mL^{-1}$. The regression equation was A=0.6832C+0.0034 (A represented the absorbance values and C represented the concentration of Sb (III) whose unit was $ng \cdot mL^{-1}$). Linear correlation coefficient was 0.999. A comparison of the main features of the proposed method with other reported methods in the literatures was shown in Table 3. This method was more effective for detecting Sb (III) with lower limit detection and had better precision than majority of the other reported methods.

3.10. Samples Analysis. A series of water samples were analyzed by the presented method. The results were shown in Table 4. The recoveries were in the ranges of 98-104% with the different standard of Sb (III) solutions spiked to the water samples. So, it demonstrated a good accuracy of this method.

4. Conclusions

The optimization method, combined with single drop microextraction using BPHA-$[C_4mim][PF_6]$ system for separation of impurities, detected by GFAAS was developed to determinate trace Sb (III) in water samples. After a series of analysis of optimization conditions, an excellent accuracy, precision, and lower limit detection were obtained by this method. The relative standard deviation of the 0.5 $\mu g \cdot L^{-1}$ Sb (III) was 4.2% (n=6). The detection limit (signal-to-noise ratio of 3) and the enrichment coefficient of Sb (III) were 0.01 $\mu g \cdot L^{-1}$ and 112, respectively. What is more, the introduction of BPHA-$[C_4mim][PF_6]$ system was not only ecofriendly comparing with traditional organic solution but also efficient for extraction. After the rapid extraction by SDME with the system of BPHA-$[C_4mim][PF_6]$, the samples could be injected and detected directly. Thus, this method was a simple, effective, and environment-friendly way to determine the trace concentration of Sb (III) in water samples.

Acknowledgments

The authors gratefully acknowledge the South China University of Technology for supporting this work and the Science and Technology Planning Project of Guangdong (Grant no. 2016A020210012) for the research fellowship.

References

[1] W. Shotyk, M. Krachler, and B. Chen, "Antimony: global environmental contaminant," *Journal of Environmental Monitoring* vol. 7, no. 12, pp. 1135-1136, 2005.

[2] X. Jiang, S. Wen, and G. Xiang, "Cloud point extraction combined with electrothermal atomic absorption spectrometry for the speciation of antimony(III) and antimony(V) in food packaging materials," *Journal of Hazardous Materials*, vol. 175, no. 1-3, pp. 146–150, 2010.

[3] Y. Li, B. Hu, and Z. Jiang, "On-line cloud point extraction combined with electrothermal vaporization inductively coupled plasma atomic emission spectrometry for the speciation of inorganic antimony in environmental and biological samples," *Analytica Chimica Acta*, vol. 576, no. 2, pp. 207–214, 2006.

[4] S. D. Richardson and T. A. Ternes, "Water analysis: emerging contaminants and current issues," *Analytical Chemistry*, vol. 83, no. 12, pp. 4616–4648, 2011.

[5] P. Smichowski, "Antimony in the environment as a global pollutant: a review on analytical methodologies for its determination in atmospheric aerosols," *Talanta*, vol. 75, no. 1, pp. 2–14, 2008.

[6] A. Gonzalvez, M. L. Cervera, S. Armenta, and M. de la Guardia, "A review of non-chromatographic methods for speciation analysis," *Analytica Chimica Acta*, vol. 636, no. 2, pp. 129–157, 2009.

[7] J. O. Vinhal, A. D. Gonçalves, G. F. B. Cruz, and R. J. Cassella, "Speciation of inorganic antimony (III & V) employing polyurethane foam loaded with bromopyrogallol red," *Talanta*, vol. 150, pp. 539–545, 2016.

[8] M. Krachler, H. Emons, and J. Zheng, "Speciation of antimony for the 21st century: Promises and pitfalls," *TrAC - Trends in Analytical Chemistry*, vol. 20, no. 2, pp. 79–90, 2001.

[9] N. Altunay and R. Gürkan, "Separation/preconcentration of ultra-trace levels of inorganic Sb and Se from different sample matrices by charge transfer sensitized ion-pairing using ultrasonic-assisted cloud point extraction prior to their speciation and determination by hydride generation AAS," *Talanta*, vol. 159, pp. 344–355, 2016.

[10] M. J. G. González, O. D. Renedo, and M. J. A. Martínez, "Simultaneous determination of antimony(III) and antimony(V) by UV-vis spectroscopy and partial least squares method (PLS)," *Talanta*, vol. 68, no. 1, pp. 67–71, 2005.

[11] A. Samadi-Maybodi and V. Rezaei, "A cloud point extraction for spectrophotometric determination of ultra- trace antimony without chelating agent in environmental and biological samples," *Microchimica Acta*, vol. 178, no. 3-4, pp. 399–404, 2012.

[12] P. Li, Y.-J. Chen, X. Hu, and H.-Z. Lian, "Magnetic solid phase extraction for the determination of trace antimony species in water by inductively coupled plasma mass spectrometry," *Talanta*, vol. 134, pp. 292–297, 2015.

[13] F. Séby, C. Gleyzes, O. Grosso, B. Plau, and O. F. X. Donard, "Speciation of antimony in injectable drugs used for leishmaniasis treatment (Glucantime®) by HPLC-ICP-MS and DPP," *Analytical and Bioanalytical Chemistry*, vol. 404, no. 10, pp. 2939–2948, 2012.

[14] H. Wu, X. Wang, B. Liu et al., "Simultaneous speciation of inorganic arsenic and antimony in water samples by hydride generation-double channel atomic fluorescence spectrometry with on-line solid-phase extraction using single-walled carbon nanotubes micro-column," *Spectrochimica Acta Part B: Atomic Spectroscopy*, vol. 66, no. 1, pp. 74–80, 2011.

[15] Y. Zhang, M. Mei, T. Ouyang, and X. Huang, "Preparation of a new polymeric ionic liquid-based sorbent for stir cake sorptive extraction of trace antimony in environmental water samples," *Talanta*, vol. 161, pp. 377–383, 2016.

[16] S. Tajik and M. A. Taher, "New method for microextraction of ultra trace quantities of gold in real samples using ultrasound-assisted emulsification of solidified floating organic drops," *Microchimica Acta*, vol. 173, no. 1-2, pp. 249–257, 2011.

[17] C. Zeng, F. Yang, and N. Zhou, "Hollow fiber supported liquid membrane extraction coupled with thermospray flame furnace atomic absorption spectrometry for the speciation of Sb(III) and Sb(V) in environmental and biological samples," *Microchemical Journal*, vol. 98, no. 2, pp. 307–311, 2011.

[18] H. H. Nadiki, M. A. Taher, and H. Ashkenani, "Ionic liquid ultrasound assisted dispersive liquid-liquid/micro-volume back extraction procedure for preconcentration and determination of ultra trace amounts of thallium in water and biological samples," *International Journal of Environmental Analytical Chemistry*, vol. 93, no. 6, pp. 623–636, 2013.

[19] S. Wen and X. Zhu, "Speciation of antimony(III) and antimony(V) by electrothermal atomic absorption spectrometry after ultrasound-assisted emulsification of solidified floating organic drop microextraction," *Talanta*, vol. 115, pp. 814–818, 2013.

[20] M. Eftekhari, M. Chamsaz, M. H. Arbab-Zavar, and A. Eftekhari, "Vortex-assisted surfactant-enhanced emulsification microextraction based on solidification of floating organic drop followed by electrothermal atomic absorption spectrometry for speciation of antimony (III, V)," *Environmental Modeling & Assessment*, vol. 187, no. 1, 2015.

[21] I. López-García, S. Rengevicova, M. J. Muñoz-Sandoval, and M. Hernández-Córdoba, "Speciation of very low amounts of antimony in waters using magnetic core-modified silver nanoparticles and electrothermal atomic absorption spectrometry," *Talanta*, vol. 162, pp. 309–315, 2017.

[22] A. R. Ghiasvand, S. Shadabi, E. Mohagheghzadeh, and P. Hashemi, "Homogeneous liquid-liquid extraction method for the selective separation and preconcentration of ultra trace molybdenum," *Talanta*, vol. 66, no. 4, pp. 912–916, 2005.

[23] S. Igarashi, N. Ide, and Y. Takagai, "High-performance liquid chromatographic-spectrophotometric determination of copper(II) and palladium(II) with 5,10,15,20-tetrakis(4N-pyridyl)porphine following homogeneous liquid-liquid extraction in the water-acetic acid-chloroform ternary solvent system," *Analytica Chimica Acta*, vol. 424, no. 2, pp. 263–269, 2000.

[24] L. Zhang, Y. Morita, A. Sakuragawa, and A. Isozaki, "Inorganic speciation of As(III, V), Se(IV, VI) and Sb(III, V) in natural water with GF-AAS using solid phase extraction technology," *Talanta*, vol. 72, no. 2, pp. 723–729, 2007.

[25] C. Dietz, J. Sanz, and C. Cámara, "Recent developments in solid-phase microextraction coatings and related techniques," *Journal of Chromatography A*, vol. 1103, no. 2, pp. 183–192, 2006.

[26] A. K. Malik, V. Kaur, and N. Verma, "A review on solid phase microextraction - High performance liquid chromatography as a novel tool for the analysis of toxic metal ions," *Talanta*, vol. 68, no. 3, pp. 842–849, 2006.

[27] R. Gürkan and M. Eser, "Application of ultrasonic-assisted cloud point extraction/flame atomic absorption spectrometry (UA-CPE/FAAS) for preconcentration and determination of low levels of antimony in some beverage samples," *Journal of the Iranian Chemical Society*, vol. 13, no. 9, pp. 1579–1591, 2016.

[28] Z. Fan, "Determination of antimony(III) and total antimony by single-drop microextraction combined with electrothermal atomic absorption spectrometry," *Analytica Chimica Acta*, vol. 585, no. 2, pp. 300–304, 2007.

[29] C. Mitani and A. N. Anthemidis, "On-line liquid phase microextraction based on drop-in-plug sequential injection lab-at-

valve platform for metal determination," *Analytica Chimica Acta*, vol. 771, pp. 50–55, 2013.

[30] C. Ye, Q. Zhou, and X. Wang, "Improved single-drop microextraction for high sensitive analysis," *Journal of Chromatography A*, vol. 1139, no. 1, pp. 7–13, 2007.

[31] E. Marguí, M. Sagué, I. Queralt, and M. Hidalgo, "Liquid phase microextraction strategies combined with total reflection X-ray spectrometry for the determination of low amounts of inorganic antimony species in waters," *Analytica Chimica Acta*, vol. 786, pp. 8–15, 2013.

[32] R. E. Rivas, I. López-García, and M. Hernández-Córdoba, "Speciation of very low amounts of arsenic and antimony in waters using dispersive liquid-liquid microextraction and electrothermal atomic absorption spectrometry," *Spectrochimica Acta Part B: Atomic Spectroscopy*, vol. 64, no. 4, pp. 329–333, 2009.

[33] S. R. Yousefi, F. Shemirani, and M. R. Jamali, "Determination of antimony(III) and total antimony in aqueous samples by electrothermal atomic absorption spectrometry after dispersive liquid-liquid microextraction (DLLME)," *Analytical Letters*, vol. 43, no. 16, pp. 2563–2571, 2010.

[34] E. M. Martinis, R. A. Olsina, J. C. Altamirano, and R. G. Wuilloud, "Sensitive determination of cadmium in water samples by room temperature ionic liquid-based preconcentration and electrothermal atomic absorption spectrometry," *Analytica Chimica Acta*, vol. 628, no. 1, pp. 41–48, 2008.

[35] H. Abdolmohammad-Zadeh and G. H. Sadeghi, "A novel microextraction technique based on 1-hexylpyridinium hexafluorophosphate ionic liquid for the preconcentration of zinc in water and milk samples," *Analytica Chimica Acta*, vol. 649, no. 2, pp. 211–217, 2009.

[36] X. Wen, Q. Deng, and J. Guo, "Ionic liquid-based single drop microextraction of ultra-trace copper in food and water samples before spectrophotometric determination," *Spectrochimica Acta Part A: Molecular and Biomolecular Spectroscopy*, vol. 79, no. 5, pp. 1941–1945, 2011.

[37] G.-T. Wei, Z. Yang, and C.-J. Chen, "Room temperature ionic liquid as a novel medium for liquid/liquid extraction of metal ions," *Analytica Chimica Acta*, vol. 488, no. 2, pp. 183–192, 2003.

[38] Z. Fan, "Speciation analysis of antimony (III) and antimony (V) by flame atomic absorption spectrometry after separation/preconcentration with cloud point extraction," *Microchimica Acta*, vol. 152, no. 1-2, pp. 29–33, 2005.

[39] H. Yiping and W. Caiyun, "Ion chromatography for rapid and sensitive determination of fluoride in milk after headspace single-drop microextraction with in situ generation of volatile hydrogen fluoride," *Analytica Chimica Acta*, vol. 661, no. 2, pp. 161–166, 2010.

Determination of Total Flavonoids Contents and Antioxidant Activity of *Ginkgo biloba* Leaf by Near-Infrared Reflectance Method

Ling-jia Zhao,[1,2] Wei Liu,[3] Su-hui Xiong,[1,2] Jie Tang,[1,2] Zhao-huan Lou,[4] Ming-xia Xie,[2] Bo-hou Xia,[1,2] Li-mei Lin,[1,2] and Duan-fang Liao[1,2]

[1]*College of Pharmacy, Hunan University of Chinese Medicine, Changsha 410208, China*
[2]*Key Laboratory for Quality Evaluation of Bulk Herbs of Hunan Province, Hunan University of Chinese Medicine, Changsha 410208, China*
[3]*School of Management of Information Engineering, Hunan University of Chinese Medicine, Changsha 410208, China*
[4]*Institute of Material Medical, Zhejiang Chinese Medical University, Hangzhou, Zhejiang, China*

Correspondence should be addressed to Bo-hou Xia; xiabohou@163.com and Li-mei Lin; lizasmile@163.com

Academic Editor: Josef Havel

Background. Total flavonoids content (TFC) is one of the most important quality indexes of *Ginkgo biloba* leaf, and it is concerned with total antioxidant activity. Near-infrared spectroscopy (NIR) method has showed its advantages in fast, accurate, qualitative, and quantitative analysis of various components in many quality control researches. In this study, a calibration model was built by partial least squares regression (PLSR) coupling with NIR spectrum to quantitatively analyze the TFC and total antioxidant activity of *Ginkgo biloba* leaf. *Results.* During the model establishing, some spectrum pretreatment and outlier diagnosis methods were optimized to establish the final model. The coefficients of determination (R^2) for TFC and total antioxidant activity prediction were 0.8863 and 0.8486, respectively; and the root mean square errors of prediction (RMSEP) were 2.203 mg/g and 0.2211 mM/g, respectively. *Conclusion.* These results showed that NIR method combined with chemometrics is suitable for quantitative analysis of main components and their activities and might be applied to quality control of relevant products.

1. Introduction

Ginkgo biloba L. (Ginkgoaceae) is an ancient tree growing in China for thousands of years. In recent decades, *Ginkgo biloba* L. occupies a prominent position among the best-selling natural products owing to its reliable and remarkable biological activities [1, 2]. Some studies and clinical trials have observed that *Ginkgo biloba* L. shows potent actions on cardiovascular system and cerebral vascular activity. Furthermore, due to its antioxidant properties, it has been used in Alzheimer's patients [3–5].

The main active components in *Ginkgo* leaf are flavonoids and terpene trilactones. Because only several kinds of terpene trilactones were found in *Ginkgo*, the pharmacological effects of them are relatively clear and the corresponding quality evaluation is simply achieved [6, 7], while more than 70 kinds of flavonoids were identified, which associate with various kinds of pharmacological actives [8–11]. Therefore, more researches have been conducted which focused on flavonoids in Ginkgo as their broad-spectrum of antioxidant and free-radical scavenging activity. Thus, total flavonoids content is often considered as an important quality index of *Ginkgo biloba* leaf. Determination of total flavonoids contents in *Ginkgo biloba* leaf and further estimating their antioxidant are of great importance to their qualities [7, 12].

In recent years, flavonoids in *Ginkgo biloba* L. have received considerable attention in various literatures, especially due to their widely recognized free-radical scavenging activity. TFC is often considered an important quality index of *Ginkgo biloba* products and samples, while DPPH radical methods are a stable N-centered radical at room temperature that is widely employed to assess the radical scavenging properties of antioxidants, such as total flavonoids. The traditional methods for determining total flavonoids in botanical materials are based on chemical extraction and couple with various analytical techniques, such as HPLC [13–15], GC [16], and ultraviolet spectrometry [17]. These methods are precise, and some methods are used as the reference methods for TFC detection. Yet, these methods all needed the sample pretreat or extraction process, which are often time-consuming and destructive. Therefore, a rapid, accurate, and even nondestructive analytical method is needed to identify the TFC and further determine their actives for the quality control of *Ginkgo biloba* L.

Near-infrared spectroscopy (NIR) has been used for various applications, such as quality estimation and quality control of various food, agriculture, and pharmaceutical products. There are also many researches based on NIR in herbal quality control and main contents analysis, and they show the advantages, including simple sample preparation and rapid and simultaneous analysis of several analytes in a large number of samples. The NIR spectra combining with appropriate mathematical models and pattern recognition techniques can be used to qualitatively and quantitatively determine quality of various products [12, 18–20].

Some studies have been published by coupling near-infrared spectroscopy with chemometrics methods to qualitative and quantitative analysis of flavonoids concentrations in various botanical leaves and relevant products, including *Ginkgo* leaf. Shi et al. determined the TFC in fresh *Ginkgo* leaves with different colors by using the NIR spectroscopy; furthermore they also analyzed the basic structure of flavonoids and relationship of wavelength regions [12]. Liu et al. published their reviews about the roles of flavonoid and its broad-spectrum free-radical scavenging activities in *Ginkgo biloba* chemical analysis and quality control [8]. Geng et al. established a quantitative near-infrared diffuse reflectance spectroscopy method for the simultaneous determination of three flavonol aglycones in *Ginkgo biloba* extracts [21]. Yet, no research has been reported to quantitatively analyze the TFC and their total antioxidant activity in *Ginkgo biloba* samples, simultaneously.

Based on these reasons, we aimed to establish a calibration model to quantitatively analyze the TFC in Ginkgo *biloba* leaves and further quantitatively estimate their antioxidant properties. During the model establishing process, some NIR signal pretreat methods were adopted to optimize the calibration model. The feasibility of combining NIR spectroscopy with chemometrics methods to rapid and nondestructive determination of TFC and their antioxidant properties was investigated.

2. Material and Methods

2.1. Chemicals and Materials. 1,1-Diphenyl-2-picrylhydrazyl (DPPH) and Trolox were purchased from Sigma-Aldrich Chemical Co. (St. Louis, MO, USA). NaNO2, NaOH, and Al (NO3)3 were analytical grade and acquired from Shanghai Macklin Biochemical Co., Ltd. (Shanghai, China). Rutin was obtained from the National Institution for Food and Drug Control (Beijing, China). Distilled water was filtered by using a Milli-Q water-purification system (Millipore, Bedford, MA, USA). 113 batches of samples were collected from Zhejiang Province. All samples were powdered by a grinder mill after dried and passed through 60-mesh place before analysis.

2.2. Total Flavonoids Content (TFC). The concentrations of flavonoids were quantified based on a colorimetric assay method [12], with slight modifications. Briefly, rutin was used as a standard to establish calibration linear with function:

$$A = 8.0045\ C + 0.0914; \quad r = 0.9959 \quad (r = \text{linear range}) \tag{1}$$

0.90-1.00 g samples were weighted, and 10 mL 60% ethanol aqueous was used to extract flavonoids from theses samples with supersonic (KQ-300DE, Kunshan Ultrasonic Equipment Co., China) for 30 min. These samples were further centrifuged at 3000 $*$ g. All the supernatant was transferred to 25 mL volumetric flask and then was fixed to 25 mL with 60% ethanol aqueous. 1.5 mL of each extracts and 4.5 mL of distilled water were pipetted into a 25 mL tube and then mixed with 1 mL 5% (w v^{-1}) $NaNO_2$ solutions. After incubation for 6 min, 1 mL of the 10% (w v^{-1}) Al $(NO_3)_3$ solutions was added to the mixture. The mixture was kept for 6 min before adding 10 mL 4% (w v^{-1}) NaOH solutions and fixed to 25 mL with 60% ethanol aqueous. Finally, the mixture was reacted for 15 min and the absorbance of the mixture solution was measured with a spectrophotometer (SP-1901, Shanghai Spectrum Instruments Co., China) at 510 nm against a blank containing 5 mL of extraction solvent. Samples were independently analyzed in triplicate times, and the mean of three tests were used and the total flavonoid content was expressed as mg rutin equivalent per g dry weight (DW).

2.3. Determination of the Total Antioxidant Activity. DPPH radical scavenging activity was determined as described by Okawa et al. [22] with a slight modification. Solutions of known Trolox concentration were used for calibration. 2 μL of samples or Trolox was mixed with 250 μL of methanolic DPPH. The homogenate was shaken vigorously and kept in darkness for 30 min. Absorption of the samples was measured on the spectrophotometer at 515 nm. Results were expressed as Trolox equivalent per g of dry weight (mM TE g dried extract^{-1}).

2.4. NIR Spectra Acquisition and Preprocessing. The NIR spectra were measured in a diffuse reflectance mode by Antaris II FT-NIR spectrophotometer (Thermo EIectron Co., USA) equipped with an integrating sphere. The spectra (4000 to 8000 cm^{-1} were analyzed, and total 4150 points/spectrum) were collected in the log (1/R) mode which was converted by the reflectance value (R). Each sample (0.5 g) was placed in the sample cup, each sample was measured three times, and the mean of the three spectra was used for further statistical analysis. The temperature was kept around at 25°C.

2.5. Multivariate Calibration Methods Establishing. The whole establishing process of calibration model has been described as follow steps.

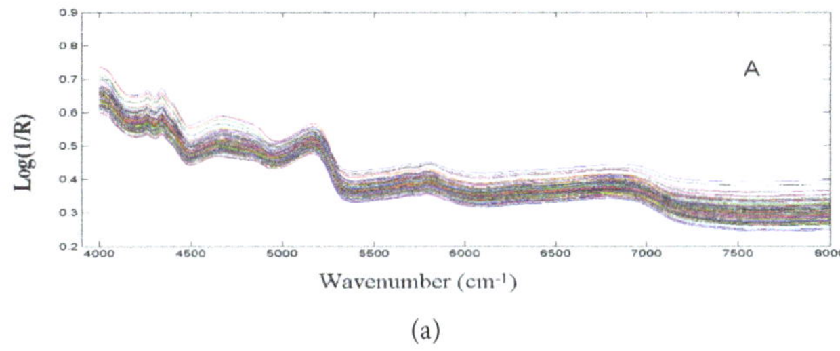

(a)

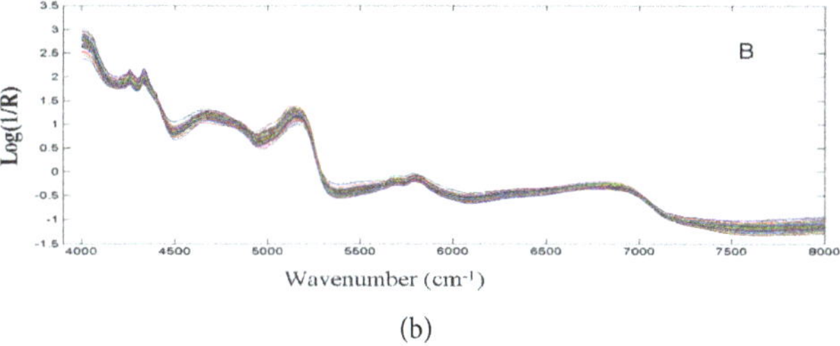

(b)

FIGURE 1: Raw NIR spectra of all samples and processed spectra.

Firstly, Kennard and Stone algorithm (K-S) [23] was adopted to split samples into calibration dataset (80%) and prediction dataset (20%), respectively. Then, the calibration dataset was used to develop calibration model for TFC and its antioxidant activity by using the partial least squares regression (PLSR). In calibration model, the number of PLS factors were optimized by 10-fold cross-validation method. It was performed as follows: (1) 90% of the calibration dataset samples were used to form the calibration model, and the remaining 10% samples were used to validate this model, and the procedure was repeated by 10 times; (2) the root mean square error of cross-validation (*RMSECV*) was then calculated as follows:

$$RMSECV = \sqrt{\frac{1}{n}\sum_{i=1}^{n}(\hat{y}_i - y_i)^2} \quad (2)$$

where n is the number of samples in the calibration set, y_i is the measured result for sample i, and $\hat{y}_i$ is the predicted value of sample i. The performances of the optimal model were evaluated according to root mean square error of calibration (*RMSEC*). *RMSEC* is calculated as follows:

$$RMSEC = \sqrt{\frac{1}{I_c}\sum_{i=1}^{I_c}(\hat{y}_i - y_i)^2} \quad (3)$$

where I_c is the number of samples in the calibration set, y_i is the measured result for sample$i$, and $\hat{y}_i$ is the predicted value of sample i.

Then, the optimal model which was validated by prediction samples in the prediction dataset *RMSEP* is calculated as follows:

$$RMSEP = \sqrt{\frac{1}{I_p}\sum_{i=1}^{I_p}(\hat{y}_i - y_i)^2} \quad (4)$$

where I_p is the number of samples in the prediction dataset, y_i is the measured result for sample i, and $\hat{y}_i$ is the predicted value of sample i.

Correction coefficient between the predicted value of PLSR model and the measurement value is calculated as follows for both the calibration and prediction set:

$$R = \sqrt{1 - \frac{\sum_{i=1}^{n}(\hat{y}_i - y_i)^2}{\sum_{i=1}^{n}(y_i - \overline{y})^2}} \quad (5)$$

where n is the number of samples in the calibration or the prediction set and $\overline{y}$ is the mean of measurement value for the calibration or the prediction set.

2.6. Spectral Signal Preprocessing. In the model establishing process, some signal pretreat need be optimized for achieving the best calibration model. In this study, some data preprocessing methods were used to process these NIR signals, such as multiplicative scattering correction (MSC), standard normal transformation (SNV), moving window smoothing, and Savitzky-Golay first derivative or second derivative (S/G 1st/2nd der). The detailed descriptions of these process methods can be found in previous researches [23–25]. All of the algorithms were implemented in MATLAB 8.0.1, and all the programs were written by own group (Mathworks).

3. Results and Discussion

3.1. TFC and Antioxidant Activity. The reduction capability of DPPH radical was determined by the decrease in absorbance by reduced DPPH to by plant antioxidants [26]. Therefore, getting systematic knowledge of flavonoids in *Ginkgo biloba* L. and their activities are highly important for the research and development of this plant. The results of TFC of these samples and their antioxidant activities were listed in Table 1.

3.2. Near-Infrared Spectra. Original NIR spectra of all samples are similar and broad; they consist of many overlapping narrow bands of different vibrational modes, as showed in Figure 1(a), the raw NIR spectra of 113 samples. It can be seen that the intensive spectral peaks are mainly in the region of 4000-8000 cm^{-1}. The multiplicative scattering correction processed spectra of all samples (Figure 1(b)) showed the most intensive band in the spectrum belonging to the vibration of the second overtone of the carbonyl group (5352 cm-1); these are caused by the stretch or deformation vibration of C-H, O-H, and N-H groups, the first two of which are abundant in the flavonoids. Also, it might be caused by the combination of stretching and deformation of the O-H group in water for the spectral peaks intense absorption bands at 6900 cm^{-1} and 5180 cm^{-1}.

3.3. NIR Calibration Model Establish. In this section, we aimed to establish a reliable and accurate calibration model for TFC and their antioxidant activities quantitative estimation. Thus, some process algorithms were taken into account.

Table 1: The total flavonoids content (TFC) and total antioxidant activity (TAA) of *Ginkgo biloba* leaves.

Samples	TFC[a]	TAA[b]	Samples	TFC[a]	TAA[b]	Samples	TFC[a]	TAA[b]
1	2.10 ± 0.18	0.47 ± 0.03	40	23.59 ± 1.79	2.98 ± 0.19	79	7.22 ± 0.46	1.01 ± 0.07
2	7.35 ± 0.42	0.99 ± 0.05	41	7.97 ± 0.48	1.06 ± 0.06	80	15.59 ± 0.95	1.96 ± 0.16
3	10.16 ± 0.76	1.93 ± 0.12	42	10.72 ± 0.98	1.14 ± 0.08	81	7.66 ± 0.68	1.64 ± 0.12
4	13.53 ± 0.87	2.12 ± 0.16	43	43.33 ± 3.16	3.97 ± 0.21	82	13.59 ± 0.97	1.93 ± 0.13
5	46.57 ± 3.25	3.48 ± 0.27	44	9.91 ± 0.69	1.19 ± 0.09	83	21.46 ± 1.46	2.94 ± 0.21
6	15.15 ± 0.93	2.33 ± 0.13	45	9.28 ± 0.64	1.75 ± 0.12	84	9.10 ± 0.54	0.83 ± 0.06
7	21.96 ± 1.98	2.67 ± 0.15	46	8.60 ± 0.63	1.00 ± 0.09	85	7.16 ± 0.58	0.57 ± 0.03
8	13.41 ± 1.15	1.52 ± 0.10	47	49.13 ± 2.67	3.84 ± 0.24	86	16.78 ± 1.07	1.24 ± 0.10
9	7.16 ± 0.63	1.51 ± 0.11	48	14.53 ± 0.93	1.99 ± 0.11	87	16.09 ± 0.99	1.39 ± 0.09
10	29.21 ± 1.92	2.45 ± 0.15	49	14.15 ± 0.96	1.47 ± 0.08	88	24.09 ± 1.54	2.81 ± 0.18
11	19.84 ± 1.82	2.34 ± 0.14	50	19.4 ± 1.65	2.51 ± 0.16	89	18.40 ± 1.48	1.68 ± 0.12
12	25.96 ± 2.29	2.83 ± 0.18	51	17.4 ± 1.34	1.68 ± 0.10	90	20.71 ± 1.35	2.79 ± 0.22
13	26.84 ± 2.32	2.59 ± 0.19	52	23.77 ± 1.87	2.69 ± 0.19	91	19.65 ± 1.28	2.62 ± 0.21
14	19.53 ± 1.75	1.82 ± 0.12	53	17.65 ± 0.95	1.99 ± 0.12	92	18.71 ± 1.23	1.66 ± 0.11
15	20.46 ± 1.69	2.08 ± 0.18	54	11.47 ± 0.73	1.19 ± 0.09	93	18.21 ± 1.01	2.23 ± 0.15
16	33.96 ± 2.98	3.26 ± 0.26	55	30.77 ± 1.98	3.15 ± 0.23	94	18.90 ± 1.42	2.18 ± 0.18
17	29.02 ± 1.92	2.33 ± 0.16	56	30.71 ± 1.85	2.84 ± 0.22	95	16.15 ± 0.97	1.60 ± 0.12
16	48.82 ± 3.44	4.42 ± 0.32	57	35.89 ± 2.45	3.18 ± 0.25	96	18.9 ± 1.12	2.78 ± 0.15
19	31.52 ± 2.76	2.60 ± 0.16	58	36.14 ± 2.24	3.50 ± 0.18	97	17.9 ± 1.53	2.91 ± 0.19
20	22.90 ± 1.54	2.04 ± 0.18	59	43.45 ± 2.89	3.57 ± 0.27	98	11.97 ± 0.98	1.22 ± 0.08
21	24.90 ± 1.59	2.67 ± 0.21	60	14.22 ± 1.07	1.47 ± 0.09	99	4.41 ± 0.35	0.72 ± 0.06
22	29.40 ± 1.97	2.96 ± 0.13	61	22.96 ± 1.48	2.64 ± 0.14	100	2.72 ± 0.17	0.32 ± 0.22
23	17.90 ± 0.98	1.61 ± 0.09	62	36.02 ± 1.91	3.59 ± 0.23	101	3.60 ± 0.23	0.52 ± 0.03
24	21.84 ± 1.89	2.57 ± 0.19	63	21.15 ± 1.58	2.62 ± 0.19	102	6.03 ± 0.51	0.84 ± 0.06
25	49.20 ± 2.92	4.65 ± 0.25	64	37.33 ± 2.68	3.78 ± 0.27	103	8.53 ± 0.64	1.40 ± 0.08
26	11.41 ± 0.70	1.19 ± 0.09	65	19.53 ± 1.39	1.72 ± 0.11	104	4.72 ± 0.36	0.55 ± 0.04
27	19.21 ± 1.06	1.80 ± 0.12	66	3.79 ± 0.27	0.71 ± 0.05	105	7.78 ± 0.42	1.35 ± 0.08
28	22.84 ± 1.42	2.93 ± 0.15	67	25.02 ± 1.52	2.07 ± 0.13	106	1.85 ± 0.12	0.97 ± 0.09
29	20.78 ± 1.39	2.50 ± 0.16	68	28.65 ± 1.86	2.31 ± 0.16	107	7.16 ± 0.58	1.91 ± 0.11
30	28.71 ± 2.35	3.41 ± 0.21	69	20.34 ± 1.53	2.17 ± 0.18	108	8.72 ± 0.63	2.41 ± 0.15
31	16.97 ± 1.04	1.65 ± 0.09	70	28.33 ± 1.94	2.19 ± 0.17	109	6.16 ± 0.51	0.55 ± 0.03
32	14.97 ± 0.78	1.32 ± 0.09	71	25.77 ± 2.05	2.42 ± 0.21	110	1.47 ± 0.93	0.24 ± 0.02
33	26.77 ± 1.85	2.99 ± 0.17	72	20.53 ± 1.86	2.18 ± 0.19	111	14.78 ± 0.54	2.81 ± 0.18
34	7.10 ± 0.55	0.81 ± 0.06	73	6.41 ± 0.51	0.88 ± 0.05	112	8.91 ± 0.61	1.42 ± 0.09
35	18.15 ± 1.07	2.63 ± 0.17	74	15.53 ± 1.27	1.46 ± 0.09	113	11.97 ± 0.85	2.42 ± 0.15
36	14.53 ± 0.96	1.39 ± 0.10	75	22.15 ± 1.75	2.28 ± 0.16			
37	29.02 ± 2.14	3.73 ± 0.21	76	8.35 ± 0.61	1.09 ± 0.07			
38	15.97 ± 0.98	1.28 ± 0.09	77	30.08 ± 1.98	3.00 ± 0.19			
39	24.02 ± 1.40	2.11 ± 0.11	78	32.27 ± 2.21	3.35 ± 0.21			

a: equivalent to rutin per g dry weight (mg g^{-1}); b: equivalent to Trolox per g of dry weight (mM g^{-1}).

The whole calibration model established process in this study contains these steps.

Firstly, the PCA method was adopted to analyze these data for exposing cluster trends in the samples information. **Secondly,** some anomalous spectra were detected by using the Mahalanobis distance and hat matrix method to detect the outliers. **Thirdly,** after removing these anomalous spectra, the remaining samples were divided into a calibration set and a prediction set by using the Kennard-Stone (K-S) algorithm. The calibration set was used to optimize the model pretreat processes and establish the calibration model; the prediction set was used as external set to validate model. In the calibration model establish process, some spectra of pretreat methods and variable selective approach were optimized.

3.4. Principle Component Analysis. Principle component analysis (PCA) can be used as a preprocessing step to show samples distribution and visually descript the samples similarities and dissimilarities. In this section, we planned

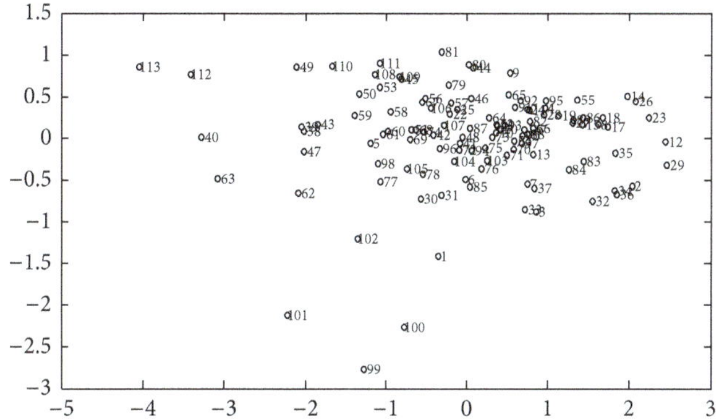

FIGURE 2: Investigation of dataset cluster trends by using PCA scores.

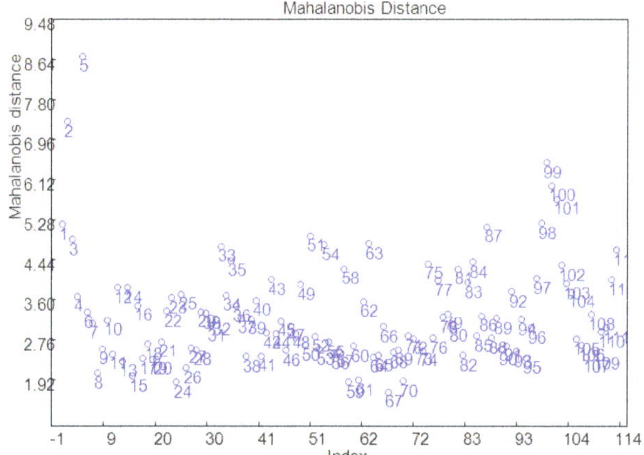

FIGURE 3: Mahalanobis distance analysis of all samples.

to use PCA to check the cluster of leaves' NIR spectra. Plotting PCA scores in two or three dimensions provides an effortless way to observe the data distribution. In the PC1-PC2 plot, the first two PCs contain about 95% (PC1: 90.32%, PC2: 5.17%) information of the raw data, and samples are unevenly distributed without obvious cluster which can be found (Figure 2). All samples were located at the whole positions of PCA scores plot. However, we still could find some samples, such as 99, 100, and 113 which are located far from other samples. These samples might be the outlier samples, which can affect the model calibration.

3.5. Anomalous Spectra Detection. In this section, two outlier measure methods were applied to accurately explore the sample information. Techniques based on the Mahalanobis distance (MD) [27] and hat matrix [28] were applied in different fields of chemometrics such as multivariate calibration, pattern recognition, and process control. In the original variable space, the MD considered the correlation in the data, since it is calculated using the inverse of the variance-covariance matrix of the data set. The "Mahalanobis distance" between all the pairs of samples was calculated. As can be seen from Figure 3, 4 samples (2, 5, 99, and 100) were defined as the outliers as the four samples were significant far away other samples.

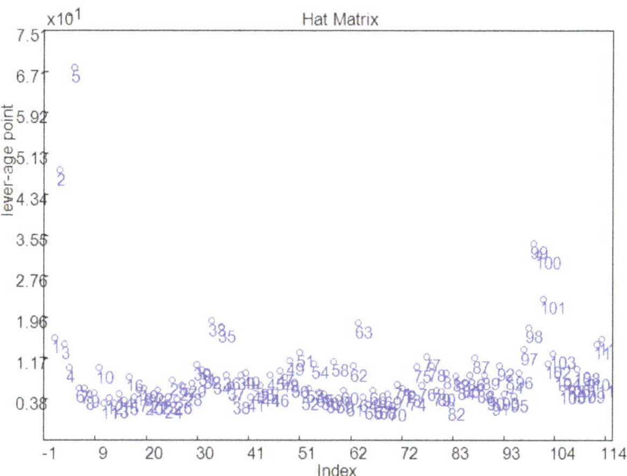

FIGURE 4: Hat matrix method analysis of all samples.

Furthermore, the hat matrix method was used to estimate the similarities of these samples. We can find that the 4 samples (2, 5, 99, and 100) were also chosen as the outliers (Figure 4). Based on these analysis, the four samples were deleted, and the remaining 108 samples were used for establishing the calibration model.

3.6. Signal Pretreatment and Prediction Model Establish. Kennard-Stone (K-S) algorithm was used to split the dataset into calibration dataset and prediction dataset with split ratio 80%. Thus, 88 samples were used to optimize the calibration model, and remaining 20 samples were used to estimate the established model. In the application of PLS algorithm, it is generally known that the spectral preprocessing methods and the number of PLS factors are critical parameters. Here, their effects on the results are discussed. The optimum number of factors is determined by the lowest root mean square error cross-validation (RMSECV).

Pretreating spectra are a procedure to optimize data and avoid disturbance due to a changing baseline. Common used pretreatments method is averaging, smoothing and normalizing with first and second derivative spectra. The first derivative can eliminate shift errors and the second derivative eliminate tilt errors. Other methods such as multiplicative scatter correction (MSC), Savizaky-Golay method (SG), and standard normal variate (SNV) are also widely used in the NIR spectra. The number of PLS factors included in the model is chosen according to the lowest RMSECV. For RMSECV, a 10-fold cross-validation was performed. Figures 5(a) and 5(b) showed RMSECV plotted versus relevant PLS factors for determining TFC and their antioxidant activity with different spectral preprocessing methods, respectively. Standard normal variate spectral preprocessing method is obviously superior to other methods with lowest RMSECV values. Therefore, standard normal variate (SNV) spectral preprocessing method and corresponding optimized factor were selected to establish the calibration model.

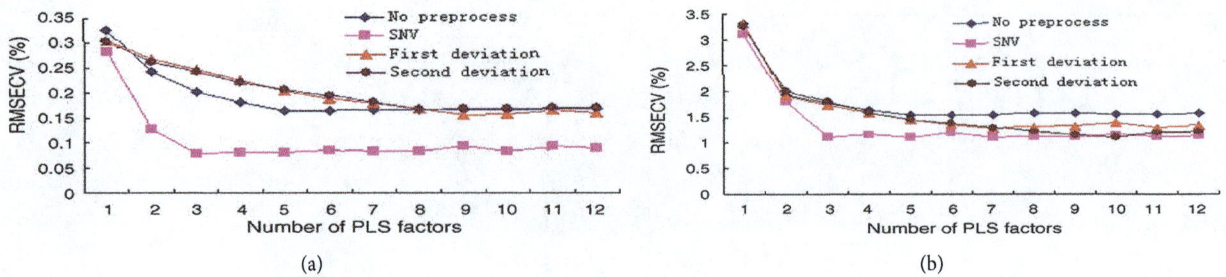

FIGURE 5: Optimized the best spectral process methods and number of PLS factors.

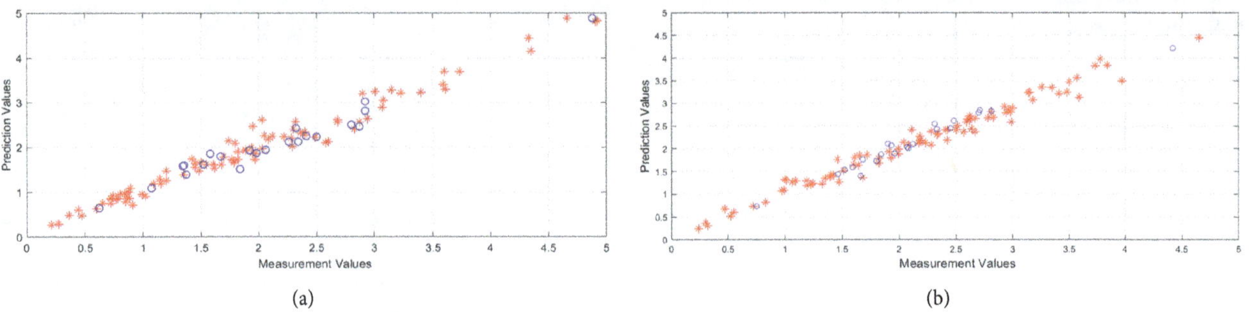

FIGURE 6: The square of correlation coefficient (R^2) and root mean square error of prediction (RMSEP) for TFC (a) and total antioxidant activity (b).

3.7. Calibration Model Validation. The robustness of the method obtained by NIR technology was validated with the 20 prediction samples. The performance of the final PLS model was evaluated in terms of root mean square error of cross-validation (RMSECV), the root mean square error of prediction (RMSEP), and the square of correlation coefficient (R^2). As can be seen from Figure 6, the coefficients of determination (R^2) for TFC and total antioxidant activity prediction were 0.8863 and 0.8486, respectively, and the root mean square errors of prediction (RMSEP) were 2.203 mg g^{-1} and 0.2211 mM g^{-1}, respectively.

4. Conclusion

In this study, a method was proposed to quantitatively analyze the TFC and their antioxidant activity of *Ginkgo biloba* L. by combining NIR spectroscopy coupled with chemometrics methods. The results verified that NIR spectroscopy was a suitable tool for quantification of TFC and their antioxidant activity, simultaneously. Comparing with other analysis methods, the NIR method has its advantages such as being simply pretreated, fast analysis speed, and being nondestructive; these make this approach has the potential of high sample throughput analysis and low costs and widely applied to products' quality control.

Authors' Contributions

Ling-jia Zhao and Wei Liu have equal contribution to this article.

Acknowledgments

This study was funded by the National Natural Science Foundation of China (81503041), Natural Science Foundation of Zhejiang Province (LY14H270008; LY15H280007), Natural Science Foundation of Hunan Province (2017JJ4045), and Project Foundation of Changsha Science and Technology Bureau (kq1701073).

References

[1] L. Marcocci, L. Packer, M.-T. Droy-Lefaix, A. Sekaki, and M. Gardes-Albert, "Antioxidant action of *Ginkgo biloba* extract EGb 761," *Methods in Enzymology*, vol. 234, pp. 462–475, 1994.

[2] J. Kobus-Cisowska, E. Flaczyk, M. Rudzińska, and D. Kmiecik, "Antioxidant properties of extracts from Ginkgo biloba leaves in meatballs," *Meat Science*, vol. 97, no. 2, pp. 174–180, 2014.

[3] A. J. Lau, G. Yang, G. Rajaraman, C. C. Baucom, and T. K. H. Chang, "Evaluation of Ginkgo biloba extract as an activator of human glucocorticoid receptor," *Journal of Ethnopharmacology*, vol. 145, no. 2, pp. 670–675, 2013.

[4] X. Yao, N. Chen, C.-H. Ma et al., "Ginkgo biloba extracts attenuate lipopolysaccharide-induced inflammatory responses in acute lung injury by inhibiting the COX-2 and NF-κB pathways," *Chinese Journal of Natural Medicines*, vol. 13, no. 1, pp. 52–58, 2015.

[5] L. Xu, Z. Hu, J. Shen, and P. M. McQuillan, "Effects of *Ginkgo biloba* extract on cerebral oxygen and glucose metabolism in elderly patients with pre-existing cerebral ischemia," *Complementary Therapies in Medicine*, vol. 23, no. 2, pp. 220–225, 2015.

[6] X. J. Xiong, W. Liu, X. C. Yang et al., "Ginkgo biloba extract for essential hypertension: a systemic review," *Phytomedicine*, vol. 21, no. 10, pp. 1131–1136, 2014.

[7] E. Pereira, L. Barros, and I. C. F. R. Ferreira, "Chemical characterization of Ginkgo biloba L. and antioxidant properties of its extracts and dietary supplements," *Industrial Crops and Products*, vol. 51, pp. 244–248, 2013.

[8] X.-G. Liu, S.-Q. Wu, P. Li, and H. Yang, "Advancement in the chemical analysis and quality control of flavonoid in Ginkgo biloba," *Journal of Pharmaceutical and Biomedical Analysis*, vol. 113, pp. 212–225, 2015.

[9] J. Deguchi, Y. Hasegawa, A. Takagi et al., "Four new ginkgolic acids from Ginkgo biloba," *Tetrahedron Letters*, vol. 55, no. 28, pp. 3788–3791, 2014.

[10] J.-Y. Qiu, X. Chen, Z. Li et al., "LC-MS/MS method for the simultaneous quantification of 11 compounds of Ginkgo biloba extract in lysates of mesangial cell cultured by high glucose," *Journal of Chromatography B*, vol. 997, pp. 122–128, 2015.

[11] G. Zhou, X. Yao, Y. Tang et al., "An optimized ultrasound-assisted extraction and simultaneous quantification of 26 characteristic components with four structure types in functional foods from ginkgo seeds," *Food Chemistry*, vol. 158, pp. 177–185, 2014.

[12] J.-Y. Shi, X.-B. Zou, J.-W. Zhao et al., "Determination of total flavonoids content in fresh Ginkgo biloba leaf with different colors using near infrared spectroscopy," *Spectrochimica Acta Part A: Molecular and Biomolecular Spectroscopy*, vol. 94, pp. 271–276, 2012.

[13] K. Ndjoko, J.-L. Wolfender, and K. Hostettmann, "Determination of trace amounts of ginkgolic acids in Ginkgo biloba L. leaf extracts and phytopharmaceuticals by liquid chromatography-electrospray mass spectrometry," *Journal of Chromatography B: Biomedical Sciences and Applications*, vol. 744, no. 2, pp. 249–255, 2000.

[14] X.-G. Liu, H. Yang, X.-L. Cheng et al., "Direct analysis of 18 flavonol glycosides, aglycones and terpene trilactones in Ginkgo biloba tablets by matrix solid phase dispersion coupled with ultra-high performance liquid chromatography tandem triple quadrupole mass spectrometry," *Journal of Pharmaceutical and Biomedical Analysis*, vol. 97, pp. 123–128, 2014.

[15] Z. Rao, H. Qin, Y. Wei et al., "Development of a dynamic multiple reaction monitoring method for determination of digoxin and six active components of Ginkgo biloba leaf extract in rat plasma," *Journal of Chromatography B*, vol. 959, pp. 27–35, 2014.

[16] Q. Zhang, G.-J. Wang, J.-Y. A et al., "Application of GC/MS-based metabonomic profiling in studying the lipid-regulating effects of Ginkgo biloba extract on diet-induced hyperlipidemia in rats," *Acta Pharmacologica Sinica*, vol. 30, no. 12, pp. 1674–1687, 2009.

[17] A. Medvedovici, F. Albu, R. D. Naşcu-Briciu, and C. Sârbu, "Fuzzy clustering evaluation of the discrimination power of UV-Vis and (±) ESI-MS detection system in individual or coupled RPLC for characterization of Ginkgo Biloba standardized extracts," *Talanta*, vol. 119, pp. 524–532, 2014.

[18] M. Jamrógiewicz, "Application of the near-infrared spectroscopy in the pharmaceutical technology," *Journal of Pharmaceutical and Biomedical Analysis*, vol. 66, pp. 1–10, 2012.

[19] S. S. Rosa, P. A. Barata, J. M. Martins, and J. C. Menezes, "Near-infrared reflectance spectroscopy as a process analytical technology tool in *Ginkgo biloba* extract qualification," *Journal of Pharmaceutical and Biomedical Analysis*, vol. 47, no. 2, pp. 320–327, 2008.

[20] M. Blanco and A. Peguero, "Analysis of pharmaceuticals by NIR spectroscopy without a reference method," *TrAC - Trends in Analytical Chemistry*, vol. 29, no. 10, pp. 1127–1136, 2010.

[21] Y. Geng and B. Xiang, "Simultaneous quantisation of flavonol aglycones in Ginkgo biloba leaf extracts applying moving window partial least squares regression models," *Journal of Near Infrared Spectroscopy*, vol. 16, no. 6, pp. 551–559, 2008.

[22] M. Okawa, J. Kinjo, T. Nohara, and M. Ono, "DPPH (1,1-diphenyl-2-Picrylhydrazyl) radical scavenging activity of flavonoids obtained from some medicinal plants," *Biological & Pharmaceutical Bulletin*, vol. 24, no. 10, pp. 1202–1205, 2001.

[23] E. Bouveresse, C. Hartmann, D. L. Massart, I. R. Last, and K. A. Prebble, "Standardization of near-infrared spectrometric instruments," *Analytical Chemistry*, vol. 68, no. 6, pp. 982–990, 1996.

[24] J. Sun, "Statistical analysis of NIR data: Data pretreatment," *Journal of Chemometrics*, vol. 11, no. 6, pp. 525–532, 1997.

[25] Y. Katsumoto, J. Jiang, R. Berry, and Y. Ozaki, "Modern pretreatment methods in NIR spectroscopy," *Near Infrared Analysis*, vol. 2, pp. 29–36, 2001.

[26] R. Ksouri, H. Falleh, W. Megdiche et al., "Antioxidant and antimicrobial activities of the edible medicinal halophyte *Tamarix gallica* L. and related polyphenolic constituents," *Food and Chemical Toxicology*, vol. 47, no. 8, pp. 2083–2091, 2009.

[27] R. De Maesschalck, D. Jouan-Rimbaud, and D. L. Massart, "The Mahalanobis distance," *Chemometrics and Intelligent Laboratory Systems*, vol. 50, no. 1, pp. 1–18, 2000.

[28] D. C. Hoaglin and R. E. Welsch, "The hat matrix in regression and anova," *The American Statistician*, vol. 32, no. 1, pp. 17–22, 1978.

21

Recent Progress of Imprinted Nanomaterials in Analytical Chemistry

Rüstem Keçili[1] and Chaudhery Mustansar Hussain[2]

[1] *Anadolu University, Yunus Emre Vocational School of Health Services, Department of Medical Services and Techniques, 26470 Eskişehir, Turkey*
[2] *Department of Chemistry and Environmental Science, New Jersey Institute of Technology, Newark, N J 07102, USA*

Correspondence should be addressed to Chaudhery Mustansar Hussain; chaudhery.m.hussain@njit.edu

Academic Editor: Stig Pedersen-Bjergaard

Molecularly imprinted polymers (MIPs) are a type of tailor-made materials that have ability to selectively recognize the target compound/s. MIPs have gained significant research interest in solid-phase extraction, catalysis, and sensor applications due to their unique properties such as low cost, robustness, and high selectivity. In addition, MIPs can be prepared as composite nanomaterials using nanoparticles, multiwalled carbon nanotubes (MWCNTs), nanorods, quantum dots (QDs), graphene, and clays. This review paper aims to demonstrate and highlight the recent progress of the applications of imprinted nanocomposite materials in analytical chemistry.

1. Introduction

Molecularly imprinted polymers (MIPs) are highly cross-linked robust materials which display excellent affinity towards target compound. For the preparation of MIPs, appropriate functional monomers and a cross-linker agent are polymerized around the target compound (template). The schematic demonstration of the molecular imprinting technique is shown in Figure 1. Due to their high affinity and selectivity for the desired compound, MIPs can be efficiently used in different application areas such as separation, catalysis, and sensor platforms [1–18]. In addition to specific molecular recognition abilities towards their target compound, MIPs can be prepared as composite nanomaterials using nanoparticles, multiwalled carbon nanotubes (MWCNTs), nanorods, quantum dots (QDs), graphene, clays in nanoscale, etc.

This paper provides the recent progress of the applications of imprinted nanocomposite materials in analytical chemistry.

2. MIPs in SPE Applications

Solid-phase extraction (SPE) is an efficient sample preparation technique which is one of the most widely applied approach in analytical chemistry. SPE has been first applied in 1940s [19]. Then, the progress for the current analytical applications was initiated in the 1970s. Different conventional materials such as silica based [20, 21], carbon based [22, 23], and clay based [24] resins were widely used in various applications of SPE. Although it is a popular sample preparation technique for the enrichment or extraction of the desired molecules from the complex matrices, the conventional SPE materials used in analytical applications exhibit lower selectivity towards the target molecules that lead to binding of other potentially interfering molecules existing in the sample matrices. This issue is very important especially for the complex biological samples such as urine and blood. MIP-based SPE materials that display great selectivity and binding affinity towards the target molecule/s can overcome the drawbacks of the conventional resins. In addition, MIPs preserve their stability under extreme

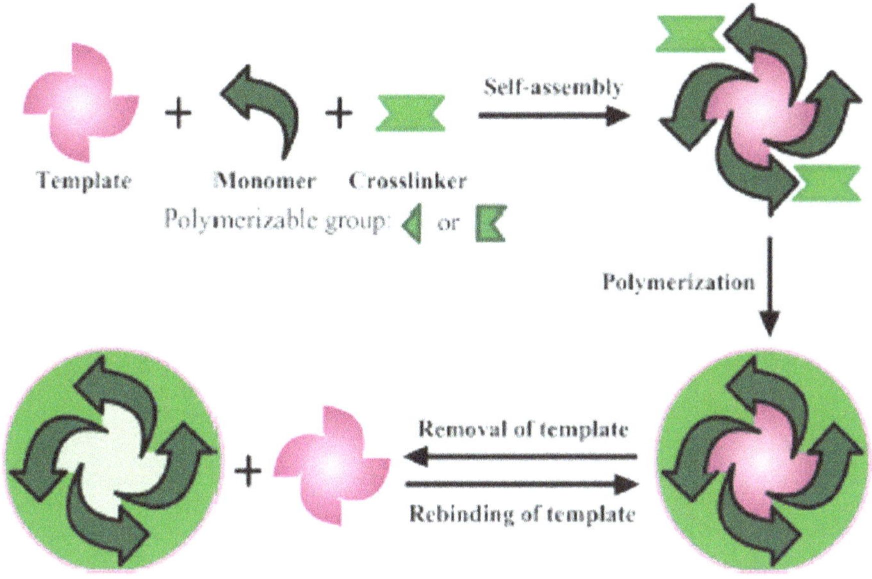

FIGURE 1: Molecular imprinting process (reproduced with permission from [25]).

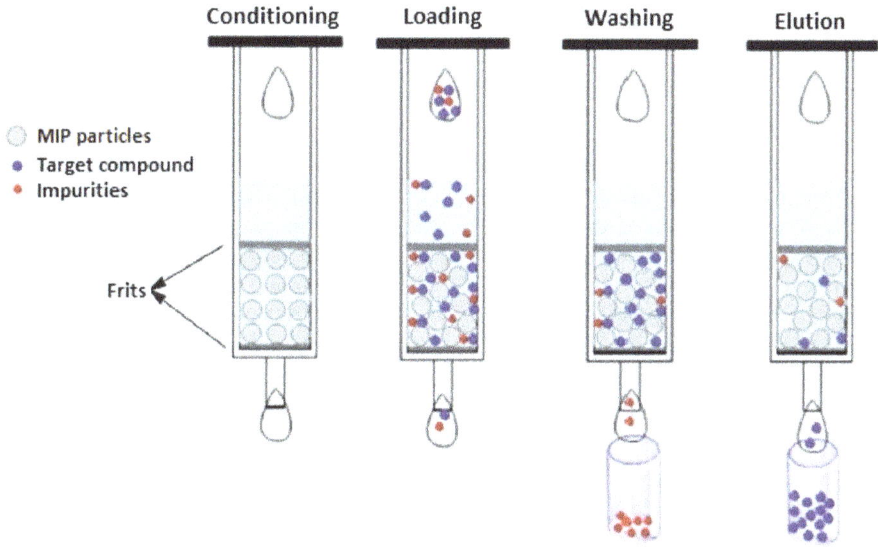

FIGURE 2: Schematic representation of SPE process (reproduced with permission from [26]).

conditions (e.g., high pressure, high temperature, and lower and higher pH).

MIP-based SPE process composed of 4 steps is schematically demonstrated in Figure 2.

Sellergren published the first SPE application of MIPs [29]. In the reported study, robust MIPs were developed for the selective extraction of drug compound pentamidine. After this successful application, many MIP-based SPE applications of various compounds in different areas were conducted and reported in the literature [30–41].

In a reported study, Su et al. developed magnetic MIP nanoparticles for the separation of bovine hemoglobin (Bhb) [114]. In their study, firstly, the preparation of magnetic $Fe_3O_4@SiO_2$-acrylic acid (AA) nanoparticles were performed. In the second step, the preparation of BHb imprinted magnetic nanoparticles was carried out by using methacrylic acid (MAA), itaconic acid (IA), and N',N-methylenebisacrylamide as functional monomers and crosslinker, respectively. The BHb imprinted magnetic nanoparticles were efficiently used for the extraction of BHb with high binding capacity (169.29 mgg^{-1}).

Viveiros et al. developed a green strategy for the preparation of selective MIPs for acetamide which is a potentially genotoxic impurity in active pharmaceutical ingredients (API) [115]. In their study, silica beads were first functionalized with 3-(Trimethoxysilyl)propyl methacrylate and then MIP layer was synthesized on the modified-silica beads using supercritical CO_2 as the green solvent. The prepared

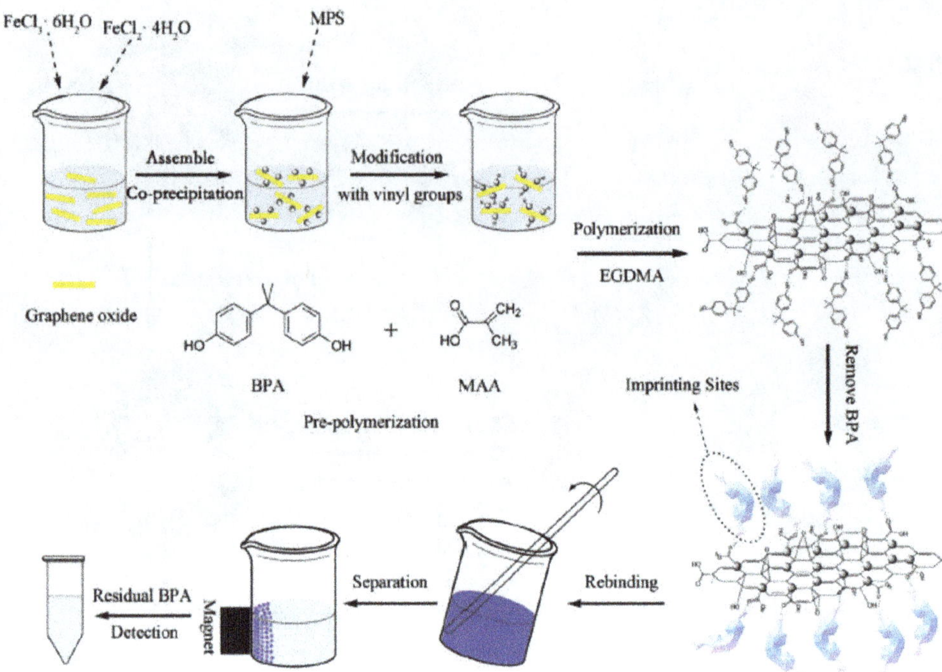

FIGURE 3: Schematic demonstration of the preparation of MIP-based magnetic graphene oxide composite towards BPA and extraction process (reproduced with permission from [27]).

acetamide imprinted polymers were successfully used for the extraction of acetamide from beclomethasone dipropionate which is the model API. The results showed that 100% of acetamide was removed by using selective MIPs with only very little loss of API (0.37%).

In another important study, Zhang and colleagues developed magnetic MIP-based-MWCNTs composite materials for the removal of Bisphenol A (BPA) from water matrices [116]. For this purpose, MAA was chosen as the functional monomer. The results from rebinding experiments for BPA in batch mode confirmed that the magnetic MIP-based MWCNTs have excellent affinity towards BPA and the obtained maximum binding capacity was 49.26 μmolg^{-1}.

Yan and colleagues demonstrated the application of MIP/silica nanocomposites for the recognition of nitrocellulose [117]. The surface of the SiO_2 particles was firstly conjugated with –OH groups and 3-(Trimethoxysilyl)propyl methacrylate (MPS) was used for the functionalization of the surface with an acrylyl groups. Then, nitrocellulose (NC) imprinted shell was synthesized on the modified-SiO_2 particles using the functional monomer MAA and cross-linkee ethylene glycol dimethacrylate (EGDMA). The results indicated that MIP/silica nanocomposites exhibited high recognition ability towards NC with a maximum capacity of 1.7 mgmg^{-1}.

In another interesting study reported by Wang and coworkers, selective extraction of BPA was successfully performed by using MIP-based magnetic graphene oxide composites [27]. For this purpose, they firstly prepared magnetic graphene oxide by using coprecipitation approach. Then, MAA (functional monomer) and BPA (template, target compound) were used for the preparation BPA imprinted magnetic graphene oxide composite. The schematic demonstration of the preparation of MIP-based magnetic graphene oxide composite towards BPA and extraction process is shown in Figure 3. The results confirmed that the prepared MIP-based magnetic graphene oxide composite displayed high selectivity towards BPA in the presence of other competing compounds such as phenol and 2,4-dichlorophenol.

Shea and his colleagues prepared imprinted hollow beads for the extraction of β-estradiol from tap water [118]. For this purpose, SiO_2 nanoparticles were used as the sacrificial support. After surface modification with 3-(Trimethoxysilyl)propyl methacrylate, selective MIP shell towards β-estradiol was synthesized on the surface of the SiO_2 nanoparticles using the functional monomer MAA and cross-linker EGDMA. The highest binding of β-estradiol was obtained within a very short time (15 min) with a maximum binding capacity of 44.5 μmolg^{-1}.

In another important study reported by Shen and colleagues [28], SiO_2 particles having MIP shell were developed for the SPE of tetrabromobisphenol A (TBBPA) from river water. For this purpose, tetrachlorobisphenol A (TCBPA) was chosen as the dummy template for the preparation of MIP towards TBBPA (Figure 4). The prepared imprinted SiO_2 particles showed fast binding kinetics (20 min) and high binding capacity (230 μmolg^{-1}) towards the target compound TBBPA.

Guo et al. reported that magnetic graphene-based MIP composite was prepared for selective recognition of bovine hemoglobin (BHb) [119]. For this purpose, magnetic graphene was prepared in the first step. Then, MIP layer

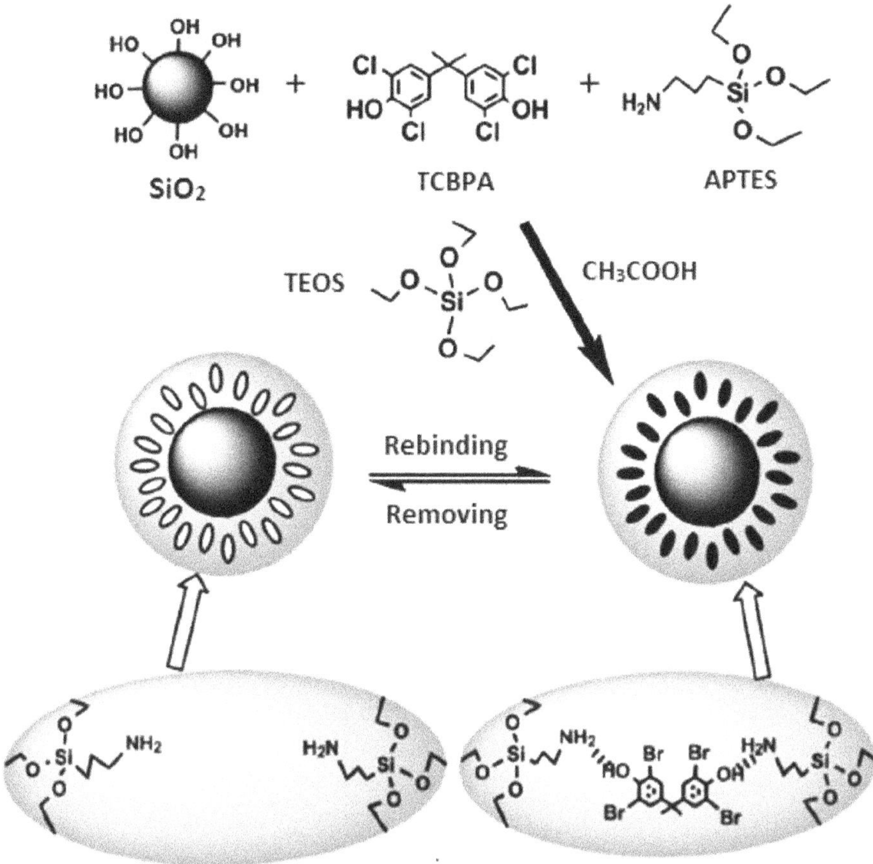

FIGURE 4: Preparation of MIP/SiO_2 composite for TBBPA (reproduced with permission from [28]).

selective to BHb was prepared using the functional monomer acrylamide (AAm) which has high affinity towards BHb and cross-linker methylene bisacrylamide (MBA). Maximum binding capacity of the magnetic graphene-based MIP composite for BHb was found to be as 186.73 mgg^{-1}.

Luo et al. developed magnetic graphene-based MIP composite for the removal of 4-nitrophenol (4-NP) from aqueous solutions [120]. Fe_3O_4 nanoparticles were immobilized on surface of graphene sheet and magnetic graphene (MGR) was prepared in the first step. Then, MGR/MIPs composite was prepared by polymerization of phenyltriethoxysilane and tetramethoxysilane in the presence of 4-NP. The preparation of the MGR/MIPs composite is demonstrated in Figure 5. The results indicated that the prepared MGR/MIP composite displayed a great binding behavior for 4-NP with an excellent binding capacity (142 mgg^{-1}).

In another research by Yang et al., core-shell magnetic MIPs were prepared for selective removal of indole from fuel oil [121]. In their research, magnetic Fe_3O_4 nanoparticles were synthesized by using coprecipitation technique. Then, surface of the prepared nanoparticles was coated with SiO_2 using 3-(Trimethoxysilyl)propyl methacrylate. In the final step, the functional monomer MAA and EGDMA (cross-linker) were polymerized on the surface of the modified magnetic nanoparticles for the preparation of selective MIP shell towards indole. The results confirmed that the prepared magnetic MIP composite displayed excellent recognition ability towards the target compound indole. The binding capacity of the composite for indole was obtained as 50.25 mgg^{-1}.

In another interesting study [122], Cao et al. prepared MIP-based-MWCNTs for the SPE of perfluorooctanoic acid from aqueous matrices. In their study, they used the functional monomer AAm for the preparation of MIP. After characterization studies, the prepared MIP-based-MWCNTs as composite SPE materials were successfully used for the selective removal of perfluorooctanoic acid from aqueous matrices. The obtained results confirmed that the binding equilibrium was obtained in 80 min. The determined binding capacity was 12.4 mgg^{-1}.

Table 1 shows the recent examples of the SPE applications of nanostructured MIP-based composites.

3. MIPs in Sensor Applications

MIP-based sensors can be categorized into 3 basic groups: electrochemical, spectroscopic, and piezoelectric sensors. In the following sections, recent examples of MIP-based sensors are briefly explained.

3.1. MIP-Based Electrochemical Sensors. In electrochemical detection, the reaction generally leads to a change of current

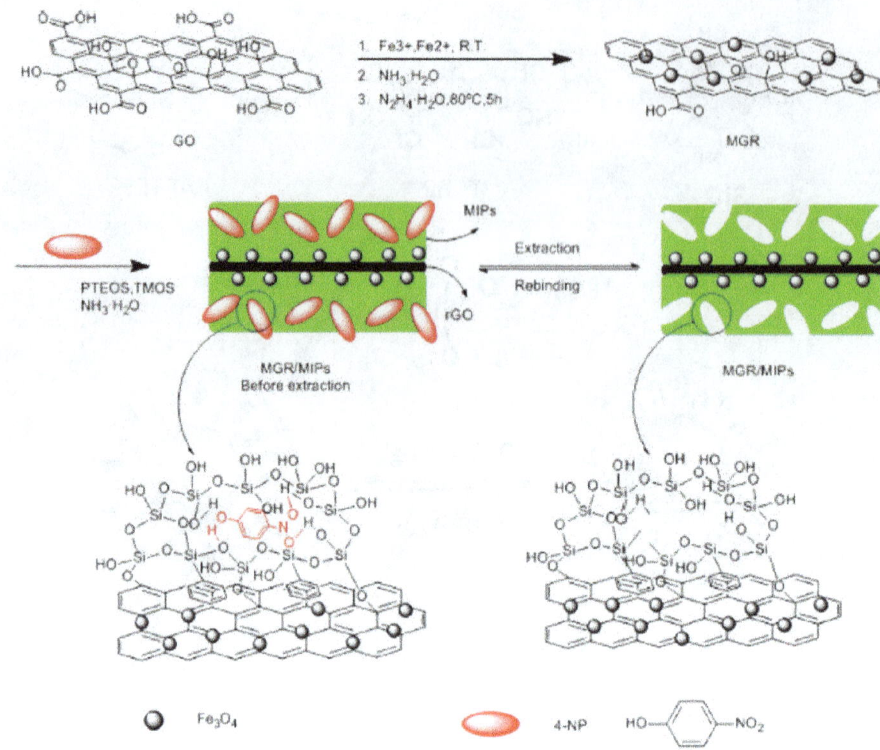

FIGURE 5: Magnetic graphene-based MIP composite towards 4-NP (reproduced with permission from [120]).

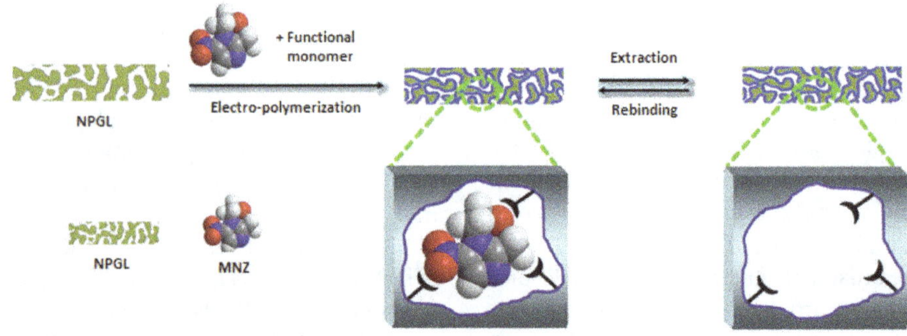

FIGURE 6: Preparation of MIP-based electrochemical sensor towards MNZ (reproduced from Li et al. (2015).

(amperometric), potential (potentiometric), or conductivity (conductometric) [129]. Selectivity and sensitivity are crucial parameter for electrochemical sensors. Surface modification of electrodes in electrochemical sensors by immobilization of recognition components is an efficient approach to obtain a high binding of target compound with good selectivity and good response. The surface modification of electrodes in the design and preparation of electrochemical sensors has firstly been reported by Itaya and Bard in 1978 [130]. Since then, many studies on the design and development of electrochemical sensors in different application areas have been reported.

In a reported study [131], an electrochemical sensor having MIP film for the theophylline recognition was prepared by Kan and colleagues. In their study, the functional monomer o-phenyldiamine was used as the functional monomer for the preparation of MIP film. After MIP film preparation on the glassy carbon electrode surface, gold nanoparticles were immobilized onto MIP film. The prepared MIP-based electrochemical sensor was characterized by SEM and binding behavior towards theophylline was tested using CV, differential pulse voltammetry, and EIS. The detection limit for theophylline was found to be as 1.0×10^{-7} $molL^{-1}$.

Li and colleagues developed an electrochemical sensor composed of nanoporous gold leaf (NPGL) electrode having selective MIP layer for the detection of metronidazole (MNZ) [132]. The preparation of the MIP-based electrochemical sensor towards MNZ is schematically shown in Figure 6. The

TABLE 1: Recent examples of nanostructured MIP-based composites in SPE applications.

Reference	Nanocomposite composition	Analyte	Sample
	Applications to environmental samples		
[42]	Magnetic nanoparticles coated with MIP having the functional monomer 4-vinyl pyridine (4-VP)	Cr^{6+}	Water
[43]	Silica-MIP composite prepared by grafting method	$[UO_2]^{2+}$	Water
[44]	Magnetic nanoparticles coated with MIP having –NH groups	Co^{2+}	Water
[45]	Chitosan-MIP magnetic nanocomposite	Ni^{2+}	Water
[46]	Silica-MIP monolithic composite column	α-cypermethrin	Soil
[47]	Cu(II)-mediated silica fiber-MIP composite	Thiabendazole	Soil
[48]	Magnetic nanoparticles coated with MIP having MAA and 4-VP as functional monomers	Methyl parathion	Soil
[49]	Magnetic nanoparticles coated with MIP prepared by using the functional monomer gelatin	17β-estradiol	Water
	Applications to clinical samples		
[50]	Magnetic SiO_2 nanoparticles having MIP shell prepared by using the functional monomer MAA	Amitriptyline	Human plasma and urine
[51]	Magnetic $SiO_{2/}$/MIP/chitosan biocomposite	Baclofen	Human urine
[52]	Magnetic nanoparticles having MIP shell prepared by using the functional monomer MAA	Rizatriptan	Human urine
[53]	Magnetic nanoparticles having MIP shell prepared by using the functional monomer MAA	Paracetamol	Human plasma
[54]	Optical fiber coated with MIP prepared by sol-gel method	Caffeine	Human serum
[55]	Magnetic nanoparticles having MIP shell prepared by using the functional monomer AAm	Protoberberine alkaloids	Rat plasma
[56]	Magnetic CNTs coated with MIP having carboxyl groups	Catecholamines	Human plasma
[57]	Magnetic nanoparticles having MIP shell prepared by using the functional monomer MAA	Tizanidine	Human urine
[58]	Magnetic nanoparticles coated with MIP having aminoimide as the functional monomer	Codeine	Human urine
[59]	Silica-MIP composite having AAm, MAA and 4-VP as functional monomers	Baicalin	Rat tissues
	Applications to food and beverage samples		
[60]	Carbon QDs-doped MIP monolithic column bearing the functional monomer MAA	Aflatoxin B1	Peanut
[61]	Magnetic nanoparticles having MIP shell bearing the functional monomer MAAm	Dimethoate	Olive oil
[62]	Magnetic MWCNTs having MIP bearing the functional monomer MAA	Melamine	Milk
[63]	Magnetic nanoparticles having MIP shell prepared by using ethyl paraoxon as the dummy template	organophosphorus pesticide	Red wine
[64]	Magnetic nanoparticles coated with MIP having AA as the functional monomer	Imidacloprid	Honey and eggplant
[65]	Magnetic nanoparticles coated with MIP having MAAm and N-3,5-bis(trifluoromethyl) phenyl-N'-4-vinylphenyl urea as functional monomers	Citrinin	Rice
[66]	Magnetic nanoparticles having MIP shell prepared by using the functional monomer MAA	Malachite green	Fish
[67]	Magnetic nanoparticles coated with MIP having oleic acid	Oxytetracycline	Honey, Egg
[68]	Carbon dots coated with MIP prepared by sol–gel method	Sterigmatocystin	Grain

Table 1: Continued.

Reference	Nanocomposite composition	Analyte	Sample
[69]	Magnetic nanoparticles coated with MIP having dopamine as the functional monomer	Gallic acid	Grape, Apple, Peach and Orange juices
[70]	Magnetic nanoparticles coated with MIP having vinyl groups	Ni(II)	Cucumber, Cantaloupe, Apple, Nectarine, Green beans, Fenugreek, Dill, Tuna fish
[71]	Silica nanoparticles having MIP shell bearing the functional monomer MAA	Ofloxacin	Milk
[72]	Magnetic nanoparticles having MIP shell bearing the functional monomer dopamine	Diethylstilbestrol	Milk
[73]	Magnetic nanoparticles having MIP shell bearing the functional monomer AAm	β-agonists	Pork
[74]	Magnetic nanoparticles having MIP shell bearing the functional monomer MAA	Chloramphenicol	Honey

experimental results confirmed that the developed electrochemical sensor has excellent binding affinity towards MNZ in fish tissue samples. The detection limit was obtained as 1.8×10^{-11} $molL^{-1}$.

In a study reported by Gupta and Goyal, a new graphene/MIP composite sensor for the determination of melatonin in biological samples was prepared [133]. For this purpose, MIP layer was prepared on the glassy carbon electrode (GCE) surface by copolymerization of 4-amino-3-hydroxy-1-naphthalenesulfonic acid and melamine around the template melatonin. The optimization studies for MIP layer formation were carried out changing the parameters such as monomer/template ratio and time. After characterization of the prepared composite electrochemical sensor for melatonin by SEM and EIS, the binding performance of the sensor towards target melatonin was carried out by using square wave voltammetry and cyclic voltammetry. The obtained results showed that efficient recognition of melatonin in plasma samples was successfully achieved. The determined detection limit was 0.006 μM.

Cui et al. prepared graphene-Prussian blue (GR-PB)/MIP-based composite electrochemical sensor for selective detection of butylated hydroxyanisole (BHA) in food samples [123]. In this study, MIP film was synthesized on the surface of GCE having GR-PB by electropolymerization of the functional monomer pyrrole and the template BHA (Figure 7). The prepared composite sensor was characterized by SEM, cyclic voltammetry (CV), electrochemical impedance spectroscopy (EIS), and chronoamperometry. The results obtained from the experiments for the sensor performance showed that immobilization of GR and PB onto the GCE increased the sensor sensitivity and the response towards target BHA. The prepared composite electrochemical sensor showed a linear response towards BHA (9×10^{-8} M to 7×10^{-5} M) and the detection limit was calculated as 7.63×10^{-8} M.

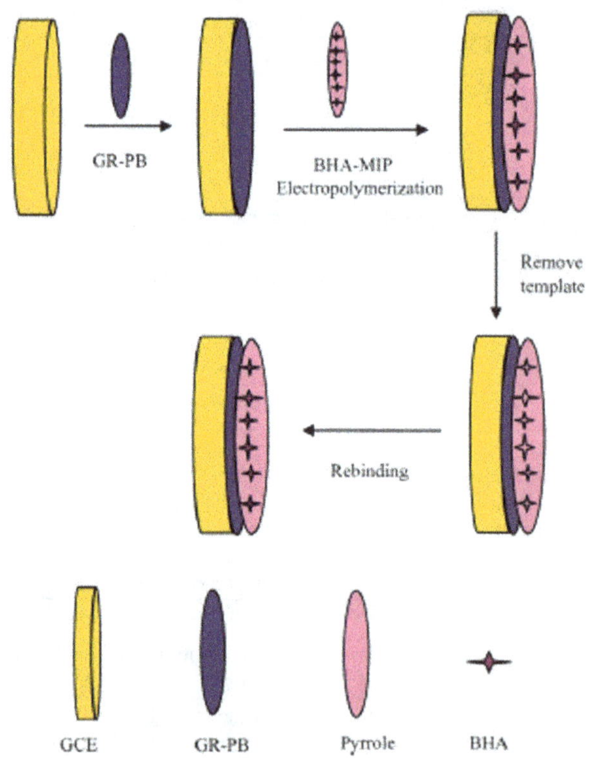

Figure 7: GR-PB/MIP-based composite electrochemical sensor towards BHA (reproduced with permission from [123]).

In an interesting study published by Prasad and colleagues, a composite electrochemical sensor composed of MIP film and MWCNTs was prepared for the detection of L-histidine [134]. MIP film selective to L-histidine was prepared by polymerization of 2-acryl amidoethyl dihydrogen phosphate (functional monomer) and EGDMA (cross-linker). In the first step, the functional monomer was interacted with Cu (II). Then, polymerization was performed in the presence of Cu (II)-functional monomer-template complex. The prepared MIP-based electrochemical sensor showed enantioselectivity towards L-histidine and the detection limit was found to be as 1.980 $ngmL^{-1}$. However, cross-reactivity studies of the prepared sensor for potentially interfering compounds in the sample such as L-phenylalanine, D- histidine,

L- and D-tryptophan, L-tyrosine, L-methionine, L-alanine, L-glycine, L-proline, urea, dopamine, creatinine, uric acid, L-glutamic acid, and L-ascorbic acid were also performed. The results confirmed that the prepared composite electrochemical sensor exhibited very low response towards these interfering compounds.

A carbon nanotube (CNT)/Graphene (GP)/MIP-based composite electrochemical sensor for the detection of bovine serum albumin (BSA) was developed by Chen and colleagues [15]. For this purpose, carbon electrode (CE) was modified with GP in the first step. Then, CNT was prepared on the surface of modified CE with GP. In the final step, MIP membrane was synthesized on the CNT/GP/CE by electrodeposition of aniline in the presence of template BSA. The prepared sensor was successfully applied for sensitive recognition of BSA in human serum with a detection limit of 6.2×10^{-11} gmL^{-1}.

Wang and coworkers reported the preparation of CdS quantum dot/graphene/MIP-based electrochemical sensor for selective recognition of 4-aminophenol in water samples [135]. In their study, fluorine-doped tin oxide (FTO) electrode was modified with CdS quantum dots and graphene (GR). Then, a MIP film selective to target compound 4-aminophenol was prepared by electropolymerization. The results confirmed that the developed electrochemical sensor specifically binds the target 4-aminophenol. The response of the sensor towards 4-aminophenol was linear in the concentration range of 5.0×10^{-8} M to 3.5×10^{-6} M and the determined detection limit was 2.3×10^{-8} M.

3.2. MIP-Based Spectroscopic Sensors. MIP-based spectroscopic sensors can be divided into 3 categories. These are MIP-based-fluorescence sensors, MIP-based-chemiluminescence sensors, and MIP-based-SPR sensors. In the fluorescence based molecular recognition of the target compound, fluorescence functional monomers are chosen for the fabrication of sensor platforms based on molecular imprinting technique [136]. When the target compound binds to the sensor, fluorescence intensity increases or decreases depending on the sensor design.

In a significant research reported by Zhang and colleagues [137], CdSe/ZnS quantum dots (QDs) coated with MIP film which shows fluorescence feature were synthesized for the sensitive recognition of carbaryl in cabbage and rice samples. For this purpose, MAA was used as the functional monomer for the synthesis of MIP layer on the QDs surface modified with the ionic liquid. The obtained results from the fluorescence measurements showed that the fluorescence sensor composed of QD_S-MIP exhibited high recognition ability towards carbaryl in the presence of metolcarb and isorcarb which are analogues of carbaryl.

Mehrzad-Samarin et al. developed a novel graphene QDs embedded silica MIP-based fluorescence sensor for the selective recognition of metronidazole [138]. The prepared sensor showed a linear response towards metronidazole in the range between 0.2 μM and 15μM. The determined detection limit was 0.15 μM.

Li and coworkers developed magnetic silica nanoparticles having selective MIP shell for the recognition of Rhodamine B from aqueous samples [139]. In this study, magnetic silica nanoparticles were coated with MIP layer using nitrobenzoxadiazole which is a fluorophore molecule. The obtained results confirmed that the efficient detection of Rhodamine B in aqueous samples was performed by using MIP-based magnetic silica nanoparticles. The maximum binding of Rhodamine B was obtained in 60 min with a high binding capacity (29.64 mgg^{-1}).

In another study reported by Jalili and Amjadi [140], MIP/green emitting carbon dot composite was prepared for the selective recognition of 3-nitrotyrosine which is a biomarker for various diseases such as rheumatoid arthritis, Alzheimer, atherosclerosis, osteoarthritis, and cardiovascular diseases. The prepared MIP-based composite fluorescence sensor was efficiently used for the selective recognition of 3-nitrotyrosine in human serum samples in the concentration range from 0.05 to 1.85 μM and the detection limit was obtained as 17 nM.

The research group of Hu was developed a ZnS QDs/MIP-based fluorescence nanosensor for the sensitive detection of sulfapyridine in tap water samples [124]. For this purpose, Mn-doped ZnS QDs was used as the fluorescence core and MIP shell was prepared on the surface of the QDs by using the functional monomer MAA, cross-linker EDMA, initiator AIBN, and template sulfapyridine (SPD). The schematic demonstration of the preparation of ZnS QDs/MIP-based fluorescence nanosensor towards sulfapyridine is shown in Figure 8. The prepared ZnS QDs/MIP-based fluorescence nanosensor exhibited high recognition ability towards SPD with a detection limit of 0.5 μM.

Chemiluminescence is another efficient approach that is used for the investigation of the recognition performance of MIP-based spectroscopic sensor systems. In this approach, a chemiluminescence system is chosen and selective MIPs are integrated to this system. When target compound binds to the MIP-based sensor, chemiluminescence emission is generated. The amount of the emission depends on the amount of bound target compound to the sensor surface.

In a study conducted by Wang and coworkers [125], a magnetic graphene oxide (GO)/MWCNTs/MIP-based chemiluminescence nanosensor was developed for the sensitive detection of lysozyme in egg samples. Figure 9 shows the schematic demonstration of the construction of the magnetic GO/MWCNTs/MIP-based chemiluminescence nanosensor towards lysozyme. The developed chemiluminescence nanosensor displayed high sensitivity towards lysozyme. The obtained detection limit was 1.9×10^{-9} gmL^{-1}.

SPR-based sensor platforms are also popular recognition systems. SPR technique relies on the measurement of the changes in refractive index of thin layer on the metal surface. The recognition element on the surface of the sensor is usually gold or silver coated with thin film. Therefore, uniform film layer is synthesized on the surface of MIP-based-SPR sensors.

Many studies were published on the development of MIP-based SPR sensors and their applications. For example, the group of Piletsky developed a molecularly imprinted nanoparticle-based SPR sensor system for the sensitive detection of diclofenac in aqueous solutions [141]. For this purpose, diclofenac imprinted nanoparticles were synthesized by using

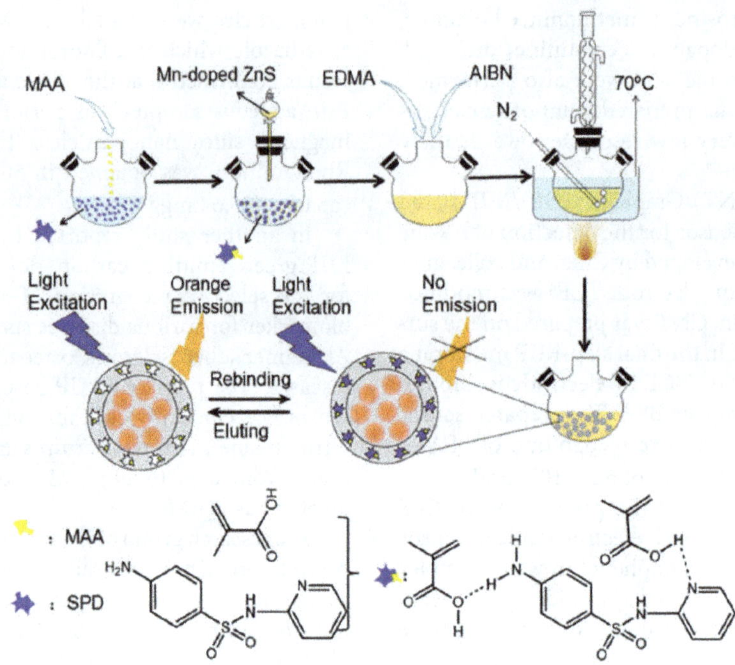

FIGURE 8: The preparation of ZnS QDs/MIP-based fluorescence nanosensor towards sulfapyridine (reproduced with permission from [124]).

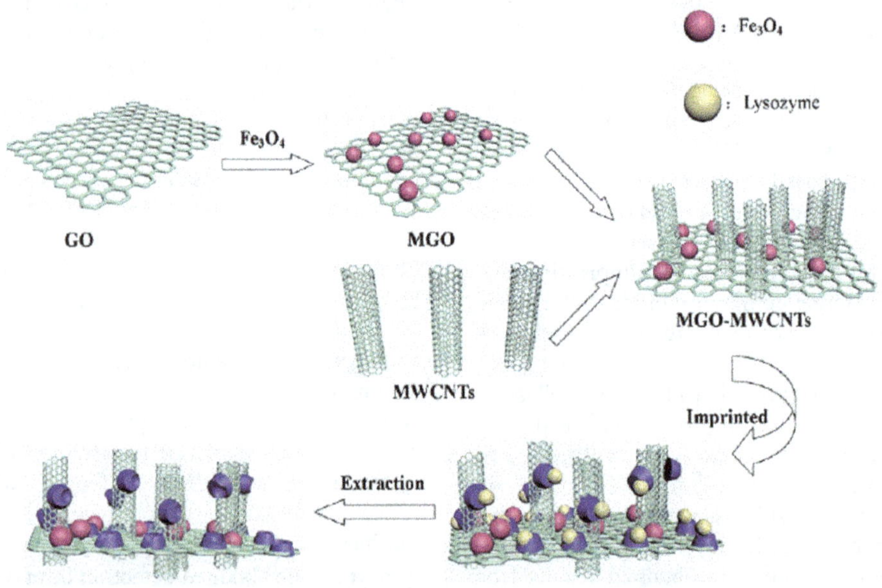

FIGURE 9: Schematic depiction of the preparation of magnetic GO/MWCNTs/MIP-based chemiluminescence nanosensor towards lysozyme (reproduced with permission from [125]).

styrene as the functional monomer, EGDMA and trimethylolpropane trimethacrylate (TRIM) as cross-linkers, and pentaerythritol tetrakis (3-mercaptopropionate) as the chain transfer agent. Then, the surface of the SPR sensor was activated by using N-Hydroxysuccinimide (NHS) and 1-Ethyl-3-(3-dimethylaminopropyl)-carbodiimide (EDC). After activation step, the prepared diclofenac imprinted nanoparticles were immobilized onto the surface of the sensor. The sensitive detection of diclofenac was successfully achieved in the concentration range from 1.24 to 80 $ngmL^{-1}$. The selectivity of the SPR sensor towards diclofenac in the presence of propranolol and vancomycin was also studied. The experimental data confirmed that the sensor exhibited high selectivity towards diclofenac.

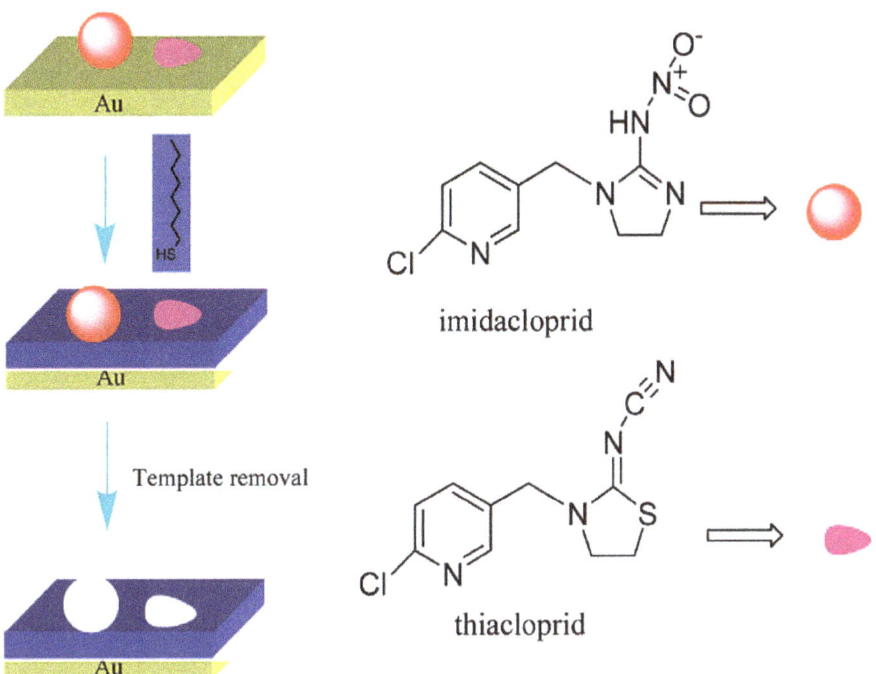

FIGURE 10: MIP-based QCM sensor towards imidacloprid and thiacloprid (reproduced with permission from [126]).

In another interesting study [142], Ashley et al. prepared a MIP-based SPR nanosensor for the sensitive detection of α-casein cleaning in place (CIP) wastewater samples. For this purpose, immobilization of the target protein α-casein (template) on the surface of glass nanobeads was carried out in the first step. Then, MIP nanoparticles were prepared by using N-(3-aminopropyl)-methacrylamide, the functional monomer acrylic acid, and cross-linker N,N′-methylenebis(acrylamide). Finally, α-casein imprinted nanoparticles were incorporated onto the SPR sensor surface. The results confirmed that the developed MIP-based SPR nanosensor showed excellent selectivity and affinity (K_D ~ 10 x 10^{-9} M) towards target protein α-casein. The detection limit was obtained as 127 $ngmL^{-1}$.

3.3. MIP-Based Piezoelectric Sensors. Quartz crystal microbalance (QCM) is another popular an analytical technique that displays high sensitivity to mass changes on the sensor surface. Many examples on different applications of QCM sensor systems have been reported in the literature and some examples are briefly described in the following.

Eren et al. [143] developed a QCM sensor system having MIP layer for the detection of lovastatin in red yeast rice. MIP layer was prepared on the surface of allyl mercaptan modified-gold electrode by the polymerization of HEMA, MAAsp as the functional monomers, and cross-linker EGDMA in the presence of template compound lovastatin. The developed QCM sensor having MIP layer was successfully applied for the sensitive recognition of lovastatin in red yeast rice samples. The limit of detection of the prepared QCM sensor towards lovastatin was found to be as 0.030 nM.

A QCM having MIP layer towards profenofos was developed by Gao and coworkers [144]. For this purpose, they used MAA as the functional monomer for the synthesis of profenofos imprinted MIP layer on the surface of gold electrode modified with 11-mercaptoundecanoic acid. The developed QCM sensor with MIP layer showed high sensitivity towards the target compound profenofos in aqueous solutions with an excellent detection limit of 2.0 x 10^{-7} $mgmL^{-1}$.

In another study [126], Bi and Yang prepared a QCM sensor platform bearing MIP layer for the detection of pesticide compounds imidacloprid and thiacloprid in celery juice. For this purpose, the immobilization of the target compounds on the surface of the gold chip was performed in the first step. Then, self-assembly of alkanethiols around the target compounds was carried out and the template removal was performed by using EtOH. The demonstration of the QCM sensor bearing MIP layer towards imidacloprid and thiacloprid is shown in Figure 10. The developed sensor system displayed good recognition behavior towards the target compounds imidacloprid and thiacloprid. It has also been noted that these sensor systems are promising and have the potential to detect pesticide residues in aqueous solutions and vegetables.

In another interesting study [145], the detection of metolcarb in food and beverage samples such as cabbage, pear, and apple juice was carried out by using MIP-based QCM sensor. The results indicated that the developed QCM sensor displayed a linear response towards metolcarb in the range between 5 and 70 μgL^{-1}. The detection limit was obtained as 2.309 μgL^{-1}.

Table 2 shows the recent examples of nanostructured MIP-based composites in sensor applications.

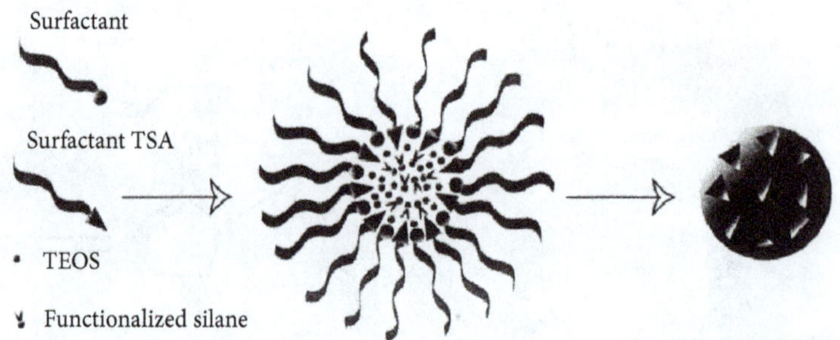

FIGURE 11: Preparation on MIP-based silica nanoparticles (reproduced with permission from [127]).

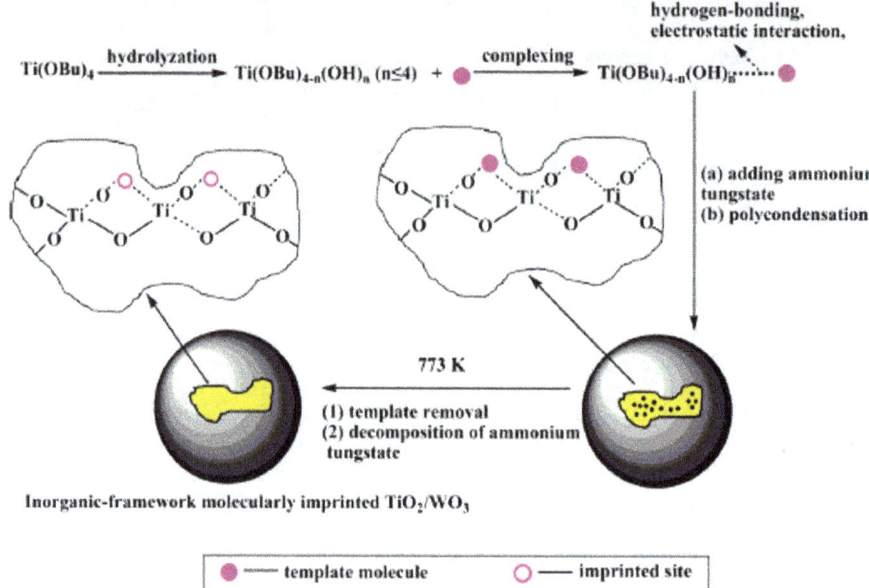

FIGURE 12: TiO_2/WO_3/MIP-based composite nanocatalyst towards 2-nitrophenol and 4-nitrophenol (reproduced with permission from [128]).

4. MIPs in Catalytic Applications

Enzyme-like catalysts are also popular application of imprinted nanomaterials in biomimetic catalysis. For the preparation of enzyme-like catalysts based on molecular imprinting approach, appropriate functional monomers are chosen and incorporated in the polymeric network by choosing the substrate of enzyme (as the template compound) or the transition state analogue (TSA) of the target reaction. After removal of the template from the polymeric network, the obtained imprinted nanomaterial behaves as enzyme-like catalyst towards the desired chemical or biochemical reaction. Some examples reported in the literature are briefly discussed in the following.

Markowitz and coworkers developed MIP-based silica nanocomposites for the selective hydrolysis of substrates of chymotrypsin and trypsin [127]. For this purpose, a TSA of α-chymotrypsin was used as the template compound for the preparation of α-chymotrypsin-like nanocatalyst (Figure 11). The silane groups conjugated with the amino acids which exist in the catalytic center of the α-chymotrypsin were used for the preparation of silica nanoparticles. The activity of the prepared MIP-based silica nanocomposites was performed by monitoring the hydrolysis of the substrates succinyl-Ala-Ala-Pro-Phe-*p*-nitroanilide and benzoyl-DL-arginine-*p*-nitroanilide. The developed imprinted nanocatalyst showed great enantioselective hydrolytic activity towards the substrate compounds.

Luo et al. developed a TiO_2/WO_3/MIP-based composite nanocatalyst for the efficient degradation of 2-nitrophenol and 4-nitrophenol [128]. One-step sol-gel method was applied for the preparation of composite nanocatalyst by using tetrabutyl orthotitanate which was chosen as the functional monomer precursor and titanium source. The schematic representation of the prepared composite nanocatalyst is shown Figure 12.

The obtained results indicated that the photocatalytic activity of the prepared TiO_2/WO_3/MIP-based composite nanocatalyst towards the target compounds is 2 times higher than its corresponding nonimprinted catalyst.

In a study reported by Bonomi et al. [146], catalytic imprinted nanogels were synthesized for the Kemp

TABLE 2: Recent examples of nanostructured MIP-based composites in sensor applications.

Reference	Nanocomposite composition	Analyte	Sample
	Electrochemical sensors		
[75]	Pencil graphite electrode coated with molecularly imprinted polypyrrole	Methylimidazole	Serum
[76]	Glassy carbon electrode modified with graphene/ Au nanoparticles/MIP composite	Colchicine	Serum and pharmaceuticals
[77]	Glassy carbon electrode modified with graphene/Ag nanoparticles/MIP composite	Creatinine	Saliva and serum
[78]	Glassy carbon electrode coated with CNT/MIP composite	Tramadol	Urine
[79]	Carbon paste electrode coated with MIP	Zn^{2+}	River water, urine and blood
[80]	Graphite electrode coated with MIP	Azithromycin	Drug
[16]	Glassy carbon electrode coated with graphene/ CNT/MIP composite	Propyl gallate	Vegetable oil
[81]	Carbon paste electrode coated with CNT/MIP nanoparticle composite	Meloxicam	Plasma
[82]	Glassy carbon electrode coated with MIP/Pd nanoparticles composite	Norepinephrine	Urine
[83]	Carbon paste electrode coated with MIP	Famciclovir	Drug
[84]	Glassy carbon electrode coated with graphene/MIP membrane composite	Artemisinin	Plant extract
[85]	Glassy carbon electrode coated with MIP/Au nanoparticles composite	Estradiol	Milk
[86]	Interdigitated electrode coated with CNT/MIP composite	Cotinine	Organic solutions
[87]	Glassy carbon electrode coated with CNT/MIP/Pt nanoparticles composite	Tartrazine	Beverages
[88]	Carbon electrode coated with graphene/MIP/Ni nanoparticles composite	Tetrabromo bisphenol A	Tap water, rain and lake water
[89]	Carbon electrode coated with graphene/MIP/Ag nanoparticles composite	Bisphenol A	Plastic samples and soil samples
[90]	Carbon paste electrode coated with MIP	Trinitrotoluene	Tap water and sea water
	Spectroscopic sensors		
[91]	CdTe QDs embedded-SiO_2 particles coated with MIP layer	Neomycin	Pork, swine liver, swine kidney, fish meat, fish liver, chicken meat, chicken kidney and milk
[92]	CdSe/ZnS QDs having MIP shell	Trichlorfon	Spinach and rape samples
[93]	Luminescent magnetic MIP nanoparticles having $LaVO_4$:Eu^{3+} nanocrystals	Diazinon	Aqueous solutions
[94]	Chemiluminescent Fe_3O_4@SiO_2 magnetic nanoparticles coated with MIP layer	Sulfadiazine	Urine
[95]	SPR sensor having MIP layer bearing the functional monomer MAA	Clenbuterol	Aqueous solutions
[96]	SPR sensor having MIP layer bearing the functional monomer MAA	Ametryn	Soybean and rice
[97]	ZnS QDs doped with Mn/MIP composite	Domoic acid	Shellfish
[98]	ZnO nanorods coated with molecularly imprinted poly(ethylene-co-vinylalcohol)	Melatonin	Urine
[99]	Magnetic nanoparticles having MIP layer bearing the functional monomer MAA	Mefenamic acid	Aqueous solutions
[100]	SPR sensor surface having MIP layer bearing the functional monomer MAA	L-nicotine	Aqueous solutions

Table 2: Continued.

Reference	Nanocomposite composition	Analyte	Sample
[101]	Graphene QDs coated with MIP layer	Dopamine	Serum and Urine
	Piezoelectric sensors		
[102]	QCM sensor surface coated with 1,3,5-pentanetricarboxylic acid imprinted film	Domoic acid	Mussel extracts
[103]	QCM sensor surface coated with MIP film having styrene/DVB copolymer	Terpenes	Herbs
[104]	QCM sensor surface coated with MIP film having 1,3,5 trisacrylamide 2,4,6 triazine as the functional monomer	Folic acid	Aqueous solutions
[105]	QCM sensor having MIP layer bearing the functional monomer MAA	Ni^{2+} and Cu^{2+}	Aqueous solutions
[106]	QCM sensor surface coated with polythiophene MIP film	Pinacolyl methyl phosphonate	Aqueous solutions
[107]	QCM sensor surface having MIP/Au nanoparticles/ poly(o-aminothiophenol) membrane	Ractopamine	Swine feed
[108]	QCM sensor having MIP layer bearing the functional monomer AA	Glucose	Aqueous solutions
[109]	QCM sensor having MIP layer bearing the functional monomer MAA	Microcystin	Lake water
[110]	QCM sensor having MIP layer bearing the functional monomer 1-Vinyl-2-pyrrolidone	Heparin	Plasma
[111]	QCM sensor having MIP layer bearing the functional monomer MAA	Methimazole	Urine
[112]	QCM sensor having MIP layer bearing zinc acrylate as the functional monomer	Human serum albumin	Human serum
[113]	QCM sensor having MIP layer bearing 3-aminopropyltriethoxysilane as the functional monomer	Enrofloxacin	Milk, egg, chicken muscle and pork

elimination reactions. The functional monomer 4-VP and template compound 5-nitro indole were used for the synthesis of imprinted nanogels. The results showed that the prepared 5-nitro indole imprinted nanogels exhibited high catalytic activity towards the substrate 1,2-benzisoxazole. Substrate selectivity of the prepared catalytic nanogels was also investigated using 5-Cl-benzisoxazole which is a substrate analogue. The catalytic nanogels displayed lower affinity towards 5-Cl-benzisoxazole compared to the substrate 1,2-benzisoxazole.

In another interesting study, Zhou and colleagues prepared a molecularly imprinted TiO_2 photocatalyst having thiol groups for the efficient removal of 2,4-dinitrophenol from wastewater [147]. MIP-based TiO_2 photocatalyst was prepared in water as a green solvent using o-phenylenediamine as the functional monomer. The results confirmed that the prepared MIP-based green photocatalyst displayed excellent selectivity and degradation activity towards 2,4-DNP in wastewater.

5. Conclusions

The growing number of published researches in which nanostructured composite MIPs have been used for different applications showed that these are promising materials for the selective extraction, sensing, and catalysis. The reported studies described in this review highlight the recent progress in SPE, sensors, and catalytic systems using nanostructured composite MIPs over the past years. Composite MIPs in nanoscale as promising materials provide a new approach for the selective SPE and sensors towards target molecules in complex matrices. On the other hand, these materials offer new routes to control aspects that determine the stereochemical outcome of a catalysis reaction.

References

[1] B. Sellergren and F. Lanza, *Molecularly Imprinted Polymers: Man-made mimics of antibodies and their application in analytical chemistry: Techniques and instrumentation in analytical chemistry*, Elsevier Science, Amsterdam, Netherlands, 2001.

[2] D. R. Kryscio and N. A. Peppas, "Critical review and perspective of macromolecularly imprinted polymers," *Acta Biomaterialia*, vol. 8, no. 2, pp. 461–473, 2012.

[3] W. J. Cheong, S. H. Yang, and F. Ali, "Molecular imprinted polymers for separation science: a review of reviews," *Journal of Separation Science*, vol. 36, no. 3, pp. 609–628, 2013.

[4] G. Vasapollo, R. D. Sole, L. Mergola et al., "Molecularly imprinted polymers: present and future prospective," *International Journal of Molecular Sciences*, vol. 12, no. 9, pp. 5908–5945, 2011.

[5] Z. Chen, M. Wang, Y. Fu, H. Yu, and D. Di, "Preparation of quercetin molecularly imprinted polymers," *Designed Monomers and Polymers*, vol. 15, no. 1, pp. 93–111, 2012.

[6] J. R. L. Guerreiro, V. Freitas, and M. G. F. Sales, "New sensing materials of molecularly-imprinted polymers for the selective recognition of Chlortetracycline," *Microchemical Journal*, vol. 97, no. 2, pp. 173–181, 2011.

[7] E. P. C. Lai and S. Y. Feng, "Molecularly imprinted solid phase extraction for rapid screening of metformin," *Microchemical Journal*, vol. 75, no. 3, pp. 159–168, 2003.

[8] Z. Zhang, D. Niu, Y. Li, and J. Shi, "Magnetic, core-shell structured and surface molecularly imprinted polymers for the rapid and selective recognition of salicylic acid from aqueous solutions," *Applied Surface Science*, vol. 435, pp. 178–186, 2018.

[9] R. Sedghi, M. Yassari, and B. Heidari, "Thermo-responsive molecularly imprinted polymer containing magnetic nanoparticles: Synthesis, characterization and adsorption properties for curcumin," *Colloids and Surfaces B: Biointerfaces*, vol. 162, pp. 154–162, 2018.

[10] W. Ji, Y. Guo, X. Wang, and D. Guo, "A water-compatible magnetic molecularly imprinted polymer for the selective extraction of risperidone and 9-hydroxyrisperidone from human urine," *Talanta*, vol. 181, pp. 392–400, 2018.

[11] X. Pan, X. He, and Z. Liu, "Molecularly imprinted mesoporous silica nanoparticles for specific extraction and efficient identification of Amadori compounds," *Analytica Chimica Acta*, vol. 1019, pp. 65–73, 2018.

[12] J. Kupai, M. Razali, S. Buyuktiryaki, R. Kecili, and G. Szekely, "Long-term stability and reusability of molecularly imprinted polymers," *Polymer Chemistry*, vol. 8, no. 4, pp. 666–673, 2017.

[13] C. Alexander, L. Davidson, and W. Hayes, "Imprinted polymers: artificial molecular recognition materials with applications in synthesis and catalysis," *Tetrahedron*, vol. 59, no. 12, pp. 2025–2057, 2003.

[14] D. Mathew, B. Thomas, and K. Devaky, "Biomimetic recognition and peptidase activities of transition state analogue imprinted chymotrypsin mimics," *Reactive and Functional Polymers*, vol. 124, pp. 121–128, 2018.

[15] H. Chen, Z. Zhang, L. Luo, and S. Yao, "Surface-imprinted chitosan-coated magnetic nanoparticles modified multi-walled carbon nanotubes biosensor for detection of bovine serum albumin," *Sensors and Actuators B: Chemical*, vol. 163, no. 1, pp. 76–83, 2012.

[16] M. Cui, J. Huang, Y. Wang, Y. Wu, and X. Luo, "Molecularly imprinted electrochemical sensor for propyl gallate based on PtAu bimetallic nanoparticles modified graphene-carbon nanotube composites," *Biosensors and Bioelectronics*, vol. 68, pp. 563–569, 2015.

[17] B. B. Prasad, R. Madhuri, M. P. Tiwari, and P. S. Sharma, "Imprinting molecular recognition sites on multiwalled carbon nanotubes surface for electrochemical detection of insulin in real samples," *Electrochimica Acta*, vol. 55, no. 28, pp. 9146–9156, 2010.

[18] A. Afzal and F. Dickert, "Imprinted Oxide and MIP/Oxide Hybrid Nanomaterials for Chemical Sensors," *Nanomaterials*, vol. 8, no. 4, p. 257, 2018.

[19] I. Liška, "Fifty years of solid-phase extraction in water analysis - Historical development and overview," *Journal of Chromatography A*, vol. 885, no. 1-2, pp. 3–16, 2000.

[20] J. Cazes, *Encyclopedia of Chromatography*, CRC Press, Germany, 2009.

[21] B. Y. Spivakov, G. I. Malofeeva, and O. M. Petrukhin, "Solid-phase extraction on alkyl-bonded silica gels in inorganic analysis," *Analytical Sciences*, vol. 22, no. 4, pp. 503–519, 2006.

[22] K. Pyrzynska, "Application of carbon sorbents for the concentration and separation of metal ions," *Analytical Sciences*, vol. 23, no. 6, pp. 631–637, 2007.

[23] J. A. Rodríguez, K. A. Escamilla-Lara, A. Guevara-Lara, J. M. Miranda, and M. E. Páez-Hernández, "Application of an activated carbon-based support for magnetic solid phase extraction followed by spectrophotometric determination of tartrazine in commercial beverages," *International Journal of Analytical Chemistry*, vol. 2015, Article ID 291827, 8 pages, 2015.

[24] M. G. Valdés, A. Pérez-Cordoves, and M. Díaz-García, "Zeolites and zeolite-based materials in analytical chemistry," *TrAC Trends in Analytical Chemistry*, vol. 25, no. 1, pp. 24–30, 2006.

[25] X. Shen, L. Zhu, N. Wang, L. Ye, and H. Tang, "Molecular imprinting for removing highly toxic organic pollutants," *Chemical Communications*, vol. 48, no. 6, pp. 788–798, 2012.

[26] P. Su, R. Wang, Y. Yu, and Y. Yang, "Microwave-assisted synthesis of ionic liquid-modified silica as a sorbent for the solid-phase extraction of phenolic compounds from water," *Analytical Methods*, vol. 6, no. 3, pp. 704–709, 2014.

[27] R.-Z. Wang, D.-L. Huang, Y.-G. Liu et al., "Selective removal of BPA from aqueous solution using molecularly imprinted polymers based on magnetic graphene oxide," *RSC Advances*, vol. 6, no. 108, pp. 106201–106210, 2016.

[28] W. Shen, G. Xu, F. Wei, J. Yang, Z. Cai, and Q. Hu, "Preparation and application of imprinted polymer for tetrabromobisphenol A using tetrachlorobisphenol A as the dummy template," *Analytical Methods*, vol. 5, no. 19, pp. 5208–5214, 2013.

[29] B. Sellergren, "Direct Drug Determination by Selective Sample Enrichment on an Imprinted Polymer," *Analytical Chemistry*, vol. 66, no. 9, pp. 1578–1582, 1994.

[30] E. Caro, R. M. Marcé, P. A. G. Cormack, D. C. Sherrington, and F. Borrull, "On-line solid-phase extraction with molecularly imprinted polymers to selectively extract substituted 4-chlorophenols and 4-nitrophenol from water," *Journal of Chromatography A*, vol. 995, no. 1-2, pp. 233–238, 2003.

[31] F. Chapuis, V. Pichon, F. Lanza, B. Sellergren, and M.-C. Hennion, "Retention mechanism of analytes in the solid-phase extraction process using molecularly imprinted polymers: Application to the extraction of triazines from complex matrices," *Journal of Chromatography B*, vol. 804, no. 1, pp. 93–101, 2004.

[32] J.-P. Lai, R. Niessner, and D. Knopp, "Benzo[a]pyrene imprinted polymers: Synthesis, characterization and SPE application in water and coffee samples," *Analytica Chimica Acta*, vol. 522, no. 2, pp. 137–144, 2004.

[33] O. Brüggemann, A. Visnjevski, R. Burch, and P. Patel, "Selective extraction of antioxidants with molecularly imprinted polymers," *Analytica Chimica Acta*, vol. 504, no. 1, pp. 81–88, 2004.

[34] F. Puoci, A. Scoma, G. Cirillo, L. Bertin, F. Fava, and N. Picci, "Selective extraction and purification of gallic acid from actual site olive mill wastewaters by means of molecularly imprinted microparticles," *Chemical Engineering Journal*, vol. 198-199, pp. 529–535, 2012.

[35] J. H. Xin, X. G. Qiao, Z. X. Xu, and J. Zhou, "Molecularly imprinted polymer as sorbent for solid-phase extraction coupling to gas chromatography for the simultaneous determination of trichlorfon and monocrotophos residues in vegetables," *Food Analytical Methods*, vol. 6, no. 1, pp. 274–281, 2013.

[36] Y. Li, C. Zheng, X. Sun, B. Ouyang, P. Ni, and Y. Zhang, "Identification of 3-chloro-1,2-propandiol using molecularly imprinted composite solid-phase extraction materials," *Analytical and Bioanalytical Chemistry*, vol. 406, no. 25, pp. 6319–6327, 2014.

[37] X. He, X. Mei, J. Wang, Z. Lian, L. Tan, and W. Wu, "Determination of diethylstilbestrol in seawater by molecularly imprinted solid-phase extraction coupled with high-performance liquid chromatography," *Marine Pollution Bulletin*, vol. 102, no. 1, pp. 142–147, 2016.

[38] S. Han, X. Li, Y. Wang, and S. Chen, "Multifunctional imprinted polymers based on CdTe/CdS and magnetic graphene oxide for selective recognition and separation of p-t-octylphenol," *Chemical Engineering Journal*, vol. 271, pp. 87–95, 2015.

[39] E. C. Morais, R. Brambilla, G. G. Correa, V. Dalmoro, and J. H. Z. Dos Santos, "Imprinted silicas for paracetamol preconcentration prepared by the sol–gel process," *Journal of Sol-Gel Science and Technology*, vol. 83, no. 1, pp. 90–99, 2017.

[40] I. Vasconcelos and C. Fernandes, "Magnetic solid phase extraction for determination of drugs in biological matrices," *TrAC Trends in Analytical Chemistry*, vol. 89, pp. 41–52, 2017.

[41] S. Ansari and M. Karimi, "Synthesis and application of molecularly imprinted polymer for highly selective solid phase extraction trace amount of sotalol from human urine samples: Optimization by central composite design (CCD)," *Medicinal Chemistry Research*, vol. 26, no. 10, pp. 2477–2490, 2017.

[42] S. Hassanpour, M. Taghizadeh, and Y. Yamini, "Magnetic Cr(VI) Ion Imprinted Polymer for the Fast Selective Adsorption of Cr(VI) from Aqueous Solution," *Journal of Polymers and the Environment*, pp. 1–15, 2017.

[43] J. Fasihi, M. Shamsipur, A. Khanchi, M. Mahani, and K. Ashtari, "Imprinted polymer grafted from silica particles for on-line trace enrichment and ICP OES determination of uranyl ion," *Microchemical Journal*, vol. 126, pp. 316–321, 2016.

[44] N. Khoddami and F. Shemirani, "A new magnetic ion-imprinted polymer as a highly selective sorbent for determination of cobalt in biological and environmental samples," *Talanta*, vol. 146, pp. 244–252, 2016.

[45] Y. Chen, X. Ma, M. Huang, J. Peng, and C. Li, "Use of a new magnetic ion-imprinted nanocomposite adsorbent for selective and rapid preconcentration and determination of trace nickel by flame atomic absorption spectrometry," *Analytical Methods*, vol. 8, no. 4, pp. 824–829, 2016.

[46] M. Zhao, X. Ma, F. Zhao, and H. Guo, "Molecularly imprinted polymer silica monolith for the selective extraction of alpha-cypermethrin from soil samples," *Journal of Materials Science*, vol. 51, no. 7, pp. 3440–3447, 2016.

[47] H. Lian, Y. Hu, and G. Li, "Novel metal-ion-mediated, complex-imprinted solid-phase microextraction fiber for the selective recognition of thiabendazole in citrus and soil samples," *Journal of Separation Science*, vol. 37, no. 1-2, pp. 106–113, 2014.

[48] S. Xu, C. Guo, Y. Li, Z. Yu, C. Wei, and Y. Tang, "Methyl parathion imprinted polymer nanoshell coated on the magnetic nanocore for selective recognition and fast adsorption and separation in soils," *Journal of Hazardous Materials*, vol. 264, pp. 34–41, 2014.

[49] Y. Hao, R. Gao, L. Shi, D. Liu, Y. Tang, and Z. Guo, "Water-compatible magnetic imprinted nanoparticles served as solid-phase extraction sorbents for selective determination of trace 17beta-estradiol in environmental water samples by liquid chromatography," *Journal of Chromatography A*, vol. 1396, pp. 7–16, 2015.

[50] K. Kamari and A. Taheri, "Preparation and evaluation of magnetic core–shell mesoporous molecularly imprinted polymers for selective adsorption of amitriptyline in biological samples," *Journal of the Taiwan Institute of Chemical Engineers*, vol. 86, pp. 230–239, 2018.

[51] A. Ostovan, M. Ghaedi, and M. Arabi, "Fabrication of water-compatible superparamagnetic molecularly imprinted biopolymer for clean separation of baclofen from bio-fluid samples: A mild and green approach," *Talanta*, vol. 179, pp. 760–768, 2018.

[52] M. Soleimani, M. Ahmadi, T. Madrakian, and A. Afkhami, "Magnetic solid phase extraction of rizatriptan in human urine samples prior to its spectrofluorimetric determination," *Sensors and Actuators B: Chemical*, vol. 254, pp. 1225–1233, 2018.

[53] S. Azodi-Deilami, A. H. Najafabadi, E. Asadi, M. Abdouss, and D. Kordestani, "Magnetic molecularly imprinted polymer nanoparticles for the solid-phase extraction of paracetamol from plasma samples, followed its determination by HPLC," *Microchimica Acta*, vol. 181, no. 15-16, pp. 1823–1832, 2014.

[54] A. Rajabi Khorrami and A. Rashidpur, "Development of a fiber coating based on molecular sol-gel imprinting technology for selective solid-phase micro extraction of caffeine from human serum and determination by gas chromatography/mass spectrometry," *Analytica Chimica Acta*, vol. 727, pp. 20–25, 2012.

[55] J. Meng, W. Zhang, T. Bao, and Z. Chen, "Novel molecularly imprinted magnetic nanoparticles for the selective extraction of protoberberine alkaloids in herbs and rat plasma," *Journal of Separation Science*, vol. 38, no. 12, pp. 2117–2125, 2015.

[56] J.-B. Ma, H.-W. Qiu, Q.-H. Rui et al., "Fast determination of catecholamines in human plasma using carboxyl-functionalized magnetic-carbon nanotube molecularly imprinted polymer followed by liquid chromatography-tandem quadrupole mass spectrometry," *Journal of Chromatography A*, vol. 1429, pp. 86–96, 2016.

[57] G. Sheykhaghaei, M. Hossainisadr, S. Khanahmadzadeh, M. Seyedsajadi, and A. Alipouramjad, "Magnetic molecularly imprinted polymer nanoparticles for selective solid phase extraction and pre-concentration of Tizanidine in human urine," *Journal of Chromatography B*, vol. 1011, pp. 1–5, 2016.

[58] T. Madrakian, F. Fazl, M. Ahmadi, and A. Afkhami, "Efficient solid phase extraction of codeine from human urine samples using a novel magnetic molecularly imprinted nanoadsorbent and its spectrofluorometric determination," *New Journal of Chemistry*, vol. 40, no. 1, pp. 122–129, 2016.

[59] X. Gu, H. He, C.-Z. Wang et al., "Synthesis of surface nano-molecularly imprinted polymers for sensitive baicalin detection in biological samples," *RSC Advances*, vol. 5, no. 52, pp. 41377–41384, 2015.

[60] G. Liang, H. Zhai, L. Huang et al., "Synthesis of carbon quantum dots-doped dummy molecularly imprinted polymer monolithic column for selective enrichment and analysis of aflatoxin B1 in peanut," *Journal of Pharmaceutical and Biomedical Analysis*, vol. 149, pp. 258–264, 2018.

[61] R. Garcia, E. P. Carreiro, J. P. Prates Ramalho et al., "A magnetic controllable tool for the selective enrichment of dimethoate from olive oil samples: A responsive molecular imprinting-based approach," *Food Chemistry*, vol. 254, pp. 309–316, 2018.

[62] M. Hashemi and Z. Nazari, "Preparation of molecularly imprinted polymer based on the magnetic multiwalled carbon nanotubes for selective separation and spectrophotometric determination of melamine in milk samples," *Journal of Food Composition and Analysis*, vol. 69, pp. 98–106, 2018.

[63] M. Wei, X. Yan, S. Liu, and Y. Liu, "Preparation and evaluation of superparamagnetic core–shell dummy molecularly imprinted polymer for recognition and extraction of organophosphorus pesticide," *Journal of Materials Science*, vol. 53, no. 7, pp. 4897–4912, 2018.

[64] N. Kumar, N. Narayanan, and S. Gupta, "Application of magnetic molecularly imprinted polymers for extraction of imidacloprid from eggplant and honey," *Food Chemistry*, vol. 255, pp. 81–88, 2018.

[65] J. L. Urraca, J. F. Huertas-Pérez, G. A. Cazorla, J. Gracia-Mora, A. M. García-Campaña, and M. C. Moreno-Bondi, "Development of magnetic molecularly imprinted polymers for selective extraction: Determination of citrinin in rice samples by liquid chromatography with UV diode array detection," *Analytical and Bioanalytical Chemistry*, vol. 408, no. 11, pp. 3033–3042, 2016.

[66] Z.-Z. Lin, H.-Y. Zhang, L. Li, and Z.-Y. Huang, "Application of magnetic molecularly imprinted polymers in the detection of malachite green in fish samples," *Reactive and Functional Polymers*, vol. 98, pp. 24–30, 2016.

[67] S. Aggarwal, Y. S. Rajput, G. Singh, and R. Sharma, "Synthesis and characterization of oxytetracycline imprinted magnetic polymer for application in food," *Applied Nanoscience*, vol. 6, no. 2, pp. 209–214, 2016.

[68] L. Xu, G. Fang, M. Pan, X. Wang, and S. Wang, "One-pot synthesis of carbon dots-embedded molecularly imprinted polymer for specific recognition of sterigmatocystin in grains," *Biosensors and Bioelectronics*, vol. 77, pp. 950–956, 2016.

[69] X. Hu, L. Xie, J. Guo et al., "Hydrophilic gallic acid-imprinted polymers over magnetic mesoporous silica microspheres with excellent molecular recognition ability in aqueous fruit juices," *Food Chemistry*, vol. 179, pp. 206–212, 2015.

[70] A. A. Asgharinezhad, N. Jalilian, H. Ebrahimzadeh, and Z. Panjali, "A simple and fast method based on new magnetic ion imprinted polymer nanoparticles for the selective extraction of Ni(ii) ions in different food samples," *RSC Advances*, vol. 5, no. 56, pp. 45510–45519, 2015.

[71] W. J. Tang, T. Zhao, C. H. Zhou, X. J. Guan, and H. X. Zhang, "Preparation of hollow molecular imprinting polymer for determination of ofloxacin in milk," *Anal. Methods*, vol. 6, no. 10, pp. 3309–3315, 2014.

[72] L. Qiao, N. Gan, F. Hu et al., "Magnetic nanospheres with a molecularly imprinted shell for the preconcentration of diethylstilbestrol," *Microchimica Acta*, vol. 181, no. 11-12, pp. 1341–1351, 2014.

[73] Y. Hu, Y. Li, R. Liu, W. Tan, and G. Li, "Magnetic molecularly imprinted polymer beads prepared by microwave heating for selective enrichment of β-agonists in pork and pig liver samples," *Talanta*, vol. 84, no. 2, pp. 462–470, 2011.

[74] L. Chen and B. Li, "Magnetic molecularly imprinted polymer extraction of chloramphenicol from honey," *Food Chemistry*, vol. 141, no. 1, pp. 23–28, 2013.

[75] A. Nezhadali, L. Mehri, and R. Shadmehri, "Determination of methimazole based on electropolymerized-molecularly imprinted polypyrrole modified pencil graphite sensor," *Materials Science and Engineering C*, vol. 85, pp. 225–232, 2018.

[76] H. Bai, C. Wang, J. Chen, Z. Li, K. Fu, and Q. Cao, "Graphene@AuNPs modified molecularly imprinted electrochemical sensor for the determination of colchicine in pharmaceuticals and serum," *Journal of Electroanalytical Chemistry*, vol. 816, pp. 7–13, 2018.

[77] Z. Zhang, Y. Li, X. Liu, Y. Zhang, and D. Wang, "Molecular Imprinting Electrochemical sensor for sensitive creatinine determination," *International Journal of Electrochemical Science*, vol. 13, pp. 2986–2995, 2018.

[78] B. Deiminiat, G. H. Rounaghi, and M. H. Arbab-Zavar, "Development of a new electrochemical imprinted sensor based on poly-pyrrole, sol–gel and multiwall carbon nanotubes for determination of tramadol," *Sensors and Actuators B: ChemicalSensors and Actuators B: Chemical*, vol. 238, pp. 651–659, 2017.

[79] A. Shirzadmehr, M. Rezaei, H. Bagheri, and H. Khoshsafar, "Novel potentiometric sensor for the trace-level determination of Zn^{2+} based on a new nanographene/ion imprinted polymer composite," *International Journal of Environmental Analytical Chemistry*, vol. 96, no. 10, pp. 1–16, 2016.

[80] M. A. Abu-Dalo, N. S. Nassory, N. I. Abdulla, and I. R. Al-Mheidat, "Azithromycin-molecularly imprinted polymer based on PVC membrane for Azithromycin determination in drugs using coated graphite electrode," *Journal of Electroanalytical Chemistry*, vol. 751, pp. 75–79, 2015.

[81] S. Azodi-Deilami, E. Asadi, M. Abdouss, F. Ahmadi, A. H. Najafabadi, and S. Farzaneh, "Determination of meloxicam in plasma samples using a highly selective and sensitive voltammetric sensor based on carbon paste electrodes modified by molecularly imprinted polymer nanoparticle-multiwall carbon nanotubes," *Analytical Methods*, vol. 7, no. 4, pp. 1280–1292, 2015.

[82] J. Chen, H. Huang, Y. Zeng, H. Tang, and L. Li, "A novel composite of molecularly imprinted polymer-coated PdNPs for electrochemical sensing norepinephrine," *Biosensors and Bioelectronics*, vol. 65, pp. 366–374, 2015.

[83] N. A. El Gohary, A. Madbouly, R. M. El Nashar, and B. Mizaikoff, "Synthesis and application of a molecularly imprinted polymer for the voltammetric determination of famciclovir," *Biosensors and Bioelectronics*, vol. 65, pp. 108–114, 2015.

[84] H. Bai, C. Wang, J. Chen, J. Peng, and Q. Cao, "A novel sensitive electrochemical sensor based on in-situ polymerized molecularly imprinted membranes at graphene modified electrode for artemisinin determination," *Biosensors and Bioelectronics*, vol. 64, pp. 352–358, 2015.

[85] X. Zhang, Y. Peng, J. Bai et al., "A novel electrochemical sensor based on electropolymerized molecularly imprinted polymer and gold nanomaterials amplification for estradiol detection," *Sensors and Actuators B: Chemical*, vol. 200, pp. 69–75, 2014.

[86] S. Antwi-Boampong, K. S. Mani, J. Carlan, and J. J. Belbruno, "A selective molecularly imprinted polymer-carbon nanotube sensor for cotinine sensing," *Journal of Molecular Recognition*, vol. 27, no. 1, pp. 57–63, 2014.

[87] L. Zhao, B. Zeng, and F. Zhao, "Electrochemical determination of tartrazine using a molecularly imprinted polymer - Multiwalled carbon nanotubes - ionic liquid supported Pt nanoparticles composite film coated electrode," *Electrochimica Acta*, vol. 146, pp. 611–617, 2014.

[88] H. Chen, Z. Zhang, R. Cai, W. Rao, and F. Long, "Molecularly imprinted electrochemical sensor based on nickel nanoparticles-graphene nanocomposites modified electrode

for determination of tetrabromobisphenol A," *Electrochimica Acta*, vol. 117, pp. 385–392, 2014.

[89] R. Cai, W. Rao, Z. Zhang, F. Long, and Y. Yin, "An imprinted electrochemical sensor for bisphenol A determination based on electrodeposition of a graphene and Ag nanoparticle modified carbon electrode," *Analytical Methods*, vol. 6, no. 5, pp. 1590–1597, 2014.

[90] T. Alizadeh, "Preparation of magnetic TNT-imprinted polymer nanoparticles and their accumulation onto magnetic carbon paste electrode for TNT determination," *Biosensors and Bioelectronics*, vol. 61, pp. 532–540, 2014.

[91] Y. Wan, Y. Liu, C. Liu et al., "Rapid determination of neomycin in biological samples using fluorescent sensor based on quantum dots with doubly selective binding sites," *Journal of Pharmaceutical and Biomedical Analysis*, vol. 154, pp. 75–84, 2018.

[92] Q. Liu, M. Jiang, Z. Ju, X. Qiao, and Z. Xu, "Development of direct competitive biomimetic immunosorbent assay based on quantum dot label for determination of trichlorfon residues in vegetables," *Food Chemistry*, vol. 250, pp. 134–139, 2018.

[93] Y. Ma, H. Li, and L. Wang, "Magnetic-luminescent bifunctional nanosensors," *Journal of Materials Chemistry*, vol. 22, no. 36, pp. 18761–18767, 2012.

[94] F. Lu, H. Li, M. Sun et al., "Flow injection chemiluminescence sensor based on core-shell magnetic molecularly imprinted nanoparticles for determination of sulfadiazine," *Analytica Chimica Acta*, vol. 718, pp. 84–91, 2012.

[95] H. Bao, T. Wei, H. Meng, and B. Liu, "Surface Plasmon resonance sensor for supersensitive detection of clenbuterol using molecularly imprinted film," *Chemistry Letters*, vol. 41, no. 3, pp. 237–239, 2012.

[96] N. Zhao, C. Chen, and J. Zhou, "Surface plasmon resonance detection of ametryn using a molecularly imprinted sensing film prepared by surface-initiated atom transfer radical polymerization," *Sensors and Actuators B: Chemical*, vol. 166-167, pp. 473–479, 2012.

[97] L. Dan and H.-F. Wang, "Mn-doped ZnS quantum dot imbedded two-fragment imprinting silica for enhanced room temperature phosphorescence probing of domoic acid," *Analytical Chemistry*, vol. 85, no. 10, pp. 4844–4848, 2013.

[98] M.-H. Lee, J. L. Thomas, Y.-L. Chen et al., "Optical sensing of urinary melatonin with molecularly imprinted poly(ethylene-co-vinyl alcohol) coated zinc oxide nanorod arrays," *Biosensors and Bioelectronics*, vol. 47, pp. 56–61, 2013.

[99] M. Ahmadi, T. Madrakian, and A. Afkhami, "Molecularly imprinted polymer coated magnetite nanoparticles as an efficient mefenamic acid resonance light scattering nanosensor," *Analytica Chimica Acta*, vol. 852, pp. 250–256, 2014.

[100] N. Cennamo, G. D'Agostino, M. Pesavento, and L. Zeni, "High selectivity and sensitivity sensor based on MIP and SPR in tapered plastic optical fibers for the detection of l-nicotine," *Sensors and Actuators B: Chemical*, vol. 191, pp. 529–536, 2014.

[101] X. Zhou, A. Wang, C. Yu, S. Wu, and J. Shen, "Facile Synthesis of Molecularly Imprinted Graphene Quantum Dots for the Determination of Dopamine with Affinity-Adjustable," *Applied Materials & Interfaces*, vol. 7, no. 22, pp. 11741–11747, 2015.

[102] W.-H. Zhou, S.-F. Tang, Q.-H. Yao, F.-R. Chen, H.-H. Yang, and X.-R. Wang, "A quartz crystal microbalance sensor based on mussel-inspired molecularly imprinted polymer," *Biosensors and Bioelectronics*, vol. 26, no. 2, pp. 585–589, 2010.

[103] N. Iqbal, G. Mustafa, A. Rehman et al., "QCM-Arrays for sensing terpenes in fresh and dried herbs via bio-mimetic MIP layers," *Sensors*, vol. 10, no. 7, pp. 6361–6376, 2010.

[104] R. Madhuri, M. P. Tiwari, D. Kumar, A. Mukharji, and B. B. Prasad, "Biomimetic piezoelectric quartz sensor for folic acid based on a molecular imprinting technology," *Advanced Materials Letters*, vol. 2, no. 4, pp. 264–267, 2011.

[105] U. Latif, A. Mujahid, A. Afzal, R. Sikorski, P. A. Lieberzeit, and F. L. Dickert, "Dual and tetraelectrode QCMs using imprinted polymers as receptors for ions and neutral analytes," *Analytical and Bioanalytical Chemistry*, vol. 400, no. 8, pp. 2507–2515, 2011.

[106] A. V. Vergara, R. B. Pernites, S. Pascua, C. A. Binag, and R. C. Advincula, "QCM sensing of a chemical nerve agent analog via electropolymerized molecularly imprinted polythiophene films," *Journal of Polymer Science Part A: Polymer Chemistry*, vol. 50, no. 4, pp. 675–685, 2012.

[107] L.-J. Kong, M.-F. Pan, G.-Z. Fang et al., "Molecularly imprinted quartz crystal microbalance sensor based on poly(o-aminothiophenol) membrane and Au nanoparticles for ractopamine determination," *Biosensors and Bioelectronics*, vol. 51, pp. 286–292, 2014.

[108] A. Mirmohseni, R. Pourata, and M. Shojaei, "Application of molecularly imprinted polymer for determination of glucose by quartz crystal nanobalance technique," *IEEE Sensors Journal*, vol. 14, no. 8, pp. 2807–2812, 2014.

[109] H. He, L. Zhou, Y. Wang et al., "Detection of trace microcystin-LR on a 20 MHz QCM sensor coated with in situ self-assembled MIPs," *Talanta*, vol. 131, pp. 8–13, 2015.

[110] M. Hussain, "Ultra-sensitive detection of heparin via aPTT using plastic antibodies on QCM-D platform," *RSC Advances*, vol. 5, no. 68, pp. 54963–54970, 2015.

[111] M. Pan, G. Fang, Y. Lu, L. Kong, Y. Yang, and S. Wang, "Molecularly imprinted biomimetic QCM sensor involving a poly(amidoamine) dendrimer as a functional monomer for the highly selective and sensitive determination of methimazole," *Sensors and Actuators*, vol. 207, pp. 588–595, 2015.

[112] X.-T. Ma, X.-W. He, W.-Y. Li, and Y.-K. Zhang, "Epitope molecularly imprinted polymer coated quartz crystalmicrobalance sensor for the determination of human serum albumin," *Sensors and Actuators B*, vol. 246, pp. 879–886, 2017.

[113] M. Pan, Y. Gu, M. Zhang, J. Wang, Y. Yun, and S. Wang, "Reproducible Molecularly Imprinted QCM Sensor for Accurate, Stable, and Sensitive Detection of Enrofloxacin Residue in Animal-Derived Foods," *Food Analytical Methods*, vol. 11, no. 2, pp. 495–503, 2017.

[114] Y. Su, B. Qiu, C. Chang et al., "Separation of bovine hemoglobin using novel magnetic molecular imprinted nanoparticles," *RSC Advances*, vol. 8, no. 11, pp. 6192–6199, 2018.

[115] R. Viveiros, F. M. Dias, L. B. Maia, W. Heggie, and T. Casimiro, "Green strategy to produce large core–shell affinity beads for gravity-driven API purification processes," *Journal of Industrial and Engineering Chemistry*, vol. 54, pp. 341–349, 2017.

[116] Z. Zhang, X. Chen, W. Rao, H. Chen, and R. Cai, "Synthesis and properties of magnetic molecularly imprinted polymers based on multiwalled carbon nanotubes for magnetic extraction of bisphenol A from water," *Journal of Chromatography B*, vol. 965, pp. 190–196, 2014.

[117] Y. Yang, X. Meng, and Z. Xiao, "Synthesis of a surface molecular imprinting polymer based on silica and its application in the identification of nitrocellulose," *RSC Advances*, vol. 8, no. 18, pp. 9802–9811, 2018.

[118] W. Chen, M. Xue, F. Xue et al., "Molecularly imprinted hollow spheres for the solid phase extraction of estrogens," *Talanta*, vol. 140, pp. 68–72, 2015.

[119] J. Guo, Y. Wang, Y. Liu, C. Zhang, and Y. Zhou, "Magnetic-graphene based molecularly imprinted polymer nanocomposite for the recognition of bovine hemoglobin," *Talanta*, vol. 144, pp. 411–419, 2015.

[120] J. Luo, Y. Gao, K. Tan, W. Wei, and X. Liu, "Preparation of a Magnetic Molecularly Imprinted Graphene Composite Highly Adsorbent for 4-Nitrophenol in Aqueous Medium," *ACS Sustainable Chemistry & Engineering*, vol. 4, no. 6, pp. 3316–3326, 2016.

[121] W. Yang, D. Niu, X. Ni, Z. Zhou, W. Xu, and W. Huang, "Core-Shell Magnetic Molecularly Imprinted Polymer Prepared for Selectively Removed Indole from Fuel Oil," *Advances in Polymer Technology*, vol. 36, no. 2, pp. 168–176, 2017.

[122] F. Cao, L. Wang, Y. Yao, F. Wu, H. Sun, and S. Lu, "Synthesis and application of a highly selective molecularly imprinted adsorbent based on multiwalled carbon nanotubes for selective removal of perfluorooctanoic acid," *Environ. Sci.: Water Res. Technol*, 2018, In press.

[123] M. Cui, S. Liu, W. Lian, J. Li, W. Xu, and J. Huang, "A molecularly-imprinted electrochemical sensor based on a graphene-Prussian blue composite-modified glassy carbon electrode for the detection of butylated hydroxyanisole in foodstuffs," *Analyst*, vol. 138, no. 20, pp. 5949–5955, 2013.

[124] Y. Hu, X. Li, J. Liu, M. Wu, M. Li, and X. Zang, "One-pot synthesis of a fluorescent molecularly imprinted nanosensor for highly selective detection of sulfapyridine in water," *Analytical Methods*, vol. 10, no. 8, pp. 884–890, 2018.

[125] Y. Wang, H. Duan, L. Li et al., "A chemiluminescence sensor for determination of lysozyme using magnetic graphene oxide multi-walled carbon nanotube surface molecularly imprinted polymers," *RSC Advances*, vol. 6, no. 15, pp. 12391–12397, 2016.

[126] X. Bi and K.-L. Yang, "On-line monitoring imidacloprid and thiacloprid in celery juice using quartz crystal microbalance," *Analytical Chemistry*, vol. 81, no. 2, pp. 527–532, 2009.

[127] M. A. Markowitz, P. R. Kust, G. Deng et al., "Catalytic silica particles via templated-directed molecular imprinting," *Langmuir*, vol. 16, pp. 1759–1765, 2000.

[128] X. B. Luo, F. Deng, L. J. Min et al., "Facile one step synthesis of inorganic-framework molecularly imprinted TiO2/WO3 nanocomposite and its molecular recognitive photocatalytic degradation of target contaminant," *Environmental Science & Technology*, vol. 47, pp. 7404–7412, 2013.

[129] N. R. Stradiotto, H. Yamanaka, and M. V. B. Zanoni, "Electrochemical sensors: A powerful tool in analytical chemistry," *Journal of the Brazilian Chemical Society*, vol. 14, no. 2, pp. 159–173, 2003.

[130] K. Itaya and A. J. Bard, "Chemically Modified Polymer Electrodes: Synthetic Approach Employing Poly(methacryl chloride) Anchors," *Analytical Chemistry*, vol. 50, no. 11, pp. 1487–1489, 1978.

[131] X. Kan, T. Liu, H. Zhou, C. Li, and B. Fang, "Molecular imprinting polymer electrosensor based on gold nanoparticles for theophylline recognition and determination," *Microchimica Acta*, vol. 171, no. 3, pp. 423–429, 2010.

[132] Y. Li, Y. Liu, J. Liu et al., "Molecularly imprinted polymer decorated nanoporous gold for highly selective and sensitive electrochemical sensors," *Scientific Reports*, vol. 5, no. 1-8, 2015.

[133] P. Gupta and R. N. Goyal, "Graphene and Co-polymer composite based molecularly imprinted sensor for ultratrace determination of melatonin in human biological fluids," *RSC Advances*, vol. 5, no. 50, pp. 40444–40454, 2015.

[134] B. B. Prasad, D. Kumar, R. Madhuri, and M. P. Tiwari, "Metal ion mediated imprinting for electrochemical enantioselective sensing of l-histidine at trace level," *Biosensors and Bioelectronics*, vol. 28, no. 1, pp. 117–126, 2011.

[135] R. Wang, K. Yan, F. Wang, and J. Zhang, "A highly sensitive photoelectrochemical sensor for 4-aminophenol based on CdS-graphene nanocomposites and molecularly imprinted polypyrrole," *Electrochimica Acta*, vol. 121, pp. 102–108, 2014.

[136] J. Tan, H.-F. Wang, and X.-P. Yan, "Discrimination of saccharides with a fluorescent molecular imprinting sensor array based on phenylboronic acid functionalized mesoporous silica," *Analytical Chemistry*, vol. 81, no. 13, pp. 5273–5280, 2009.

[137] C. Zhang, H. Cui, J. Cai, Y. Duan, and Y. Liu, "Development of Fluorescence Sensing Material Based on CdSe/ZnS Quantum Dots and Molecularly Imprinted Polymer for the Detection of Carbaryl in Rice and Chinese Cabbage," *Journal of Agricultural and Food Chemistry*, vol. 63, no. 20, pp. 4966–4972, 2015.

[138] M. Mehrzad-Samarin, F. Faridbod, A. S. Dezfuli, and M. R. Ganjali, "A novel metronidazole fluorescent nanosensor based on graphene quantum dots embedded silica molecularly imprinted polymer," *Biosensors and Bioelectronics*, vol. 92, pp. 618–623, 2017.

[139] H. Li, N. Li, J. Jiang et al., "Molecularly imprinted magnetic microparticles for the simultaneous detection and extraction of Rhodamine B," *Sensors and Actuators B: Chemical*, vol. 246, pp. 286–292, 2017.

[140] R. Jalili and M. Amjadi, "Bio-inspired molecularly imprinted polymer–green emitting carbon dot composite for selective and sensitive detection of 3-nitrotyrosine as a biomarker," *Sensors and Actuators B: Chemical*, vol. 255, pp. 1072–1078, 2018.

[141] Z. Altintas, A. Guerreiro, S. A. Piletsky, and I. E. Tothill, "NanoMIP based optical sensor for pharmaceuticals monitoring," *Sensors and Actuators B: Chemical*, vol. 213, pp. 305–313, 2015.

[142] J. Ashley, Y. Shukor, R. D'Aurelio et al., "Synthesis of molecularly imprinted polymer nanoparticles for αcasein detection using surface plasmon resonance as a milk allergen sensor," *ACS Sensors*, vol. 3, no. 2, pp. 418–424, 2018.

[143] T. Eren, N. Atar, M. L. Yola, and H. Karimi-Maleh, "A sensitive molecularly imprinted polymer based quartz crystal microbalance nanosensor for selective determination of lovastatin in red yeast rice," *Food Chemistry*, vol. 185, pp. 430–436, 2015.

[144] N. Gao, J. Dong, M. Liu et al., "Development of molecularly imprinted polymer films used for detection of profenofos based on a quartz crystal microbalance sensor," *Analyst*, vol. 137, no. 5, pp. 1252–1258, 2012.

[145] G. Fang, Y. Yang, H. Zhu et al., "Development and application of molecularly imprinted quartz crystal microbalance sensor for rapid detection of metolcarb in foods," *Sensors and Actuators B: Chemical*, vol. 251, pp. 720–728, 2017.

[146] P. Bonomi, A. Servant, and M. Resmini, "Modulation of imprinting efficiency in nanogels with catalytic activity in the Kemp elimination," *Journal of Molecular Recognition*, vol. 25, no. 6, pp. 352–360, 2012.

[147] X. Zhou, C. Lai, D. Huang et al., "Preparation of water-compatible molecularly imprinted thiol-functionalized activated titanium dioxide: Selective adsorption and efficient photodegradation of 2,4-dinitrophenol in aqueous solution," *Journal of Hazardous Materials*, vol. 346, pp. 113–123, 2018.

Permissions

All chapters in this book were first published in IJAC, by Hindawi Publishing Corporation; hereby published with permission under the Creative Commons Attribution License or equivalent. Every chapter published in this book has been scrutinized by our experts. Their significance has been extensively debated. The topics covered herein carry significant findings which will fuel the growth of the discipline. They may even be implemented as practical applications or may be referred to as a beginning point for another development.

The contributors of this book come from diverse backgrounds, making this book a truly international effort. This book will bring forth new frontiers with its revolutionizing research information and detailed analysis of the nascent developments around the world.

We would like to thank all the contributing authors for lending their expertise to make the book truly unique. They have played a crucial role in the development of this book. Without their invaluable contributions this book wouldn't have been possible. They have made vital efforts to compile up to date information on the varied aspects of this subject to make this book a valuable addition to the collection of many professionals and students.

This book was conceptualized with the vision of imparting up-to-date information and advanced data in this field. To ensure the same, a matchless editorial board was set up. Every individual on the board went through rigorous rounds of assessment to prove their worth. After which they invested a large part of their time researching and compiling the most relevant data for our readers.

The editorial board has been involved in producing this book since its inception. They have spent rigorous hours researching and exploring the diverse topics which have resulted in the successful publishing of this book. They have passed on their knowledge of decades through this book. To expedite this challenging task, the publisher supported the team at every step. A small team of assistant editors was also appointed to further simplify the editing procedure and attain best results for the readers.

Apart from the editorial board, the designing team has also invested a significant amount of their time in understanding the subject and creating the most relevant covers. They scrutinized every image to scout for the most suitable representation of the subject and create an appropriate cover for the book.

The publishing team has been an ardent support to the editorial, designing and production team. Their endless efforts to recruit the best for this project, has resulted in the accomplishment of this book. They are a veteran in the field of academics and their pool of knowledge is as vast as their experience in printing. Their expertise and guidance has proved useful at every step. Their uncompromising quality standards have made this book an exceptional effort. Their encouragement from time to time has been an inspiration for everyone.

The publisher and the editorial board hope that this book will prove to be a valuable piece of knowledge for researchers, students, practitioners and scholars across the globe.

List of Contributors

Pappula Nagaraju
Department of Pharmaceutical Analysis, Hindu College of Pharmacy, Guntur 522002, Andhra Pradesh, India

Balaji Kodali
College of Pharmaceutical Sciences, Acharya Nagarjuna University, Nagarjuna Nagar 522510, Guntur, Andhra Pradesh, India

Peda Varma Datlas
Clinical Pharmacology and Bio Sciences Division, RA Chem Pharma, Hyderabad, India

Surya Prakasarao Kovvasu
College of Pharmacy,Western University of Health Sciences, Pomona, CA 91733, USA

Rania Abtroun
Department of Toxicology, Bab-El-Oued hospital, Avenue Lamine Debaghine, 16009, Algiers, Algeria

Abderrezak Khelfi and Berkahoum Alamir
National Center of Toxicology, Avenue Petit Staouali Delly Brahim, 16062, Algiers, Algeria

Mohammed Azzouz and Mohammed Reggabi
Department of Biology and Toxicology, Ait-Idir hospital, Avenue Abderrezak Hahad Casbah, 16017, Algiers, Algeria

Alexander Chernonosov
Institute of Chemical Biology and Fundamental Medicine, Siberian Branch of Russian Academy of Sciences, Academician Lavrentiev Avenue 8, Novosibirsk 630090, Russia

Sasiprapa Choochuay, Jutamas Phakam and Natthasit Tansakul
Department of Veterinary Pharmacology and Toxicology, Faculty of Veterinary Medicine, Kasetsart University, 10900, Bangkok,Thailand

Sasiprapa Choochuay, Jutamas Phakam and Natthasit Tansakul
Center for Advanced Studies for Agriculture and Food, KU Institute for Advanced Studies, Kasetsart University (CASAF, NRU-KU), 10900, Bangkok, Thailand

Prakorn Jala
Faculty of Veterinary Medicine, Kamphaengsaen Campus, 73140, Nakhon Pathom, Thailand

Thanapoom Maneeboon
Kasetsart University Research and Development Institute, 10900, Bangkok, Thailand

Cátia Salvador, and A. Teresa Caldeira
HERCULES Laboratory, Évora University, Largo Marquês de Marialva 8, 7000-809 Évora, Portugal

M. Rosário Martins
Departamento de Quimica, School of Sciences and Technology, Évora University, Rua Romão Ramalho 59, 7000-671 Évora, Portugal

Henrique Vicente
Centro ALGORITMI, Universidade do Minho, Braga, Portugal

Young Chel Kwun
Department of Mathematics, Dong-A University, Busan 49315, Republic of Korea

Adeel Farooq
Department of Mathematics, COMSATS University Islamabad, Lahore Campus, Lahore 54000, Pakistan

Waqas Nazeer
Division of Science and Technology, University of Education, Lahore 54000, Pakistan

Zohaib Zahid
Department of Mathematics, University of Management and Technology, Lahore 54000, Pakistan

Saba Noreen
Department of Mathematics and Statistics, The University of Lahore, Lahore, Pakistan

Shin Min Kang
Department of Mathematics and RINS, Gyeongsang National University, Jinju 52828, Republic of Korea
Center for General Education, China Medical University, Taichung 40402, Taiwan

Yang Tian, Yue Wang, ShanshanWu, Zhian Sun and Bolin Gong
College of Chemistry and Chemical Engineering, North Minzu University, Yinchuan, 750021, China

Rasdin Ridwan and Wan Mazlina Md Saad
Centre of Medical Laboratory Technology, Faculty of Health Sciences, Universiti Teknologi MARA, Puncak Alam Campus, 42300 Bandar Puncak Alam, Selangor, Malaysia

Hairil Rashmizal Abdul Razak
Centre of Medical Imaging, Faculty of Health Sciences, Universiti Teknologi MARA, Puncak Alam Campus,42300 Bandar Puncak Alam, Selangor, Malaysia

Mohd Ilham Adenan
Faculty of Applied Sciences, Universiti Teknologi MARA, 40450 Shah Alam, Selangor, Malaysia
Atta-ur-Rahman Institute for Natural Product Discovery, Level 9, Bangunan FF3, Universiti Teknologi MARA, Puncak Alam Campus, 42300 Bandar Puncak Alam, Selangor, Malaysia

Hurmus Refiker and Ruveyde Sezer Aygun
Department of Chemistry, Middle East Technical University, Ankara, Turkey

Hurmus Refiker
Institute of Applied Sciences, University of Kyrenia, Sehit Bakir Yahya Street, Kyrenia, North Cyprus, Mersin 10, Turkey

Melek Merdivan
Department of Chemistry, Dokuz Eylul University, Izmir, Turkey

Rüstem Keçili
Anadolu University, Yunus Emre Vocational School of Health Services, Department of Medical Services and Techniques, 26470 Eskisehir, Turkey

Ying Chen, Kunze Du, Jin Li, Yun Bai and Yan-xu Chang
Tianjin State Key Laboratory of Modern Chinese Medicine, Tianjin University of Traditional ChineseMedicine, Tianjin 300193, China
Key Laboratory of Formula of Traditional Chinese Medicine, Tianjin University of Traditional Chinese Medicine, Ministry of Education, Tianjin 300193, China

Mingrui An and Zhijing Tan
Department of Surgery, University of Michigan, Ann Arbor, MI 48109, USA

Claudia Invernizzi, Tommaso Rovetta, Maurizio Licchelli and Marco Malagodi
Arvedi Laboratory of Non-Invasive Diagnostics, CISRiC, University of Pavia, Via Bell'Aspa 3, 26100 Cremona, Italy

Claudia Invernizzi
Department of Mathematical, Physical and Computer Sciences, University of Parma, Parco Area delle Scienze, 7/A, 43124 Parma, Italy

Tommaso Rovetta
Department of Physics, University of Pavia, Via Bassi 6, 27100 Pavia, Italy

Maurizio Licchelli
Department of Chemistry, University of Pavia, via Taramelli 12, 27100 Pavia, Italy

Marco Malagodi
Department of Musicology and Cultural Heritage, University of Pavia, Corso Garibaldi 178, 26100 Cremona, Italy

Saoussen Hammami and Adel Megriche
Université de Tunis El Manar, Faculté des Sciences de Tunis, UR11ES18 Unité de recherche de Chimie Minérale Appliquée. Campus Universitaire Farhat Hached, 2092 Manar 1. Tunis, Tunisia

Nassira Chniba Boudjada
Institut Néel, BP 166, 38042 Grenoble Cedex 9, France

Dongxiao Wen, Ying Cui and Huaixia Yang
Pharmacy College, Henan University of Chinese Medicine, Zhengzhou 450008, China

Qianrui Liu and Jinming Kong
School of Environmental and Biological Engineering, Nanjing University of Science and Technology, Nanjing 210094, China

S. Sowjanya
Department of Pharmaceutical Chemistry, Vishwa Bharathi College of Pharmaceutical Sciences, Perecherla, Guntur 522 009, Andhra Pradesh, India

Ch. Devadasu
Department of Pharmaceutical Analysis & Quality Assurance, Vignan Pharmacy College, Vadlamudi, Guntur 522 213, Andhra Pradesh, India

Elizabeth Mary Mathew and Sudheer Moorkoth
Department of Pharmaceutical Quality Assurance, Manipal College of Pharmaceutical Sciences, Manipal Academy of Higher Education, Manipal 576104, India

Leslie Lewis
Department of Paediatrics, Kasturba Medical College, Manipal Academy of Higher Education, Manipal 576104, India

Pragna Rao
Department of Biochemistry, Kasturba Medical College, Manipal Academy of Higher Education, Manipal 576104, India

Joanna McFarlane and Michelle K. Kidder
Oak Ridge National Laboratory, USA

Marissa E. Morales-Rodriguez
The Bredesen Center at the University of Tennessee Knoxville, USA

Young Hye Hahm, Sung Ho Hahm and Hyoun Young Jo
Rophibio Inc., Cheongju 28160, Republic of Korea

Yeong Hee Ahn
Department of Biomedical Science, Cheongju University, Cheongju 28160, Republic of Korea

Xiaoshan Huang, Mingxin Guan, Zhuliangzi Lu and Yiping Hang
School of Chemistry and Chemical Engineering, South China University of Technology, Guangzhou 510640, China

Ling-jia Zhao, Su-hui Xiong, Jie Tang, Ming-xia Xie, Bo-hou Xia, Li-mei Lin and Duan-fang Liao
College of Pharmacy, Hunan University of Chinese Medicine, Changsha 410208, China
Key Laboratory for Quality Evaluation of Bulk Herbs of Hunan Province, Hunan University of Chinese Medicine, Changsha 410208, China

Wei Liu
School of Management of Information Engineering, Hunan University of Chinese Medicine, Changsha 410208, China

Zhao-huan Lou
Institute of Material Medical, Zhejiang Chinese Medical University, Hangzhou, Zhejiang, China

Rüstem Keçili
Anadolu University, Yunus Emre Vocational School of Health Services, Department of Medical Services and Techniques,26470 Eskişehir, Turkey

Chaudhery Mustansar Hussain
Department of Chemistry and Environmental Science, New Jersey Institute of Technology, Newark, N J 07102, USA

Index